ELECTRON FLOW
IN ORGANIC CHEMISTRY

ELECTRON FLOW
IN ORGANIC CHEMISTRY

Paul H. Scudder
New College
of the University of South Florida

JOHN WILEY & SONS, INC.

New York • Chichester • Brisbane • Toronto • Singapore

ACQUISITIONS EDITOR   Nedah Rose
PRODUCTION SUPERVISOR   Charlotte Hyland
MANUFACTURING MANAGER   Lorraine Fumoso
COPY EDITING SUPERVISOR   Richard Blander

This book was set in Times Roman by the author and printed and bound by Hamilton Printing.   The cover was designed by Ann Marie Renzi and printed by Phoenix Color Corp.

Library of Congress Cataloging in Publication Data:

Scudder, Paul H.
   Electron Flow in Organic Chemistry / Paul H. Scudder

      p.   cm.
   Includes index.
   ISBN 0-471-61381-9 (pbk.)
   1. Chemistry, Organic. 2. Chemical bonds. 3. Reaction mechanisms.
   I. Title

QD251.2.S39   1992
547.1- - dc20   91-40620
                                                                 CIP

Printed and bound by the Hamilton Printing Company.
Printed in the United States of America
10 9 8 7 6 5 4 3 2

Preface

An important part of the scientific method is the ability to postulate a reasonable hypothesis, to make a very good educated guess. This text teaches how to write reasonable reaction mechanisms, and assumes only a general chemistry background. It shows how to develop a good chemical intuition for organic chemistry. Good intuition arises from the automatic use of general trends to guide the decision process.

The ability to formulate a reasonable hypothesis can be taught. We must extract the essence of the field: the conceptual tools, the rules of thumb, the trends, the modes of analysis, and everything that one would use to construct an expert system. First we must review the knowledge base: the basics of bonding, stability, and reactivity trends. Then we need to learn how to classify into generic groups so that the size of the field is manageable. Next, we establish the rules and understand the tools. Finally, we must make explicit, simplify, and teach the intuitive decision process used by an experienced organic chemist.

As with any foreign language, we must learn grammar rather than memorize phrases. In this text, we are concerned with the processes by which reactions occur, the "how and why" of the field. We get a command of organic chemistry based much more upon understanding the basics than upon memorization. Too often a science course becomes an exercise in trivia collection rather than one in reasoning.

Organic chemistry is approached from a mechanistic viewpoint. Each mechanistic process is divided into its individual units. These basic units, the twenty electron flow pathways, become the building blocks of all the common mechanistic processes. This text is centered around the understanding and proper combination of these pathways.

We use the concept of electron flow together with the **rigorous** use of curved arrows as an electron bookkeeping device. The use of curved arrows without mechanistic restraint has rightfully been criticized. Because a process can be drawn with arrows has no bearing on its correctness. It often seems that novice "electron pushers" are just trying to rearrange the lines and dots of the reactant structure into the lines and dots of product with a minimum number of intermediates. They may finish in a few steps but with more arrows than there were at Custer's Last Stand. There is much less tendency to write nonsense with arrows if you are assembling proven mechanistic units, the electron flow pathways.

I have attempted to keep the use of technical jargon to a bare minimum to enhance the readability of the text. In the discussion of various reactions, I drew out the structures rather than used the names of the compounds. I've included several concepts and tools new to most undergraduate organic texts to aid in understanding the most difficult sections of the typical organic course. Hard-soft acid base theory is used to guide decisions and to explain and predict the dual reactivity of many species. There is also much use of energy diagrams and surfaces so that the student has a physical model to aid with the more complex decisions that involve multiple variables.

An optional level of explanation is included that makes use of frontier molecular orbital theory to explain reactivity. A beginning student who has difficulty with molecular orbital concepts can skip these sections without penalty.

I envision the text to be used in several ways. It can be used in the major's sophomore undergraduate organic chemistry course as a short, highly mechanistic text to act as a supplemental "yellow pages" to any synthetically organized "white pages" text.

Second, it can be used as the primary text in an advanced undergraduate or beginning graduate course in organic reaction mechanisms. Third, it can be used as a supplemental review text for graduate courses in physical organic chemistry, enzymatic reaction mechanisms, or biochemistry. This text is a product of fourteen years of teaching and seven years of class testing.

Finally, I would like to thank all that have helped to bring this book to fruition: my father, Prof. Harvey I. Scudder who has helped me greatly refine a concept and algorithm based instructional approach, and my mentors, Prof. Barry Trost and Prof. Virgil Boekelheide. I am indebted to my students who helped me work through the many versions of this text. I appreciate the comments of the reviewers of this manuscript. I would like to thank my colleagues at New College, Prof. Jane Stephens, Prof. David Rohrbach, and Prof. Suzanne Sherman. I thank Nathan Pfluger, who wrote the program for the energy surfaces and the many others at New College who contributed their help and encouragement. I gratefully acknowledge the encouragement of my parents and my wife, son, and daughter, who inspired me to keep writing in the face of an ever growing project. In addition, I would like to thank the University of South Florida and the New College Foundation for their support of the New College program. Finally, I would like to thank all those at John Wiley & Sons who made the publication of this book possible.

This book is dedicated to my students, who have taught me to question everything.

Contents in Brief

Contents

1

BONDING AND ELECTRON DISTRIBUTION

1.1   INTRODUCTION

Organizational Structures

Before setting out to spend a year learning a complex subject like organic chemistry you should take a little time to examine the overall organizational structure that is imposed upon the material. The limitations of one organizational structure are not usually apparent until you see the advantages of another. These limitations make it useful to have different organizational structures for the same material, for example, the White and Yellow Pages of a telephone book. It is rather difficult to find the number of a plumber in the alphabetical White Pages or the number of a friend in the functionally organized Yellow Pages. Any complex mass of data, concepts, and processes must be organized to be taught. However, access to the material is then limited to routes that the organizational structure makes easy.

Because the majority of organic chemists do synthesis – the creation of complex organic compounds from simpler compounds – most authors of organic chemistry textbooks organize the field to facilitate the synthetic transformation of one class of compounds into another. This "White Pages" synthetic organizational structure consists of a description of how each class of compounds can be synthesized and how they can be converted into others.

In this text we organize the field of organic chemistry in a "Yellow Pages" manner; reactions are grouped by similar mechanisms, by how the reactions actually work. The similarity of the mechanism-grouped reactions allows them to be discussed in a more generic fashion. Each mechanistic process can be divided into its individual steps. These steps, called electron flow pathways, become the building blocks of all the common mechanistic processes. The purpose of this text is centered around the understanding and proper combination of these building blocks. The good news is that there appear to be only 20 of them.

We are concerned with the mechanistic processes by which reactions occur. Our approach is to understand why and when a reaction will occur and to establish a command of organic chemistry based more upon understanding the basics than upon memorization. The most impressive result of organizing the material by mechanistic process is that you will develop a sound chemical intuition and will be capable of making good educated guesses. In addition, the how and why of organic chemistry is far more interesting than the memorization of its components. We will make use of trends, generalities, and rules of thumb as tools to aid us in predicting what will happen in an organic reaction. These tools have their exceptions and limitations, but they allow us to develop an overall "feel" for organic reactions. We can worry about the exceptions, if necessary, after the bulk of the material is mastered.

Molecular orbital interaction diagrams and frontier molecular orbital theory are used in an additional, more advanced level of explanation (indicated by a sidebar) that the less prepared student can skip without penalty.

In the first few chapters we review the necessary basics of bonding and reactivity. In the middle chapters we establish the trends within each generic grouping of reactive partners and discuss the mechanistic building blocks. In the last chapters we describe the decision-making process of how to combine those partners to predict products.

The Principle of Electron Flow

A reaction will occur when there is an energetically favorable path by which electrons can flow from the electron source to the electron acceptor (sink). A reaction is a flow of electron density from an electron-rich region to an electron-deficient region (the obvious exceptions to this are free radical reactions, treated separately in Chapter 10). To know which regions are electron rich or electron deficient, we must be able to predict the distribution of electron density over a molecule.

A *nucleophile* ("nucleus-loving," Nu:⁻) is a *Lewis base* (electron pair donor) that acts as an electron source. Nucleophiles can be negatively charged or neutral. An *electrophile* ("electron-loving," E⁺) is a *Lewis acid* (electron pair acceptor) that acts as an electron sink. Electrophiles can be positively charged or neutral.

Any bond formed is the combination of a nucleophile and an electrophile. The most probable product of a reaction results from the best electron source attacking the best electron acceptor. The curved-arrow notation (Section 1.3) allows us to describe the flow of electrons from source to sink.

The concept of flow is very important. **Just as water flows under the influence of gravity, electrons flow under the influence of charge: from electron-rich atoms to electron-deficient atoms.** From *Coulomb's law* we know that **unlike charges attract.** A poor electron source will not react with a poor electron sink within a useful length of time. One might consider this a case of no pull and no push, so no appreciable electron movement occurs. A poor electron sink requires combination with a good electron source for a reaction to occur.

In the first chapters of this text we discuss the basics; in the remaining chapters we answer the questions, What properties distinguish a good electron source, a good electron acceptor, and a good pathway for electron flow? There are relatively few pathways through which the common electron sources and sinks react. The use of generic electron sources and sinks, and generic electron flow pathways makes the similarities and interrelationships of the major reactions in organic chemistry become more obvious. The electron flow pathways become the building blocks of even the most complex organic reaction mechanisms; all of the mechanisms seem to "flow" from first principles.

1.2 LEWIS STRUCTURES AND RESONANCE FORMS

To use electron flow to predict reaction products, an accurate method for electron bookkeeping is necessary. Lewis dot structures are used to keep track of all electrons, and curved arrows are used to symbolize electron movement. You must be able to draw a proper Lewis structure complete with formal charges accurately and quickly. Your command of curved arrows also must be automatic. These two points cannot be overemphasized since all explanations of how a reaction occurs will be expressed in the language of Lewis structures and curved arrows. A Lewis structure must contain the proper number of electrons, the correct distribution of those electrons over each of the atoms, and the correct formal charge. We will show all valence electrons; lone pairs will be shown as darkened dots, bonds by lines.

An atom is most stable if it can achieve the electronic configuration of the nearest inert gas, thus having a completely filled valence shell. Hydrogen with two electrons around it, a duet, achieves the configuration of helium. Second-row elements achieve the inert gas configuration of neon with eight valence electrons, an octet. Third-row elements achieve an octet but may expand their valence shell and have up to twelve electrons.

Procedure for Drawing Lewis Structures

Add up the valence electrons contributed from each atom; add an additional electron for a negative charge, or subtract one to account for a positive charge to get the **number of valence electrons.** Sum the number of electrons needed to fulfill the duets for the hydrogens and the octets for all other elements to get the **full shell count.** Subtract the total number of valence electrons from the full shell count to obtain the **number of electrons shared** in bonds. Since a bond is the sharing of two electrons, the **number of bonds** is half of the number of electrons shared. (For some electron-deficient species, the number of electrons shared is calculated to be greater than the number of valence electrons; in this case, the number of bonds is just half the number of valence electrons.)

Full shell No. - Valence No. = Shared No.
Shared No. ÷ 2 = No. of bonds

Draw single bonds between all connected atoms to establish a skeleton, or preliminary, structure. You need to know the pattern in which the atoms are connected (the most symmetrical structure is often correct, if you have to guess). Count the number of bonds used in the structure and compare this with the total number of bonds that you calculated. **Place any additional bonds between atoms with incomplete octets** to satisfy the following **general bonding trends**. Only third-row elements like phosphorus or sulfur can exceed four bonds.

One bond: H, F, Cl, Br, I
Two bonds: O, S
Three bonds: N, P
Four bonds: C, Si

Subtract the number of shared electrons from the number of valence electrons to get the **number of unshared electrons**. Place those unshared electrons as lone pairs on atoms that need them to complete octets.

Valence No. - Shared No. = Unshared No. of electrons

Assign formal charges. Formal charge is a comparison of the number of electrons an atom "owns" in the Lewis structure with the number it would have if it were free. The atom is assigned only half of the electrons that it shares in a bond, but all of its unshared electrons.

Formal charge = free atom valence - (No. of bonds + Unshared No.)

The sum of the formal charges must be equal to the total charge of the species. As a cross-check, usually if an atom has more than the number of bonds listed in the general bonding trends, it will have a positive formal charge; if it has less, it will bear a negative formal charge. Formal charge is a good indicator of the electron polarization in a molecule and thus is helpful in identifying electron-rich and electron deficient regions.

Example: Methoxide ion, $CH_3O^\ominus$

To satisfy the duets and octets we need 2 for each H and 8 each for the C and O, a full shell count of 22. The total number of valence electrons is 14; we get 6 from O, 4 from C, 1 from each of 3 H's, and 1 for the minus charge. The number of shared electrons is 8, thus the number of bonds is 4; the skeleton uses all of them.

```
Full shell  22          H
 - Valence  14          |
 - Shared    8     H-C-O
Unshared     6          |
                        H
```

We have used eight electrons, and there are six electrons left to be added as lone pairs to complete oxygen's octet. All that is left to do is assign formal charge. Oxygen started with six valence electrons, and in this structure has one bond to it and six unshared electrons, so -1 must be its formal charge. A check shows that the shells are correct for all the atoms, all the valence electrons have been used, and that the sum of the formal charges equals the total charge. The final structure is

$$\begin{array}{c} H \\ | \\ H{-}C{-}\overset{..}{\underset{..}{O}}{:}^{\ominus} \\ | \\ H \end{array}$$

Example: Acetaldehyde, CH_3CHO

To satisfy the duets and octets we need 2 for each H and 8 each for the C and O, a full shell count of 32. The total number of valence electrons is 18; we get 6 from the O, 4 from each of 2 C's, 1 from each of 4 H's. The number of shared electrons is 14, thus the number of bonds is 7; the skeleton uses 6.

$$\begin{array}{rl} \text{Full shell} & 32 \\ \text{- Valence} & 18 \\ \hline \text{- Shared} & 14 \\ \hline \text{Unshared} & 4 \end{array}$$

$$\begin{array}{c} H \qquad O \\ | \qquad / \\ H{-}C{-}C \\ | \qquad \backslash \\ H \qquad H \end{array}$$

Another bond must be made. This bond should go between two atoms that have incomplete octets, C and O, giving the structure

$$\begin{array}{c} H \qquad O \\ | \qquad \mathbin{/\!/} \\ H{-}C{-}C \\ | \qquad \backslash \\ H \qquad H \end{array}$$

Seven bonds used 14 electrons; the 4 unshared valence electrons are added as lone pairs to complete octet of oxygen. The formal charges are all zero; the final structure is

$$\begin{array}{c} H \qquad \overset{..}{O}{:} \\ | \qquad \mathbin{/\!/} \\ H{-}C{-}C \\ | \qquad \backslash \\ H \qquad H \end{array}$$

Electronegativity and Dipoles

The two extremes of bonding are ionic and covalent. In a covalent bond, a pair of electrons is shared between two atoms. In an ionic bond, that pair of electrons resides primarily on one of the two atoms, producing two oppositely charged ions that are attracted to each other.

Pauling defined *electronegativity* as the power of an atom in a molecule to attract electrons to itself. The electronegativity difference of two bonded atoms is an indication of how polarized, how ionic, the bond is. In homonuclear diatomic molecules like H_2 there is pure covalency, no polarization, for the electronegativity difference is zero. Common table salt, $Na^{\oplus} Cl^{\ominus}$, contains a good example of an ionic bond in which the electronegativity difference is large. Table 1.1 is a partial periodic table of the average electronegativities of commonly encountered elements. Elements more electronegative than carbon make up the upper right corner in the periodic table and are in boldface.

The electronegativity of an atom will vary depending on what else is bonded to it. For example, the carbon atom of a CF_3 group is more electronegative than the carbon atom of a CH_3 group. In the CF_3 group, the three highly electronegative fluorines withdraw electron density from the CF_3 carbon atom, which in turn will withdraw more electron density from whatever it's bonded to.

Both the degree and the direction of the polarization of a bond can be predicted by the electronegativity difference. The bonding electron pair is more likely to be found around the more electronegative atom. Carbon can be either partially plus, $\partial+$, or partially minus, $\partial-$, depending on the electronegativity of the group bonded to it. The dipole moment arrow, which points from the positive to the negative end, is shown above the bond in the following structures. The polarization degree and direction of the bonding electrons will become very important in understanding how two species interact since like charges repel and unlike charges attract.

$$\overset{\longmapsto}{\underset{\partial+ \quad \partial-}{>\!C-Cl}} \qquad\qquad \overset{\longleftarrow}{\underset{\partial- \quad \partial+}{>\!C-Li}}$$

Table 1.1 Average Electronegativities of Selected Elements

H 2.21									
Li 0.98					B 2.04	C 2.55	N 3.04	O 3.44	F 3.98
Na 0.93	Mg 1.31				Al 1.61	Si 1.90	P 2.19	S 2.58	Cl 3.16
K 0.82			Cu (I) 1.90	Zn 1.65				Se 2.55	Br 2.96
				Cd 1.69					I 2.66
				Hg 2.00	Tl(III) 2.04	Pb(IV) 2.33			

Resonance Hybrids

Often one Lewis structure is not sufficient to describe the electron distribution in a molecule. In many cases, the use of a *resonance hybrid* is necessary. All resonance forms are valid Lewis structures. In resonance forms, **only the electrons move and not the atoms.** Each resonance form does not have a separate existence but is a part of a hybrid whole. The use of a double-headed arrow, $\leftrightarrow$, between the forms reinforces the notion of a hybrid representation of a single structure. The forms are not in equilibrium with each other (equilibrium is denoted by $\rightleftharpoons$). The polarized bond in HCl can be represented by using resonance structures to show the partial ionic nature of the bond.

$$H - \ddot{\underset{\cdot\cdot}{C}l}\!: \quad \longleftrightarrow \quad H^{\oplus} \; :\!\ddot{\underset{\cdot\cdot}{C}l}\!:^{\ominus}$$

Only one hybrid exists, not individual equilibrating resonance forms. We are trying to describe a molecule's rather diffuse electron cloud with lines and dots; it is not surprising that quite often one Lewis structure is insufficient. The acetate ion, $CH_3COO^{\ominus}$, requires two resonance forms to describe it. Both structures are of equal importance; the carbon–oxygen bond lengths are equal. Generally, the more resonance forms of similar energy an ion has (implied by similar structure), the more stable it is.

$$\underset{H}{\overset{H \quad :\ddot{O}:^{\ominus}}{H-\overset{|}{\underset{|}{C}}-\overset{}{C}=\ddot{\underset{\cdot\cdot}{O}}}} \quad \longleftrightarrow \quad \underset{H}{\overset{H \quad :O:}{H-\overset{|}{\underset{|}{C}}-\overset{\|}{C}-\ddot{\underset{\cdot\cdot}{O}}:^{\ominus}}}$$

Frequently the hybrid cannot be adequately represented by equally weighted resonance forms. The hybrid may be more like one form than the other, and thus the resonance forms are denoted as major and minor. For formaldehyde, $H_2C=O$, the resonance forms represent the uneven electron distribution of a polarized multiple bond; the charge-separated minor resonance form places a negative charge on the more electronegative atom. The major resonance form will tend to have the most covalent bonds, complete shells, and the least amount of charge separation; it will be the most stable of the possible resonance forms. Resonance structures with an incomplete shell are usually minor. Resonance forms having an electronegative atom with an incomplete octet are insignificant.

$$
\begin{array}{ccc}
\text{:O:} & \text{:O:}^{\ominus} & \text{:O:}^{\oplus} \\
\| & | & | \\
\text{H--C--H} \longleftrightarrow & \text{H--C--H} & \text{H--C--H} \\
& \oplus & \ominus \\
\text{Major} & \text{Minor} & \text{Not significant}
\end{array}
$$

Resonance forms for acetamide, CH_3CONH_2, reveal the polarization of the amide group. The partially negative oxygen, not the partially positive nitrogen, will be the reactive site when an amide is used as an electron source. The third resonance form requires the p orbital of the nitrogen lone pair to line up with the p orbital on carbon to get the proper overlap for the double bond (see Section 1.7 and Figure 1.23 for elaboration).

$$
\begin{array}{ccc}
\text{H :O:} & \text{H :O:}^{\ominus} & \text{H :O:}^{\ominus} \\
| \ \| & | \ | & | \ | \\
\text{H--C--C--N--H} \longleftrightarrow & \text{H--C--C--N--H} \longleftrightarrow & \text{H--C--C=N--H} \\
| \ \ \ | & | \ \oplus \ | & | \ \ \ \oplus \\
\text{H} \ \ \text{H} & \text{H} \ \ \text{H} & \text{H} \ \ \text{H} \\
\text{Major} & \text{Minor} & \text{Minor}
\end{array}
$$

1.3 CURVED-ARROW NOTATION

A **full-headed curved arrow indicates the movement of two electrons from the tail of the arrow to the head.** A half-headed curved arrow indicates the shift of one electron likewise. The two ways in which a bond can break are *heterolytic* (two electron) or *homolytic* (one electron).

$$
\begin{array}{cc}
\text{A--B} \rightarrow \text{A}^{\oplus} + {}^{\ominus}\text{:B} & \text{A--B} \rightarrow \text{A·} + \text{·B} \\
\text{Heterolytic cleavage} & \text{Homolytic cleavage}
\end{array}
$$

Arrows indicate a movement or flow of electrons that **must come from a site of electron density,** either a lone pair or a bond, and move to a site that can accept additional electron density. If an arrow comes from a bond, that bond is broken. If an arrow comes from a lone pair, the lone pair is removed. If the head of the arrow points between two atoms, it forms a new bond between them; if it points to an atom, it forms a new lone pair on that atom.

A source of confusion for beginning students is that for intermolecular bond-forming reactions, some authors will point the arrow between the two atoms, whereas others will point it directly at the second atom. We will try a compromise — when an arrow goes between two molecules, the head of the arrow is drawn close to the appropriate atom on the second molecule. The following are slightly different ways to show the formation of a bond between Nu:$^{\ominus}$ and E$^{\oplus}$ to give Nu—E; we will use the arrow notation on the right. A useful generalization is that an arrow that comes from a lone pair will always form a bond, not another lone pair.

$$
\ddot{\text{Nu}}^{\ominus} \downarrow \text{E}^{\oplus} \quad \text{or} \quad \ddot{\text{Nu}}^{\ominus} \rightarrow \text{E}^{\oplus} \quad \text{or} \quad \ddot{\text{Nu}}^{\ominus} \searrow \text{E}^{\oplus} \quad \text{gives} \quad \text{Nu—E}
$$

The bond or lone pair from which the first arrow in an electron flow originates is called the *electron source*. The head of the last arrow in an electron flow points to the *electron sink*. **Arrows will always point away from negative charges and toward positive ones.** Sometimes it is useful to use arrows to interconvert resonance structures, but those arrows do not really indicate electron flow. Some specific examples will help illustrate the correct use of arrows:

$$H-\overset{\ominus}{\underset{..}{\overset{..}{O}}}: \curvearrowright H-\overset{..}{\underset{..}{O}}\overset{\overset{:O:}{\parallel}}{\diagdown}\overset{C}{\diagdown}CH_3 \quad \rightarrow \quad H-\overset{..}{\underset{..}{O}}-H + :\overset{\ominus}{\underset{..}{O}}:\overset{\overset{:O:}{\parallel}}{\diagdown}\overset{C}{\diagdown}CH_3$$

The first arrow on the left comes from the lone pair on the electron-rich hydroxide anion and makes a bond between the hydroxide oxygen and the hydrogen. The second arrow breaks the O—H bond and makes a new lone pair on oxygen. Note that with correct electron bookkeeping **the charge on one side of the transformation arrow will be the same as on the other side** (in this case one minus charge). Charge is conserved. If the electron movement signified by the curved arrows is correct, the products will also be valid Lewis structures.

$$CH_3-\overset{:\overset{\ominus}{\underset{..}{O}}:}{\underset{\underset{CH_3}{|}}{C}}-\overset{..}{\underset{..}{O}}-H \quad \rightarrow \quad CH_3-\overset{\overset{:O:}{\parallel}}{\underset{\underset{CH_3}{|}}{C}} + \overset{\ominus}{:}\overset{..}{\underset{..}{O}}-H$$

In this second example, the flow comes from the lone pair on the negative oxygen and forms a double bond. The flow continues by breaking the carbon–oxygen bond to form a new lone pair on the oxygen of the hydroxide ion. The electron source in this example changes from negative to neutral because the flow removes electrons from it; the sink becomes negative in accepting the electron flow.

$$H\overset{\ominus}{\underset{..}{\overset{..}{S}}}: \qquad \qquad H\overset{..}{\underset{..}{S}}: $$
$$H_2C=C\diagup^H \qquad \rightarrow \qquad H_2C-C\diagup^H$$
$$CH_3\overset{..}{\underset{..}{O}}\diagup C=\overset{..}{\underset{..}{O}} \qquad \qquad CH_3\overset{..}{\underset{..}{O}}\diagup C-\overset{..}{\underset{..}{O}}\overset{\ominus}{:}$$

In this last example, the flow starts with the electron-rich sulfur anion and forms a carbon–sulfur bond with the CH_2 group. The pi bond breaks and forms a new pi bond. The flow finishes by breaking the carbon–oxygen pi bond and forming a new lone pair on the electronegative oxygen.

Exercise: Cover the right side of the previous reactions and draw the product.

A good way to see if you have mastered arrows and the concept of electron flow is to provide the arrows given the reactants and products. In these one-step mechanism problems, you must decide which bonds were made and broken and which direction the electron flow went. Here is an example:

$$H-\overset{..}{\underset{..}{O}}: \qquad H-\overset{..}{\underset{..}{Br}}: \qquad \rightarrow \qquad H-\overset{..}{\overset{\oplus}{O}}\diagup^H + :\overset{..}{\underset{..}{Br}}:^{\ominus}$$
$$\diagdown H \qquad \qquad \qquad \qquad \diagdown H$$

Notice the bonding changes: oxygen has one less lone pair and has formed a new bond to the hydrogen from HBr; the HBr bond is broken and bromine now has another lone pair. Now look at the charges: oxygen is now positive and bromine is now negative. Electron flow must have come from oxygen (the source) and ended up on bromine (the sink) to account for the change in charge. Only one set of arrows could be correct: the first arrow must come from the oxygen lone pair and form a new O—H bond; the second arrow must break the H—Br bond and form a new lone pair on bromine.

A more complex example might be useful.

Oxygen again has lost a lone pair and formed a bond to hydrogen. A carbon–carbon double bond has formed, the carbon–chlorine bond is broken, and a new lone pair is on chlorine. The minus charge on oxygen in the reactants is now on chlorine in the products; the flow must have come from oxygen (the source) to chlorine (the sink). Again only one set of arrows could be correct: the first arrow must come from the lone pair on the negative oxygen and form an O—H bond; the second arrow must break the C—H bond and form a double bond; the third arrow must break the C—Cl bond and form a lone pair on chlorine.

Not all electron sources are charged, as this last example shows:

The lone pair on nitrogen is gone and a new carbon–nitrogen bond has formed; the carbon–iodine bond is broken and iodine now has an extra lone pair. The charges indicate that electron density has been drained away from nitrogen (which has become plus in the product) and deposited on iodine to give the negative iodide anion. Only one set of arrows could be correct: the first arrow starts from the nitrogen lone pair and forms the N—C bond; the second arrow breaks the C—I bond and forms a new lone pair on iodine.

Exercise: Cover the answers in these previous examples and draw the arrows.

Precautions: It is very important that you pay strict attention to the Lewis structures and arrow positions. Lack of care can lead to some rather absurd structures and proposals. Arrows always point in the same direction as the electron flow, never against it. Never use a curved arrow to indicate the motion of atoms;

arrows are reserved for electron flow only. Be forewarned that some texts may combine several steps on one structure to avoid redrawing a structure; others may show a partial set of arrows and expect you to fill in the rest mentally. In your study and practice always draw out each electron flow step completely, for errors that would otherwise be easy to find may become difficult to locate if several steps are jumbled together.

1.4 NOMENCLATURE AND ABBREVIATIONS

Organic chemistry is like a foreign language: it is cumulative and requires that vocabulary be learned in addition to grammar. You must be able to count in organic and know names of the common functional groups (see Table 1.2). A much larger functional-group glossary is in the Appendix. You should learn these words now even though some will not be used until later. Use the Appendix as a reference. Vocabulary is best learned as you need it, but there is so much to learn that a head start is helpful.

Table 1.2 Common Functional Groups

Name	Functional Group	Example
Acyl halide	–C̈–X (with =O)	CH_3–C̈–Cl̈ (with =O)
Alcohol	C–Ö–H	H_3C–Ö–H
Aldehyde	–C̈–H (with =O)	H_3C–C̈–H (with =O)
Alkane	C–C	H_3C–CH_3
Alkene	C=C	H_2C=CH_2
Alkyl halide	C–X	H_3C–B̈r̈
Alkyne	–C≡C–	HC≡CH
Amide	–C̈–N (with =O)	H_3C–C̈–NH_2 (with =O)
Amine	C–N	H_3C–N̈H_2
Aromatic ring	 ⬡	H_3C–⬡ 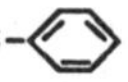
Carboxylic acid	–C̈–Ö–H (with =O)	CH_3–C̈–Ö–H (with =O)
Diene	C=C–C=C	H_2C=CH–HC=CH_2
Ester	–C̈–Ö–C (with =O)	CH_3–C̈–Ö–CH_3 (with =O)
Ether	C–Ö–C	CH_3–Ö–CH_3
Ketone	C–C̈–C (with =O)	CH_3–C̈–CH_3 (with =O)
Nitrile	C–C≡N̈	CH_3–C≡N̈
Organometallic	C–M	CH_3–Li

Exercise: Cover the right side of the page and draw Lewis structures for all functional groups. Cover the left side and name all functional groups from the Lewis structures.

List of the First Ten Alkanes

1	Methane	CH_4
2	Ethane	CH_3CH_3
3	Propane	$CH_3CH_2CH_3$
4	Butane	$CH_3CH_2CH_2CH_3$
5	Pentane	$CH_3CH_2CH_2CH_2CH_3$
6	Hexane	$CH_3CH_2CH_2CH_2CH_2CH_3$
7	Heptane	$CH_3CH_2CH_2CH_2CH_2CH_2CH_3$
8	Octane	$CH_3CH_2CH_2CH_2CH_2CH_2CH_2CH_3$
9	Nonane	$CH_3CH_2CH_2CH_2CH_2CH_2CH_2CH_2CH_3$
10	Decane	$CH_3CH_2CH_2CH_2CH_2CH_2CH_2CH_2CH_2CH_3$

Abbreviations Used in the Text

Ac	Acetyl $CH_3C=O$
Ar	Any aryl (aromatic) group
b	Brønsted base, proton acceptor
$\partial+$	A partial positive charge
$\partial-$	A partial negative charge
don	Electron-donating group
E	Electrophile, Lewis acid
Et	Ethyl CH_3CH_2
ewg	Electron-withdrawing group
G	Unspecified group
HA	Brønsted acid, proton donor
HOMO	Highest occupied molecular orbital
i-Pr	Isopropyl $(CH_3)_2CH$
L	Leaving group
LUMO	Lowest unoccupied molecular orbital
M	Metal atom
Me	Methyl CH_3
MO	Molecular orbital
n-Bu	Normal-butyl group $CH_3CH_2CH_2CH_2$
Nu	Nucleophile
Ph	Phenyl group, C_6H_5, a monosubstituted benzene
R	Any alkyl chain
t-Bu	Tertiary butyl $(CH_3)_3C$
Ts	Toluenesulfonyl, $CH_3C_6H_4SO_2$
X	Chlorine, bromine, or iodine
Y,Z	Heteroatoms, commonly oxygen, nitrogen, or sulfur
±	Racemic mixture
→→	Multistep process
‡	Transition state
......	Partially broken bond (or weak complexation)

Line Structure

Line structure is a very easy way to represent organic structures. Each corner in the line corresponds to a carbon atom, and the hydrogen atoms are not drawn. Line structure is fast and convenient to use but can lead to rather hard to locate errors. Any portion of a molecule that is participating in the reaction must be drawn out, showing all the carbons and hydrogens. It is much too easy to forget about the hydrogen atoms omitted in line structure. Groups not participating in the reaction may be shown in line structure or abbreviated. In the tables, dashes off of a carbon will be used to denote a bond to R or H. In the generic classes, Y (or Z) is used to symbolize any heteroatom group: $Y^{\ominus}$, YH, YR.

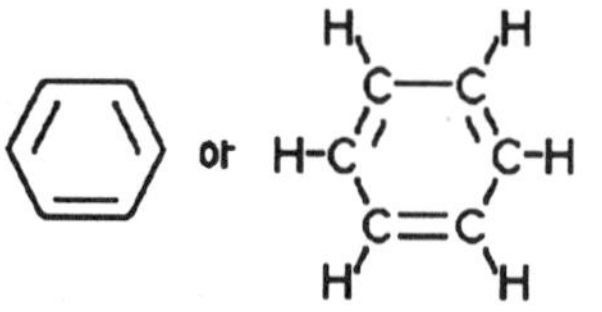

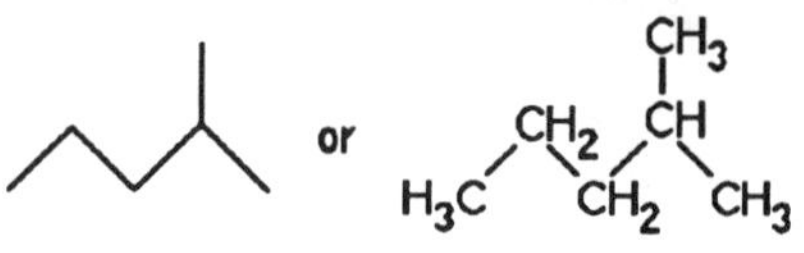

1.5 HYBRIDIZATION AND BONDING

To review, an atomic orbital can hold two electrons and is described by a mathematical expression called a *wave function*. From the wave functions, we can get the energies of the orbitals and the probability distribution in space of any electrons occupying each orbital. A three-dimensional probability distribution of electron density is difficult to display graphically. Figure 1.1 shows several ways of representing a $2p$ orbital. We arbitrarily shade one lobe of the $2p$ orbital to indicate that its mathematical sign is different from the other lobe. The $2p$ orbital has two lobes and a planar node through the center. A *node* is the region in space where the mathematical expression goes through zero.

Figure 1.1 Representations of a $2p$ orbital. The nucleus is indicated by a heavy dot (*a*) Dot density cross section. (*b*) Three-dimensional surface. (*c*) Simplified version for text.

Mathematical combinations of atomic orbitals on the same atom give hybridized orbitals useful for the description of bonding with other atoms. Figure 1.2 shows how a $2s$ orbital can be added to and subtracted from a $2p$ orbital to give the two sp hybrid orbitals. To subtract, we change the sign (shading) of the $2s$ orbital then add it to the $2p$ orbital. The two orbital wave functions reinforce where the mathematical signs are the same (the lobe gets larger) and cancel out where the sign is different (the lobe gets smaller). The number of orbitals is conserved: when we combine two atomic orbitals, we get two hybrid orbitals.

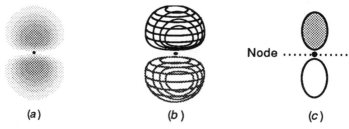

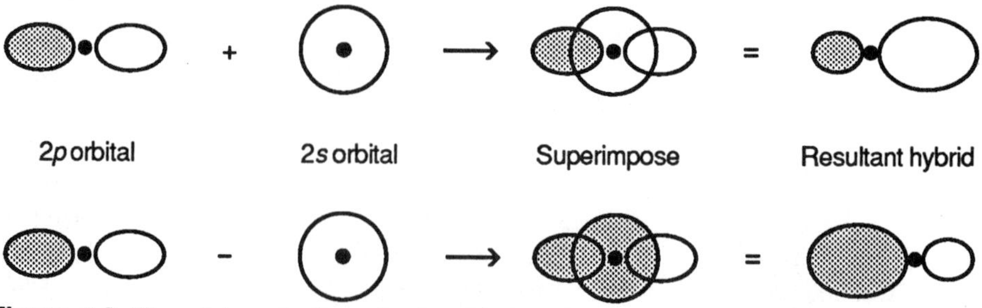

Figure 1.2 The mixing of a $2p$ and a $2s$ orbital on the same atom giving two sp hybrids.

Carbon has one $2s$ and three $2p$ orbitals that can be used for hybridization (Table 1.3). Combining the $2s$ and one of the $2p$ orbitals to get two sp hybrids leaves two $2p$ orbitals remaining. Because the $2s$ orbital is lower in energy and closer to the nucleus than a $2p$ orbital, **hybrid orbitals that contain a higher %s character will form bonds that will be shorter, stronger, and lower in energy.** To determine the hybridization of a carbon atom, count the atoms it is bonded to: an sp^3 carbon bonds to four other atoms, an sp^2 carbon bonds to three, and an sp carbon bonds to only two. Use the Lewis structure rather than the line structure because the latter does not show all the hydrogens.

Table 1.3 Summary of the Properties of Hybrid Orbitals

	Hybridization of Carbon		
	sp	sp^2	sp^3
Number of orbitals	2	3	4
Interorbital angle	180°	120°	109.5°
% s character	50	33	25
% p character	50	67	75
Orientation	Linear	Trigonal	Tetrahedral
Electronegativity of C	3.29	2.75	2.48
Remaining p orbitals	2	1	0

Single Bonds

Molecular orbitals involve the combination of orbitals on different atoms. Figure 1.3 shows how two $1s$ orbitals can be combined by addition and subtraction to give the molecular orbitals for hydrogen. Again, **if we combine two atomic orbitals, we get two molecular orbitals.** A covalent bond is the sharing of an electron pair between two nuclei. The *bonding molecular orbital* has no nodes (the highest probability of finding the electrons is between the two nuclei) and is lower in energy. The *antibonding molecular orbital* has a node between the two nuclei (the probability goes through zero there) and is a higher-energy orbital. Orbitals are filled from lowest energy upwards. Since the hydrogen molecule has two electrons, one from each atom, the bonding molecular orbital is filled and the antibonding orbital is empty.

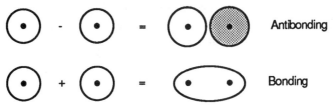

Figure 1.3 The bonding and antibonding molecular orbitals of hydrogen.

Similarly we can combine two sp^3 orbitals to produce a sigma bond. Sigma, σ, bond orbitals are cylindrical along the axis of the bond. We obtain a *bonding* and an *antibonding* combination (denoted by *). **Strong bonds are due to good overlap of the bonding orbitals.** A carbon–carbon single bond is about 1.54 Å long.

Because an sp^3 carbon is tetrahedral, organic molecules are usually drawn in three dimensions. The solid wedge in Figure 1.4 symbolizes that the bond is coming out of the paper, and the dashed wedge symbolizes that the bond is going back into the paper. Although single bonds are drawn as if they were "frozen in space," rotation has no significant barrier. However, a large barrier may arise if two large groups can bump into one another when rotation about the bond occurs.

Sigma bonding   Sigma antibonding

Figure 1.4 The bonding and antibonding orbitals of a carbon–carbon single bond are made from the combination of two sp^3 hybrid orbitals.

Stereochemistry is the description of how the atoms of a molecule are arranged in space. *Conformational isomers* differ only in rotations about single bonds. The eclipsed and staggered conformational isomers of ethane, CH_3CH_3, are shown in Figure 1.5.

The staggered conformation is the lowest energy; the eclipsed is the highest since the C—H bonds are in the process of passing each other, and their electron clouds repel one another. Ethane's rotational barrier is only 3 kcal/mol (13 kJ/mol). One kilocalorie per mol is equal to 4.184 kJ/mol. The best way to learn what organic compounds actually look like is by working with molecular models.

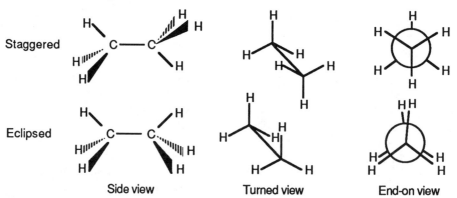

Side view   Turned view   End-on view

Figure 1.5 Three views of two conformational isomers of ethane.

Cyclohexane exists in a chair conformation in which all the bonds are staggered. All the C—H bonds roughly in the plane of the ring are called equatorial; all those perpendicular are called axial. Cyclohexane can easily flip between two chair conformations, as shown in Figure 1.6; the barrier is only 10.8 kcal/mol (45 kJ/mol). All the axial positions become equatorial and vice versa upon ring flip. With substituted rings, the equilibrium favors the larger groups in the less crowded equatorial position.

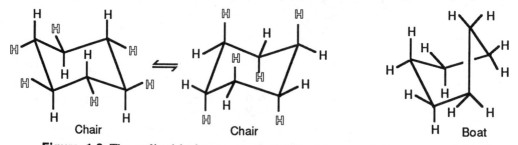

Chair   Chair   Boat

Figure 1.6 The outlined hydrogens on the left cyclohexane chair are equatorial and upon ring flip become axial. The boat conformation of cyclohexane is estimated to be 6.4 kcal/mol (26.7 kJ/mol) higher in energy than the chair.

A very important implication of the three-dimensional nature of organic compounds is that they can be *chiral*, occurring in left- and right-handed forms. A compound is chiral if it does not superimpose on its mirror image (Figure 1.7). A molecule is not chiral if it contains an internal mirror plane. A carbon atom with four different groups bonded to it is called a chiral center. A molecule can be chiral without having a chiral center if it has a chiral shape such as a propeller or helix. A compound can contain chiral centers and not be chiral if there is an internal mirror plane. Therefore, whenever you are uncertain, build the model of the compound and try to superimpose it on a model of its mirror image.

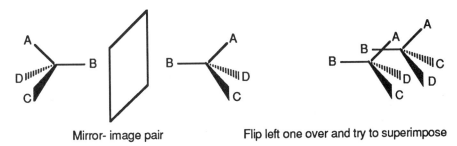

Mirror- image pair   Flip left one over and try to superimpose

Figure 1.7 A chiral compound will not superimpose on its mirror image.

Stereoisomers are isomers that have the same sequence of bonds and differ only in the arrangement of atoms in space. *Enantiomers* are mirror image pairs that are not superimposable (if the pair superimposes, then the two are the same). Stereoisomers that are not enantiomers are called *diastereomers*. Figure 1.8 shows the relationships between the different types of isomers.

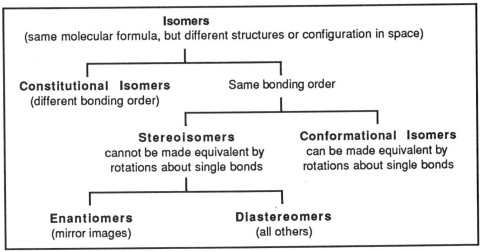

Figure 1.8 The isomer family tree.

Figure 1.9 shows that reactions at a chiral center can affect stereochemistry in three ways: retention, inversion, or racemization. It is important to heed the changes in stereochemistry because any proposed reaction mechanism must account for them. A reaction that produces a racemic mixture from a chiral reactant must proceed through an intermediate that is not chiral. A reaction that produces a predominance of one stereochemistry of product is *stereoselective*. If starting material stereochemistry determines the stereochemistry of product, the reaction is *stereospecific*.

Weak bonds are a site of reactivity. The average bond strength for a C—C single bond is 83 kcal/mol (347 kJ/mol). Rings containing three and four atoms are strained by 28 and 26 kcal/mol (117 and 109 kJ/mol) because the orbitals can no longer be directed along a line between the atoms, and their overlap and bond strength decreases (see Figure 1.10). This "*ring strain*" makes small rings easy to break. Rings of five, six, and seven atoms are relatively strain-free. Bonds between atoms that both contain lone pairs of electrons (for example, Br_2) are weak because of lone pair–lone pair repulsion. Orbitals of different sizes overlap poorly and therefore form weak bonds.

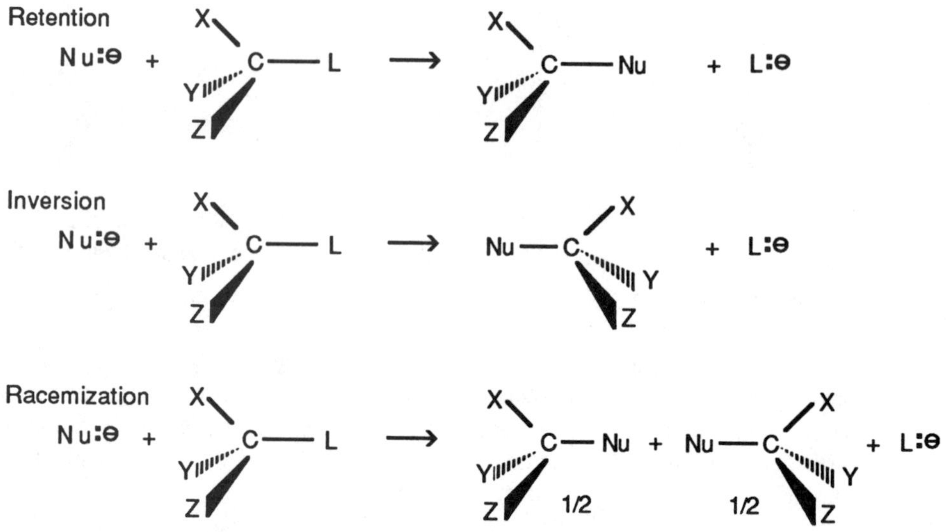

Figure 1.9 The three possible fates of a chiral center undergoing a chemical reaction: retention, inversion, or racemization (producing a 1:1 mixture of enantiomers, shown by ±).

In general, the overlap with a carbon-based orbital decreases as one goes down a column in the periodic table. For example, the strengths of a $2p$, $3p$, $4p$, and $5p$ halogen to carbon bond are, respectively, C—F, 116 kcal/mol (485 kJ/mol); C—Cl, 81 kcal/mol (339 kJ/mol); C—Br, 68 kcal/mol (285 kJ/mol); and C—I, 51 kcal/mol (213 kJ/mol).

Figure 1.10 The larger six-membered ring on the left does not distort the sigma bond; the three-membered ring on the right does.

Double Bonds

Pi bonds are made by two orbitals interacting side by side in the same plane, and are relatively reactive due to less effective overlap. The combination of two $2p$ orbitals to yield the molecular orbitals for ethene, $CH_2=CH_2$, is shown in Figure 1.11. Double bonds result from a sigma bond and a pi bond between the two bonding atoms. Since the sigma bond lies in the nodal plane of the pi bond, the sigma and pi bonds of a double bond are independent.

The requirement that the *p* orbitals overlap causes a very large barrier to rotation about the double bond, about 63 kcal/mol (264 kJ/mol). *Cis* (two groups on the same side) and *trans* (two groups on opposite sides) double-bond isomers do not interconvert at any reasonable temperatures. The *trans* isomer tends to be slightly more stable than the *cis* isomer because the groups may bump into one another when they are *cis*. More alkyl substitution (replacing an H on the double bond by an R) on the double bond makes the alkene slightly more stable. For example, an equilibrium mixture of butenes is found to contain 3% 1-butene, 23% *cis*-2-butene, and 74% *trans*-2-butene.

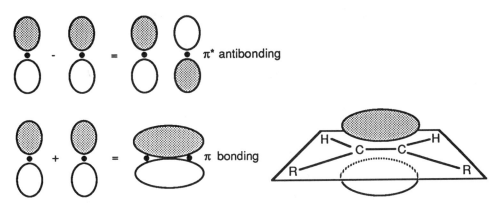

Figure 1.11 The bonding and antibonding orbitals of a carbon–carbon pi bond are made from the combination of two 2*p* orbitals by subtraction and addition. Atoms adjacent to the pi bond lie in a plane, as shown on the right. The R groups in this example are *cis* to each other.

The average bond strength of a double bond is 146 kcal/mol (611 kJ/mol); after subtraction of the sigma bond strength, the pi bond is worth about 63 kcal/mol (264 kJ/mol). The overlap of a pi bond is greatly diminished if one of the ends is twisted so that the two *p* orbitals are no longer coplanar (in the same plane). The greater the amount of this twist, the less stable the pi bond. At a 90° twist angle the pi bond no longer exists, for the two *p* orbitals are perpendicular to each other and no longer interact. Double bonds with as little as a 30° twist are reactive and difficult to make. A *trans* double bond in a ring is twisted as the ring gets smaller (Figure 1.12). The smallest ring that contains a *trans* double bond and is still stable enough to be isolated at room temperature is *trans*-cyclooctene, which is about 11 kcal/mol (46 kJ/mol) more strained than *cis*-cyclooctene. Cyclobutene, with an untwisted *cis* double bond, is stable at room temperature. Pi bonds between orbitals of different sizes are very weak because of poor overlap; a good example is the 2*p*–3*p* carbon–sulfur double bond (Figure 1.13).

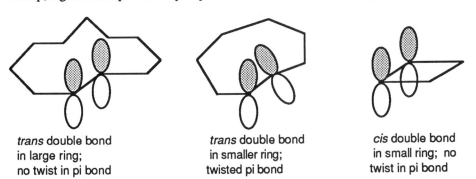

trans double bond
in large ring;
no twist in pi bond

trans double bond
in smaller ring;
twisted pi bond

cis double bond
in small ring; no
twist in pi bond

Figure 1.12 The distortion of a pi bond with varying ring size.

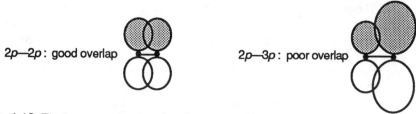

Figure 1.13 The better overlap in a $2p$—$2p$ versus a $2p$—$3p$ pi bond.

Triple Bonds

A triple bond is composed of one sigma bond and two pi bonds. Since the two pi bonds are perpendicular to each other, they are treated as two separate, noninteracting pi bonds. **Perpendicular orbitals do not interact.** The triple bond is linear and atoms bonded to it lie in a straight line (Figure 1.14).

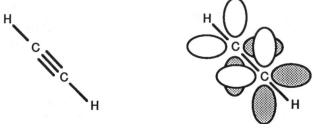

Figure 1.14 The pi orbitals of a triple bond.

The average bond strength of a triple bond is 200 kcal/mol (837 kJ/mol); therefore each pi bond is worth about 59 kcal/mol (247 kJ/mol). If placed in a ring, the triple bond can be bent from colinearity and become weaker and much more reactive. Cyclooctyne is the smallest ring that contains a triple bond and is still stable enough to be isolated.

Cumulenes

Cumulenes are compounds with two adjacent perpendicular double bonds (Figure 1.15). Each double bond can be considered separately. In heterocumulenes, heteroatoms have replaced one or more of the carbons, for example, carbon dioxide, $O=C=O$.

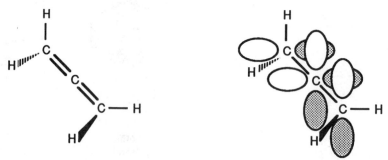

Figure 1.15 The pi orbitals of a cumulene. The central carbon is sp hybridized and the two CH_2's lie in perpendicular planes.

Hydrogen Bonding

Hydrogen bonding is the attraction of the permanent $\partial+$ of a polarized hydrogen–heteroatom bond with the $\partial-$ of a heteroatom lone pair. Although the hydrogen bond strength averages about 5 kcal/mol (21 kJ/mol), the importance of hydrogen bonding should not be underestimated. Hydrogen bonding causes water to be a liquid rather than a gas at room temperature. With rare exceptions, the carbon–hydrogen bond is not significantly polarized, and therefore does not hydrogen bond. Hydrogen bonds with a linear arrangement of all three atoms are most common. We can represent a hydrogen bond as a minor resonance form, signifying a weak bond (the nuclei do not move).

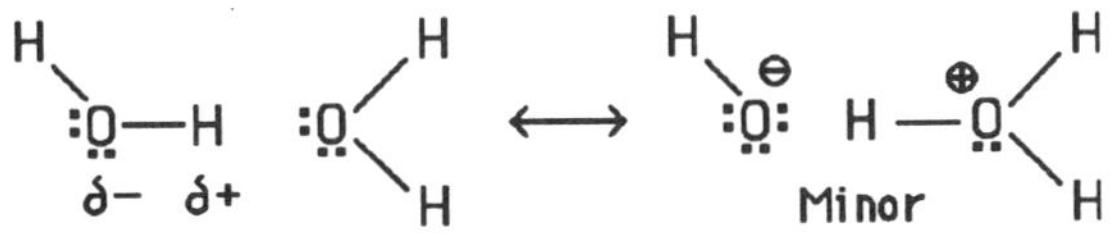

Pi-complexes

Pi-complexes, also called donor–acceptor complexes, are a weak association of an electron-rich molecule with an electron-poor molecule. The donor is commonly the electron cloud of a pi bond or aromatic ring; the acceptor can be a metal ion, a halogen, or another organic compound (Figure 1.16).

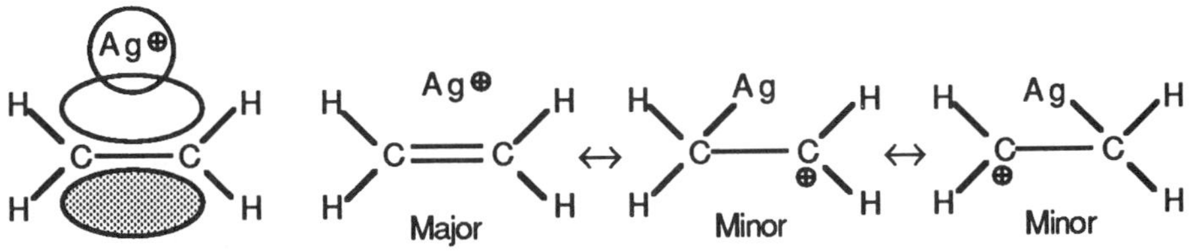

Figure 1.16 An example of a pi-complex between a silver ion and an alkene. The resonance forms on the right are an attempt to describe the orbital overlap shown on the left.

Often a very weak bond like a hydrogen bond or a complexation is indicated by a dotted line, ·······, which is also used to symbolize a partially broken bond.

1.6 ORBITAL INTERACTION DIAGRAMS
(A Supplementary, More Advanced Explanation)

An understanding of interaction diagrams is not absolutely necessary for using the principle of electron flow to predict organic reaction products. However, it is useful for understanding reactivity trends and the stability of reactive intermediates.

Another way to describe the mixing of orbitals shown in the previous section is to use an orbital interaction diagram. The orbitals to be interacted are drawn on each side of the diagram and their interaction is shown in the center of the diagram. When two orbitals interact, the result is two new orbitals. Shown on the left side of Figure 1.17 is the interaction of two hydrogen $1s$ orbitals containing one electron each to form a bonding orbital into which both electrons are placed. The two electrons in the bonding orbital have a lower energy than they had before the atomic orbitals interacted to form the bonding and antibonding molecular orbitals. The electrons are stabilized; energy is released, and a bond is formed. Shown on the right side of Figure 1.17 is the interaction of two helium $1s$ orbitals containing two electrons each. Now both the bonding and antibonding orbitals must be filled. The two electrons in the antibonding orbital are destabilized more than the two electrons in the bonding orbital are stabilized, resulting in net destabilization and no bond between the helium atoms.

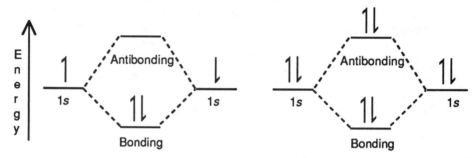

Figure 1.17 The left side shows the orbital interaction diagram for two hydrogen atoms and the right side shows the interaction of two helium atoms.

The better the overlap between two orbitals, the greater their interaction. If the orbitals are oriented such that their antibonding overlap exactly cancels their bonding overlap, then there is no interaction (Figure 1.18). If the orbitals are too far apart to overlap, there is no interaction.

Figure 1.18 Examples of orbitals that do not interact because their bonding and antibonding overlap exactly cancels.

The closer in energy the two interacting orbitals are, the greater their interaction. A greater interaction results in both a larger raising of the antibonding and a larger lowering of the bonding molecular orbitals formed. When a full orbital interacts with an empty orbital, a large stabilization of the bonding electron pair is possible. A good example of the lowering of the bonding molecular orbital as the energy difference between the interacting orbitals changes is shown by Figure 1.19.

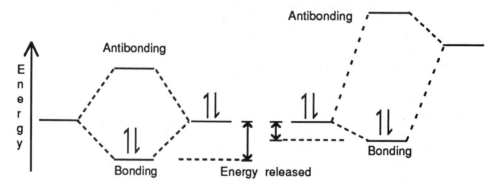

Figure 1.19 The energy released upon interaction of a full with an empty orbital. On the left side the two orbitals are of identical energy and the interaction is large. On the right is the small interaction of two orbitals of greatly differing energies.

Electronegativity Effects

Electronegativity affects orbital energies and electron distribution. Because oxygen has a higher nuclear charge than carbon, its $2p$ orbital is of lower energy than the carbon $2p$ orbital. As illustrated in Figure 1.20, the lower-energy oxygen $2p$ orbital tends to pull down the energy of the π and π^* molecular orbitals. The lowered π^* orbital energy will have an important influence on reactivity (Section 2.7).

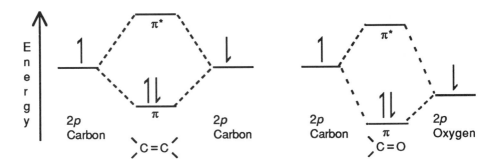

Figure 1.20 The effect of electronegativity on the interaction of two $2p$ orbitals.

When two orbitals of differing energy interact, the molecular orbital contains a greater percentage of the atomic orbital closest to it in energy. This is easily demonstrated (Figure 1.21) by examining the two extremes. As we have seen for a carbon–carbon pi bond, when the two orbitals are of equal energy each contributes equally to the molecular orbital. The other extreme is when the orbitals are so far apart in energy that they do not interact: the higher orbital would contain only the higher atomic orbital. An intermediate case in which the energies of the two orbitals differ but the orbitals still interact is shown in the center.

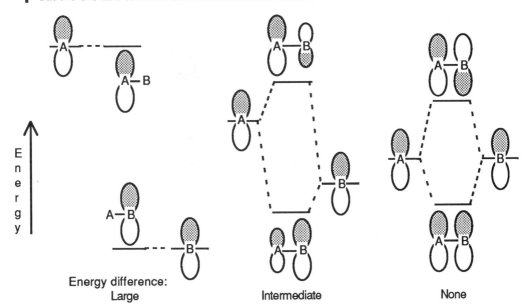

Figure 1.21 The interaction of orbitals of differing energy. The orbital with the larger contribution is drawn larger.

In the π molecular orbitals of a *carbonyl*, the carbon–oxygen double bond, the lower-energy oxygen $2p$ orbital contributes more to the π than to the π^* molecular orbital, and the higher-energy carbon $2p$ orbital contributes more to the π^* than to the π molecular orbital (Figure 1.22). Because of the distortion of the bonding molecular orbital, the electrons in the carbonyl pi bond have a greater probability of being found around the oxygen. The valence bond resonance forms are just another way of describing this electron distribution in the molecular orbital.

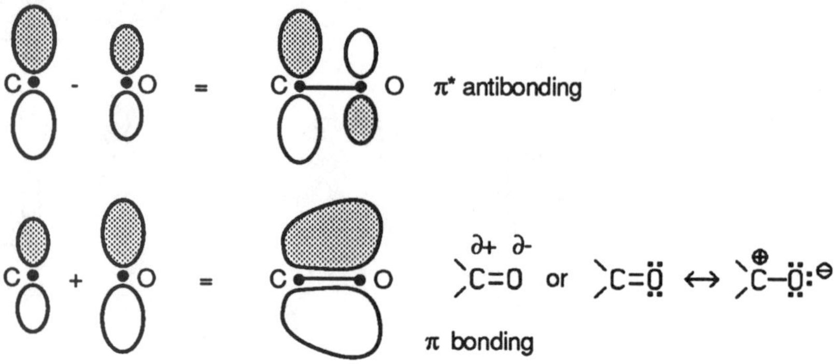

Figure 1.22 The pi molecular orbitals of a carbonyl group.

1.7 CONJUGATION

Two pi bonds connected by a single bond behave as one *conjugated* system. The p orbitals of the two pi units are then close enough to have good overlap and additional pi bonding. Figure 1.23 shows how a double bond and a carbonyl group can be placed in a molecule to form both a conjugated and unconjugated system. The partial plus of the carbonyl is delocalized by conjugation. Because of the additional pi bonding, a conjugated system is more stable than the corresponding unconjugated system by about 3.5 kcal/mol (14.6 kJ/mol); any equilibrium between the two favors the conjugated system.

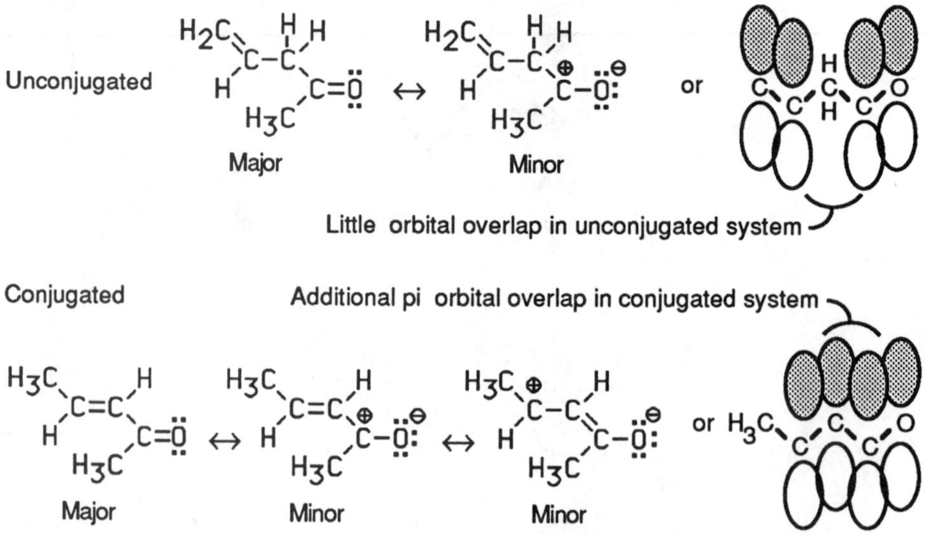

Figure 1.23 The resonance forms and pi overlap of an unconjugated system (top) and a conjugated system (bottom) of a carbon–carbon pi bond and a carbon–oxygen pi bond.

Another example of conjugation is the nitrogen lone pair joined by resonance with a carbonyl group shown in Figure 1.24 for amides. The nitrogen lone pair is in a *p* orbital that overlaps the *p* orbitals of the pi bond to make a three *p* orbital pi system. Any three *p* orbital pi system is called an *allylic system*. The conjugated system will have properties different from its individual parts because it now acts as a mixture and not as two independent functional groups.

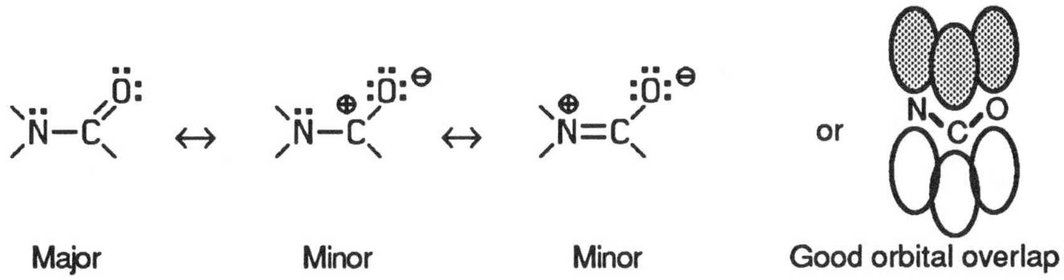

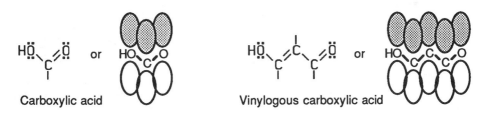

Major   Minor   Minor   Good orbital overlap

Figure 1.24 Conjugation of a nitrogen lone pair with a carbonyl to give an amide group.

Still another example is the carboxylic acid group in which an alcohol is conjugated with a carbonyl group. The resonance forms of the carboxylic acid explain the increased acidity of the hydrogen:

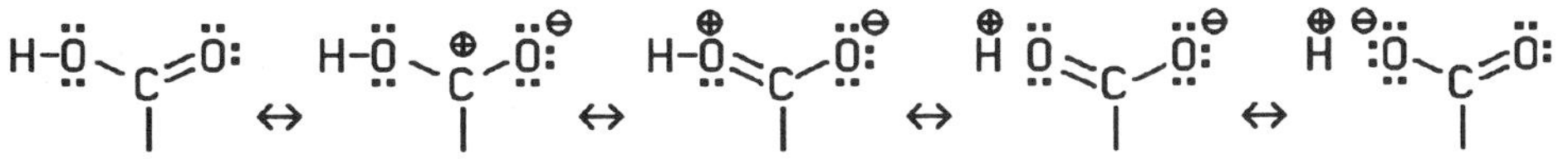

Vinylogy is the extension of the properties of a system by the insertion of a carbon–carbon double bond. The conjugated system increases in length, but the properties remain approximately the same. A good example of this is the comparison of a carboxylic acid and a vinylogous carboxylic acid (Figure 1.25). The alcohol group is still conjugated with the carbonyl through the intervening pi bond.

Carboxylic acid   Vinylogous carboxylic acid

Figure 1.25 The insertion of a carbon–carbon double bond between the OH and the C=O of the carboxylic acid on the left the creates a vinylogous carboxylic acid.

Similar resonance forms in a vinylogous carboxylic acid also account for an acidic hydrogen.

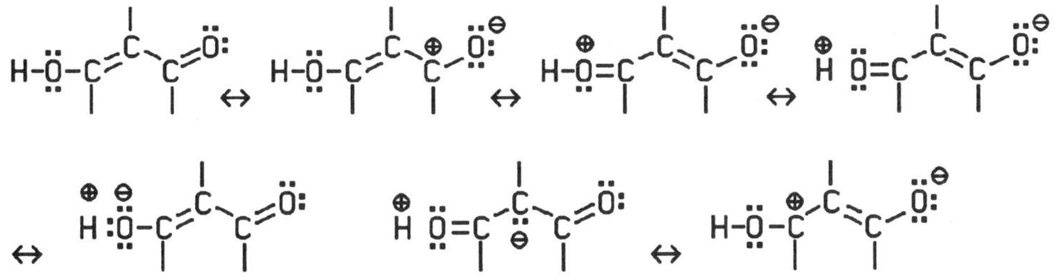

The following structures compare several other vinylogous systems with their respective parent systems.

Ester

Vinylogous ester

Amide

Vinylogous amide

Acyl chloride

Vinylogous acyl chloride

1.8   MOLECULAR ORBITAL THEORY FOR ACYCLIC SYSTEMS
(A Supplementary, More Advanced Explanation)

Since the electron has a wavelike behavior, the electrons in a molecule will have properties similar to a wave confined to a limited space. The stable configurations will be standing waves. We can view *molecular orbitals* (MOs) like standing waves on a guitar string because they too are waves confined to a limited space. As the energy of the molecular orbital increases, the number of nodes increases, like the higher harmonics on a vibrating string (Figure 1.26). The nodes are the regions where the wave goes through zero, indicated by vertical dotted lines in the figure. As the frequency becomes higher, the energy becomes higher.

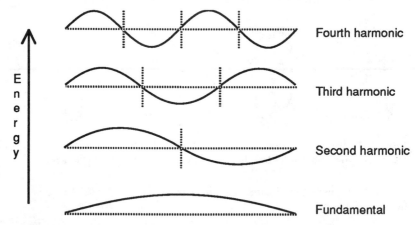

Figure 1.26 The standing waves of a vibrating string. The fundamental (first harmonic) has no nodes; the second harmonic is twice the frequency of the fundamental and has one node; the third harmonic is three times the frequency of the fundamental and has two nodes, and the fourth harmonic is four times the frequency of the fundamental and has three nodes.

Two *p* Orbitals, a Simple Pi Bond

Since the pi bonding portion of a molecule is perpendicular to the sigma bonded framework, it can be treated independently. In a double bond, **two *p* orbitals produce** two molecular orbitals, π (bonding) and π* (antibonding), shown again in Figure 1.27. The node in the wave corresponds to the mathematical sign change. The lowest-energy molecular orbital always has no nodes between the nuclei (we ignore the node in the plane of the atoms common to all pi systems). The electron pair in the bonding MO is shared between the nuclei. The antibonding MO has a node between the nuclei.

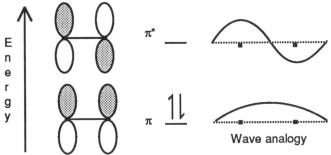

Figure 1.27 The pi molecular orbitals of ethene. The dots in the wave analogy approximate the nuclei position. The standing wave extends beyond the nuclei because the orbitals do.

Three *p* Orbitals, the Allyl Unit

Three *p* orbitals, all aligned parallel to one another, make up the allyl unit and **produce three molecular orbitals.** The MOs for the following systems differ from the completely carbon allylic system only in the previously discussed electronegativity effects (the bonding MOs will have a larger contribution from the heteroatom *p* orbital and the antibonding MOs will have a lesser contribution).

To determine how many electrons occupy any pi system, start by counting each double bond as contributing two electrons, but do not count any pi bonds that are perpendicular to the pi system of interest, for example, the second pi bond of a triple bond. Because the maximum that a single atom can contribute is a single filled *p* orbital containing two electrons, if a heteroatom has several lone pairs it can contribute only two electrons. Only one *p* orbital on an atom can align itself with the pi system. If an atom is doubly bonded and contains a lone pair, the lone pair cannot be counted since it must be in an orbital that is perpendicular to the double bond. All the examples above have four electrons in a three *p* orbital pi system.

Shown in Figure 1.28 are several different ways to describe the MOs for C=C—C:$^\ominus$, the allyl anion. The MOs are labelled from the lowest energy to the highest and are drawn with the lowest-energy MO at the bottom. The lowest-energy MO, Ψ_1, has no nodes and will always be bonding. The highest-energy molecular orbital, Ψ_3, has a node between each nucleus and will always be antibonding. In pi systems containing an odd number of *p* orbitals there is a *nonbonding* MO, in this case Ψ_2, in which the wave puts a node through the central atom. Nodes are always symmetrical about the center; in waves with an odd number of nodes, one of the nodes must go through the center.

For C=C—C$^\oplus$, the allyl cation, there would be two electrons in the pi system, filling just Ψ_1, leaving the others empty. The shape and energy of the MOs are independent of the occupancy. The allyl anion has four electrons in the pi system, filling both Ψ_1 and Ψ_2. The valence bond resonance forms, C=C—C:$^\ominus$ ↔ $^\ominus$:C—C=C, describe the electron distribution in Ψ_2 of the anion. The two electrons of the anion can be found on the ends of the pi system and not in the middle.

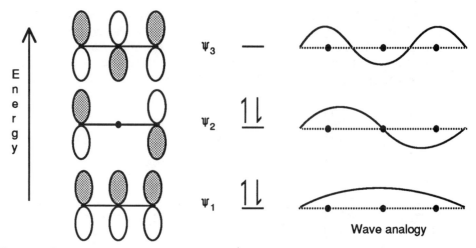

Figure 1.28 The pi molecular orbitals of the allyl unit.

Four p Orbitals, the Diene Unit

Figure 1.29 gives the MOs for 1,3-butadiene, $H_2C=CH-HC=CH_2$, which has **four p orbitals** in the pi system and thus **four MOs**. The four electrons are placed in the two bonding MOs, leaving the two antibonding MOs empty. Again note the similarity to a vibrating string. The *highest occupied molecular orbital* (HOMO) is Ψ_2, and the *lowest unoccupied molecular orbital* (LUMO) is Ψ_3.

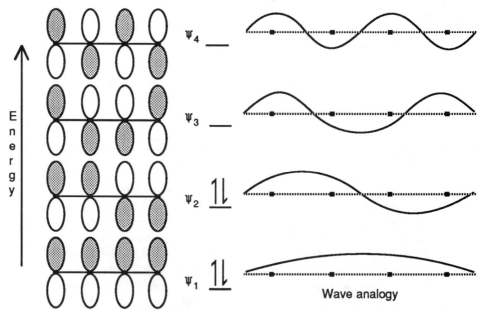

Figure 1.29 The pi molecular orbitals of the diene unit.

An alternative way to order the MOs is to count the bonding and antibonding overlap within each MO. Ψ_1 is bonding between the first p orbital and the second, bonding between the second and the third, and bonding between the third and the fourth. Ψ_2 is bonding between the first p orbital and the second, antibonding between the second and the third, and bonding between the third and the fourth. Ψ_3 is antibonding between the first p orbital and the second, bonding between the second and the third, and antibonding between the third and the fourth. Ψ_4 is antibonding between all, therefore is the highest energy of all of them.

Figure 1.30 shows how the relative energy of the pi orbitals changes as a second pi bond is added to form a conjugated system. As can be seen from comparing the energy of the MOs of ethene and 1,3-butadiene, additional conjugation raises the HOMO in energy and lowers the LUMO in energy.

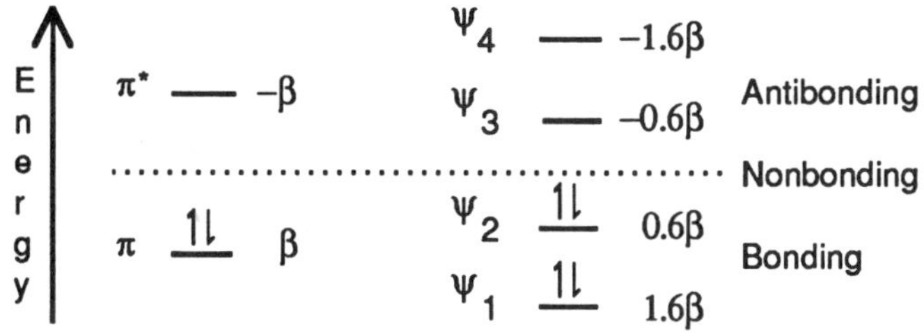

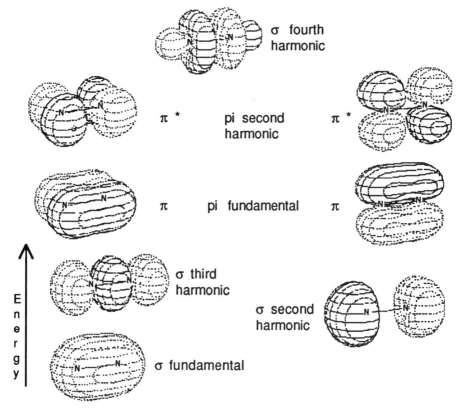

Figure 1.30 A comparison of the relative orbital energies of ethene (left) and 1,3-butadiene (right). The stabilization of an electron in the π of ethene is defined as β .

Delocalization of Sigma Electrons (More Advanced)

Advanced molecular orbital calculations for the nitrogen molecule (:N≡N:) give a simple set of molecular orbitals (Figure 1.31). Two perpendicular pairs of π and π^* molecular orbitals arise from the nitrogen–nitrogen triple bond.

Figure 1.31 The molecular orbitals of nitrogen. The lowest five orbitals are filled. Modified with permission from W. L. Jorgensen and L. Salem *The Organic Chemist's Book of Orbitals*, page 79, Copyright ©1973 by Academic Press.

The sigma system from these calculations may be a bit of a surprise; the sigma bonding electrons are not constrained to individual bonds but are delocalized over the entire molecule. Instead of four localized orbitals, sigma bonding, sigma antibonding, and two lone pair orbitals, there are four sigma molecular orbitals: the fundamental, the second harmonic, the third harmonic, and the fourth harmonic.

Sigma molecular orbitals pose a difficult representational problem, for they usually are as large and as complex as the framework of the molecule itself. Commonly we treat the sigma framework as if it were completely localized, chopping up the representational problem into intellectually convenient little pieces, each piece corresponding to a particular bond or electron pair. Toward this end, the atomic orbitals are mixed to give hybridized orbitals that are more convenient for the discussion of these localized bonding orbitals. These localized bonding orbitals are easy to use and correspond to the lines and dots of our Lewis formula. There are very few penalties for using hybrid orbitals and treating the framework as localized, when it actually is not.

1.9   AROMATICITY, CYCLIC SYSTEMS

Benzene, C_6H_6, is much less reactive than simple alkenes and is described by two neutral resonance structures:

Any cyclic conjugated compound will be especially stable when the ring contains **4 n + 2 pi electrons**, where n is any integer (*Hückel's rule*). This extra stability, "resonance stabilization," of the 2, 6, 10, 14, 18, ... electron systems is termed *aromaticity*. The resonance stabilization of benzene is commonly estimated to be about 36 kcal/mol (151 kJ/mol), but this is not an experimentally verifiable number. The $4n$ pi electron systems (4, 8, 12, 16, ...) not only are less stable than the 4 n + 2 systems but are destabilized relative to their open-chain analog and are called *antiaromatic*. To summarize, aromatic stabilization occurs in rings that have an unbroken loop of p orbitals (a p orbital on each atom) and 4 n + 2 electrons in the loop.

If the molecule has an unbroken loop of p orbitals, you will be able to draw at least two resonance structures that shift the double bonds around the loop, as we did for benzene. If any atom in the ring is sp^3 hybridized, there can be no p orbital on that atom; the loop is broken, and the ring cannot be aromatic.

To determine how many electrons occupy the loop, count each double bond as contributing two electrons. A lone pair on an atom in the loop is counted as contributing two electrons if there is no double bond to that atom. (If an atom is doubly bonded and contains a lone pair, the lone pair cannot be counted since it must be in an orbital that is perpendicular to the loop.) If an atom has two lone pairs, just one is counted (only one can align with the loop). Count a triple bond as contributing only two electrons (one of the two pi bonds must be perpendicular to the loop).

To understand this alternating stability and instability for cyclic conjugated polyenes we need to look at the molecular orbitals. Although the derivation of these molecular orbitals is beyond the scope of this text, a qualitative understanding is possible.

Molecular Orbital Theory for Cyclic Systems
(A Supplementary, More Advanced Explanation)

The molecular orbitals for benzene are shown in Figure 1.32; the view is from the side. The ring is planar, and every atom in the ring bears a p orbital; we have a closed loop of overlapping *p* orbitals. As with the linear MOs, the more nodes the MO has, the higher its energy. We have basically taken our linear system and brought the ends together so that now we have standing waves around a loop. Note that two pairs of MOs, called *degenerate pairs*, have the same energy, for they have the same number of nodes. When we place the six pi electrons in the MOs for benzene, a degenerate pair is completely filled; benzene has a filled shell.

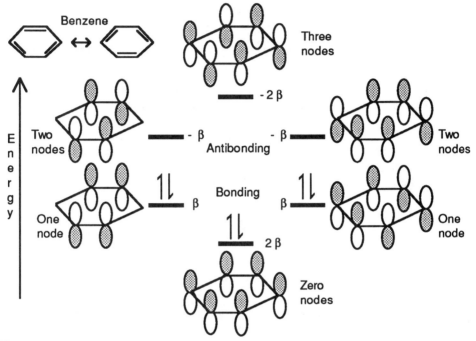

Figure 1.32 The pi molecular orbitals of benzene. The nodes are planes that go through the center and cut both sides of the loop.

Cyclopentadienyl cation, on the other hand, is a very unstable molecule. Its MO energy levels are shown in Figure 1.33. When the four pi electrons are placed in the MOs for cyclopentadienyl cation, the degenerate pair is only half-filled; cyclopentadienyl cation has an incomplete shell. The instability of cyclopentadienyl cation comes as a direct result of these unpaired electrons in the half-filled shell.

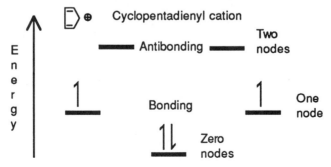

Figure 1.33 The pi molecular orbitals of cyclopentadienyl cation.

The MO energy levels of cyclopentadienyl radical are shown in Figure 1.34. Cyclopentadienyl radical has one more electron than the cation, thus five electrons total.

It has one unpaired electron and is a reactive radical. Radical reactions will be covered in Chapter 10.

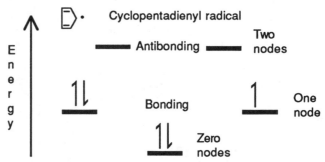

Figure 1.34 The pi molecular orbitals of cyclopentadienyl radical.

The MO energy levels of cyclopentadienyl anion are shown in Figure 1.35. Cyclopentadienyl anion has two more electrons than the cation, thus six electrons total, and therefore has a filled degenerate pair and is predicted correctly to be aromatic and stable; it has enough electrons to fill the shell.

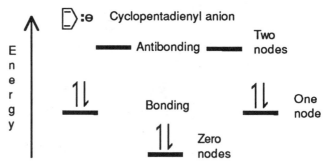

Figure 1.35 The pi molecular orbitals of cyclopentadienyl anion.

The stabilization of an aromatic can be placed on a semiquantitative scale. If we define the stabilization of one electron in the pi orbital of ethene as β, then the stabilization of two electrons in a double bond is just 2β. The stabilization of six pi electrons in three double bonds is 6β. Benzene's MO energy levels are at 2β, 1β, 1β, -1β, -1β, and -2β; the first three are filled with two electrons each for a stabilization of 8β total $(2 \times 2\beta + 4 \times 1\beta)$. The six electrons of benzene pi loop are then 2β more stabilized than the six electrons in three isolated double bonds.

The MO energy levels for any simple cyclic conjugated polyene can be estimated by drawing the regular polygon of the ring point down inscribed in a circle of 4β diameter (Figure 1.36). The vertices of the polygon will correspond to the energies of the MOs on a vertical scale. The center of the circle corresponds to the nonbonding level (zero β) with the bonding levels below it and the antibonding above.

Figure 1.36 The pi molecular orbital energies of cyclic systems.

Since the degenerate pair must be completely filled or completely empty in order for the molecule to be aromatic, stable arrangements arise when 2, 6, 10, ... or $4n + 2$ pi electrons are in the cyclic conjugated pi system. Unstable half-filled degenerate shells will occur with 4, 8, 12, ... or $4n$ pi electrons. The $4n$ systems tend to distort from planarity to diminish this destabilizing conjugation.

ADDITIONAL EXERCISES

1.1 Draw Lewis structures with resonance for the following neutral compounds.

$O-O-O$ H_2C-N-N $O-C-O$ $H-O-N-O$ $H-O-NO_2$

1.2 Draw Lewis structures with resonance for the following charged species.

$H_2CNH_2^{\oplus}$ $OCN^{\ominus}$ $HOCO_2^{\ominus}$ $H_2COH^{\oplus}$ $H_2CCHCH_2^{\oplus}$

1.3. What is the polarization of the indicated bond?

$Br-Br$ $HO-Cl$ H_2B-H $H_2C=O$ $I-Cl$

1.4 Circle the electrophiles and underline the nucleophiles in the following group.

BF_3 $H^{\oplus}$ Ne NH_3 $^{\ominus}C\equiv N$

1.5 Draw the Lewis structure(s) that would be the product of the arrows.

(a)

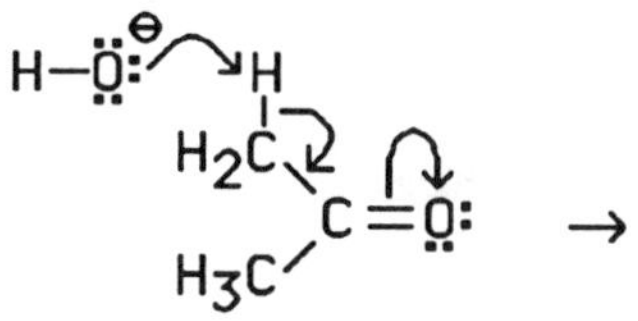

(b)

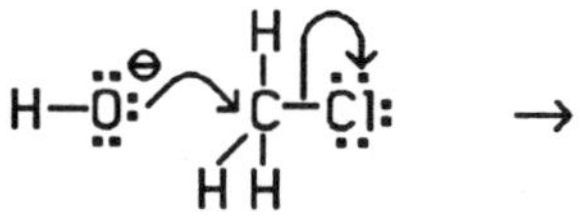

(c)

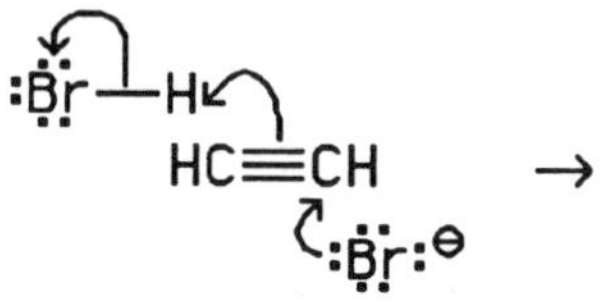

(d)

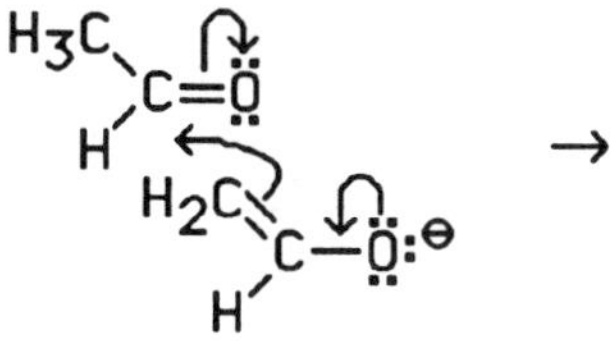

(e)

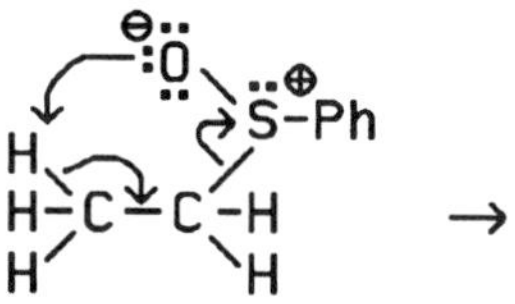

1.6 Give the curved arrows necessary for the following reactions.

(a)

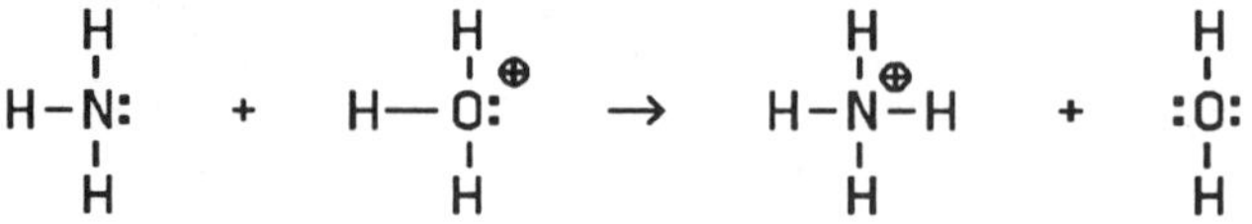

$$\underset{\overset{|}{H}}{\overset{\overset{H}{|}}{H-N:}} \;+\; \underset{\overset{|}{H}}{\overset{\overset{H}{|}}{H-\overset{\oplus}{O}:}} \;\rightarrow\; \underset{\overset{|}{H}}{\overset{\overset{H}{|}}{H-\overset{\oplus}{N}-H}} \;+\; \underset{\overset{|}{H}}{\overset{\overset{H}{|}}{:O:}}$$

(b)

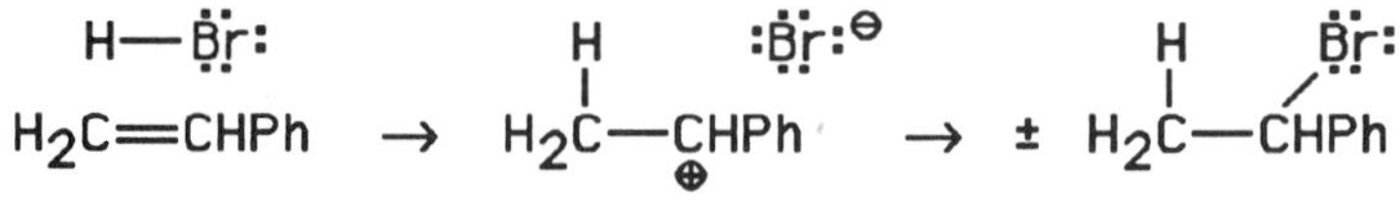

$$\begin{array}{ccc}
H-\ddot{B}r: & H & :\ddot{B}r:^{\ominus} & H \quad \ddot{B}r: \\
H_2C{=}CHPh & \rightarrow & H_2\overset{|}{C}{-}\underset{\oplus}{CHPh} & \rightarrow \; \pm \; H_2\overset{|}{C}{-}CHPh
\end{array}$$

(c)

$$\begin{array}{ccc}
:N{\equiv}C:^{\ominus} & :N{\equiv}C\diagdown & :N{\equiv}C\diagdown \\
\underset{H_3C}{\overset{H_3C}{}}{C}{=}\ddot{O}: & \underset{H_3C}{\overset{H_3C}{}}C{-}\ddot{O}:^{\ominus} & \underset{H_3C}{\overset{H_3C}{}}C{-}\ddot{O}{-}H \\
\qquad H{-}C{\equiv}N: & \qquad H{-}C{\equiv}N: & \qquad {}^{\ominus}:C{\equiv}N:
\end{array}$$

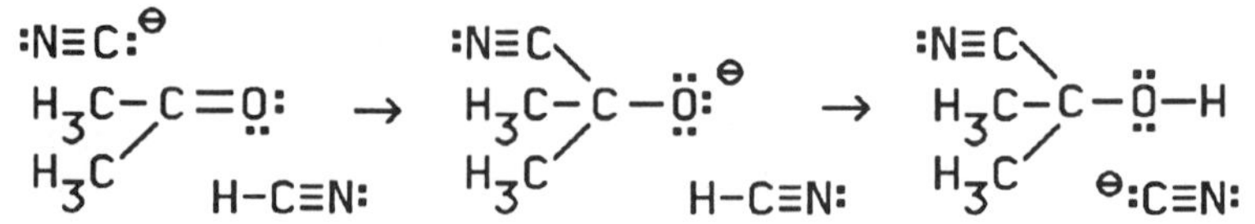

(d)

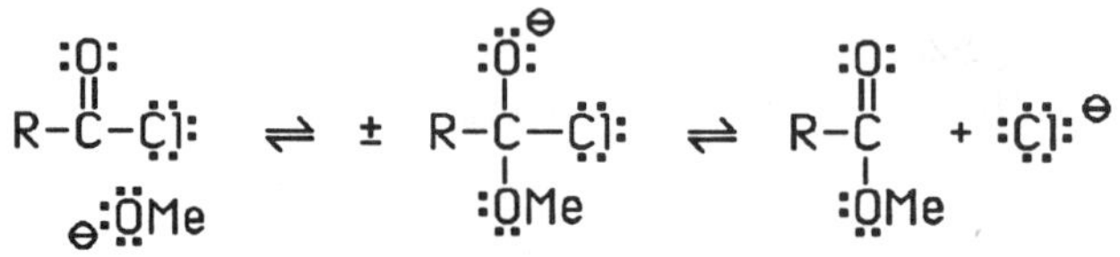

$$\begin{array}{ccc}
\overset{\textstyle :O:}{\underset{}{\parallel}} & \overset{\textstyle :\ddot{O}:^{\ominus}}{\underset{}{|}} & \overset{\textstyle :O:}{\underset{}{\parallel}} \\
R{-}\overset{|}{C}{-}\ddot{C}l: & \rightleftharpoons \pm R{-}\overset{|}{C}{-}\ddot{C}l: & \rightleftharpoons R{-}C \;+\; :\ddot{C}l:^{\ominus} \\
\underset{\ominus}{:}\ddot{O}Me & \quad :\ddot{O}Me & \quad :\ddot{O}Me
\end{array}$$

(e)

$$^{\ominus}A: \qquad \underset{(CH_3)_3C\oplus}{} \qquad \rightarrow \qquad ^{\ominus}A: \qquad \underset{(CH_3)_3C}{} \qquad \rightarrow \qquad A{-}H \qquad (CH_3)_3C{-}$$

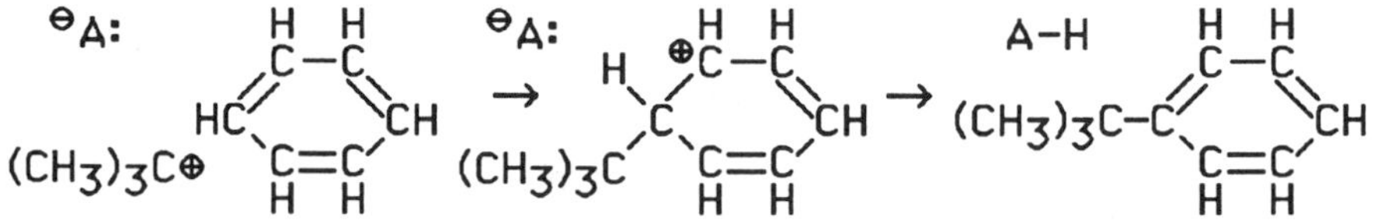

1.7 Draw full Lewis structures for the following line structures.

1.8 In the following structures circle any carbon atom that bears a significant partial positive charge. (Hint: look for electronegative atoms.)

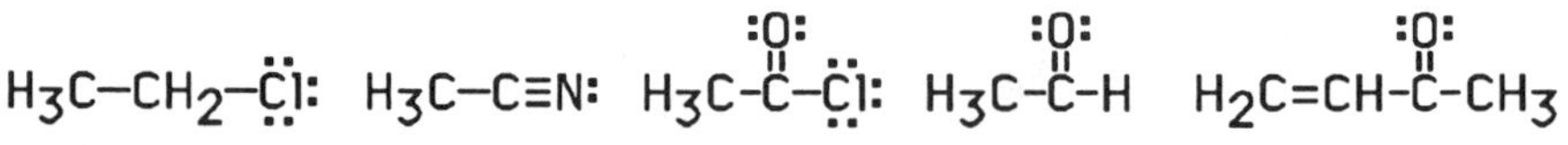

$$H_3C{-}CH_2{-}\ddot{C}l: \quad H_3C{-}C{\equiv}N: \quad H_3C{-}\overset{\overset{\textstyle :O:}{\parallel}}{C}{-}\ddot{C}l: \quad H_3C{-}\overset{\overset{\textstyle :O:}{\parallel}}{C}{-}H \quad H_2C{=}CH{-}\overset{\overset{\textstyle :O:}{\parallel}}{C}{-}CH_3$$

1.9 In the following structures circle any carbon atom that bears a negative or significant partial negative charge. (Hint: draw some resonance forms.)

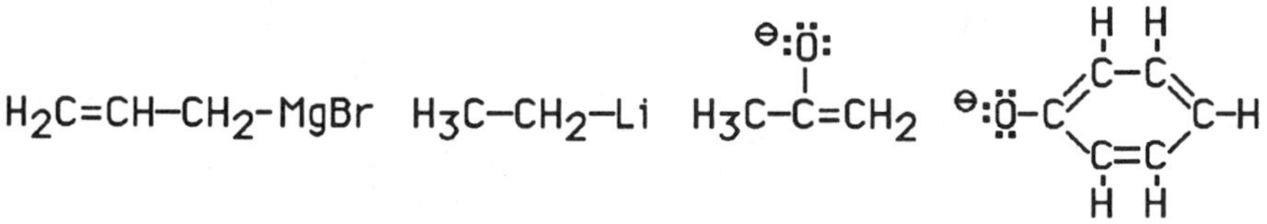

$$H_2C{=}CH{-}CH_2{-}MgBr \quad H_3C{-}CH_2{-}Li \quad H_3C{-}\overset{\overset{\textstyle ^{\ominus}:\ddot{O}:}{|}}{C}{=}CH_2 \quad ^{\ominus}:\ddot{O}{-}C$$

1.10 Give the hybridization of the carbons in these structures.

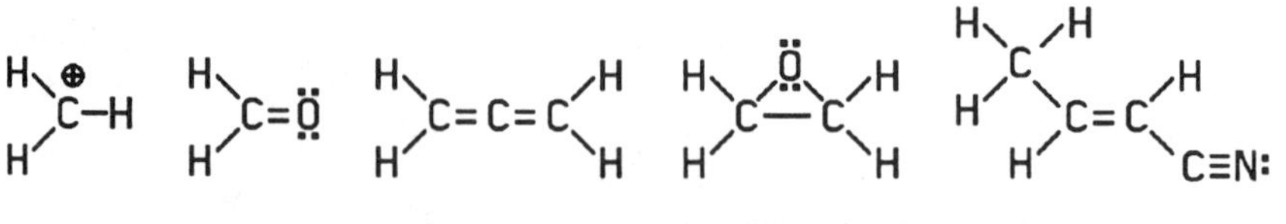

$$\underset{H}{\overset{H}{}}\overset{\oplus}{C}{-}H \quad \underset{H}{\overset{H}{}}C{=}\ddot{O} \quad \underset{H}{\overset{H}{}}C{=}C{=}\underset{H}{\overset{H}{}}C \quad \underset{H}{\overset{H}{}}\overset{\overset{\textstyle :\ddot{O}:}{}}{C}{-}\underset{H}{\overset{H}{}}C \quad \underset{H}{\overset{H}{}}C{=}\underset{C{\equiv}N:}{\overset{H}{}}C$$

1.11 Which of the following structures are chiral?

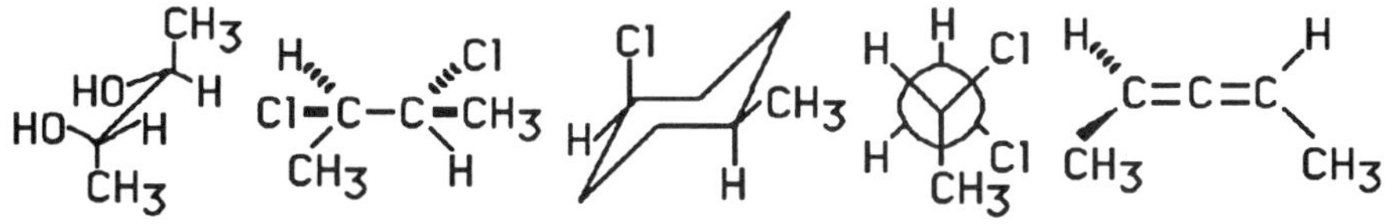

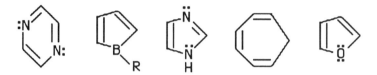

1.12 Draw the pi overlap for an amide, carbon dioxide, an ester, and a vinylogous amide.

1.13 Circle the aromatic compounds in the following list.

1.14 Circle the conjugated systems in the following list.

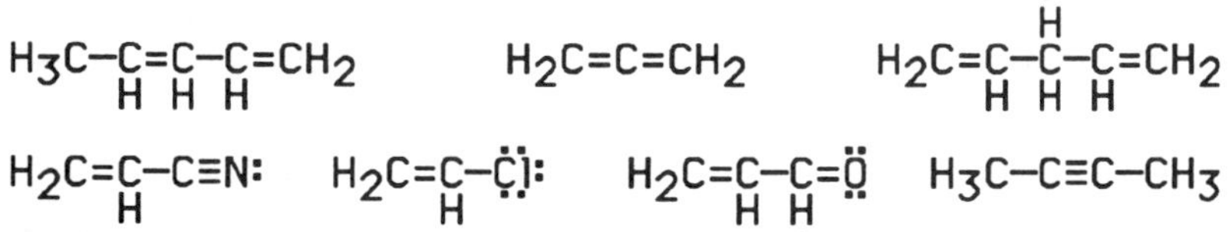

$H_3C-\underset{H}{C}=\underset{H}{C}-\underset{H}{C}=CH_2$ $H_2C=C=CH_2$ $H_2C=\underset{H}{C}-\overset{H}{\underset{H}{C}}-\underset{H}{C}=CH_2$

$H_2C=\underset{H}{C}-C\equiv N\colon$ $H_2C=\underset{H}{C}-\ddot{\underset{\cdot}{C}}\colon$ $H_2C=\underset{H}{C}-\underset{H}{C}=\ddot{O}$ $H_3C-C\equiv C-CH_3$

1.15 Indicate the number of electrons in the pi system of the following compounds.

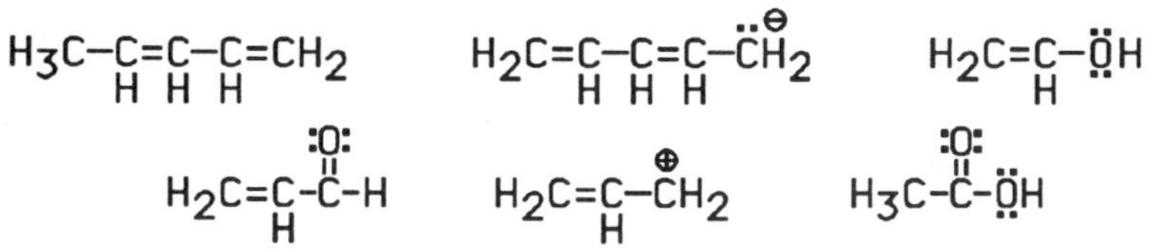

$H_3C-\underset{H}{C}=\underset{H}{C}-\underset{H}{C}=CH_2$ $H_2C=\underset{H}{C}-\underset{H}{C}=\underset{H}{C}-\overset{..\ominus}{C}H_2$ $H_2C=\underset{H}{C}-\ddot{O}H$

$H_2C=\underset{H}{C}-\overset{:O:}{\overset{\|}{C}}-H$ $H_2C=\underset{H}{C}-\overset{\oplus}{C}H_2$ $H_3C-\overset{:O:}{\overset{\|}{C}}-\ddot{O}H$

2

THE PROCESS
OF BOND
FORMATION

A reaction will occur whenever it is energetically reasonable for it to do so. If the reactant can overcome the energy barrier to reaction and produce a product that is lower in energy, the reaction will always proceed. Therefore, to be able to determine when a reaction will occur, we need to understand some basics of energetics.

2.1 THERMODYNAMICS, POSITION OF EQUILIBRIUM

Energy diagrams are a plot of energy versus the reaction coordinate, a measure of the degree of a molecule's progress toward complete reaction. The energy diagram is actually a slice along the lowest energy path on an energy surface. Although the actual energy surface for all but the simplest of reactions is very complex, we can learn much from a simple three-dimensional surface that plots only the most important variables. The vertical axis would be energy and the two horizontal axes would each be the distance between atoms undergoing bond breaking or bond making.

For example, for a simple reaction, $Nu:^{\ominus} + Y{-}L \rightarrow Nu{-}Y + L:^{\ominus}$, one horizontal axis would be the distance between Nu and Y; the other horizontal axis would be the distance between Y and L. As the reaction progresses, the distance between Nu and Y decreases while that between Y and L increases. The reaction would go through a point of highest energy (a *transition state*, symbolized by $\ddagger$) in which Y is partially bonded to both Nu and L.

For simplicity, we can start with two dimensions by looking down at the surface from above. The reaction's simple energy surface as viewed from the top is shown in Figure 2.1. The dark line from reactants, R, to products, P, is the path of lowest energy that the reacting species might follow. As the reactants proceed along this diagonal lowest energy path, the Y—L bond begins to break (any movement on the surface to the right stretches the Y—L bond), and the Nu—Y bond begins to form (any movement on the surface in the downward direction brings Nu and Y closer together). The midpoint of the diagonal corresponds to a point where the Y—L bond is half-broken, and the Nu—Y bond is half-formed.

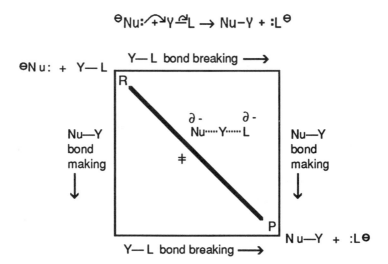

Figure 2.1 A simplified energy surface viewed from the top.

The third dimension, energy, becomes apparent in Figure 2.2 as we view the surface from off to the side. The lowest energy path, again shown by the dark line, starts back at

R and goes up through the "mountain pass," the transition state, and then down to point P in the front corner of the surface (the path is shaded when it is behind the surface). It is downhill to the products from the transition state.

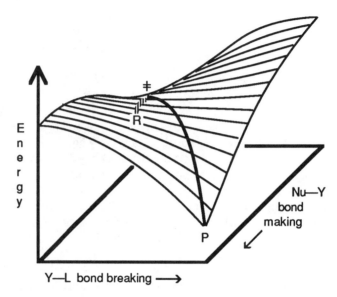

Energy

Y—L bond breaking ⟶

Nu—Y bond making

Figure 2.2 A simplified energy surface viewed from the side.

The slice through the surface from point R to P along the lowest energy path is the energy diagram (Figure 2.3). For this particular simple surface, the reaction coordinate was the diagonal in the top view of the surface. Energy diagrams are easier to visualize than the lowest energy path displayed on a more complex surface; there the path may weave around, and fall into and climb out of the energy minima of several intermediates along the way to product.

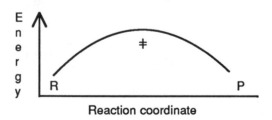

Energy

R   P

Reaction coordinate

Figure 2.3 The lowest energy slice of the simplified energy surface.

The example in Figure 2.3 with the slight modification that P is more stable than R would produce an energy diagram shown in Figure 2.4 (energy is expressed as G, free energy). The standard free-energy difference between the reactants and the products is $\Delta G°$ and is related to the equilibrium constant, K_{eq}, by the formulas given below:

$$\Delta G° = -RT \ln K_{eq} = -2.303\,RT \log K_{eq}$$

or $\log K_{eq} = -\Delta G°/2.303RT$ or $K_{eq} = 10^{-\Delta G°/2.303RT}$

where $R = 1.99 \times 10^{-3}$ kcal/mol-K (8.33×10^{-3} kJ/mol-K)

and T = temperature in K.

At a common room temperature of 25°C, $T = 298\,K$ so $2.303RT = 1.37$ kcal/mol (5.73 kJ/mol), and therefore if $\Delta G°$ is expressed in kcal/mol, $K_{eq} = 10^{-\Delta G°/1.37}$. The formula for $\Delta G°$ expressed in kJ/mol is $K_{eq} = 10^{-\Delta G°/5.73}$.

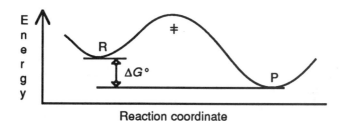

Reaction coordinate

Figure 2.4 An energy diagram in which the products are more stable than the reactants

As the standard free-energy difference, $\Delta G°$, gets more negative, K_{eq} gets larger, and the more the lower-energy compound P will predominate in the equilibrium mixture. If $\Delta G°$ is a positive value, then the product P is uphill in energy from the reactants. Table 2.1 gives representative values for room temperature, 25°C, with P becoming more stable than R. **At room temperature, every 1.37 kcal/mol (5.73 kJ/mol) changes the equilibrium constant by a factor of 10.** Thus, we have a quick way to interconvert $\Delta G°$ in kcal/mol to K_{eq} at room temperature: divide $\Delta G°$ by -1.37 to obtain the exponent of K_{eq}; conversely, multiply the exponent of K_{eq} by -1.37 to get $\Delta G°$ in kcal/mol. For kJ/mol use -5.73 instead.

Table 2.1 $\Delta G°$ and K_{eq} Values for 25°C

$\Delta G°$ kcal/mol	K_{eq}	Reactant	Product	$\Delta G°$ (kJ/mol)
+5.46	0.0001	99.99	0.01	+22.84
+4.09	0.001	99.9	0.1	+17.11
+2.73	0.01	99	1	+11.42
+1.37	0.1	91	9	+5.73
+1.0	0.18	85	15	+4.18
+0.5	0.43	70	30	+2.09
0	1	50	50	0
-0.5	2.33	30	70	-2.09
-1.0	5.41	15	85	-4.18
-1.37	10	9	91	-5.73
-2.73	100	1	99	-11.42
-4.09	1,000	0.1	99.9	-17.11
-5.46	10,000	0.01	99.99	-22.84
-9.56	10^7	Essentially complete		-40.00

Exercise: Cover the value of K_{eq} in Table 2.1 and give the approximate K_{eq} from the $\Delta G°$; then cover the $\Delta G°$ and give the approximate $\Delta G°$ from the K_{eq}.

A reaction can have a positive $\Delta G°$ and still proceed. For example, if the reaction had an unfavorable equilibrium constant with $\Delta G°$ as +2.73 kcal/mol (+11.42 kJ/mol), the equilibrium mixture would contain a 99:1 R to P mixture. If we started the reaction with 100 mol of pure R, the reaction would be spontaneous to produce 1 mol of P to give the equilibrium mixture.

If a following reaction removed the small amount of P that was formed, then more P would be produced to restore the equilibrium mixture. That following reaction would act as a *driving force* to overcome an unfavorable R to P equilibrium.

The standard free energy change of a reaction can be separated into its components:

$$\Delta G° = \Delta H° - T\Delta S°$$

The standard *enthalpy* change, $\Delta H°$, **is a measure of the heat absorbed or evolved in a reaction** and bears a negative sign if heat is given off (exothermic reactions). All exothermic reactions break weak bonds and make strong ones. (If heat is absorbed, the reaction is endothermic and $\Delta H°$ is positive). A bond-strength table is included in the Appendix. The heat of reaction can be calculated from the difference in the heats of formation of the reactants and products, but often those heats of formation are not known. A method to get an approximate heat of reaction is to consider just the bonds being formed and broken. Since bond breaking requires heat and bond making releases it, the heat of reaction can be approximated as:

$$\Delta H° = \Delta H_{(bonds\ broken)} - \Delta H_{(bonds\ made)}$$

Another use for the bond-strength table is to predict which of two products has most likely a lower heat of formation and is the preferred product of the reaction. The product that overall has the strongest bonds is preferred. Only the bonds that differ in each structure need be considered. For example, for two possible products, A and B, we are just finding out which way is exothermic for the reaction: product A $\rightleftharpoons$ product B.

The standard *entropy* change, $\Delta S°$, **is a measure of the change in the randomness within the system.** If there is more disorder after the reaction, for example, one molecule breaking in two (with no complicating factors such as solvation), $\Delta S°$ will be positive. The entropy term arises from a combination of translational, rotational, and vibrational motions.

Example problem

Calculate $\Delta H°$ for $H_2C=CH_2 + HBr \rightarrow H_3C-CH_2Br$

Answer: Bonds broken = C=C 146 kcal/mol (611 kJ/mol) + H—Br 87 kcal/mol (364 kJ/mol) = 233 kcal/mol (975 kJ/mol)
Bonds made = C—C 83 kcal/mol (347 kJ/mol) + C—H 99 kcal/mol (414 kJ/mol) + C—Br 68 kcal/mol (285 kJ/mol) = 250 kcal/mol (1046 kJ/mol)
Since $\Delta H°$ = bonds broken - bonds made = 233 - 250 = -17 kcal/mol (-71 kJ/mol)
Because the sign is negative, the reaction is exothermic; it gives off heat. It is important to notice that we broke the double bond, at 146 kcal/mol (611 kJ/mol), and replaced it by making a single bond, at 83 kcal/mol (347 kJ/mol).

2.2   KINETICS, RATE OF REACTION

A reaction may be downhill in energy and still not proceed at a reasonable rate. **The rate of reaction is controlled by the barrier height to reaction, $\Delta G^{\ddagger}$, the free energy of activation.** If the barrier is too high to be overcome using the available energy, no reaction occurs. If enough energy is added to overcome the barrier, then the reaction proceeds. Paper is capable of sitting in air without burning for quite a long time at room temperature, but if the temperature is raised high enough it will burst into flames. In Figure 2.5, note that the $\Delta G^{\ddagger}$ for the forward and reverse reactions is different. Since B starts out at a lower energy it must climb more of an energy hill than A to get to the same transition state, denoted by the symbol ‡.

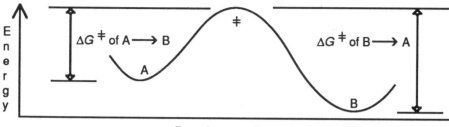

Reaction coordinate

Figure 2.5 The free energy of activation.

If the barrier for B to A is very large, then the reaction is an irreversible *unimolecular* (first-order) process, A → B, and the rate of B formation is $d[B]/dt = k_f[A]$ where k_f is the forward rate constant. If the reaction is reversible, then the overall rate is the reverse rate, $k_r[B]$, subtracted from the forward rate, $k_f[A]$. Overall rate = $d[B]/dt = k_f[A] - k_r[B]$. The initial rate (when [B] = 0) is just $k_f[A]$, but as the concentration of B grows, so does the extent of reverse reaction. At equilibrium, the concentration of B does not change, so $d[B]/dt$ is zero, making $k_f[A] = k_r[B]$ which rearranges to $[B]/[A] = k_f/k_r = K_{eq}$.

The rate of an irreversible *bimolecular* (two molecule) reaction in which A collides with B to produce C is expressed as a rate constant, k, times the concentrations.

$$A + B \rightarrow C \qquad Rate = k[A][B]$$

Concentrations for reactions vary, but the average organic reaction is run at 1 M or less. As an upper limit on concentration recall that pure water is 55.5 M. If B is a reactive intermediate, its concentration may be very small (if [B] goes from 10^{-3} to 10^{-6} M, the rate would drop by a factor of a thousand). Reactive intermediates often have large rate constants which can more than make up for their low concentrations.

The largest value that the bimolecular rate constant can have is 10^{10} L/mol-sec, which corresponds to a reaction upon every collision. This limit, the rate at which the two molecules collide, is called the *diffusion controlled limit*.

The rate constant, k, is related to the barrier height, $\Delta G^{\ddagger}$, by the formula

$k = (\kappa T/h)e^{-\Delta G^{\ddagger}/RT}$
R = gas constant = 1.99×10^{-3} kcal/mol-K (8.33×10^{-3} kJ/mol-K)
T = temperature in K
κ = Boltzmann's constant = 1.38×10^{-23} J/K
h = Planck's constant = 6.63×10^{-34} J-sec

Using the above equation, we can calculate the rate constant k for various values of $\Delta G^{\ddagger}$, shown in Table 2.2. The half-life of a unimolecular reaction is equal to $\ln 2/k$ or $0.693/k$, and a reaction is 97% complete after five half-lives. A unimolecular reaction with a $\Delta G^{\ddagger}$ of 15 kcal/mol (63 kJ/mol) would be complete in 0.05 seconds. If the barrier is 25 kcal/mol (105 kJ/mol), then completion would take almost 2 weeks. **At room temperature, dropping the $\Delta G^{\ddagger}$ by 1.37 kcal/mol (5.73 kJ/mol) increases the rate tenfold.** The number of molecules in the reaction mixture at room temperature that have enough energy to traverse a 20 kcal/mol (84 kJ/mol) barrier is small; the average energy of a molecule at room temperature is only 1 kcal/mol (4.2 kJ/mol). However, if we lower the barrier many more molecules will have sufficient energy to react.

Table 2.2 $\Delta G^{\ddagger}$, Rate Constants, Half-lives, and Completion Times of a Unimolecular Reaction at Room Temperature, 298 K (25°C)

$\Delta G^{\ddagger}$ in kcal/mol	k_{298} in sec^{-1}	Half-Life	97% complete
15 (63 kJ/mol)	64.1	0.01 sec	0.05 sec
16 (67 kJ/mol)	11.9	0.06 sec	0.3 sec
17 (71 kJ/mol)	2.2	0.3 sec	1.6 sec
18 (75 kJ/mol)	4.1×10^{-1}	1.7 sec	8.5 sec
19 (79 kJ/mol)	7.5×10^{-2}	9.2 sec	45.9 sec
20 (84 kJ/mol)	1.4×10^{-2}	49.6 sec	4.1 min
21 (88 kJ/mol)	2.6×10^{-3}	4.5 min	22.3 min
22 (92 kJ/mol)	4.8×10^{-4}	24 min	2 hr
23 (96 kJ/mol)	8.9×10^{-5}	130 min	10.8 hr
24 (100 kJ/mol)	1.6×10^{-5}	11.7 hr	2.4 days
25 (105 kJ/mol)	3.0×10^{-6}	63.2 hr	13.2 days

Table 2.2 used the formula $\ln(c^{\circ}/c) = kt$ to relate the rate constant, k, and the original concentration, c°, to the concentration, c, at time t for a unimolecular reaction. Reactant concentrations are very important in a bimolecular reaction. If reactant B in a bimolecular reaction is in a large excess, then its concentration will not change significantly over the course of the reaction and can be considered a constant, making the reaction *pseudo-first order*. We can then use the above formula to calculate the concentration of A at time t by substituting the pseudo-first-order rate constant, $k' = k[B]$.

We can estimate the half-life and 97% completion times for a bimolecular reaction in which both reactants are at the same concentration by using the k from Table 2.2 and the formula $(1/c) - (1/c^{\circ}) = kt$. For example, a reaction with a 20 kcal/mol (84 kJ/mol) barrier and both reactants at 1 M has a 97% completion time of 38.5 min at 25°C. A common landmark value is the $\Delta G^{\ddagger}$ **of a reaction that proceeds at a reasonable rate at room temperature is about 20 kcal/mol (84 kJ/mol).**

Example problem

If the concentration of both reactants of a bimolecular reaction were at a common value of 0.1 M, how long would it take to be 97% complete at 25°C if $\Delta G^{\ddagger}$ is 20 kcal/mol (84 kJ/mol)?

Answer: 97% complete means 3% starting material remains, so c is 0.003 M if c° was 0.1 M. Substitution of these values and a k of 0.014 from Table 2.2 into the formula: $(1/c) - (1/c^{\circ}) = kt$, gives the time as 6.4 hr.

We can see that our landmark value for a reasonable reaction rate at room temperature is highly dependent on $\Delta G^{\ddagger}$, the molecularity of the reaction, the reactant concentrations, and of course, on how long the experimenter considers a reasonable time to wait. This value still is useful as a reference point. The ring flip of cyclohexane that has a $\Delta G^{\ddagger}$ of 11 kcal/mol (46 kJ/mol) is very fast at room temperature. A reaction with a $\Delta G^{\ddagger}$ of 40 kcal/mol (167 kJ/mol) does not proceed at room temperature and has to be heated to over 300°C before it proceeds quickly. An approximate rule of thumb for the temperature dependence of a reaction is that the rate doubles for every 10°C increase in temperature.

If the reaction goes by a stepwise mechanism, it may have an energy diagram similar to that shown in Figure 2.6, in which intermediates lie in minima, energy valleys, and transition states are always at maxima, the top of the energy hills. The *lifetime* of intermediate C is defined as the inverse of its rate constant for unimolecular decomposition. As the height of the lowest barrier to decomposition of C drops, the lifetime of C gets shorter. When the lifetime of the intermediate gets less than 10^{-12} sec (only long enough for several molecular vibrations), then A + B → D is considered concerted, occurring in one step. Still, the various bond-forming and -breaking processes need not be exactly synchronous.

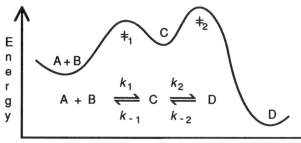

Reaction coordinate

Figure 2.6 The energy diagram for a reaction that proceeds through an intermediate, C. Over the equilibrium arrows are the rate constants for each forward and reverse step.

The highest energy hill, $\ddagger_2$, is a major factor in determining the rate of reaction along with the concentration of intermediate C, which may not be known. Any proposed reaction mechanism must fit the experimentally derived rate expression. The rate expression can be be obtained in simplified form if two assumptions are made: the second step is irreversible (k_{-2} is negligible), and the second step is slow compared to the first (k_2 is much smaller than k_1 or k_{-1}). The reaction rate for the formation of D is $k_2[C]$, where [C] is determined by the equilibrium constant, $k_1/k_{-1} = [C]/[A][B]$.

Solving for [C] gives: $[C] = k_1[A][B]/k_{-1}$

Substitution into the rate expression gives

$$\text{rate} = \frac{d[D]}{dt} = k_2[C] = \frac{k_2 k_1[A][B]}{k_{-1}}$$

The rate expression without any restrictions on the rate constants can be obtained by assuming that intermediate C does not build up, the *steady-state approximation*. The rate of formation of C, $k_1[A][B] + k_{-2}[D]$, equals its rate of destruction, $k_{-1}[C] + k_2[C]$. Solving for [C] gives

$$[C] = \frac{k_1[A][B] + k_{-2}[D]}{k_{-1} + k_2}$$

Substitution for [C] into the rate expression $d[D]/dt = k_2[C] - k_{-2}[D]$ gives

$$\frac{d[D]}{dt} = \frac{k_2(k_1[A][B] + k_{-2}[D])}{k_{-1} + k_2} - k_{-2}[D]$$

This simplifies to rate expression with a forward and a reverse term just like our simple equilibrium example

$$\text{rate} = \frac{d[D]}{dt} = \frac{k_1 k_2[A][B]}{k_{-1} + k_2} - \frac{k_{-1}k_{-2}[D]}{k_{-1} + k_2}$$

If C → D is irreversible, $k_{-2} = 0$, the second term drops out giving

$$\text{rate} = \frac{d[D]}{dt} = \frac{k_1 k_2[A][B]}{k_{-1} + k_2}$$

If the second step is slow compared to the first, $k_{-1} \gg k_2$, the expression becomes the same as our one with the two assumptions.

If the first step in a multistep process is the slowest, this "bottleneck" step will be the *rate-determining step* and not any of the faster steps that follow in the overall process.

A transition state contains partially broken and formed bonds, but the degree of this varies and depends on how close to the starting materials or products the point of transition is. The *Hammond postulate*, illustrated by Figure 2.7, states that the transition states of exothermic reactions resemble starting materials in energy and geometry whereas transition states of endothermic reactions resemble products. Transition states close to starting materials, A → B, have rather little bond breaking or making. Transition states close to products, C → D, have almost complete bond breaking and making.

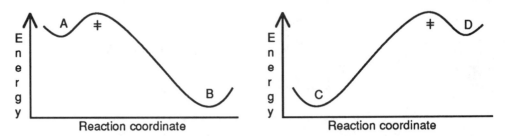

Figure 2.7 The transition state shifts position as the overall energy of the process changes.

Exercise

Hold a strip of paper in your hand so that both ends are at the same level and bow it up as in Figure 2.4. Notice that the high point, our transition state, is in the exact center. Now lower the end of the strip held in your right hand and notice that the transition point moves toward your left hand. Now raise your right higher than your left and notice how the transition state shifts and compares with that shown in Figure 2.7.

As shown in Figure 2.8, the stability of the product has a great influence on the energy of the transition states of endothermic reactions (D → E and F) but has very little influence in the energy of the transition states of exothermic reactions (A → B and C).

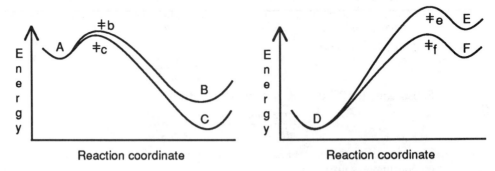

Figure 2.8 The degree that the product stability influences the transition state.

Therefore, with few exceptions, the more exothermic a reaction is, the less the product stability matters in the energy of the transition state and the less selective the reaction is. **The more reactive a species is, the less selective it is.** A general rule of reactivity is, **the more stable a compound is, the less reactive it is.** The converse is also true: The less stable a compound is, the more reactive it is. Basically, a highly reactive, unstable compound might react with almost anything, whereas a less reactive, more stable compound is more limited in what it can react with. It would react at the best site for reaction in preference to other, less favorable sites.

Some reactions can produce two different products, and it is necessary to determine whether the reaction is reversible to be able to predict the product. For example, in Figure 2.9, if the reaction is completely reversible, an equilibrium is established in which the major species in the reaction pot is the most stable, lowest-energy one, G. A reaction that produces the more stable product is under *thermodynamic control*. The product ratio is determined by the difference in free energy of the two products, ΔG° (Table 2.1).

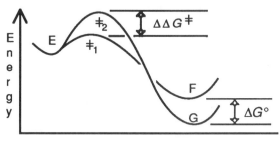

Reaction coordinate

Figure 2.9 The energy diagram for two competing reactions. Reaction E to F has the lower barrier to reaction (kinetic), whereas reaction E to G produces the more stable product (thermodynamic).

If the reaction is not reversible, then the product that is formed faster, F, will be the major one. When the relative rates of reaction (barrier heights) determine the product ratio, the reaction is under *kinetic control*. As the temperature drops and the molecules have less energy, the reaction takes the lower barrier pathway.

When the relative rates of reaction control the product ratio, how much effect does a small difference in the $\Delta G^\ddagger$ make? The equation for the relative rate of formation of two products at the same temperature varies exponentially with the difference in the barrier heights, $\Delta\Delta G^\ddagger$.

$$\text{Relative rate of formation} \quad G/F = k_2/k_1 = e^{-\Delta\Delta G^\ddagger/RT}$$

At room temperature, a 1.37 kcal/mol (5.73 kJ/mol) difference in barrier heights means that F is formed about ten times as fast as G. This would mean that the product mixture would contain ten times as much F, the product of traversing the lower barrier, as it would G.

2.3 THE BASICS OF BARRIERS

What causes a barrier? The $\Delta G^\ddagger$ can be broken into two terms:

$$\Delta G^\ddagger = \Delta H^\ddagger - T\Delta S^\ddagger$$

The $\Delta H^\ddagger$ term arises because bond breaking must usually lead bond making to a slight degree. The old bond must be partially broken before the new one can begin to form. Since it costs energy to break a bond and we have not yet gained any energy from bond formation, extra energy in the form of $\Delta H^\ddagger$ is required to start the reaction. Any reaction that requires the complete breakage of a bond before any bond formation can begin will have a larger $\Delta H^\ddagger$ than a reaction whose bond breaking and making occur at almost the same time. The larger the $\Delta H^\ddagger$ is, the larger the barrier is.

The $\Delta S^{\ddagger}$ term is a measure of the degree of disorder in the transition state. If the transition state for the reaction requires that the reacting molecules be aligned relative to each other in a specific organization, then disorder is sacrificed and the value of $\Delta S^{\ddagger}$ is a negative number. A negative $\Delta S^{\ddagger}$ makes the $-T\Delta S^{\ddagger}$ term a large positive number, which increases the value $\Delta G^{\ddagger}$ (raises the barrier). The reaction proceeds slowly because few collisions have the proper alignment. As the temperature increases and molecular motion gets more violent, the alignment problem gets worse.

There are many factors that will increase the barrier height. As an intermediate increases in energy, the transition states leading up to it must also increase in energy. Therefore, the stability of any intermediates will directly affect the barrier height. As shown in Figure 2.10, anything that tends to stabilize the intermediate relative to the reactant lowers the barrier, and conversely, anything that stabilizes the reactant relative to the intermediate raises the barrier. As an example, if the reactant has additional bonding that is lost in going to the transition state, the barrier is raised. If the transition state has additional bonding that is not present in the reactant, the barrier is lowered. We will spend the next chapter discussing the stability of intermediates.

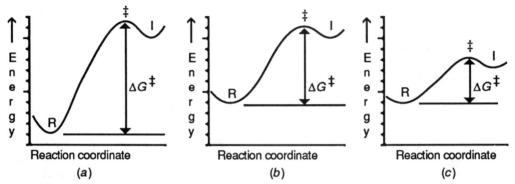

Figure 2.10 Use diagram b for reference. In a compared to b the reactant R is stabilized relative to the intermediate I; therefore the barrier increases. In c compared to b the intermediate is stabilized relative to reactant R; the barrier decreases.

Several generalizations on barriers are useful: Formation of charge from neutral reactants tends to raise the barrier, whereas neutralization of charge lowers the barrier. If reactant groups have trouble achieving the proper alignment for reaction, the barrier is raised. If the site for reaction is not very accessible, the barrier is raised. If the intermediate and the transition state leading up to it are better stabilized by solvation than the reactants, the barrier is lowered by this solvent stabilization (see Section 3.1).

A *catalyst* is a substance that speeds up a reaction and can be recovered unchanged. The most common catalysts for organic reactions are acids or bases. An acid can protonate a reactant, giving it a full positive charge and making it a better electron sink. A base can remove a proton, making the reactant anionic and a better electron source. On mechanism problems, beginners will sometimes try to do both acid-catalyzed and base-catalyzed processes at the same time, forgetting that a strong acid and base present in the same vessel would neutralize each other.

2.4   ORBITAL OVERLAP IN BOND FORMATION

Electron flow must occur through overlapping orbitals. If the orbital overlap is poor, the path is less likely to occur. **The formation of sigma bonds**

requires approximate colinearity of the reacting orbitals because that produces the best overlap, thus the strongest bond. **The formation of pi bonds requires approximate coplanarity of the reacting orbitals.**

The orbital alignment requirements have some slack; an error of 10° off the proper angle appears to have little effect. Pi-bond-forming reactions that involve deformations from coplanarity of up to 30° occur but are rare. Pi overlap falls off with the cosine of the twist angle. A 30° twist has 87% of the maximum overlap, a loss of about 8.4 kcal/mol (35 kJ/mol); a 60° twist has 50% of the maximum overlap, a loss of about 31.5 kcal/mol (132 kJ/mol). The extreme, a double bond with a 90° twist, would have no pi bonding at all, a loss of about 63 kcal/mol (264 kJ/mol) bonding energy stabilization because perpendicular orbitals do not interact.

Deformed sigma bonds can also be made, as illustrated by three-membered ring formation, but again at a cost; the strain energy of a three-membered ring is 27 kcal/mol (113 kJ/mol). Deformed sigma and pi bonds are less stable, and so the paths forming them are necessarily of higher energy; this is important since we will be attempting to predict the lowest energy path on the energy surface.

Generally, reactions break weak bonds and form strong bonds. Good orbital overlap must occur in the product (thus strong bonds) and also all along the path of electron flow to the product. Reactions that occur in two or more steps have the opportunity between steps to rotate groups into orbital alignment for the next step. Reactions that occur in one step (concerted reactions) do not have this option; their pathways may be of higher energy because of rigid orientation requirements in the transition state.

It is relatively easy to forget the three-dimensional nature of organic compounds when they are written flat on a page. **Whenever you have the slightest suspicion that the orbital alignment may be poor, build the molecular model.**

Enzymes use the orbital alignment requirements of a reaction to achieve selectivity. Depending on the enzyme, the reactant will be held in such a way that the appropriate bonds are rotated into the alignment required for the desired reaction. In the following example, the aromatic ring conjugated to the C=N pi bond forms the pi system of the electron sink. The N—C bond is free to rotate to bring either the C—R, C—H, or C—COO$^\ominus$ bond into alignment with the p orbitals of the pi system of the electron sink. One particular enzyme holds the molecule such that only the C—COO$^\ominus$ bond, indicated in bold, aligns with the pi system of the electron sink and is selectively broken. With minor deviations, the entire molecule lies in the plane of the paper except the bond in bold which comes up out of the plane. In this conformation, only the bond in bold is properly aligned to form the N=C pi bond in the product.

Pi system of sink

2.5   MICROSCOPIC REVERSIBILITY, STEREOELECTRONICS

The most basic rule of energetics is **the principle of microscopic reversibility: the forward and reverse reactions follow the same lowest energy route but in opposite directions. There is only one energetically**

best pathway. If A to B is the lowest energy path forward, then B to C to A cannot be the lowest energy path back. In deciding the lowest energy path, remember to look in both directions. The best pathway for addition of a nucleophile is also the best pathway for expulsion of it.

Stereoelectronic effects occur when the position of an orbital in space affects the course of a reaction. These orbital-position effects are a direct consequence of the principle of microscopic reversibility. For example, when a nucleophile attacks a three p orbital system such as an ester on the central carbon atom, it must approach the central p orbital along its axis, as shown in Figure 2.11.

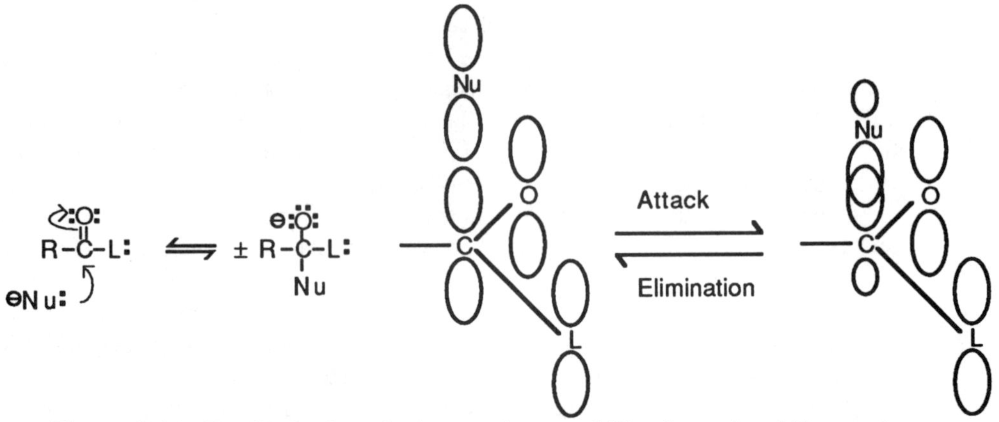

Figure 2.11 The illustration of microscopic reversibility for nucleophilic attack.

The position in space of the remaining p orbitals in the product are parallel to the newly formed bond. For the reverse reaction, which follows the same path but in the opposite direction, the lone pair orbitals of O and L must be aligned parallel to each other and to the breaking bond at the transition state. The orbitals are then lined up so that they can easily become the allylic pi system of the product. If one of the lone pairs were not lined up, the allylic system could not be established at the transition state, and that transition state would be much higher in energy. Allylic stabilization amounts to about 14 to 25 kcal/mol (59 to105 kJ/mol).

Another example is the attack by an electrophile on the carboxylate anion, an allylic system bearing more than one lone pair (Figure 2.12). Since the allylic system stabilizes both starting material and product, the lowest energy route would maintain that allylic system throughout the reaction path. In general, **the transition state of lowest energy maximizes the extent of bonding.** The loss of the electrophile (a proton for example) would create a new lone pair in the plane of the carboxylate, perpendicular to the allylic pi system. The reverse reaction must be electrophilic attack on a lone pair that

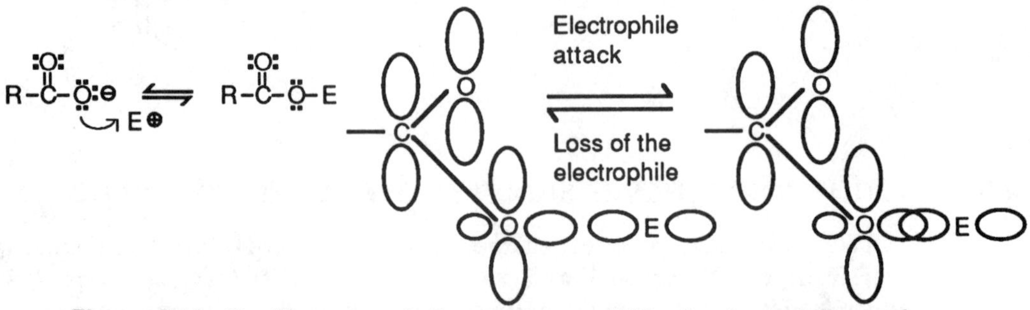

Figure 2.12 The illustration of microscopic reversibility for electrophilic attack.

is not part of the allylic pi system. Because of the principle of microscopic reversibility, we can study either the forward or reverse reaction, whichever is the easiest.

2.6 STERIC LIMITATIONS

All groups on a molecule take up space, and this occupied space is called *steric bulk*. Interactions that attempt to force two groups to occupy the same space are very unfavorable. The like-charged electron clouds on the two groups repel each other, and as they are forced together the extra energy required is called *steric hindrance*, or nonbonded repulsion. If the path that the nucleophile must follow to satisfy orbital overlap requirements has too much steric hindrance, the reaction will not occur. The nucleophile will simply bounce off the group that is in the way.

Steric hindrance caused by the interaction of the site for attack with the attacking species is a function of the steric bulk of both of them. Bulky nucleophiles need very open sites for attack, and hindered sites can be attacked only by small nucleophiles. The steric interactions between site and nucleophile play an important part in the competition between an anion acting as a nucleophile or acting as a base (substitution vs. elimination, Section 8.5).

How close is too close? If the distance between two groups is less than the sum of their van der Waals radii, then strong repulsion begins. For most groups this distance is between 3 and 4 Å, which is about twice the normal C–C single bond length of 1.54 Å. Models are very useful for comparing the relative sizes of groups. Some common groups ordered by increasing size are hydrogen, methyl, ethyl, isopropyl, phenyl, *tert*-butyl.

2.7 HOMO–LUMO INTERACTIONS
(A Supplementary, More Advanced Explanation)

All nucleophile–electrophile interactions involve the overlap and interaction of the highest occupied molecular orbital (HOMO) of the nucleophile with the lowest unoccupied molecular orbital (LUMO) of the electrophile to yield the new bonding orbital and a new antibonding orbital. **The closer in energy the two interacting orbitals are, the greater their interaction. The better the overlap is between the two orbitals, the greater the interaction.** Just as we used interaction diagrams in Section 1.6 to explain bonding stabilization, we can use them now to predict the bonding stabilization that occurs in a transition state when two different molecules react to form a new bond. The full orbital of the nucleophile interacts with the empty orbital of the electrophile to form a new bonding orbital and also a new antibonding orbital. The electrons of the nucleophile flow into the new bonding orbital and fill it. The nucleophile and electrophile approach each other so as to maximize the overlap of the HOMO with the LUMO, resulting in greater interaction and greater stabilization of the transition state. A generalized example, is shown in Figure 2.13.

A specific example is shown in Figure 2.14. When a nucleophile attacks a carbonyl group, it will approach the carbonyl group so as to achieve the best overlap of its HOMO with the carbonyl's LUMO whose largest lobe is on carbon. We have interacted a full orbital with an empty one to achieve stabilization.

Likewise, when an electrophile attacks a carbonyl group, it will approach the carbonyl group so as to achieve the best overlap of its LUMO with the carbonyl's HOMO, the lone pair on oxygen (Figure 2.15). The molecular orbitals are telling us what we already knew from the minor resonance form of the carbonyl group. Nucleophiles attack the $\partial+$ end, and electrophiles attack the $\partial-$ end of the carbonyl group.

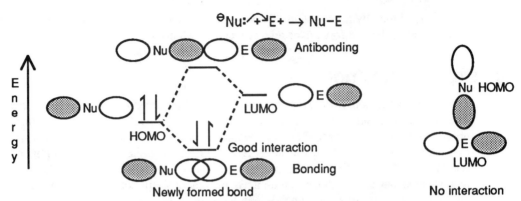

Figure 2.13 The interaction of the HOMO of the nucleophile and the LUMO of the electrophile to produce a sigma bond. The orbital arrangement on the right of the figure gives no interaction because the orbitals have equal positive and negative overlap (recall Figure 1.18).

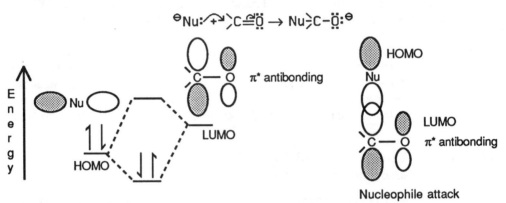

Figure 2.14 The interaction of the HOMO of the nucleophile and the LUMO of the carbonyl to produce a sigma bond.

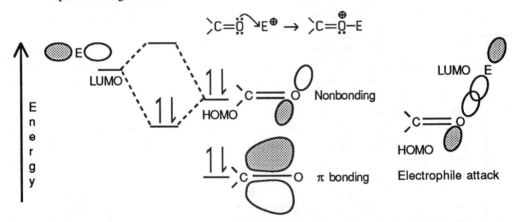

Figure 2.15 The interaction of the LUMO of the electrophile and the HOMO of the carbonyl to produce a sigma bond. The carbonyl HOMO is the lone pair and not the pi bond.

When a nucleophile displaces a leaving group in a substitution reaction, the nucleophile HOMO interacts with the LUMO of the leaving group bond as shown in Figure 2.16. Nucleophile attack on a single bond to a leaving group produces the transition state wherein the nucleophile overlaps the largest lobe of the LUMO to get the greatest stabilization possible. The nucleophile displaces the leaving group from the backside, and therefore the configuration about the carbon attacked is inverted in this type of process. Frontside displacement would involve much poorer interaction with the LUMO because of nearly equal positive and negative overlap.

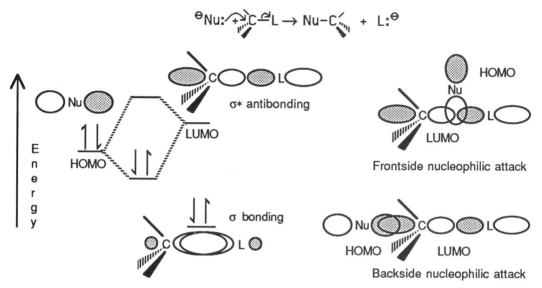

Figure 2.16 The interaction of the HOMO of the nucleophile and the LUMO of the leaving group bond to carbon to produce a new sigma bond.

Although resonance structures generally give enough information about the partial pluses and minuses to explain a molecule's reactivity (on the basis of opposite charges attracting each other), we occasionally need to refer to HOMO–LUMO interactions for additional help. Few explanations in this text rely entirely on HOMO–LUMO arguments.

2.8 HARD AND SOFT ACID–BASE THEORY

Hardness or *softness* is a property of an acid or base that is independent of its strength. An important aspect of electron availability is **polarizability, the ease of distortion of the valence electron shell of an atom by an adjacent charge.** For example, the valence electrons of iodide are shielded from the nuclear charge by all of the core electrons and thus are capable of being distorted toward a partially positive reactive site much more easily than the valence electrons of fluoride. **Polarizability increases going down a column in the periodic table.**

Hard base — It is most often a small negatively charged ion of a strongly electronegative element with **low polarizability** that is difficult to oxidize.

Soft base — It is often a large, neutral species, or one with a diffuse charge of a weakly electronegative element with **high polarizability** that is easy to oxidize. **More resonance forms of an ion indicate a more diffuse charge.**

Examples of some hard and soft bases are the following:

Hard, $F^{\ominus}$, H_2O, $HO^{\ominus}$, $Cl^{\ominus}$, NH_3
Borderline, $Br^{\ominus}$
Soft, $R^{\ominus}$, $CN^{\ominus}$, $HS^{\ominus}$, $H_2C=CH_2$, $I^{\ominus}$, $H^{\ominus}$.

For our purposes it is more important to be able to compare the relative hardness of bases; several trends are helpful:

The less electronegative bases are softer: hard $F^\ominus$ > $HO^\ominus$ > $H_2N^\ominus$ > $H_3C^\ominus$ soft.

The more polarizable the base, the softer it is: hard $F^\ominus$ > $Cl^\ominus$ > $Br^\ominus$ > $I^\ominus$ soft.

The less charged or more diffuse the charge, the softer the base is:

$$R_3C{:}^\ominus > R{-}\overset{\overset{\displaystyle :O:}{\|}}{C}{-}\overset{\ominus}{C}H_2 \leftrightarrow R{-}\overset{\overset{\displaystyle :\ddot{O}{:}^\ominus}{\|}}{C}{=}CH_2 > R{-}\overset{\overset{\displaystyle R_2N{:}}{\|}}{C}{=}CH_2 \leftrightarrow R{-}\overset{\overset{\displaystyle R_2\overset{\oplus}{N}}{\|}}{C}{-}\overset{\ominus}{C}H_2 > H_2C{=}CH_2$$

Hard   Soft

Hard acid — It is most often a small positively charged ion of an element with **low polarizability.**

Soft acid — It is often a large, neutral, or diffusely charged species or element with **high polarizability.**

Some examples of hard and soft acids are the following:

Hard, $Al^{3\oplus}$, $Mg^{2\oplus}$, $Li^\oplus$, $H^\oplus$, $Na^\oplus$, BF_3, $AlCl_3$
Borderline, $Cu^{2\oplus}$, $Zn^{2\oplus}$, $R^\oplus$
Soft, R-X, $Cd^{2\oplus}$, $Ag^\oplus$, Br_2, $Tl^{3\oplus}$, $Hg^{2\oplus}$, BH_3.

Again, the hard-to-soft trends for acids are most important.

The more polarizable acid is softer: hard $Mg^{2\oplus}$ > $Cu^{2\oplus}$ > $Cd^{2\oplus}$ > $Hg^{2\oplus}$ soft.

The less charge it has, the softer the acid is: harder $Al^{3\oplus}$ > $Mg^{2\oplus}$ > $Na^\oplus$ softer (but still reasonably hard).

The less charged or more diffuse the charge, the softer the acid is:

$$\text{hard} \quad R{-}\overset{\oplus}{C}{=}O > R_3C^\oplus > \overset{\overset{\displaystyle \delta+}{}}{RH_2C}{-}OSO_2Ar > RH_2C{-}Br \quad \text{soft}$$

A good example of a hard-to-soft trend is the increase in softness of a carbon–halogen sigma bond as the halogen gets less electronegative. Since the electronegativity of the halogen decreases in going from fluorine to iodine, the partial plus on the carbon of the carbon-halogen sigma bond also decreases. The carbon–iodine sigma bond is very soft because there is little difference in electronegativity between carbon and iodine. Softness is important to consider in reactions that break the carbon–halogen bond.

The HSAB principle: Hard bases favor binding with hard acids; soft bases favor binding with soft acids.

The soft–soft interaction is a covalent bond. It is favored by good overlap of orbitals that are relatively close in energy. The hard–hard interaction is an ionic bond. It is favored by highly charged species that have small radii so that they can get close together to form a strong ionic bond. The soft–hard interaction is relatively weak. The HSAB principle simply expresses the tendency of reactions to form strong bonds, both ionic and covalent.

An organometallic can react with a metal salt to produce a new organometallic and a new salt in the process called *transmetallation*. As expected from the HSAB principle, this equilibrium favors the formation of the more covalent organometallic from the softer pair and the more ionic salt from the harder pair. Transmetallation allows the conversion of a reactive organometallic into a more covalent, less reactive organometallic:

$$2\,RMgCl \;+\; CdCl_2 \;\rightarrow\; R_2Cd \;+\; 2\,MgCl_2$$

R soft–Mg hard + Cd soft–Cl hard $\rightarrow$ R soft–Cd soft + Mg hard–Cl hard

This section builds up to a very important take-home message:

When a pair of molecules collide, two attractive forces lead to reaction: the hard–hard attraction (opposite charges attracting each other), and the soft–soft attraction (the interaction of filled orbitals with empty orbitals).

Hard and Soft from a HOMO–LUMO Perspective
(A Supplementary, More Advanced Explanation)

Hard and soft are best explained in terms of the HOMO and LUMO energies and their separation (Figure 2.17). Soft species have high-energy HOMOs and a small energy separation between the HOMO and LUMO. The soft–soft interaction, a covalent bond, is favored by a good interaction between the HOMO of the base and the LUMO of the acid. The closer in energy these two orbitals are, the larger the interaction, and the more stable the covalent bond formed is.

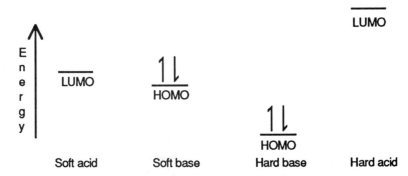

Figure 2.17 Representative HOMO and LUMO levels for hard and soft acids and bases.

Hard species have stable low-energy HOMOs and a large energy separation between the HOMO and LUMO. This large energy separation between the HOMO and LUMO means that there is little covalent interaction. The hard–hard interaction is the electrostatic attraction of opposite charges.

The hard–soft interaction has two problems: the more diffuse charge (if any) of the soft species makes a poor ionic bond. The large energy gap between the LUMO of the hard species and the HOMO of the soft species significantly decreases their interaction resulting in a weak covalent bond (Figure 2.18).

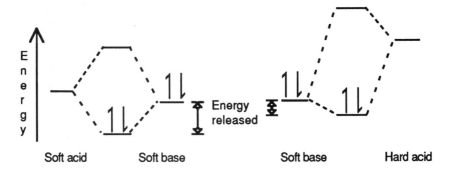

Figure 2.18 The soft–soft versus the soft–hard interaction.

2.9   PERTURBATION OF THE HOMO AND LUMO
(A Supplementary, More Advanced Explanation)

When two groups are conjugated with each other, the new conjugated system has properties that are different from the original two groups (Section 1.7). In molecular-orbital terms, the groups are said to perturb each other, and this perturbation can be viewed as the change of one of the groups when the other is connected to it.

For example, we can look at the change that occurs in the molecular orbitals of a carbon–carbon pi bond as we attach a group that conjugates with it. One extreme would be to attach an atom with a fully occupied *p* orbital, an excellent pi donor. In general, **any group that conjugates with a pi system and donates electron density to it is a pi donor.** The other extreme would be to attach an atom with an empty *p* orbital, an excellent pi acceptor. **Any group that conjugates with a pi system and withdraws electron density from it is a pi acceptor.**

The right side of Figure 2.19 shows what happens if a good pi donor is attached to a double bond. The pi donor HOMO interacts with both the double-bond HOMO and LUMO to produce a new allylic system with a new HOMO and LUMO, which are raised from those of the double bond, displayed in the center of the figure. By raising the HOMO of the system, a pi donor bonded to a double bond makes the combination more nucleophilic.

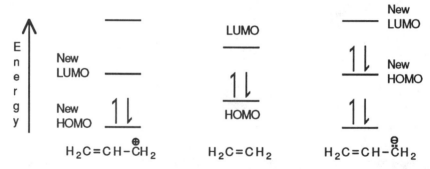

Figure 2.19 The center of the figure shows the HOMO and LUMO of an unperturbed pi bond. The left side shows that both the HOMO and LUMO drop in energy when a pi acceptor is attached. The right side shows that both the HOMO and LUMO are raised in energy when a pi donor is attached.

The left side of Figure 2.19 shows what happens if a good pi acceptor is attached to a double bond. The LUMO of the pi acceptor interacts with both the double-bond HOMO and LUMO to produce a new allylic system with a new HOMO and LUMO, which are lowered from those of the double bond. By lowering the LUMO of the system, a pi acceptor bonded to a double bond makes the combination more electrophilic. When the energy of the LUMO of the electrophile is lower, then it is softer, and the stabilization resulting from overlap with the average nucleophile's HOMO is much greater (recall Figure 2.18). Alternately, replacement of a carbon atom in a double bond by a more electronegative heteroatom will lower both the HOMO and LUMO of the pi bond but without making an allylic system. In this way carbon-oxygen double bonds are more electrophilic than carbon-carbon double bonds.

A good covalent bond is a soft–soft interaction formed between an electrophile with a low lying LUMO and a nucleophile with a high-lying HOMO (Section 2.8). Optimally the HOMO and LUMO are close in energy and therefore have a large interaction, releasing much energy, and forming a strong bond. It is possible to minimize the energy gap between the nucleophile HOMO and electrophile LUMO by raising the HOMO of the nucleophile and likewise lowering the LUMO of the electrophile, making both more reactive.

For example, consider the case of two neutral reactants, A and B, which initially have no difference in their HOMO and LUMO levels, as shown on the left in Figure 2.20. If the HOMO and LUMO levels in each are far enough apart, there is no reaction. The soft–soft interaction is poor, and since the reactants are not charged there is no hard–hard interaction. As an example, two molecules of $H_2C=CH_2$ do not normally react with each other. However, if a pi donor is added to reactant A, raising both HOMO and LUMO, and a pi acceptor is added to reactant B, lowering both HOMO and LUMO, then reaction occurs easily. The HOMO of A' is now close in energy to the LUMO of B', producing a very good soft–soft interaction. Reactant A' is now a good electron source, and reactant B' is now a good electron sink. In essence, by raising the source and lowering the sink we have made bond formation, the flow of electrons from A' to B', a good process.

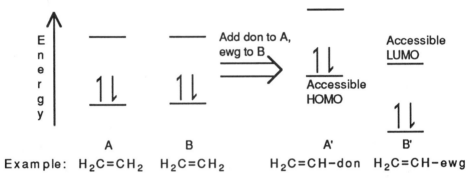

Figure 2.20 The perturbation of two reactants by adding a pi donor (don) to one and a pi acceptor (ewg) to the other, thus producing a more reactive pair.

The best electron sources have an accessible high-energy HOMO. When the energy of the HOMO of the nucleophile is high, the nucleophile is softer, and the stabilization resulting from overlap with the electrophile's LUMO is much greater. The best electron sinks have an accessible low-energy LUMO.

ADDITIONAL EXERCISES

2.1 According to the following energy diagram,

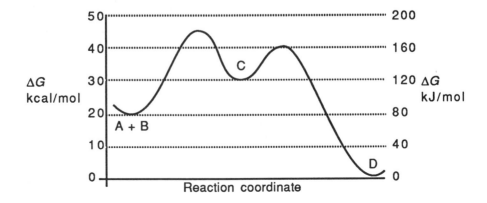

2.1 (continued)
   (a) What is the rate-determining step of A → D?
   (b) What is the $\Delta G°$ of C → D?
   (c) What is the $\Delta G^{\ddagger}$ of A → C?
   (d) What is the $\Delta G^{\ddagger}$ of D → C?
   (e) Is C stable enough to be isolated at room temperature?

2.2 According to the following energy diagram,

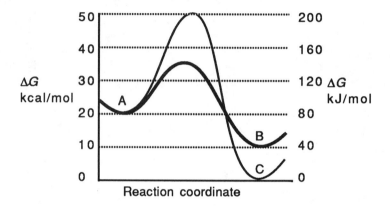

   (a) What is the kinetic product from A?
   (b) What is the thermodynamic product from A?
   (c) What is the $\Delta G^{\ddagger}$ for A → B?
   (d) What is the $\Delta G^{\ddagger}$ for A → C?
   (e) What is the $\Delta G°$ for A → C?
   (f) At 0°C, what is the most likely product from A?
   (g) Give the approximate **equilibrium** ratio of C/B at room temperature.

2.3 There are two ways that a compound can be stable: one due to $\Delta G°$ and the other due to $\Delta G^{\ddagger}$. Draw energy diagrams to explain.

2.4 Explain why possibility B and not possibility A occurs even though they are technically the same process. Models may help.

Possibility A   Possibility B

2.5 State the law of microscopic reversibility.

2.6 Use a ΔH calculation to determine which of the two possible products is the more stable.

$$CH_2=C\begin{smallmatrix}O^{\ominus}\\CH_3\end{smallmatrix} + CH_3-I \rightleftharpoons CH_3-\overset{O}{\overset{||}{C}}-CH_2CH_3 + I^{\ominus} \text{ or } CH_2=C\begin{smallmatrix}OCH_3\\CH_3\end{smallmatrix} + I^{\ominus}$$

2.7 State the relationship between selectivity and reactivity.

2.8 In each molecule which of the two boldface carbon atoms is softer?
$I-CH_2CH_2CH_2CH_2-OSO_2Ph$ $H_2C=CH-CHO$

2.9 Using models, explain why a methyl on a cyclohexane ring prefers to be equatorial.

2.10 Using the bond-strength table in the Appendix, calculate the ΔH of reaction for each of the following processes.

(a)

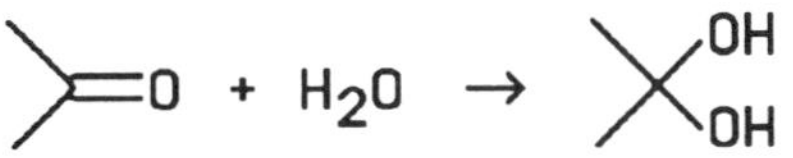

(b)

2.11 At room temperature, a reversible reaction equilibrates A $\rightleftharpoons$ B to an 85:15 ratio. What is the free-energy difference between A and B?

2.12 A reaction equilibrates C $\rightleftharpoons$ D with a $\Delta G°$ of -1 kcal/mol (-4.2 kJ/mol). Assuming that the reaction remains reversible and the $\Delta G°$ does not change, what would be the D:C ratio if this reaction were run at -50°C? At 25°C? At 100°C?

2.13 If the concentration of one reactant of a bimolecular reaction is at 1.0 M and the other at 0.1 M, how long would it take to be 97% complete at 25°C if $\Delta G^{\ddagger}$ is 20 kcal/mol (84 kJ/mol)? Use Table 2.2 to obtain a value for the rate constant.

2.14 The relative rates of two related unimolecular reactions is 10,000:1 under identical conditions at 25 °C. What is the relative difference in barrier heights?

2.15 For the following allylic species, would attack by an electrophile at carbon occur in the plane of the allylic system (the plane in which the atoms of $O-C=C$ lie) or perpendicular to it? Would attack by an electrophile at oxygen occur in the plane of the allylic system or perpendicular to it?

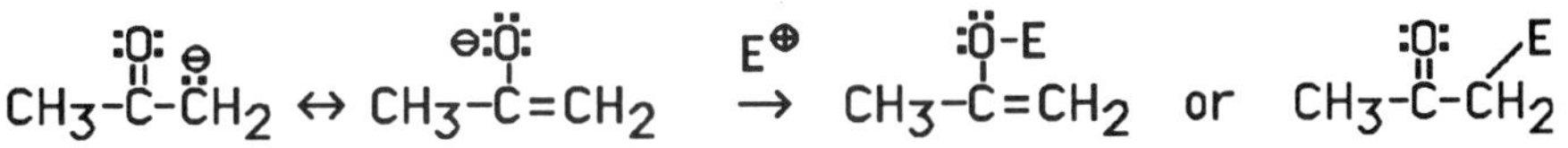

3

STABILITY AND REACTIVITY OF INTERMEDIATES

3.1 ION STABILITY, SOLVATION, AND MEDIA EFFECTS

Because most common organic reactions involve several steps and intermediates, you must be able to judge the stability of intermediates so that you can choose among different routes in predicting the product of a reaction.

The old phrase "a snowball's chance in hell" illustrates the fact that the media can play a critical part in the probability of existence of any species. For example, in water at pH 1 the concentration of hydroxide ion is 10^{-13} M. The probability that a reactive species will encounter hydroxide in a medium this acidic is very, very low.

Acidic media contain powerful electrophiles and rather weak nucleophiles, whereas basic media contain excellent nucleophiles and weak electrophiles. Similarly, no medium can possibly be both strongly acidic and strongly basic.

You could prepare two different solutions, one with an excellent electrophile and the other with an excellent nucleophile, and pour them together. However, as in tossing a snowball into a blast furnace, you must be prepared for the process to be rapid and probably violent. Forgetting the media restrictions is one of the most common mistakes of students learning to do mechanisms.

Since many nucleophiles, electrophiles, and reactive intermediates are charged, it is important to understand how charged species are stabilized. Nature abhors an isolated, localized charge. There are two ways, intramolecular and intermolecular, that charges are spread out and thereby stabilized.

Charge can be distributed over the molecule by appropriate interaction with other groups on the molecule. This distribution is most effective through pi bonds by *resonance* . The more resonance forms an ion has (which indicates pi delocalization), the more stable it will tend to be. Antiaromatic ions are the exceptions to this rule (Section 1.9). A second but less efficient mode of charge distribution is through the polarization of the sigma bonds, by *induction*. Unlike resonance, induction drops off rapidly with distance. A third way that groups on a molecule can influence a charge is through space, by a *field effect*. Since these last two effects are difficult to separate, any nonresonance effects are commonly grouped together as inductive/field effects.

Charge can be shielded by ion pairing. Bringing an ion of opposite charge into close proximity forms a stabilizing ionic bond due to the Coulombic attraction of opposite charges. If a reaction that produces a charged intermediate or product is run in a less polar solvent that cannot offer sufficient solvation (discussed next), the reaction mechanism would require ion pairing for the reaction to proceed at a reasonable rate.

Charge can be partially stabilized by polar solvent molecules. This stabilization results from the clustering of the solvent dipoles of the opposite charge around the charged molecule and becomes, in effect, an intermolecular sharing of the charge (Figure 3.1). Water can solvate cations and anions well; it has lone pairs with which to stabilize cations and polarized O—H bonds, which can hydrogen-bond to anions. A proton in water is commonly represented in organic mechanisms as the hydronium ion, $H_3O^{\oplus}$, or rarely as hydrogen bonded to four other water molecules as $H_9O_4^{\oplus}$, but the aggregate is certainly much larger.

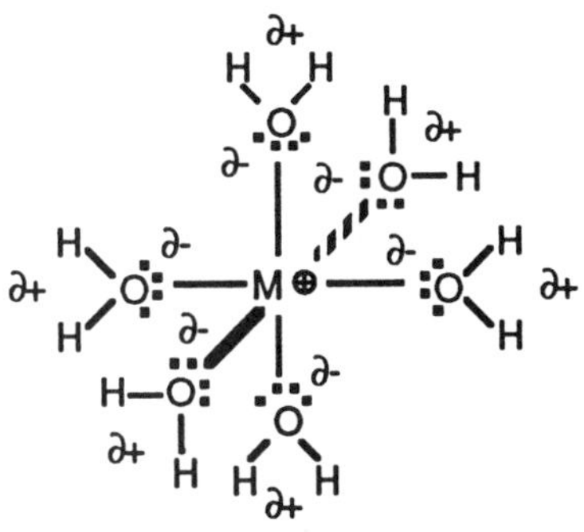

Figure 3.1 The solvation of an inorganic cation in water.

The arrangement of common solvents in the following list is from most to least polar. Solvents are also classified as either *protic* (solvents with acidic hydrogens) or *aprotic* (no acidic hydrogens). The most polar protic solvent, water, has two polarized O—H bonds and two lone pairs. Hexane, a nonpolar aprotic solvent, has no polarized bonds (C—H bonds are not polarized) and no lone pairs. **The more alkyl "grease" a solvent has, the less polar it is.** Remember that **"like dissolves like."** Ionic compounds such as salts dissolve in polar solvents and not in nonpolar ones. Nonpolar compounds dissolve in nonpolar solvents and not in polar ones.

> **Polar**
> Water
> Methanol, ethanol, acetic acid,
> Isopropanol, dimethyl sulfoxide (DMSO), dimethylformamide (DMF), acetonitrile
> Acetone, dichloromethane, pyridine, ethyl acetate
> Tetrahydrofuran (THF), diethyl ether, carbon tetrachloride
> Benzene, hexane
> **Nonpolar**

Solvation is a hard–hard interaction, and therefore the smaller and harder the ion, the more energy is released when it is solvated. The more energy released, the more stabilized and tightly solvated the ion is. Tightly solvated ions will be less useful in reactions because to react they must break out of this stabilizing solvation. The tightness of solvation will have a major effect in determining reactivity. The importance of solvation is often underestimated in chemical reactions.

There are "highly polar aprotic" solvents, such as dimethyl sulfoxide (DMSO) and hexamethylphosphoric triamide (HMPT) that can stabilize cations much better than anions. The ∂- dipole end of these solvents is sterically accessible and therefore can nestle in close to a cation, solvating it well. However, their ∂+ dipole end is highly hindered and can not get very close to an anion, therefore solvating it poorly. Because the anions are less stabilized by these solvents, they are more reactive. Bases become much more basic in these solvents.

$$(CH_3)_2\overset{\oplus}{\ddot{S}}-\overset{..}{\underset{..}{O}}:^{\ominus} \qquad\qquad ((CH_3)_2\overset{..}{N})_3\overset{\oplus}{P}-\overset{..}{\underset{..}{O}}:^{\ominus}$$
$$\text{DMSO} \qquad\qquad\qquad\quad \text{HMPT}$$

Solvent polarity can restrict the possibilities for reaction paths. Whenever charged species are present in a reaction, the reaction barrier is highly dependent on the polarity of the solvent (Figure 3.2). Solvent stabilization of the reactants more than the products decreases reactivity. Solvent stabilization of the products more than the reactants increases reactivity. In addition to stabilizing ions, polar protic solvents can also allow proton transfer and equilibration between the various ionic species in solution.

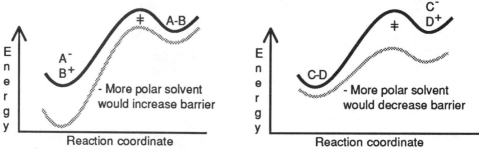

Figure 3.2 The effect of a more polar solvent on the barrier to reaction depends on which the polar solvent stabilizes more: the reactants (left) or products (right).

Example problem

Which of the following solvents, ethanol, hexane, diethyl ether, would dissolve the most KOH? The least?

Answer: KOH is a salt and therefore very polar. It will dissolve the most in a polar solvent, the least in a nonpolar. The most polar solvent of the three is ethanol. Hexane is nonpolar; diethyl ether is slightly polar.

3.2 RANKING OF ACIDS AND BASES, THE pK_a CHART

The pK_a chart serves as a good ranking of acids. For the reaction $H-A \rightleftharpoons H^{\oplus} + A^{\ominus}$, the proton donor $H-A$ is the Brønsted acid, and the anion of the acid, $A^{\ominus}$, is its conjugate base. Because the pK_a chart (see Appendix) is a composite of many experimental observations, it has the greatest extrapolation errors at its extremes and is most accurate in the center, between -2 and 16. **Stronger acids have lower pK_a values.** A low pK_a corresponds to a large value for K_a, the dissociation constant for $H-A \rightleftharpoons H^{\oplus} + A^{\ominus}$. The larger the value of the K_a, the more ionized is the acid. **An acid with a negative pK_a is a strong acid;** a pK_a of zero gives a K_a of one. When $[HA] = [A^{\ominus}]$, these terms cancel out of the K_a expression, leaving $K_a = [H^{\oplus}]$; therefore the pK_a of the acid matches the solution pH when $[HA] = [A^{\ominus}]$.

$$H-A \rightleftharpoons H^{\oplus} + A^{\ominus} \qquad K_a = \frac{[H^{\oplus}][A^{\ominus}]}{[HA]}$$

$$pK_a = -\log K_a \qquad K_a = 10^{-pK_a}$$

Several factors contribute to acidity: the strength of the $H-A$ bond, the stability of $A^{\ominus}$, and the polarity of the solvent. If the $H-A$ bond is stronger, it will be more difficult to break, and the acid will be weaker (HF is a relatively weak acid for this reason). In general, **strong acids have weak, stable conjugate bases** (a less stable conjugate base will discourage dissociation). Polar solvents favor dissociation by stabilizing charged species; an acid will dissociate less in a nonpolar solvent.

Example problem

Which of the following compounds, HNO_3, HF, HCl, is the strongest acid? The weakest acid?

Answer: The acid with the lowest pK_a is the strongest. The strongest acid is HCl at pK_a = -7; the next is HNO_3 at -1.4; the weakest is HF at 3.2.

The pK_a for compounds not on the pK_a chart must be estimated. Since the hydrogens in the middle of an alkyl chain are not acidic, we need concern ourselves only with the hydrogens on electronegative heteroatoms or hydrogens immediately adjacent to functional groups. Functional groups that contain a polarized multiple bond (C=Y or C≡Y) can delocalize the electron pair formed upon loss of $H^{\oplus}$, and greatly increase the acidity of adjacent hydrogens.

First identify the functional groups in the compound, and then find the pK_a for hydrogens adjacent to these functional groups. In this way the pK_a for $(CH_3)_2C=O$ on the chart must do for all simple ketones. For compounds of similar structure, the error from this estimation process is usually only one to two pK_a units.

Example problem

What is the pK_a of $CH_3CH_2CH_2CH_2CH_2C{\equiv}N$?

Answer: The functional group is a nitrile, $R-C{\equiv}N$. The only nitrile on the pK_a chart is $CH_3C{\equiv}N$ at a pK_a of 25. Therefore in $CH_3CH_2CH_2CH_2CH_2C{\equiv}N$ the CH_2 next to the nitrile will have a pK_a close to 25. The other CH_2s are not acidic.

Finding the most acidic hydrogen of a compound is not always easy, especially if the compound has more than one functional group. Hydrogens on electronegative atoms are reasonably acidic. Hydrogens on atoms adjacent to the $\partial+$ end of a polarized multiple bond are acidic. A hydrogen on a carbon situated between two polarized multiple bonds, for example, $N{\equiv}C-CH_2-C{\equiv}N$, would be even more acidic because the electron pair formed upon loss of $H^\oplus$ would be very delocalized. In conclusion, check all hydrogens on electronegative atoms and those adjacent to functional groups; watch for any H whose loss would produce a highly delocalized system.

Example problem

What is the most acidic hydrogen of the following compound?

$$CH_3CH_2\overset{\overset{O}{\|}}{C} CH_2\overset{\overset{O}{\|}}{C}O\,CH_3$$

Answer: The functional groups are a ketone and an ester. Both CH_2s are adjacent to the $\partial+$ end of a polarized multiple bond and are acidic. The pK_a's of each set of hydrogens are given below.

$$CH_3CH_2\overset{\overset{O}{\|}}{C} CH_2\overset{\overset{O}{\|}}{C}O\,CH_3$$

$$50 \quad 19.2 \quad 10.7 \quad {>}50$$

Neither methyl group is acidic, for both lack any sort of stabilization of their conjugate base. The CH_2 between the two carbonyls is the most acidic; it has the lowest pK_a because the anion is delocalized.

$$CH_3CH_2\overset{\overset{:\ddot{O}:}{|}}{C}{-}\overset{\overset{\ominus}{|}}{\underset{H}{C}}{-}\overset{\overset{:\ddot{O}:}{\|}}{C}\ddot{O}CH_3 \leftrightarrow CH_3CH_2\overset{\overset{:\ddot{O}:}{|}}{C}{=}\overset{\ominus}{\underset{H}{C}}{-}\overset{\overset{:\ddot{O}:}{\|}}{C}\ddot{O}CH_3 \leftrightarrow CH_3CH_2\overset{\overset{:\ddot{O}:}{\|}}{C}{-}\overset{}{\underset{H}{C}}{=}\overset{\overset{:\ddot{O}:^{\ominus}}{|}}{C}\ddot{O}CH_3$$

The previous examples illustrated the estimation of the pK_a for compounds that were structurally close to ones on the pK_a chart. It is necessary to recognize general structural features that affect acidity so that crude estimations of acidity can be made for compounds that do not have close relatives on the pK_a chart. Structural features that affect acidity are many, and examples can be found on the pK_a chart.

Second only to the previously discussed resonance effects are the inductive/field effects. Most groups, being more electronegative than carbon, tend to increase acidity. A good relative ranking of these inductive/field effects for various groups, G, is found in the acidity of a series of carboxylic acids, $G-CH_2-COOH$. A group such as $(CH_3)_3N^\oplus$ can increase the acidity of acetic acid (in which G is H, p$K_a = 4.8$) by three pK_a units, a factor of 1,000. Inductive/field effects fall off rapidly with distance; the effect of the group in $G-CH_2-CH_2-COOH$ is only 40% of that for $G-CH_2-COOH$.

Inductive/field effects are approximately additive. For example, the acidity of $Cl_2CH-COOH$ is approximately halfway between that of $ClCH_2-COOH$ and $Cl_3C-COOH$.

Besides resonance and inductive/field effects, solvation can affect acidity. Bulky alkyl groups can sterically hinder solvation and will decrease acidity; for example, acidity decreases with $CH_3OH > (CH_3)_2CHOH > (CH_3)_3COH$. Internal hydrogen bonding to the conjugate base will definitely increase acidity.

Although it is beyond the scope of this text, there exist tables of substituent constants, σ, that allow the approximate calculation of an unknown pK_a from a related parent pK_a via equations of the form $pK_a(\text{unknown}) = pK_a(\text{parent}) - \sigma\rho$, in which ρ is an experimentally derived parameter for the particular system of interest. As might be expected, the error in such a method is inherently greater than estimations based upon close structural analogs of known pK_a.

The pK_a chart also serves as a good ranking of bases. For the reaction $b^{\ominus} + H^{\oplus} \rightleftharpoons b-H$, the proton acceptor is the base $b^{\ominus}$ and $b-H$ is the conjugate acid. **The basicity of the base increases as the pK_{abH} increases** (pK_{abH} is the pK_a of the conjugate acid). **Strong bases have weak conjugate acids.** Poor solvation of an anionic base will increase its basicity. For example, the conjugate acid pK_a's of alcohol anions are 10 to 15 units higher in DMSO than in water. The following list gives in rank order by pK_{abH} of common bases used in organic reactions.

$$pK_{abH} = \quad \begin{array}{ll} \textbf{Strong base} & \\ \approx 50 & CH_3Li, \text{ n-BuLi} \\ \approx 36 & Et_2NLi, NaNH_2, NaH, KH \\ \approx 26 & [(CH_3)_3Si]_2NLi \\ \approx 19 & t\text{-BuOK} \\ \approx 16 & EtONa, CH_3ONa, NaOH \\ \approx 10 & Na_2CO_3, Et_3N \\ \approx 5 & CH_3COONa, \text{ pyridine} \\ \textbf{Weak base} & \end{array}$$

Example problem

Which of the following, $EtO^{\ominus}$, NH_3, $NH_2^{\ominus}$, is the strongest base? Which is the weakest?

Answer: The strongest base has the highest pK_{abH}. This means that $NH_2^{\ominus}$ with a pK_{abH} of 35 is the strongest base. The next strongest is $EtO^{\ominus}$, with a pK_{abH} of 16. The weakest base is NH_3, with a pK_{abH} of 9.2. Do not forget to look up the conjugate acid of the base: add a proton, then find it on the pK_a chart: for NH_3 as a base, look up $^{\oplus}NH_4$ at 9.2 and not NH_3.

3.3 CALCULATION OF K_{eq} FOR PROTON TRANSFER

Proton transfers go toward the formation of the weaker base.

$$HO^{\ominus} + HCl \rightleftharpoons HOH + Cl^{\ominus}$$

The reaction goes from stronger base, hydroxide, pK_{abH} 15.7, to the weaker base, chloride, pK_{abH} -7. There are four charge types of the proton transfer reaction since the base can be negatively charged or neutral, and the acid can be positively charged or neutral.

Calculation of the equilibrium constant for the proton transfer reaction is straightforward. Any proton transfer reaction can be considered to be the sum of two K_a equilibria, one written in the forward direction and the second written in the reverse. The equilibrium constant for the reverse reaction is the reciprocal of K_a.

K_a forward	$H{-}A \rightleftharpoons H^\oplus + A^\ominus$
K_a reverse	$H^\oplus + b^\ominus \rightleftharpoons b{-}H$
Sum (the $H^\oplus$ on both sides cancels):	$H{-}A + b^\ominus \rightleftharpoons A^\ominus + b{-}H$

When equilibria are added the equilibrium constants are multiplied.

$$K_a \text{ forward} = K_{aHA} = \frac{[H^\oplus][A^\ominus]}{[HA]} \qquad K_a \text{ reverse} = \frac{1}{K_{abH}} = \frac{[bH]}{[H^\oplus][b^\ominus]}$$

$$K_{eq} = (K_{aHA})(\frac{1}{K_{abH}}) = \frac{K_{aHA}}{K_{abH}} = \frac{[H^\oplus][A^\ominus]}{[HA]} \frac{[Hb]}{[H^\oplus][b^\ominus]} = \frac{[A^\ominus][bH]}{[HA][b^\ominus]}$$

$$\log K_{eq} = \log \frac{K_{aHA}}{K_{abH}} = \log K_{aHA} - \log K_{abH} = (-\log K_{abH}) - (-\log K_{aHA})$$

$$\boxed{\log K_{eq} = pK_{abH} - pK_{aHA} \qquad K_{eq} = 10^{\{pK_{abH} - pK_{aHA}\}}}$$

Simply take the pK_{abH} of the base and subtract from it the pK_a of the acid to get the exponent of the K_{eq}.

How unfavorable can a proton transfer be and still be useful in a reaction?

$$CH_3CH_2\overset{\ominus}{\underset{\ddot{}}{O}}{:} \curvearrowright H \overset{\curvearrowleft}{-} \overset{:\overset{..}{O}:}{\overset{\|}{CH_2COCH_2CH_3}} \rightleftharpoons CH_3CH_2\overset{..}{O}{-}H + \overset{\ominus}{\underset{\ddot{}}{C}}H_2\overset{:\overset{..}{O}:}{\overset{\|}{C}}OCH_2CH_3$$

$b^\ominus$	+	$H{-}A$	$b{-}H$	+	$A^\ominus$
Base		Acid	Conjugate acid		Conjugate base

In the example just given, since the pK_a of bH is 16 and that of HA is 24, the K_{eq} would be very small, $10^{(16-24)}$ or 10^{-8}, indicating a strong preference for $b^\ominus$ and HA. For illustration, if $b^\ominus$ and HA are initially equimolar, $[b^\ominus] = [HA]$, then it follows from the stoichiometry of the reaction that $[bH] = [A^\ominus]$. Substitution into the K_{eq} expression produces

$$K_{eq} = \frac{[A^\ominus][bH]}{[HA][b^\ominus]} = \frac{[A^\ominus]^2}{[b^\ominus]^2} \quad \text{or} \quad \sqrt{K_{eq}} = \frac{[A^\ominus]}{[b^\ominus]} = \sqrt{10^{-8}} = 10^{-4} \quad \text{or} \quad \frac{1}{10,000} = \frac{[A^\ominus]}{[b^\ominus]}$$

Therefore the concentration of $b^\ominus$ at equilibrium is 10,000 times greater than $A^\ominus$. On occasion a K_{eq} as unfavorable as 10^{-8} can still be useful. Products arising from $A^\ominus$ will be significant only if $b^\ominus$ produces a less favored product that easily reverts back to reactants, or if $A^\ominus$ is much more reactive than $b^\ominus$. Consumption of $A^\ominus$ will drive the equilibrium to replenish it.

If the proton transfer K_{eq} is greater than 10^{+8}, it can for all practical purposes be considered irreversible. An irreversible proton transfer is often the last step in a reaction and the driving force for the entire process. Each pK_a unit converts to a 1.37 kcal/mol (5.73 kJ/mol) contribution to the $\Delta G°$ of the reaction at room temperature.

Proton transfer is the first step in many important reactions. The ability to predict which protons will be pulled off in basic media or which sites will be protonated in acidic media is very important. The reaction solvent often behaves as the actual acid or base for proton transfer because the strongest acid that can occur in any media is the protonated solvent, and the strongest base is the deprotonated solvent. Thus, the strongest acid in water is $H_3O^{\oplus}$ and the strongest base is $OH^{\ominus}$.

The speed of proton transfer is usually very fast, most often at the diffusion controlled limit. The exception for this is proton transfer in which a carbanion deprotonates a C—H bond. This reaction is so slow that other reactions can easily compete with it.

Example problem

Calculate the K_{eq} for proton transfer for the following reaction.

$$EtO^{\ominus} + PhOH \rightleftharpoons EtOH + PhO^{\ominus}$$

Answer: The base in the forward reaction is $EtO^{\ominus}$ at a $pK_{abH} = 16$; the acid is PhOH at $pK_a = 10$. The K_{eq} equals $10^{(pK_{abH} - pK_{aHA})}$ or $10^{(16 - 10)} = 10^{+6}$. The equilibrium favors products.

3.4 CARBOCATION STABILITY AND TRENDS

There are two common hybridizations for carbocations, sp^2 for alkyl carbocations and sp for vinyl carbocations. The carbon bearing the positive charge hybridizes to place the valence electrons in the most stable arrangement. Since hybrid orbitals with a higher percentage of s character (%s) are more stable, these are filled first, and an empty p orbital is left over. Alkyl carbocations are relatively easy to form and are sp^2 at the cationic center. Vinyl carbocations, however, are more difficult to form because the cation is on a more electronegative sp hybridized carbon. Figure 3.3 gives the geometries of the alkyl and vinyl carbocations.

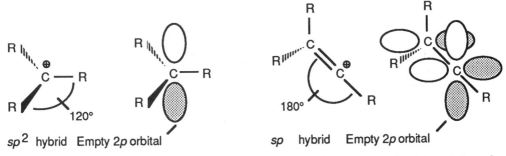

sp^2 hybrid Empty $2p$ orbital sp hybrid Empty $2p$ orbital

Figure 3.3 The hybridization of an alkyl carbocation is shown on the left and that of a vinyl carbocation is shown on the right.

Carbocation formation is very rare in solvents that cannot stabilize the cation. Clustering of the lone pairs on solvent molecules around the cationic center is an intermolecular donation of electron density that stabilizes the carbocation.

A lone pair on an atom directly bonded to the cationic center serves as a pi-donor group and stabilizes the carbocation by donating some of the lone pair electron density. A pi bond is formed between the pi-donor lone pair and the empty $2p$ orbital of the carbocation (Figure 3.4). The less electronegative the atom bearing the lone pair, the greater will be its capacity to donate. A lone pair in a nonbonding $2p$ orbital will be able to stabilize a carbocation better than a lone pair in a $3p$ or $4p$ orbital because of better overlap with the empty $2p$ carbon orbital. **The better the electron donor group is, or the more donors attached to the cationic center, the more stable the carbocation is.**

Carbon–carbon double bonds can stabilize the carbocation by donating some electron density from the pi bond by resonance. The electrons in a pi bond are less available because they are in a lower-energy bonding molecular orbital compared to a nonbonding orbital of a lone pair. Resonance with a pi bond forms a conjugated system, distributing the charge over more atoms (Figure 3.4). Stabilization of a carbocation by one double bond gives the allylic cation (Section 1.8).

Alkyl groups stabilize the carbocation to a much lesser extent by donating some electron density from the sigma framework via *hyperconjugation* (Figure 3.4). The more substituted a carbocationic center is, the more stable it is (tertiary, $R_3C^\oplus$ > secondary, $R_2HC^\oplus$ > primary, $RH_2C^\oplus$ > methyl, $H_3C^\oplus$).

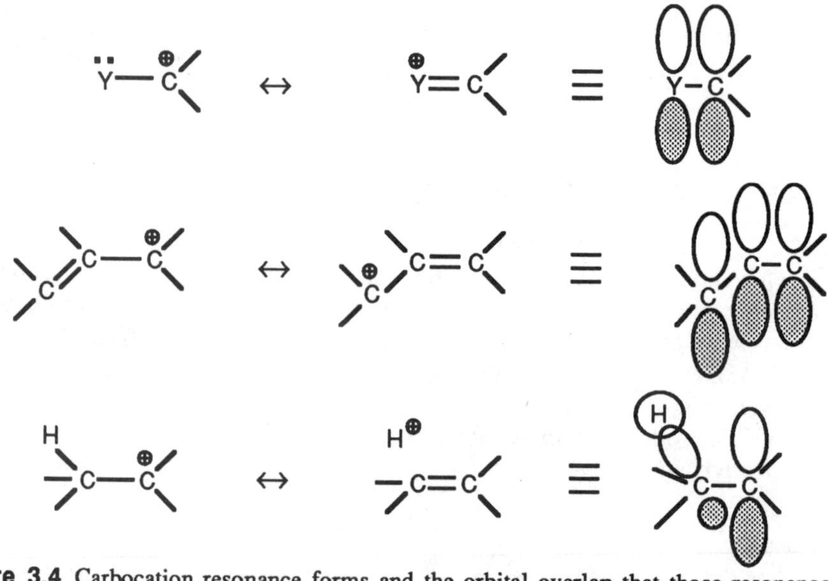

Figure 3.4 Carbocation resonance forms and the orbital overlap that those resonance forms represent. *Top*: Resonance with a lone pair is best. *Middle*: Resonance with a pi bond is next. *Bottom*: sigma bond hyperconjugation is poorest; the $H^\oplus$, pi bond resonance form is minor.

We can then state a general trend for carbocation stability. **Carbocations stabilized by resonance with a lone pair are more stable than those stabilized only by resonance with a carbon–carbon double bond, which in turn are more stable than those stabilized only by alkyl group substitution.** The following ranking of carbocations by stability illustrates these trends (Table 3.1).

There are areas of overlap where three of a lesser type of stabilization are at least as good as one of the better type. Some cations are stable enough to be on the pK_a chart. The more stable cations are weaker acids. Recall that when the pH equals the pK_a, the conjugate acid and base concentrations are equal. For example, a protonated amide requires very acidic pH 0.5 water before it is half-cation and half-neutral. Vinyl carbocations are less stable than the corresponding alkyl carbocation of similar substitution (secondary vinyl cation is less stable than the secondary cation, but more stable than the primary cation). Aromaticity greatly stabilizes carbocations; an example is tropylium cation, $C_7H_7^\oplus$ (Section 1.9). Antiaromatic ions such as cyclopentadienyl cation (Section 1.9) are highly destabilized.

Table 3.1 The Carbocation Stability Ranking

Carbocation name	Structure	Comments on Stabilization
Stable carbocations		
Guanidinium ion	$(H_2\overset{..}{N})_3\overset{\oplus}{C}$	Three N lone pairs stabilize, pK_a = 13.6
Tropylium cation	(ring structure)	Stabilized by aromaticity
Protonated amide	$H_3C-\overset{\overset{\displaystyle :\overset{..}{O}H}{}}{\underset{}{C}}{}^{\oplus}NH_2$	N and O lone pairs and one R, pK_a = 0.5
Moderately stable		(Easily formed)
Protonated carboxylic acid	$H_3C-\overset{\overset{\displaystyle :\overset{..}{O}H}{}}{\underset{}{C}}{}^{\oplus}\overset{..}{O}H$	Two O lone pairs and one R, pK_a = −6
Triphenylmethyl cation	$(\text{Ph}-)_3\overset{\oplus}{C}$	Resonance with three phenyl groups
Protonated ketone	$H_3C-\overset{\overset{\displaystyle :\overset{..}{O}H}{}}{\underset{}{C}}{}^{\oplus}CH_3$	An O lone pair donor and two R's, pK_a = −7
Diphenylmethyl cation	$(\text{Ph}-)_2\overset{\oplus}{C}-H$	Resonance with two phenyl groups
Average stability		(Common, of similar stability)
Tertiary cation	$(H_3C)_3\overset{\oplus}{C}$	Three alkyl groups stabilize
Benzyl cation	$\text{Ph}-\overset{\oplus}{C}H_2$	Resonance with phenyl group
Primary allyl cation	$H_2C=CH-\overset{\oplus}{C}H_2$	Resonance with a pi bond only
Acylium ion	$H_3C-\overset{\oplus}{C}=\overset{..}{O}$	A vinyl cation stabilized by an O lone pair
Moderately unstable		(Infrequent)
Secondary cation	$(H_3C)_2\overset{\oplus}{C}-H$	Stabilized by two alkyl groups
Secondary vinyl cation	$H_2C=\overset{\oplus}{C}-CH_3$	*sp* hybrid. Less stable than secondary cation
Unstable		(Very rare)
Primary cation	$H_3C-\overset{\oplus}{C}H_2$	Only one alkyl group stabilizes
Primary vinyl cation	$H_2C=\overset{\oplus}{C}H$	Less stable than primary cation due to sp hybrid
Phenyl cation	(ring) $\overset{\oplus}{C}$	A bent vinyl cation
Methyl cation	$\overset{\oplus}{C}H_3$	No stabilization

Example problem

Which of the following cations, $(CH_3)_2HC^\oplus$, $Ph(CH_3)_2C^\oplus$, $(CH_3)_3C^\oplus$, is the most stable? The least stable?

Answer: The delocalized carbocation, $Ph(CH_3)_2C^\oplus$, is the most stable. The *tert*-butyl carbocation, $(CH_3)_3C^\oplus$, stabilized by three methyl groups, is of intermediate stability, whereas the isopropyl carbocation, $(CH_3)_2HC^\oplus$, stabilized by only two methyl groups, is the least stable.

Carbocation Stabilization from a HOMO–LUMO Perspective
(A Supplementary, More Advanced Explanation)

Another way to look at the stabilization of a carbocation is to consider the interaction of the empty p orbital (LUMO) with the full donor orbital (HOMO). The best donors to an empty p orbital will be full orbitals close to the same energy (the stabilizing interaction is greater). For lone pair donors, as the atom bearing the lone pair becomes more electronegative, the pi-donor effect decreases because the energy of the HOMO drops. Stronger bonds have lower-energy HOMOs, farther away in energy from the empty carbon $2p$ orbital. A carbon–carbon pi bond will be a better donor than a carbon–hydrogen sigma bond (Figure 3.5).

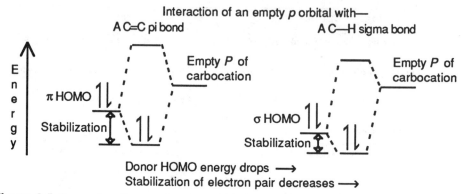

Figure 3.5 Interaction diagrams for the stabilization of a carbocation by a C=C pi bond and a C—H sigma bond.

The strained C—C sigma bonds of a three-membered ring have a HOMO much higher in energy than unstrained C—C sigma bonds and are the exception to sigma bonds' being poor at stabilizing cations. In fact, a three-membered ring can stabilize an adjacent carbocation (called a cyclopropylcarbinyl system, C_3H_5—$CH_2^{\oplus}$) better than a phenyl group can, C_6H_5—$CH_2^{\oplus}$.

Nonclassical Carbocations

Certain primary or secondary carbocations can enter into a type of bonding with a pi bond or strained sigma bond that involves three orbitals overlapping to give a bonding MO that is occupied by two electrons, similar to a pi-complex. These bridged species can be transition states, intermediates, or sometimes the lowest-energy structure of the carbocation. The 2-norbornyl cation (Figure 3.6) is a nonclassical ion, but much experimentation was required to determine that the bridged ion was the lowest-energy species and not a transition state between two interconverting secondary carbocations.

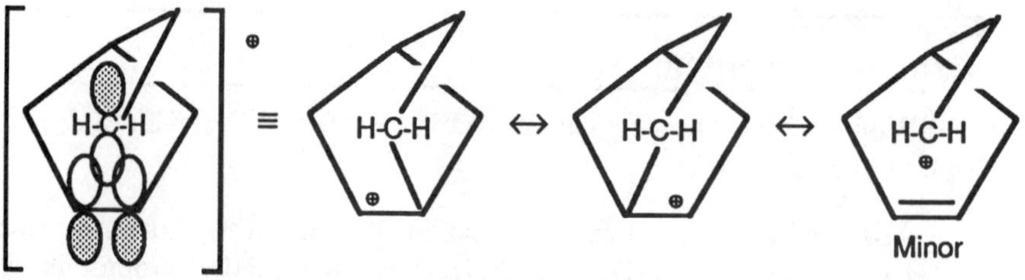

Figure 3.6 The 2-norbornyl nonclassical ion. The resonance forms on the right describe this unusual three-center, two-electron bonding molecular orbital.

3.5   RANKING OF ELECTRON-DONOR GROUPS

The ranking of an electron-donor group (don) is very important. The electron richness of electron sources will depend on this trend. The ranking is based on how available the electrons of the donor are; the more stabilized they are, the less available they are. Electrons in nonbonding orbitals are the most available. The best donors are anions that can donate an electron pair by resonance. The better donors all contain lone pairs that can form a strong pi bond to the empty p orbital of the cation, and are pi donors for that reason. The following is a ranked list of electron donor groups.

Excellent donors	
$-\overset{\ominus}{\ddot{C}}H_2$	Anionic
$-\overset{\ominus}{\ddot{N}}H$	Anionic, more electronegative than C
$-\ddot{\ddot{O}}{:}^{\ominus}$	Anionic, more electronegative than N
Good donors	
$-\ddot{N}(CH_3)_2$	Neutral
$-\ddot{N}H_2$	Neutral
$-\ddot{\ddot{O}}H$	Neutral, more electronegative than N
$-\ddot{\ddot{O}}CH_3$	Neutral, more electronegative than N
$-\ddot{N}H-\overset{\overset{:O:}{\|}}{C}-CH_3$	Amide, carbonyl decreases lone pair availability
$-\ddot{\ddot{S}}CH_3$	Sulfur, poor $2p$–$3p$ pi bond
Poor donors	
$-Ar$	Delocalization of charge into the aromatic ring
$-R$	Hyperconjugation only, no lone pairs to donate
$-H$	No substituent at all
Very poor donors	(pi donors whose overall effect is electron withdrawing)
$-\ddot{\ddot{C}}F$	Cl, electronegative, poor $2p$–$3p$ pi bond
$-\ddot{\ddot{O}}-\overset{\overset{:O:}{\|}}{C}-CH_3$	Electronegative, carbonyl decreases lone pair availability

For first-row elements, as the atom containing the lone pair becomes more electronegative, it is less able to donate electron density. For this reason, carbon anions are better donors than nitrogen anions, which are in turn better donors than oxygen anions. Anions are the best donors because they donate both by resonance and by inductive/field effects. Neutral heteroatoms are more electronegative than carbon, so inductive electron withdrawal competes with resonance electron donation, making them poorer lone pair donors than anions.

An electron-withdrawing group attached to a donor decreases the electron availability of the donor. For example, the attachment of an electronegative carbonyl group to a heteroatom that bears a lone pair diminishes the electron donation ability of that lone pair; an amide nitrogen is a much poorer donor than an amine nitrogen. Delocalization of the lone pair with the carbonyl stabilizes the lone pair, makes the nitrogen partially positive, and more electronegative.

The formation of a strong pi bond is responsible for the stabilization afforded by the pi donor. However, pi bonds between orbitals from different shells necessarily have poor overlap and are weak. A sulfur atom is a poor donor because there is poor overlap of the sulfur $3p$ orbital with the carbon $2p$ orbital. The two worst lone pair donors are poorer donors than alkyl groups because their high electronegativity is not compensated by their weak pi donation ability. For example, a chlorine atom is very electronegative and is a poor resonance electron donor because of poor $2p$-$3p$ overlap.

Resonance effects will usually dominate over inductive/field effects. When the resonance effect is blocked, the only groups that will still be donors are anions and alkyl groups. The electronegativity of all the other groups will cause them to be inductive electron-withdrawing groups. Electron-withdrawing groups destabilize carbocations. Similarly, electron-donor groups destabilize anions.

Example problem

Which of the following, OCH_3, CH_3, NH_2, is the strongest donor? Which is the weakest?

Answer: Both the ether and the amine are lone pair donors and so are better donors than a methyl group, which is the weakest. The amine is the best donor because the amine nitrogen is less electronegative than the ether oxygen.

3.6 CARBANION STABILITY AND TRENDS

Since the major factor governing the acidity of the carbon–hydrogen bond is the stability of the resultant carbanion, a useful generalization is that **the more acidic the carbon acid (lower pK_a), the more stable the carbanion.** A pK_a chart then becomes a handy reference for carbanion stability. Table 3.2 illustrates the pK_a, electronegativity, and hybridization trends for some carbanions.

Table 3.2 Hybridization and Stability of Carbanions

Decreasing stability $\longrightarrow$	$HC\equiv C{:}^{\ominus}$	$H_2C={\overset{\ominus}{C}}H$	$H_3C-{\overset{\ominus}{C}}H_2$
Hybridization	sp	sp^2	sp^3
%s character	50	33	25
pK_{abH}	25	44	50
Electronegativity of C	3.29	2.75	2.48

Hybridization has a dramatic effect on carbanion stability. Since the $2s$ orbital is closer to the nucleus and lower in energy than the $2p$ orbital, an anion residing in a hybrid orbital with greater %s (therefore a more electronegative orbital) is more stable. **Anions on more electronegative atoms are more stable.**

The most common mode of solvent stabilization is hydrogen bonding to the anion, which works well for less basic ($pK_{abH} \leq 25$) carbanions in protic media like water or alcohols. The more basic anions are quenched by irreversibly deprotonating the protic solvent.

As with cations, conjugation with double bonds or aromatic rings stabilizes anions by distributing the charge over more atoms. An electron-withdrawing group directly bonded to the carbon bearing the negative charge can stabilize the anion best. The better the electron withdrawal, the greater the stabilization. The stabilization is greatest by resonance (through pi bonds) to a pi acceptor. Inductive (through sigma bonds) and field effects (through space) to electronegative atoms can also stabilize the negative charge. Aromaticity greatly stabilizes anions; the only common example is cyclopentadienyl anion $C_5H_5^{\ominus}$ (Section 1.9).

The more electron-withdrawing groups on the carbanion, the more stable it will be. However, additional groups do not have as much an effect as the first one. The attachment of one carbonyl drops the pK_a of methane from 48 to 19.2; the attachment of a second carbonyl group drops the pK_a less, from 19.2 to 9.0.

Example problem

Which of the following, $^{\ominus}C{\equiv}CR$, $^{\ominus}CH_2Ph$, $^{\ominus}CH(C{\equiv}N)_2$, is the most stable carbanion? Which is the least?

Answer: The most stable carbanion will have the lowest pK_{abH}. This means that $^{\ominus}CH(C{\equiv}N)_2$, with a pK_{abH} of 11.2, is the most stable. The next in stability is $^{\ominus}C{\equiv}CR$, with a pK_{abH} of 25. The least stable carbanion is $^{\ominus}CH_2Ph$, with a pK_{abH} of 41.

Carbanion Stabilization from a HOMO–LUMO Perspective
(A Supplementary, More Advanced Explanation)

The best electron withdrawal occurs when the energy of the pi acceptor's LUMO is close to that of the full $2p$ orbital of the carbanion. For example, compare how two groups, the C=C and the C=O, stabilize an adjacent carbanion. The pK_a of methane is about 48; if we attach a C=C, which can stabilize the anion by delocalization, the pK_a drops to 43. If we instead attach a C=O, which can stabilize by delocalization and is a more electronegative group, the pK_a drops all the way to 19.2, indicating a much greater stabilization of the anion. As illustrated by Figure 3.7, the LUMO of the carbonyl is lower in energy than that of a simple carbon–carbon double bond (recall Section 1.6) and therefore interacts more with the full $2p$ orbital of the carbanion.

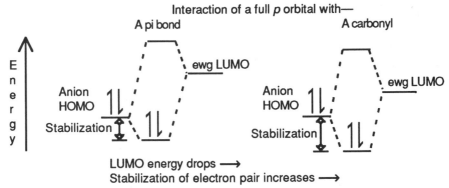

Figure 3.7 Interaction diagrams for the stabilization of an anion by an empty orbital.

The lower in energy the LUMO of the electron-withdrawing group (ewg) is, the better the group will be at withdrawing electrons. The lower-energy LUMO can interact better with filled orbitals. The better electron-withdrawing groups contain a low-lying empty p orbital or π^* orbital that can form a strong pi bond to the full p orbital of the anion.

3.7   RANKING OF ELECTRON-WITHDRAWING GROUPS

The acidity of CH_3—ewg increases as groups better at withdrawing electrons more effectively stabilize the anionic conjugate base, $^{\ominus}CH_2$—ewg. If the pK_a's of CH_3—ewg's are compared, the pK_a chart becomes a guide (good enough for our purposes) for the approximate ranking of the common electron-withdrawing groups. **The more acidic (lower pK_a) the CH_3 — ewg is, the better the ewg group is at withdrawing electrons.** The following is a list of electron-withdrawing groups ranked by the pK_a of CH_3—ewg.

Best ewg

pK_a		
10.2	$CH_3-\overset{\oplus}{N}(=\ddot{O})\ddot{O}{:}^{\ominus}$	
19–23	$CH_3-\overset{:O:}{\underset{\|}{C}}-R$	$CH_3-\overset{:O:}{\underset{:O:}{\overset{\|}{S}}}-CH_3$
24–25	$CH_3-\overset{:O:}{\underset{\|}{C}}-OR$	$CH_3-C\equiv N{:}$
30–35	$CH_3-\overset{:O:}{\underset{\|}{C}}-NR_2$	$CH_3-\overset{:O:}{\underset{\|}{\ddot{S}}}-R$

Worst ewg

The best electron-withdrawing groups can stabilize the anion by resonance, for this reason they are called pi acceptors. They delocalize the anion and distribute the charge onto more electronegative atoms. The two most common electron-withdrawing groups are the nitrile and the carbonyl. The ability of the carbonyl to withdraw electron density depends on the nature of the group bound on the other side of it. As the group attached to the other side of the carbonyl becomes a better donor, the carbonyl becomes a poorer electron-withdrawing group.

Example problem

Which of the following, CN, Ph, NO_2, is the best electron-withdrawing group? Which is the poorest?

Answer: The best electron-withdrawing group will have the lowest pK_a for CH_3—ewg. Therefore CH_3NO_2, with a pK_a of 10.2, makes the nitro group the best. The next is the nitrile, CH_3CN, with a pK_a of 25. The worst electron-withdrawing group is phenyl because $PhCH_3$ has a pK_a of 41. The phenyl group is not really an electron-withdrawing group; its field/inductive effect is mildly withdrawing. It stabilizes the charge simply by delocalization; something it does almost equally well for either anions or cations.

Groups that are not pi acceptors stabilize an adjacent anion less effectively, and their CH_3—ewg pK_a's are not available. Very electronegative groups, such as trimethylammonium cation, $^{\oplus}N(CH_3)_3$, and trifluoromethyl, CF_3, can withdraw electron density through field/inductive effects. The halogens are resonance electron donors, but their inductive effect is withdrawing. The field/inductive effects of an electron-withdrawing group (resonance effects blocked) is occasionally needed and can be ranked by the extent that the ewg lowers the pK_a of the carboxylic acid group, ewg—CH_2COOH (Section 3.2).

Example problem

Which of the following is the most acidic phenol?

$$CH_3O \text{---} \langle \rangle \text{---} OH \qquad \langle \rangle \text{---} OH \qquad O_2N \text{---} \langle \rangle \text{---} OH$$

Answer: Stabilization of the anionic conjugate base by a good ewg will increase the acidity of the phenol. A lone pair donor will destabilize the conjugate base and decrease the acidity of the phenol. The lowest pK_a (most acidic) is 7.2, the nitrophenol. The unsubstituted phenol at pK_a 10.0 is next. Least acidic at pK_a 10.2 is the methoxyphenol.

3.8 RANKING OF LEAVING GROUPS

$$-\overset{|}{C} \!-\! L \rightarrow -\overset{|}{C}^\oplus \quad L:^\ominus \qquad or \qquad -\overset{|}{C} \!-\! L^\oplus \rightarrow -\overset{|}{C}^\oplus \quad L:$$

A group that separates from a compound and takes with it its bonding pair of electrons is called a *leaving group* (L) or nucleofuge, and can depart as an anion or as a neutral species. It is common to rank leaving groups by their conjugate acid pK_a, the pK_{aHL}. **A good leaving group, L, has a negative pK_{aHL}.** The pK_{aHL}'s of poor leaving groups range from 13 to 30. The loss of a group whose pK_{aHL} is greater than 30, like $NH_2^\ominus$, is exceedingly rare.

Some groups do not fit this highly simplistic scheme for ranking leaving groups. If the L is highly electronegative, or if the carbon-leaving group bond is weaker or more polarizable, the bond is more easily broken. In ranking leaving groups by their pK_{aHL}, we are drawing a parallel between two reactions that are close but not identical:

$$\overset{|}{\underset{|}{C}} \!-\! L \rightleftharpoons \overset{|}{\underset{|}{C}}^\oplus + :L^\ominus \qquad vs. \qquad H \!-\! L \rightleftharpoons {}^\oplus H + :L^\ominus$$

Only three types of leaving groups fall significantly out of order when ranked by pK_{aHL}: sulfonates, $RSO_3^\ominus$ and $ArSO_3^\ominus$, are much better leaving groups than predicted by their pK_{aHL}'s (they are better than iodide); fluoride and cyanide are much poorer leaving groups than predicted.

When using a pK_a chart to rank leaving groups, always look up the conjugate acid of the leaving group on the chart. If EtO$^\ominus$ falls off, look up EtOH, pK_a +16; if EtOH falls off, look up EtOH$_2$$^\oplus$, pK_a -2. Protonation of a group before it falls off makes it cationic, more electronegative, and a much better leaving group. Poor neutral leaving groups are rare. The following is a ranked list of common leaving groups and their conjugate acid pK_a.

Good	pK_{aHL}	Fair to Poor	pK_{aHL}
N_2	<-10	RCOO$^\ominus$	+4.8
$CF_3SO_3^\ominus$	<-10	NH_3	+9.2
ArSO$_3^\ominus$	-6.5	RS$^\ominus$	+10.6
I$^\ominus$	-10	HO$^\ominus$	+15.7
Br$^\ominus$	-9	EtO$^\ominus$	+16
Cl$^\ominus$	-7		
EtOH	-2.4		
H_2O	-1.7		
CF$_3$COO$^\ominus$	+0.5		

Example problem

Which of the following, R—Cl, R—Br, R—OH, bears the best leaving group? Which bears the worst?

Answer: The best leaving group is bromide, pK_{aHL} of -9. The next-best leaving group is chloride, pK_{aHL} of -7. The worst leaving group is hydroxide, $pK_{aHL} = 15.7$.

3.9 THE ΔpK_a RULE

If the reactants and the products of a reaction are not charged, we can calculate the ΔH (Section 2.1) and thereby know whether the reaction is uphill or downhill in energy. Unfortunately, this calculation is of limited use for mechanisms, since most reaction steps involve charged species as nucleophiles, electrophiles, or intermediates. We need a way to tell whether a step involving a charged species is energetically uphill or down.

For example, the displacement of a leaving group by a nucleophile is an equilibrium, $Nu:^\ominus + Y–L \rightleftharpoons Nu–Y + L:^\ominus$. Therefore a rule of thumb is needed to determine whether the reaction will proceed to the left or to the right. We can use the relative basicity of the Nu and the L as a guide, because **usually the reaction will proceed to form the weaker base**, as was seen in the proton transfer process. Sometimes a slightly stronger base (one stronger by about eight pK_a units at the most) can be formed if the bond broken is weaker than the one formed; the formation of the stronger bond would counterbalance the formation of a slightly stronger base. Even a 10 unit climb in pK_{abH}, corresponding to a 13.7 kcal/mol (57.3 kJ/mol) climb in energy, could occasionally be compensated for by bond formation. Therefore the following is a very helpful and general rule of thumb:

The ΔpK_a **rule: The leaving group or anion produced should be no more than about eight pK_a units more basic than the incoming nucleophile or base.**

As with any rule of thumb, this one has many exceptions. However, the ΔpK_a rule is exceedingly helpful in predicting the energetics of a reaction. We need some rule, no matter how approximate, to tell us whether a particular step or alternative is uphill or downhill in energy. The mechanism for basic hydrolysis of an ester can be used to illustrate the ΔpK_a rule.

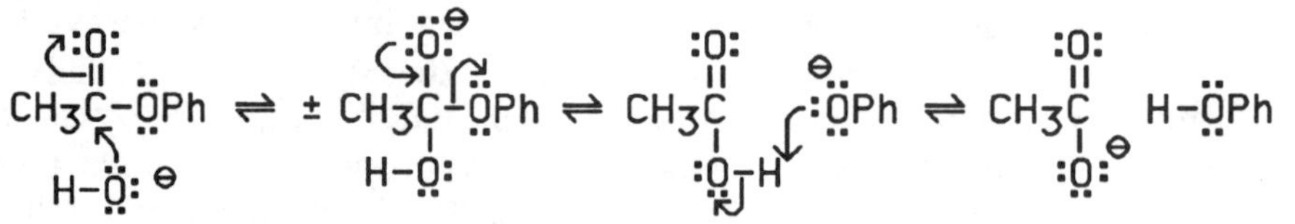

A reasonable energy diagram for the process is given in Figure 3.8. The rate-determining step is the attack of hydroxide ion on the ester. This is expected because the best leaving group from the first intermediate is $PhO^\ominus$, and expulsion of the better leaving group should have a lower barrier (viewed from the well of the intermediate). The three possible groups that could be lost are $PhO^\ominus$ (pK_{aHL} 10), $HO^\ominus$ (pK_{aHL} 15.7), or $CH_3^\ominus$ (pK_{aHL} 48), and clearly the first is the best. Methyl anion is not a leaving group.

The ester and the acid are of approximately the same energy, but the $PhO^\ominus$ anion has a more distributed charge than hydroxide. The K_{eq} for proton transfer between $PhO^\ominus$ and the acid is calculated (Section 3.3) to be $10^{(10 - 4.8)} = 10^{+5.2}$ corresponding to a ΔG° at room temperature of approximately $(5.2)(-1.37) = -7.1$ kcal/mol (29.7 kJ/mol), almost irreversibly downhill (Section 2.1). This last proton transfer step is the driving force for the reaction.

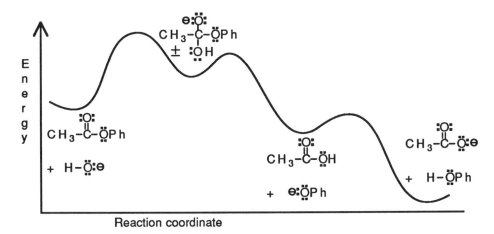

Figure 3.8 Energy diagram for the basic hydrolysis of phenyl acetate.

The plot of pK_{abH}, the pK_a of conjugate acid of the most basic partner in each step versus the reaction coordinate, is shown in Figure 3.9 and has a profile similar to the energy diagram (Figure 3.8). Several points can be made. Over the course of a reaction the pK_{abH} of the intermediates may increase slightly, but they will either hover about the initial pK_{abH} or very likely fall to yield a weaker base at the end of the reaction. The breaking of a weak bond and formation of a stronger bond can compensate for the formation of an intermediate with a higher pK_{abH}. **No reaction has huge jumps upward in the pK_{abH} of its intermediates.** There can be great downward jumps as strongly basic intermediates are neutralized.

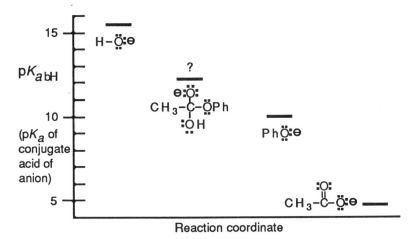

Figure 3.9 The pK_{abH} of the most basic species in each step of the basic hydrolysis of phenyl acetate plotted against the reaction coordinate. The pK_{abH} must be estimated for the first intermediate.

Obviously, the match is not perfect. In the energy diagram (Figure 3.8) the energy of all species in each step is taken into account, whereas in the pK_{abH} diagram, only the most basic partner was considered. The loss of entropy when the two reactants join to form the first intermediate causes a significant error. However, overall we can draw a rough association between a reaction proceeding energetically downhill and the fact that the pK_{abH} of the products is almost always less than that of the reactants. Consequently, the following is an extremely useful, although very crude, approximation. **The energy drops if the pK_{abH} drops significantly.** This gross generalization will allow us to choose which of two alternative reaction routes is probably of lower energy.

Summary

The pK_a chart can be used for
> Ranking the strength of acids
> Ranking the strength of bases
> Calculating a proton transfer K_{eq}
> Ranking the stability of carbanions
> Approximate ranking of leaving groups
> Approximate ranking of electron-withdrawing groups
> Estimating the energetics of reaction.
> Helping to rank nucleophiles (in conjunction with softness) (Section 4.2)

Example problem

Does the following reaction favor products or reactants?

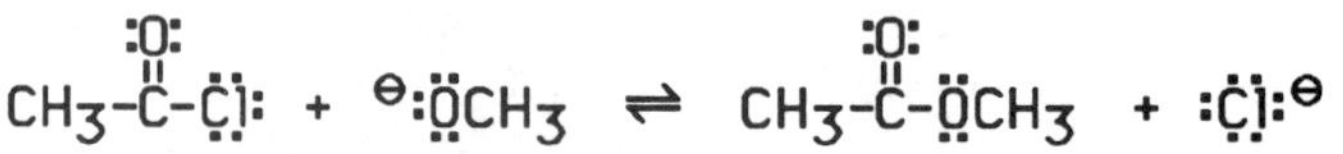

$$CH_3-\overset{:O:}{\overset{||}{C}}-\ddot{\overset{..}{C}}\overset{..}{l}: \; + \; {}^{\ominus}:\ddot{O}CH_3 \; \rightleftharpoons \; CH_3-\overset{:O:}{\overset{||}{C}}-\ddot{O}CH_3 \; + \; :\ddot{\overset{..}{C}}\overset{..}{l}:^{\ominus}$$

Answer: Since the pK_{abH} of methoxide is 15.5 and that of chloride is -7, chloride is obviously the weaker base. The reaction proceeds to the right to form the weaker base.

ADDITIONAL EXERCISES

3.1 Using the pK_a chart, calculate the numerical value of K_{eq} for the following:

PhONa + PhSH $\rightleftharpoons$ PhSNa + PhOH

NaH + =$\ddot{O}$ $\rightleftharpoons$ H$_2$ + —$\ddot{\overset{..}{O}}$:$^{\ominus}$

$(CH_3)_3COK + HC{\equiv}CH \rightleftharpoons (CH_3)_3COH + HC{\equiv}CK$

3.2 Trends: with the help of the pK_a chart when needed, rank all species, beginning with the numeral 1 to designate:
The most stable carbanion

$R\overset{\ominus}{\overset{..}{C}}H_2$:$\ominus$:$\ominus$:$\ominus$ $R-\overset{:O:\,\ominus}{\overset{||}{C}}-\overset{..}{C}H_2$

The most stable carbocation

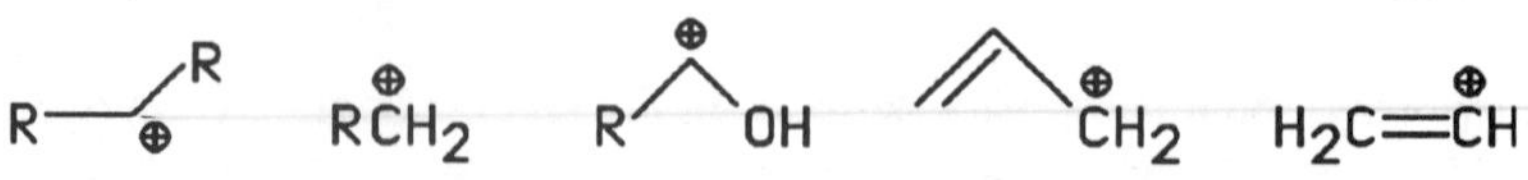

$R\overset{}{\overset{\oplus}{\diagdown}}\diagup^{R}$ $R\overset{\oplus}{C}H_2$ $R\diagup\overset{\oplus}{\diagdown}OH$ $\diagup\!\!=\!\!\diagdown\overset{\oplus}{C}H_2$ $H_2C{=}\overset{\oplus}{C}H$

The best electron-withdrawing group

$-\overset{:O:}{\overset{||}{C}}-\overset{..}{N}R_2$ $-\overset{:O:}{\overset{||}{C}}-\overset{..}{O}R$ $-\overset{:O:}{\overset{||}{C}}-CH_3$ $-NO_2$ $-\overset{:O:}{\overset{||}{S}}-CH_3$

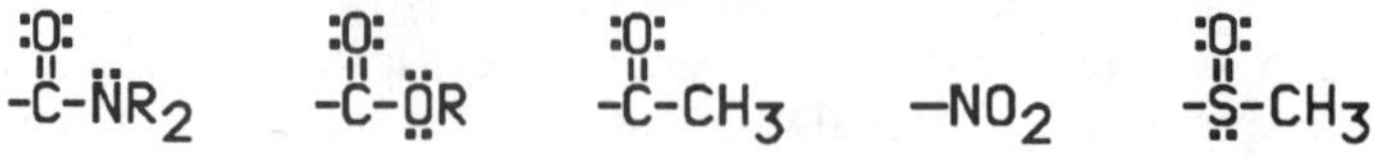

The best base
$^{\ominus}$CN   PhNH$_2$ $^{\ominus}$NH$_2$ HO$^{\ominus}$ (CH$_3$)$_3$CO$^{\ominus}$

The best electron donor
—O$^{\ominus}$ —OR   —R   —NR$_2$ —Cl

The best leaving group attached to R
R—OH$_2$$^{\oplus}$ R—Br   R—Cl   R—OCH$_2$CH$_3$ R—NR$_2$

The best acid
H$_3$O$^{\oplus}$ HF   MeOH   HCN   NH$_3$

The most polar solvent
CH$_3$CH$_2$CH$_2$CH$_2$CH$_3$ EtOEt   H$_2$O   CH$_3$OH   CH$_3$CH$_2$CH$_2$OH

3.3 Using the ΔpK_a rule, predict whether each of the following equilibria would favor reactants or products.

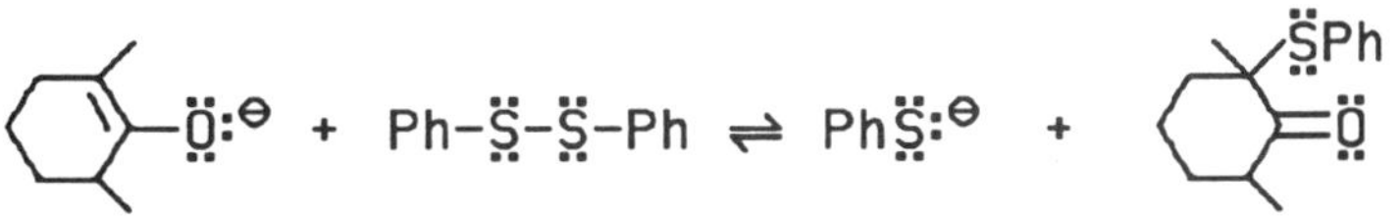

$^{\ominus}$OCH$_3$ + H$_3$CCOCl $\rightleftharpoons$ Cl$^{\ominus}$ + H$_3$CCOOCH$_3$

HS$^{\ominus}$ + H$_3$C-I $\rightleftharpoons$ H$_3$C-SH + I$^{\ominus}$

3.4 With the help of the pK_a chart, circle the most acidic H in each of the following compounds.

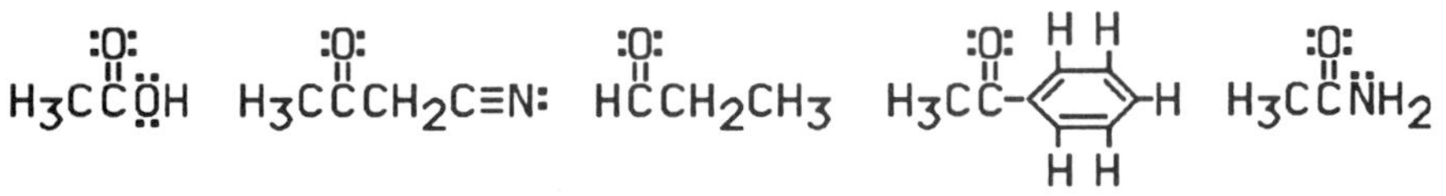

3.5 Use the ΔpK_a rule to determine which of the two following alternatives is the lower-energy process.

CH$_3$COOEt + HO$^{\ominus}$ reacting to form H$_2$O + $^{\ominus}$CH$_2$COOEt
or CH$_3$COOEt + HO$^{\ominus}$ reacting to form EtOH + CH$_3$COO$^{\ominus}$

3.6 Use the leaving group trend to decide which of the following two alternatives is the lower-energy process.

ClCH$_2$CH$_2$CH$_2$Br + CN$^{\ominus}$ displacing bromide to form ClCH$_2$CH$_2$CH$_2$CN
or displacing chloride to form BrCH$_2$CH$_2$CH$_2$CN

3.7 Use the carbocation stability trend to decide which of the following two alternatives is the lower-energy process.

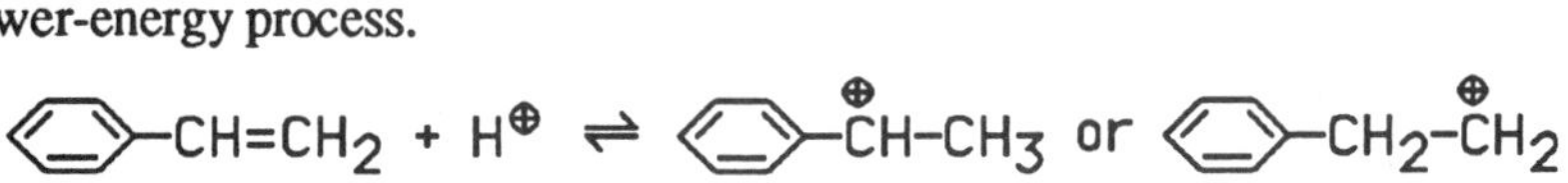

3.8 Use your knowledge of anion stability and resonance to decide which of the following two alternatives is the lower-energy process.

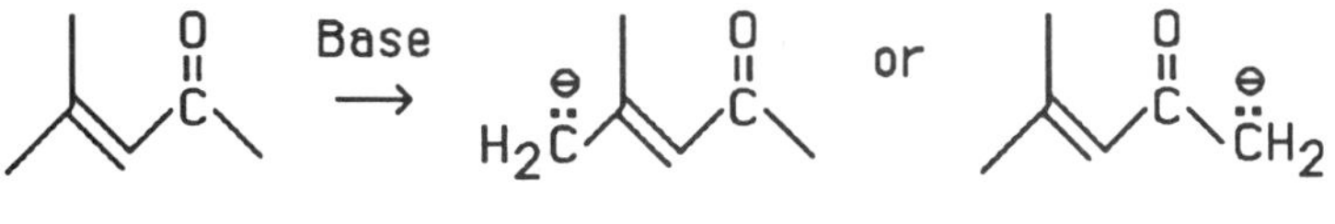

3.9 Use your knowledge of proton transfer equilibria to decide which of the following two alternatives is the lower-energy process.

$$H_3C-\overset{:O:}{\overset{\|}{C}}-\overset{:O:}{\overset{\|}{C}}-\overset{..}{O}H \; + \; {}^{\ominus}:\overset{..}{O}H \; \rightarrow \; H_2\overset{:O:}{\overset{\|}{C}}-\overset{:O:}{\overset{\|}{C}}-\overset{..}{O}H \; \text{ or } \; H_3C-\overset{:O:}{\overset{\|}{C}}-\overset{:O:}{\overset{\|}{C}}-\overset{..}{O}:^{\ominus}$$

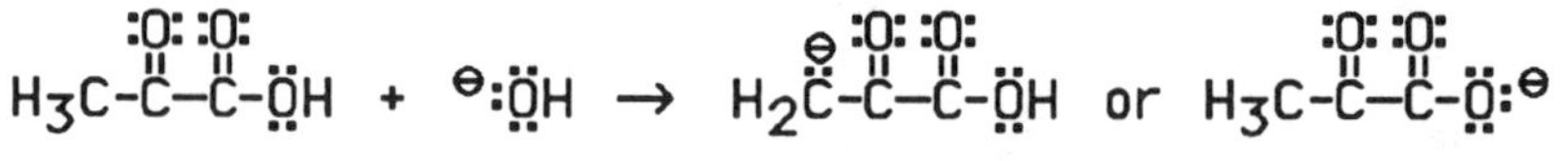

3.10 Decide which of the following two alternatives is the lower-energy process.

$$H_2\overset{\ominus}{\underset{}{C}}-\overset{:O:}{\overset{\|}{C}}-\overset{\ominus}{\underset{}{C}}H-\overset{:O:}{\overset{\|}{C}}-\overset{..}{O}CH_3 \; + \; H_3CCH_2-\overset{..}{\underset{..}{B}r}: \; \rightarrow \; H_3CCH_2-H_2C-\overset{:O:}{\overset{\|}{C}}-\overset{\ominus}{\underset{}{C}}H-\overset{:O:}{\overset{\|}{C}}-\overset{..}{O}CH_3 \quad :\overset{..}{\underset{..}{B}r}:^{\ominus}$$

$$\text{or} \; H_2\overset{\ominus}{\underset{}{C}}-\overset{:O:}{\overset{\|}{C}}-\overset{\ominus}{\underset{}{C}}H-\overset{:O:}{\overset{\|}{C}}-\overset{..}{O}CH_3 \; + \; H_3CCH_2-\overset{..}{\underset{..}{B}r}: \; \rightarrow \; \pm \; H_2\overset{\ominus}{\underset{}{C}}-\overset{:O:}{\overset{\|}{C}}-\underset{\underset{H_3CCH_2}{|}}{C}H-\overset{:O:}{\overset{\|}{C}}-\overset{..}{O}CH_3 \quad :\overset{..}{\underset{..}{B}r}:^{\ominus}$$

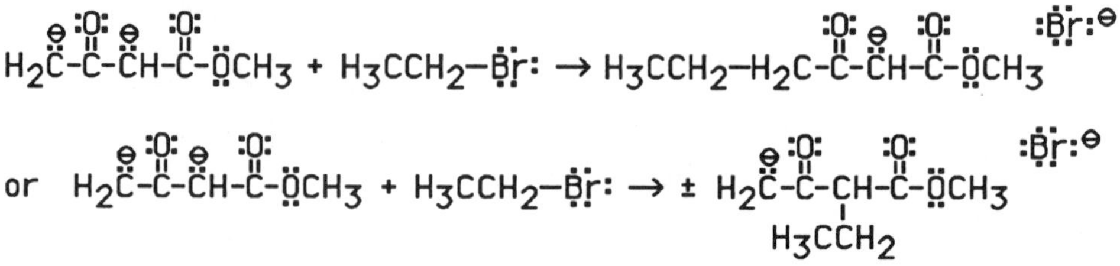

3.11 Decide which of the following two alternatives is the lower-energy process.

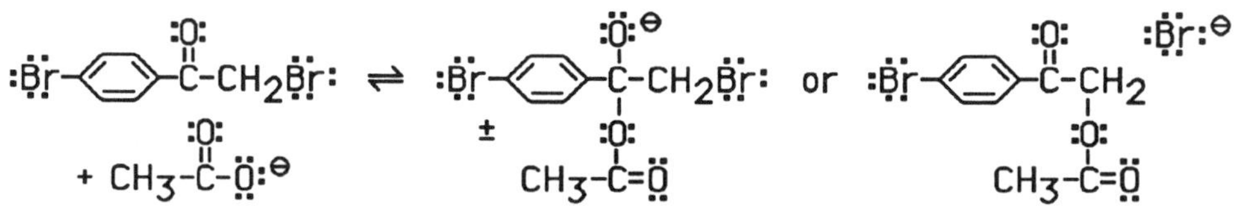

3.12 Ascorbic acid, vitamin C, has one very acidic H ($pK_{a1} = 4.1$), and all the others are not very acidic ($pK_{a2} = 11.8$). Use your knowledge of anion stability and resonance to find the acidic H and explain why it is so acidic.

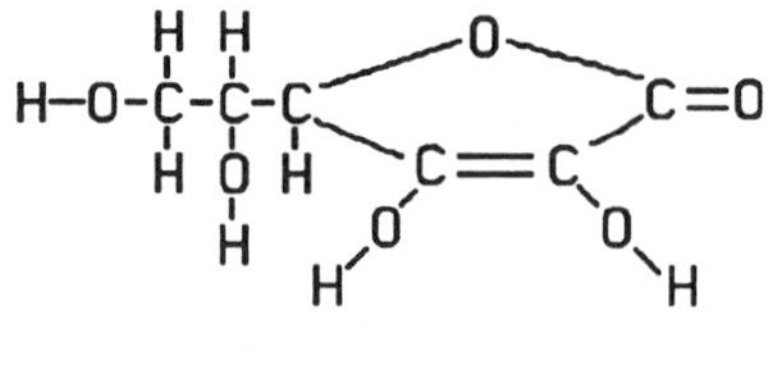

3.13 Using the pK_a chart, rank the following by acidity and explain the reason for the order you find.

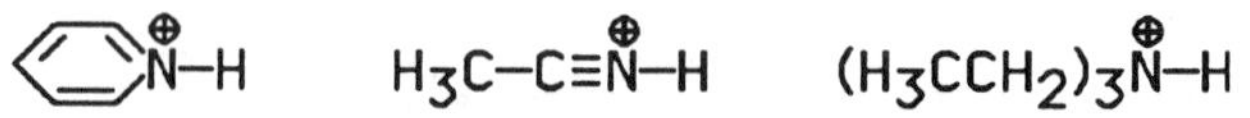

3.14 Account for the different rates in the ionization of the leaving group X. *Hint:* look at the stability of the resultant carbocation.

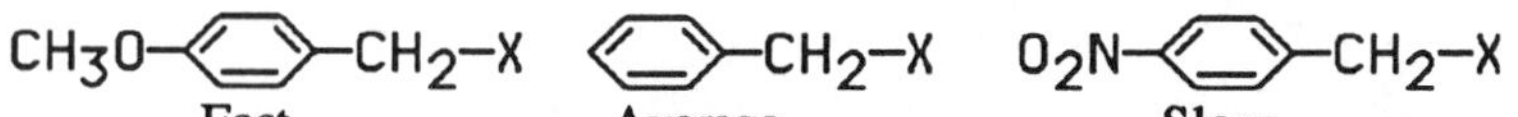

3.15 Use your knowledge of carbocation stability to explain why the loss of the leaving group for

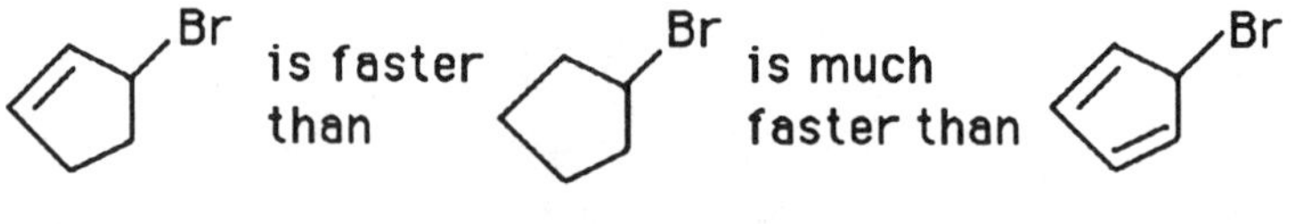

4

CLASSIFICATION OF ELECTRON SOURCES

4.1 GENERALIZED RANKING OF ELECTRON SOURCES

In this chapter we shall explore the classification and ranking of electron sources into the 12 common "generic" classes of electron sources. To repeat for emphasis, **there are only 12 generic classes from which a set of arrows, an electron flow, can start.** Since some classes may be quite rare and others may have several dozen examples, it is very important that you learn to classify according to generic class. **It is much easier to handle multiple examples of 12 familiar classes than**

hundreds of different special cases. The first thing that you should do in learning a new reaction is to classify the electron source into one of these 12 classes. Three questions must be answered about each class. What characterizes the members of the class? What is the most electron-rich site? What determines reactivity within each class?

Before discussing the individual classes, we need to get an overview of all of them. Since most reactions have a hard–hard and a soft–soft component, it is important to notice both how much negative charge a source bears and how soft the source is. **Electrons that reside in more stable, lower-energy bonding orbitals are less available for use as electron sources in reactions.** We can divide good electron sources into three groups: nonbonding electrons, electron-rich sigma bonds, and electron-rich pi bonds.

The best electron sources are usually nonbonding electron pairs. They are electron rich, and no bonds need be broken to use them as electron sources. Other excellent electron sources are highly ionic sigma bonds and also pi bonds highly polarized by excellent electron-donor groups.

The most stabilized electrons, such as those in strong sigma bonds or aromatic pi bonds, make poor electron sources. For example, carbon–carbon single bonds are generally poor sources. There are exceptions of course; an aromatic bearing an excellent electron donor is an improved but still mediocre source.

Although an electron flow path is shown for each source in this chapter, it is included only to point out the electron pair from which the electron flow must start. The detailed discussion of the electron flow paths is presented in Chapter 6.

4.2   NONBONDING ELECTRONS

Lone Pairs as Brønsted Bases

A nonbonding electron pair can serve as a Lewis base and attack an electron-deficient carbon or serve as a Brønsted base and pull off a proton. As a Brønsted base, the electron flow starts from the lone pair of the base.

$$b: \overset{\curvearrowright}{} H \overset{\curvearrowleft}{} A \rightarrow b–H + A:^{\ominus}$$

Section 3.2 showed how to rank Brønsted bases. For the calculation of the K_{eq} for proton transfer see Section 3.3. Nucleophilicity is a gauge of the rate of a Lewis base's attack on an electron-deficient carbon. A proton is much harder than an electron-deficient carbon, so the relative hardness of the nonbonding electron pair is important in any discussion of basicity versus nucleophilicity (see Section 8.5). To rank the nucleophilicity of a nonbonding electron pair, we must consider the major factors of charge, size, solvation, and the minor factors of electronegativity and strength of the bond formed.

Lone Pairs on Heteroatoms as Nucleophiles

$$R–\overset{\ominus}{\underset{}{Z}}: \overset{\curvearrowright}{} E^{\oplus} \rightarrow R–Z–E$$

The electron flow starts from the lone pair. Nonbonding electron pairs are good electron sources, especially the lone pairs of anions. The following list gives the relative nucleophilicity toward methyl iodide of many lone pair nucleophiles. A charged ion is always a better nucleophile than its neutral counterpart. Note that methoxide anion is almost 2 million times more nucleophilic than methanol.

Relative nucleophilicity toward CH₃I
(in CH₃OH as solvent)

CH₃OH	1
F$^{\ominus}$	500
CH₃COO$^{\ominus}$	20,000
Cl$^{\ominus}$	23,000
Et₂S	220,000
NH₃	320,000
PhO$^{\ominus}$	560,000
Br$^{\ominus}$	620,000
CH₃O$^{\ominus}$	1,900,000
Et₃N	4,600,000
CN$^{\ominus}$	5,000,000
I$^{\ominus}$	26,000,000
Et₃P	520,000,000
PhS$^{\ominus}$	8,300,000,000
PhSe$^{\ominus}$	50,000,000,000

We might suspect that the more basic species would also be more nucleophilic, but nucleophilicity does not parallel basicity exactly because softness must be considered. To verify this fact we need only to look at the previous list and see that iodide at pK_{abH} -10 is over a thousand times more nucleophilic than the more basic acetate, pK_{abH} 4.8. **For lone pairs of the *same element*, the more basic the lone pair is, the more nucleophilic it is.** To rank nucleophiles in this way, remember to look up the conjugate acid on the pK_a chart. The following is a ranked list of the pK_{abH} of only the oxygen nucleophiles from the previous list.

Most nucleophilic and most basic O	**pK_{ab}H**
CH₃O$^{\ominus}$	15.5
PhO$^{\ominus}$	10
CH₃COO$^{\ominus}$	4.8
CH₃OH	-2.4
Least nucleophilic and least basic O	

Steric hindrance can decrease nucleophilicity. As the nucleophile becomes larger, its bulk tends to get in the way of its acting as a nucleophile, especially when the electrophile is also large. For this reason the larger *tert*-butoxide, (CH₃)₃CO$^{\ominus}$, is a much poorer nucleophile than the smaller methoxide, CH₃O$^{\ominus}$.

Electron availability decreases as the orbital that the electrons reside in is made more stable. As the hybridization of the nitrogen lone pair goes from sp^3 to sp^2 to sp, the lone pair gets less basic and less nucleophilic as indicated by the following trend.

Most nucleophilic and most basic N
Et₃N: pK_{abH} = 10.7; lone pair hybridization = sp^3

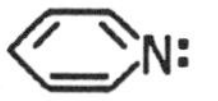

N: pK_{abH} = 5.2; lone pair hybridization = sp^2

CH₃C≡N: pK_{abH} = -10; lone pair hybridization = sp
Least nucleophilic and least basic N

However, there are exceptions to this hybridization trend. An ester's carbonyl oxygen sp^2 lone pairs are much more basic and more nucleophilic than its ether oxygen sp^3 lone pairs. When an ester's carbonyl lone pair is protonated, a delocalized cation is formed. When an ester's ether oxygen lone pair is protonated, the cation is less stable because it is not delocalized and the electron withdrawal by the adjacent carbonyl destabilizes it.

$$\text{R-}\overset{\overset{\displaystyle :\ddot{O}:}{\|}}{\text{C}}\text{-}\ddot{\ddot{O}}\text{-Et} + \text{H}^{\oplus} \quad \rightleftharpoons \quad \text{R-}\overset{\overset{\displaystyle \overset{\oplus}{:}\text{O-H}}{\|}}{\text{C}}\text{-}\ddot{O}\text{-Et} \quad + \quad \text{R-}\overset{\overset{\displaystyle :\ddot{O}:}{\|}}{\text{C}}\text{-}\underset{\underset{\displaystyle H}{|}}{\overset{\oplus}{\ddot{O}}}\text{-Et}$$

Ester + Acid   Major   Minor

Solvation is a hard–hard interaction and therefore **solvent effects drastically alter nucleophilicity.** Small, hard, highly shielded ions must "break out" of the solvent cage to be available as nucleophiles, and thus nucleophilicity will vary greatly with solvation. Solvation, in fact, appears to be mostly responsible for the following halide nucleophilicity trend in methanol. The ion most strongly solvated is fluoride, the weakest iodide. In polar aprotic solvents, the halide anion is poorly solvated, and the nucleophilicity order reverses, being dominated by the strength of the bond formed (the C−F bond is strongest).

Best halide nucleophile $I^{\ominus} > Br^{\ominus} > Cl^{\ominus} > F^{\ominus}$ Worst

A pK_a chart can be used as a reference for nucleophilicity only if the difference in softness is considered. A partially plus carbon atom is a much softer electrophile than a proton. Soft ions are more nucleophilic in protic solvents because tighter solvation greatly decreases the nucleophilicity of the hard ions.
We can restate the HSAB principle. Hard nucleophiles favor binding with hard electrophiles; soft nucleophiles favor binding with soft electrophiles. Most nucleophilicity charts show relative rates of nucleophilic attack with methyl iodide as the electrophile. A carbon–iodine bond is very soft because the electronegativity difference is nearly zero (Section 2.8). Therefore softer (less electronegative, more polarizable) atoms have lone pairs that are better electron sources toward soft electrophiles.
In summary, to rank the nucleophilicity of nonbonding electron pairs reacting in protic solvents with soft electrophiles such as R−X, first rank by softness, then by basicity (within the same attacking atom). However, for nonbonding electron pairs reacting with harder electrophiles such as a proton or a carbonyl, rank by basicity. Very reactive electrophiles like carbocations are not selective and react with the most abundant nucleophile (commonly the solvent).

Example problem

Which of the following, PhO$^{\ominus}$, PhS$^{\ominus}$, CH$_3$COO$^{\ominus}$, is the best nucleophile to react with CH$_3$I in methanol? Which is the worst?

Answer: Since the electrophile is soft and the solvent is protic, the softness of the nucleophile must be considered. Because sulfur is softer than oxygen, the softest and most nucleophilic is PhS$^{\ominus}$, thiophenoxide. The remaining oxygen anions can then be ranked by basicity. Of the oxygen anions, phenoxide (PhO$^{\ominus}$) at a pK_{abH} of 10, is more basic than acetate (CH$_3$COO$^{\ominus}$), whose pK_{abH} is 4.8. Therefore the next best nucleophile is phenoxide, and the least nucleophilic is acetate.

Metals

$$M: \; + \; R\!-\!\ddot{\underset{..}{Br}}: \; \rightarrow \; M^{2+} \; + \; R:^{\ominus} \; + \; :\ddot{\underset{..}{Br}}:^{\ominus}$$

The electron flow starts from the metal electrons. The valence electrons of a reactive metal such as lithium, sodium, magnesium, or zinc are a good electron source. Reactions of metals occur by *electron transfer* from the metal surface to the electron sink. This process oxidizes the metal producing the metal ion and reduces the electron sink. The electron transfer very likely occurs one electron at a time (Chapter 10). In the *halogen metal exchange reaction* shown, the R–Br bond is broken in the reduction step, producing an ionic or partially ionic organometallic salt.

Metals differ greatly in how easily they are oxidized. The most reactive metals are those with a very negative standard electrode potential. Of the commonly used metals, the alkali metals lithium, sodium, and potassium are highly reactive; the next most reactive is magnesium, then zinc.

4.3   ELECTRON-RICH SIGMA BONDS

We can consider a partially ionic sigma bond as an already partially "broken" bond. An organometallic bond, for example, has electrons that are very available. Conversely, a carbon–carbon covalent bond is completely covalent; its electrons reside in a very stable sigma bonding orbital and are not available.

Other partially broken single bonds are acidic hydrogen–heteroatom bonds. Hydrogen bonding to an appropriate solvent weakens them further. Since a proton is never dropped off by itself (Section 3.1) but rather picked off by a base or by a hydrogen bonded solvent molecule acting as a base, we will not consider acidic hydrogen–heteroatom bonds separately but regard them as already covered under the discussion of Brønsted bases. Although mechanisms are commonly written with the role of the solvent omitted, the electron flow actually starts with the nonbonding electron pair on the base or solvent.

Organometallics

$$M\!-\!R \; \longleftrightarrow \; M^{\oplus} \; {}^{\ominus}R:\!\curvearrowright\!E^{\oplus} \; \rightarrow \; M^{\oplus} \; + \; R\!-\!E$$

The electron flow starts from the polarized sigma bond. Organometallics vary in their ability to act as electron sources. Very ionic organometallics are excellent electron sources, whereas the most covalent organometallics are poor sources. Therefore checking the electronegativity difference between the metal and carbon will allow us to rank the reactivity of the organometallic when the R group is the same. **A good source will have a large electronegativity difference.** The following is a list of several organometallics ranked by their electronegativity differences (R is the same for all).

	Most reactive
1.62	R-Na
1.57	R-Li
1.24	R-MgX
0.9	R_2Zn
0.86	R_2Cd
0.65	$R_2Cu^{\ominus}Li^{\oplus}$
0.55	R_2Hg
0.22	R_4Pb
	Least reactive

To determine the reactivity of an organometallic on the basis of R when the metal is the same, compare the carbanion stability (Section 3.6). The less stable the carbanion, the more reactive it is as an electron source. For stabilized carbanions, the identity of the metal is of less importance than the stabilization of the carbanion.

Simple alkyl carbanions, like CH_3Li, are sp^3 hybridized at the carbanionic center and commonly exist as the tetramer, $(CH_3Li)_4$. This tetramer exists as a tetrahedron of lithium ions with a carbanion snuggled in between the three lithium atoms of each face. The degree of aggregation affects reactivity; lithium-chelating tetramethylethylenediamine, $(CH_3)_2NCH_2CH_2N(CH_3)_2$, is used to dissociate this aggregate to the more reactive dimer. As the number of electron-donating alkyl groups on the carbanion increases, the carbanion gets less stable and more reactive. The alkyllithium general reactivity trend is tertiary > secondary > primary > methyl.

Example problem

Which of the following, CH_3Li, CH_3MgBr, CH_3CdCl, is the most reactive organometallic? Which is the least?

Answer: The most ionic is the most reactive. The least electronegative metal in this group is lithium, so CH_3Li is the most reactive. The next most reactive is CH_3MgBr; the least reactive and the most covalent organometallic is CH_3CdCl.

Metal hydrides

$$M{-}H \longleftrightarrow M^{\oplus}\ {}^{\ominus}H\!:\!\overset{\curvearrowright}{+}E^{\oplus} \rightarrow M^{\oplus} + H{-}E$$

The electron flow starts from the polarized sigma bond. Metal hydrides can serve as good electron sources, but their behavior varies with the electronegativity of the metal. Alkali metal hydrides, NaH and KH, are ionic and function primarily as bases reacting with acidic protons to form hydrogen gas; little reduction occurs. The complex metal hydrides, $AlH_4^{\ominus}$ and $BH_4^{\ominus}$, donate hydride to reduce the substrate and are not as good as bases. To summarize, the harder alkali metal hydrides are good bases and poor nucleophiles, whereas the softer complex metal hydrides are good nucleophiles and weaker bases. As with the organometallics in the previous section, the electronegativity difference governs the reactivity of the metal hydrides. The metal hydride will become somewhat less reactive if the metal is made more electronegative by attaching an inductively withdrawing group. As the hydride source becomes less reactive, it also becomes more selective. The complex metal hydride reactivity trend is $AlH_4^{\ominus}$ > $HAl(OR)_3^{\ominus}$ > $BH_4^{\ominus}$ > $H_3BCN^{\ominus}$.

4.4   ELECTRON-RICH PI BONDS

Allylic Sources $:Z{-}C{=}C \leftrightarrow {}^{\oplus}Z{=}C{-}\overset{..}{C}{}^{\ominus}$

Pi bonds tend to be weaker than sigma bonds and are therefore usually easier to break. We can make a pi bond more reactive by attaching a good lone-pair donor. The electron flow can come from either end of the allylic system. When pi bonds are used as electron sources, an attached lone pair electron-donor group can greatly increase the availability of the pi bond's electrons and stabilize the cationic product usually formed upon an electrophilic attack on the double bond. We have in essence "extended" via delocalization with a double bond the properties of the lone pair donor group (vinylogy). These sources contain three adjacent *p* orbitals (Section 1.8), and almost all are much more reactive than a simple alkene.

Resonance structures show us that either the double bond or the heteroatom can serve as an electron source. Because allylic sources can "bite" at two different sites they are called *ambident nucleophiles*. Usually the double bond is the attacking nucleophile; the decision about which end of the allylic source attacks the electrophile will be discussed in Section 8.4.

We will see later that some electron sinks are in equilibrium with a species that can serve as an allylic electron source, $Z=C-C-H \rightleftharpoons H-Z-C=C$; this equilibrium is called *tautomerization* and is discussed in Chapter 6 under path combinations.

The rankings of allylic electron sources reflect the electron-donor group trends already covered in Section 3.5 and thus do not present new trends for you to learn. **The better the donor is, the better the allylic electron source is.** The following are several allylic electron sources, ordered by their rank as sources:

Best			Worst
$:\ddot{O}:^{\ominus}$ R–C=CH$_2$	$\ddot{N}(CH_3)_2$ R–C=CH$_2$	$:\ddot{O}CH_3$ R–C=CH$_2$	$:\ddot{X}:$ R–C=CH$_2$
Enolate	Enamine	Enol ether	Vinyl halide

The last one in the series, the vinyl halide, is actually a poorer electron source than a simple alkene because halogens are poorer donors than alkyl groups.

Example problem

Which of the following, $CH_2=CH-O^{\ominus}$, $CH_2=CH-NR_2$, $CH_2=CH-Cl$, is the most reactive allylic source? Which is the least?

Answer: The best donor is the oxygen anion, so the best allylic source of the three is the enolate, $CH_2=CH-O^{\ominus}$; the next best is the enamine, $CH_2=CH-NR_2$. The least is the vinyl chloride, $CH_2=CH-Cl$, because chlorine is a very poor donor.

Vinylogous allylic sources, $:Z-C=C-C=C$ or $:Z-C=C-C=O$, commonly react at the center atom and therefore will be treated as a subset of this source. A few examples of these vinylogous allylic electron sources are, respectively, extended enolates, acetoacetates, and malonates.

Allylic Alkyne Sources $\ddot{Z}-C\equiv C-$

If we add a second pi bond perpendicular to an allylic system we create this rare electron source that should just be treated as a subset of the allylic sources.

4.5   SIMPLE PI BONDS

$$\overset{\nearrow E^{\oplus}}{\underset{}{\rangle}C\overset{|}{=}C\langle} \quad \rightarrow \quad \overset{}{\underset{}{\rangle}\overset{\oplus}{C}\overset{E}{\underset{|}{-}}C\langle}$$

The pi bond is the electron source. For allylic sources, the electron pair donor stabilizes the resultant cation. The simple alkene reactivity trend reflects the stabilization of the resultant carbocation by alkyl substitution and delocalization with other double bonds. The ranking of simple double bonds as electron sources is

Best ══╱   ╲╱╲╱   ╲╱╲   ╲╱╲   ══╲ Worst

When a substituted double bond is used as an electron source, often two carbocations, which can differ in stability, are possible. **The formation of the more stable carbocation will be the lower-energy process and will determine the site of electrophilic attack** (a modern version of Markovnikov's rule). Figure 4.1 shows that if the group on the double bond acts as an electron donor, the carbocation adjacent to and stabilized by the donor will be formed. Donors stabilize cations, and electron-withdrawing groups destabilize cations. Avoid forming cations adjacent to electron-withdrawing groups.

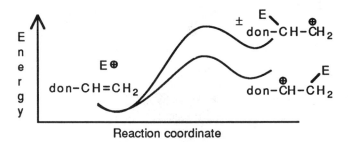

Figure 4.1 Energy diagrams for electrophilic addition to substituted alkenes. The cation adjacent to the donor is stabilized.

To make the choice, draw out both carbocations and then pick the more stable one according to the trends you learned in Section 3.4. This is one of many times that we will use the trends to make a decision about the route a reaction will take.

Example problem

Predict where an electrophile would attack the following compound.

⋀⋀

Answer: Draw out all four possible carbocations that would be formed upon electrophilic attack and rank their stability.

| Disubstituted allylic carbocation | Primary carbocation | Secondary carbocation | Monosubstituted allylic carbocation |

The ranking of these carbocations from most to least stable is disubstituted allylic > monosubstituted allylic > secondary > primary. Since the electrophile will add to form the most stable cation, the electrophile will add on the end to form the disubstituted allylic carbocation.

Alkyne Sources

The two perpendicular pi bonds of a triple bond can be considered separately, and thus this source becomes merely a subset of the double-bond sources. Alkynes are slightly poorer electron sources than alkenes because the vinyl cation formed upon electrophile attack is relatively unstable. Acetylide anions, $RC\equiv C^{\ominus} M^{\oplus}$, are treated as organometallic reagents (Section 4.3) and not as alkyne sources.

Example problem

Which of the following, $CH_3C\equiv CCH_3$, $RC\equiv C-NR_2$, $HC\equiv CH$, is the most reactive triple-bond electron source? Which is the worst?

Answer: The best donor attached to the triple bond is the NR_2, so the best triple-bond electron source of the three is $RC\equiv C-NR_2$; the next best is $CH_3C\equiv CCH_3$ because methyl is a weak donor. The worst is $HC\equiv CH$, because bearing no donors the cation formed upon electrophilic attack lacks any form of stabilization.

The least common of the simple pi bond sources are allenes, $R_2C=C=CR_2$, whose two perpendicular double bonds must again be considered separately. Allenes tend to be less reactive than simple alkenes because the intermediate carbocation formed upon electrophilic attack is less stable. When an electrophile adds to the center of an allene, the initially formed carbocation is perpendicular to the second double bond and therefore is not delocalized, a loss of about 14 to 25 kcal/mol (59 to105 kJ/mol) of allylic stabilization. Electrophilic addition at the end of the allene gives the unstable secondary vinyl cation.

4.6 AROMATIC RINGS

The pi cloud of the aromatic ring reacts with an electrophile to produce a carbocation that then loses an electrophile, commonly $H^{\oplus}$, to restore the aromatic stabilization of the ring. Almost all aromatic rings are poor electron sources because the electrons in an aromatic ring are very stabilized (Section 1.9) and are therefore not very available. However, an aromatic ring can be made more reactive by the attachment of one or more good electron-donor groups. Any group that has a resonance interaction with an aromatic ring has the greatest effect at the ortho and para positions.

An electron-donor group attached on the ring increases the electron availability at the ortho and para positions and stabilizes the carbocation formed upon attack at those positions. Aromatic rings substituted with donors are more reactive than the unsubstituted compound and react with electrophiles at the ortho and para positions.

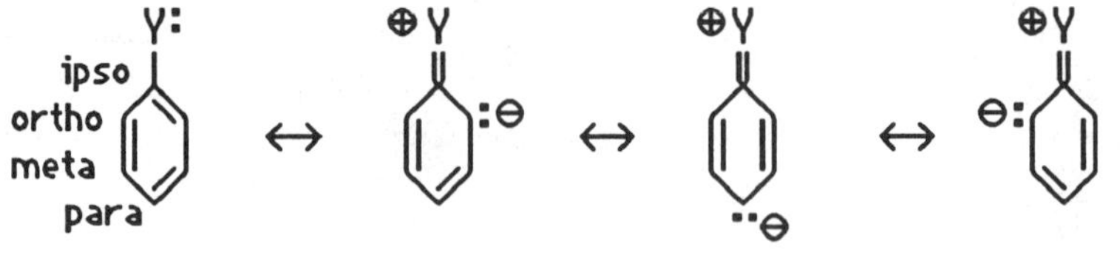

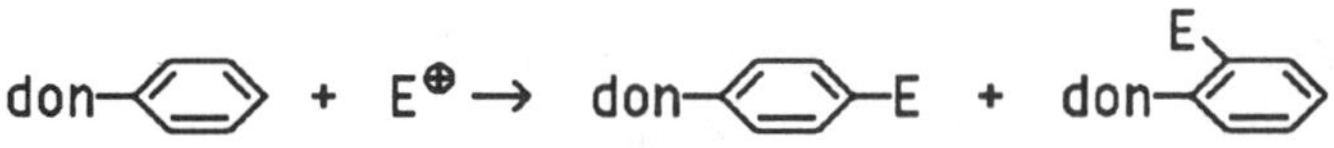

An electron-withdrawing group attached to the ring decreases the electron availability at the ortho and para positions and destabilizes the carbocation formed upon attack at those positions. Aromatic rings substituted with electron-withdrawing groups are less reactive than the unsubstituted compound and react with electrophiles at the meta position.

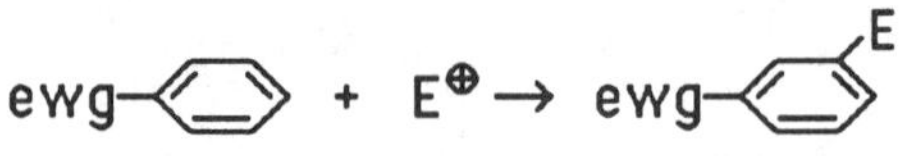

The position of attachment is the most electron-rich site for groups significantly less electronegative than carbon, like R_3Si. Ipso electrophilic attack replaces the R_3Si group with the electrophile, E.

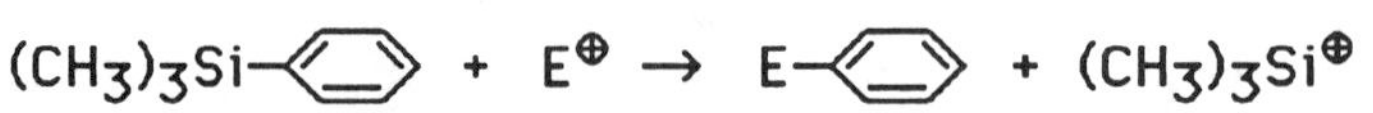

Example problem

Which of the following, $PhCH_3$, $PhNR_2$, PhCOR, is the most reactive aromatic ring toward electrophilic attack? Which is the least?

Answer: The best donor is the NR_2, so the most reactive ring of the three is $PhNR_2$; the next best donor is methyl, so the next most reactive ring is $PhCH_3$. The least reactive ring is the PhCOR, because a ketone is an electron withdrawing group and deactivates the ring.

4.7   SUMMARY OF GENERIC ELECTRON SOURCES

Table 4.2 lists the 12 generic electron sources from which an electron flow in a reaction may start. Conspicuous in their absence are covalent sigma bonds. **Regular sigma bonds are just too stabilized to act as electron sources** (only when destabilized in a highly strained three-membered ring may a carbon–carbon single bond behave as a source). The only single bonds that serve as sources are highly ionic ones, single bonds to metals.

The remaining task is to come up with a method for classification of a given structure into its most appropriate generic class. Often several functional groups are present in a reactant, and several reactants may be present in a reaction. One functionality may possibly fall into two classes, so we need to pick the most applicable. We need to recognize the generic sources in a way such that we find the most reactive first. A flow chart of the classification process for electron sources can be found in the Appendix.

1. Metals, organometallics, and metal hydrides are usually highly reactive.
2. Next look for lone pair donors and the electron rich pi bonds.
3. Then search for the simple pi bonds.
4. The least reactive, aromatics, are sorted out last.

Table 4.1 The Twelve Generic Electron Sources

Generic Class	Symbol	Examples
Nonbonding Electrons (4.2)		
Lone pairs on heteroatoms (Nu / base dual behavior)	Z:	$I^{\ominus}$, $HO^{\ominus}$, H_2O, H_3N, t-$BuO^{\ominus}$ $CH_3COO^{\ominus}$
Metals	M:	Na, Li, Mg, Zn metals
Electron-rich sigma bonds (4.3)		
Organometallics	R$-$M	RMgBr, RLi, $R_2Cu^{\ominus}$ $Li^{\oplus}$
Metal hydrides	$MH_4^{\ominus}$ (Nu)	$LiAlH_4$, $NaBH_4$,
	MH (Bases)	NaH, KH
Electron-rich pi bonds (4.4)		
Allylic sources	C=C$-$Z:	Enols: C=C$-$OH Enolates: C=C$-$O$^{\ominus}$ Enamines: C=C$-$NR$_2$
Allylic alkyne sources	C≡C$-$Z:	$EtC≡C$-NEt_2
Simple pi bonds (4.5)		
Alkenes	C=C	$H_2C=CH_2$
Dienes	C=C$-$C=C	$H_2C=CH$-$HC=CH_2$
Alkynes	C≡C	HC≡CH
Allenes	C=C=C	$H_2C=C=CH_2$
Aromatic rings (4.6)		
Aromatics	ArH	

Example problem

To which generic classes do these belong, Ph$-$NH$_2$, CH$_2$=CH$-$Li ?

Answer: We notice that in Ph$-$NH$_2$ there is a nitrogen lone pair and an aromatic ring. Following our classification scheme, we drop it into the lone pair class, Z, rather than into the less reactive aromatic class. Likewise for CH$_2$=CH$-$Li , we classify it as an organometallic, R$-$M, rather than the much less reactive alkene.

ADDITIONAL EXERCISES

4.1 Give the generic class of each of the following electron sources.

$LiAlH_4$	MeLi	NH_3	MeOCH=CH$_2$	Et$_2$NLi
Mg	MeCOO$^{\ominus}$	EtMgBr	Me$_2$NCH=CH$_2$	H_2O
NaH	CH$_3$C≡CH	Me$_2$C=CH$_2$	t-BuO$^{\ominus}$	PhCH$_3$

4.2 Give the generic class of each of the following electron sources.

PhNH$_2$	$^{\ominus}$C≡N	Cl$^{\ominus}$	$^{\ominus}$CH$_2$COOCH$_3$	CH$_3$OH
NaBH$_4$	$^{\ominus}$O-CH=CH$_2$	HO$^{\ominus}$	Zn	NaOEt
$^{\ominus}$SC≡N	$H_2C=C=CH_2$	PhSH	$H_2C=CHCH=CH_2$	F$^{\ominus}$

4.3 Draw Lewis structures for the species in problem 4.1 and indicate on the structures which electron pair(s) would be the start of the electron flow in a reaction.

4.4 Draw Lewis structures for the species in problem 4.2 and indicate on the structures which electron pair(s) would be the start of the electron flow in a reaction.

4.5 Circle the one atom on each molecule that is best for electrophilic attack.

$$
\langle \bigcirc \rangle\text{-CH=CH-CH=CH}_2 \quad (H_3C)_2\overset{..}{N}\overset{\overset{:\overset{..}{O}:}{||}}{C}CH_3 \quad H_2C=CH\overset{\overset{:\overset{..}{O}:}{||}}{C}CH_3 \quad (H_3C)_2C=CH_2
$$

4.6 Rank all species beginning with the numeral 1 to designate:

The best nucleophile to react with CH_3I in CH_3OH

PhSe$^\ominus$ PhO$^\ominus$ PhS$^\ominus$ $CH_3COO^\ominus$ CH_3OH

The most reactive organometallic
Me-K   Me-MgBr   Me$_2$Hg   Me$_2$Cd   MeLi

The best allylic electron source

$$
\overset{}{=\!\!\!\diagup}\diagdown\overset{..}{\underset{\ominus}{N}}R \quad =\!\!\!\diagup\diagdown\overset{..}{N}R_2 \quad =\!\!\!\diagup\diagdown\overset{..}{\underset{..}{O}}R \quad =\!\!\!\diagup\diagdown\overset{..}{\underset{\ominus}{O}}: \quad =\!\!\!\diagup\diagdown\overset{..}{\underset{..}{C}}:
$$

The most reactive aromatic ring toward electrophilic attack

$$
\langle\bigcirc\rangle\text{-}\overset{\overset{:\overset{..}{O}:}{||}}{C}\text{-}\overset{..}{\underset{..}{O}}R \quad \langle\bigcirc\rangle\text{-}\overset{..}{\underset{..}{O}}H \quad \langle\bigcirc\rangle\text{-NO}_2 \quad \langle\bigcirc\rangle\text{-H} \quad \langle\bigcirc\rangle\text{-CH}_3
$$

4.7 State Markovnikov's rule.

4.8 At which position(s) on the aromatic ring would you expect electrophilic attack to occur for the following compounds?

PhCF$_3$ PhOH   PhNO$_2$ PhCOOH   PhCl

4.9 Use the nucleophilicity trend to decide which of the following two alternatives is the faster process in methanol solvent.

$CH_3CH_2CH_2Br + CH_3O^\ominus$ displacing bromide to form $CH_3CH_2CH_2OCH_3$ or
$CH_3CH_2CH_2Br + CH_3S^\ominus$ displacing bromide to form $CH_3CH_2CH_2SCH_3$

4.10 When an acyl chloride is added to aqueous ammonia two products are possible. Use the nucleophilicity trend to decide which of the following two alternatives is the faster process.

$CH_3COCl + H_2O$ displacing chloride to form CH_3COOH or
$CH_3COCl + NH_3$ displacing chloride to form CH_3CONH_2

5

CLASSIFICATION OF ELECTRON SINKS

5.1   GENERALIZED RANKING OF ELECTRON SINKS

In this chapter we will study the 18 generic classes of electron sinks into which an electron flow ends. Because most reactions have a hard–hard and a soft–soft component, again it is important to notice both how much positive charge the sink bears and how soft it is. **The larger the partial plus is, the better the electron sink can attract a negatively charged nucleophile.**

The three general classes of electron sinks are an empty orbital, a weak single bond to a leaving group, and a polarized multiple bond. Excellent electron sinks are electron-deficient cationic species having an empty orbital, such as protons and carbocations, that can easily form a new bond with an electron source. Similarly, strong Lewis acids like $AlCl_3$ or BF_3 are also good electron sinks. Medium electron sinks are groups that accept the electron flow from the electron source by breaking a weak bond and forming a stable anion. The bond that is broken is either a pi bond of a polarized multiple bond or a sigma bond to a leaving group. Almost all of these electron sinks have a partial or inducible partial plus charge on the atom that gets attacked by the electron source, the nucleophile. Here are a few examples listed with their bond polarization:

$$H-L \qquad Y-L \qquad \underset{/}{\overset{|}{C}}-L \qquad \underset{/}{\overset{|}{C}}=Y \qquad -C{\equiv}Y \qquad \underset{/}{\overset{|}{C}}=C{\overset{ewg}{\underset{\diagdown}{}}}$$

The poorer electron sinks would be any of the above species with poorer leaving groups, less electronegative atoms, or poorer electron-withdrawing groups.

Although an electron flow path is shown for each sink in this chapter, it is included only to point out the atom that would be attacked by the electron source. A detailed discussion of the electron flow paths is presented in Chapter 6.

5.2 ELECTRON-DEFICIENT SPECIES

Carbocations

$$^{\ominus}Nu{:}\curvearrowright\overset{\oplus}{\underset{/}{C}}- \quad \rightarrow \quad Nu-\overset{|}{\underset{|}{C}}$$

The best electron sinks, occurring almost exclusively in acidic media, are reactive cations such as carbocations (Section 3.4). Most are such good electron sinks that they can react with even very poor electron sources like aromatic rings. The most stable carbocations are the least reactive. Highly stabilized carbocations like $^{\oplus}C(NH_2)_3$ are so stable that they can exist in basic media and make very poor electrophiles. The following are some of the more common reactive carbocations.

$$H_3C-\overset{\oplus}{C}\overset{\ddot{O}H}{\underset{\ddot{O}H}{\cdots}} \qquad H_3C-\overset{\oplus}{C}\overset{\ddot{O}H}{\underset{CH_3}{\cdots}} \qquad Ph-\overset{\oplus}{C}\overset{H}{\underset{H}{}} \qquad H_3C-\overset{\oplus}{C}\overset{CH_3}{\underset{CH_3}{}} \qquad H_2C{=}\overset{H}{\underset{}{C}}-\overset{\oplus}{C}\overset{H}{\underset{H}{}}$$

Lewis Acids

$$^{\ominus}Nu{:}\curvearrowright\overset{}{\underset{/}{A}}- \quad \rightarrow \quad Nu-\overset{|}{\underset{|}{A}}{}^{\ominus}$$

Electron-deficient Lewis acids like $AlCl_3$ or BF_3 behave much like carbocations. Like the $^{\oplus}CR_3$ carbocation, BR_3 has an empty $2p$ orbital, exactly the same orbital configuration (*isoelectronic*), and will tend to react in a similar way. The strength of a Lewis acid varies with substituents; attached donors reduce the acidity. A very crude ranking of Lewis acid strengths is $BX_3 > AlX_3 > FeX_3 > SbX_5 > SnX_4 > ZnX_2 > HgX_2$. Another strong Lewis acid is $TiCl_4$.

Metal Ions

$$M^{2\oplus}$$

Although not really electron deficient, metal ions may bear a high positive charge and can act as electrophilic catalysts because they can complex with many electron sinks and make them more positive, and thus more prone to nucleophilic attack. Examples are $Ag^{\oplus}$, $Hg^{2\oplus}$, and $Mg^{2\oplus}$.

Transition metal ions such as $MnO_4^{\ominus}$ and $CrO_4^{2\ominus}$ can serve as electron acceptors in oxidation reactions, for they can have many available oxidation states.

5.3 WEAK SINGLE BONDS

Acids

$$^{\ominus}Nu\text{:} \frown H\text{--}A \rightarrow Nu\text{--}H \ + \ A\text{:}^{\ominus}$$

All Brønsted acids belong to this classification. The more stable the conjugate base is (Section 3.2), the more acidic the acid is. Strong acids will protonate the solvent; therefore the strongest acid that can exist in a solvent is the protonated solvent. See Section 3.3 for the calculation of the proton transfer K_{eq}.

Weak Single Bonds Between Heteratoms

$$^{\ominus}Nu\text{:} \frown Y\text{--}L \rightarrow Nu\text{--}Y \ + \ L\text{:}^{\ominus}$$

Bonds between heteroatoms and good leaving groups are generally weak. Lone pair–lone pair repulsion is responsible for weak bonds in RO—OR and RS—SR. These weak bonds are broken easily by an electron source that attacks Y and displaces the leaving group. These reactions are usually exothermic since the bond formed is almost always stronger than the weak Y—L bond. In compounds like I—Cl, it is very important to note that the $\partial+$ end of the molecule gets attacked by the nucleophile (in this case iodine because Cl is more electronegative). A subset of the Y—L class that will be discussed in Sections 6.3 and 7.5 is the :Nu—L class in which Y bears a nucleophilic lone pair. Example: halogens like Br_2 and Cl_2

Leaving Group Bound to an sp^3 Hybridized Carbon

$$Nu\text{:} \frown C\text{--}L \rightarrow Nu\text{--}C \ + \ L\text{:}^{\ominus}$$

The sp^3 bound leaving groups can be directly displaced by the nucleophile as shown, or the leaving group can ionize off first to form a carbocation. As the leaving group gets better, the following equilibrium begins to produce more of the carbocation, an excellent electron sink. The weaker electron sources must wait until the carbocation is formed before the reaction can occur. The extent of this equilibrium depends heavily on the stability of the carbocation produced (Section 3.4) in addition to the quality of the leaving group (Section 3.8). (See Section 8.3 for a complete discussion of the ionization of leaving groups.) We will find in Section 5.5 that vinylic, C=C—L, sp^2 bound leaving groups behave much differently from sp^3 bound leaving groups.

$$\overset{\cdot}{C}\text{--}L \ \rightleftharpoons \ \left[\overset{\cdot}{C}^{\oplus} \ \text{:}L^{\ominus} \right] \ \rightleftharpoons \ \overset{\cdot}{C}^{\oplus} \ + \ \text{:}L^{\ominus}$$
$$\text{Ion pair}$$

Any electron sink that contains a leaving group will become a much better sink as the leaving group improves. One common way to improve the quality of the leaving group is with acid catalysis. Acid catalysis is the complexation of a proton or strong Lewis acid (for example, BF_3 or $AlCl_3$) with an electron sink, which increases the electron sink's ability to accept electrons.

Poor leaving group   Good leaving group

In the previous acid catalysis example, the pK_{aHL} of the leaving group improves over 17 pK_a units (hydroxide loss at a pK_{aHL} of 15.7 compared to water loss at a pK_{aHL} of -1.7). In fact, no reaction at all would occur without catalysis because hydroxide is a poor leaving group. **With few exceptions, no reaction occurs when the electron sink bears a poor leaving group (pK_{aHL} greater than 12).** Because diethyl ether, $CH_3CH_2OCH_2CH_3$, is so unreactive, it is a common solvent for reactions involving very reactive nucleophiles. Strained ring ethers (three- and four-membered) are exceptions. Ring strain raises the energy of the reactant relative to a product in which the ring is broken, and the strain is relieved. The five-membered ring ether (tetrahydrofuran, THF) is not strained and is unreactive.

Reactive   Unreactive

The best sinks of the sp^3 bound leaving group class are allylic leaving groups, $CH_2=CH-CH_2-L$, and benzylic leaving groups, $Ar-CH_2-L$, because both are especially activated toward direct displacement and ionization (see Section 6.1). Some additional examples of sp^3 bound leaving group sinks are

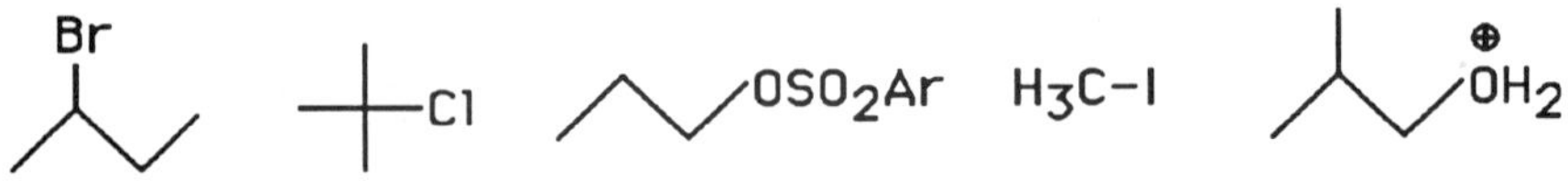

Compounds with more than one leaving group are just a subset of this class. Ketals, $R_2C(OR)_2$, and orthoesters, $RC(OR)_3$, are common examples.

Example problem

Which of the following, CH_3OCH_3, CH_3I, CH_3Cl, is the most reactive toward nucleophilic attack? Which is the least?

Answer: The most reactive bears the best leaving group. The most reactive is CH_3I, with a pK_{aHL} of -10. CH_3Cl is next, with a pK_{aHL} of -7. The least reactive is the ether, CH_3OCH_3, with a pK_{aHL} of 15.5.

5.4 POLARIZED MULTIPLE BONDS WITHOUT LEAVING GROUPS

Heteroatom–Carbon Double Bonds

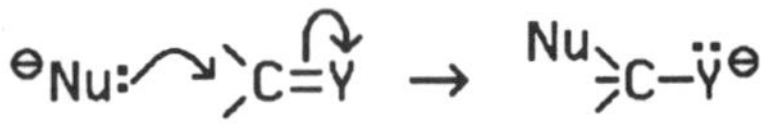

The electronegativity of Y polarizes the double bond to put a partial plus charge on carbon, making it susceptible to nucleophilic attack. The electronegativity of Y stabilizes the anionic product of the nucleophilic attack. The atom Y is most commonly oxygen, occasionally nitrogen. Steric effects are important; the less hindered aldehydes are more reactive than ketones. Conjugation stabilizes the carbonyl and thus makes it less reactive; for example, $PhCOCH_3$ is less reactive than CH_3COCH_3.

Acid catalysis, often by proton transfer or hydrogen bonding to Y, enhances the ability of this electron sink to accept electrons. Since the lone pair on an sp^2 nitrogen is easier to protonate than a lone pair on an sp^2 oxygen (see pK_a chart), acidic catalysis occurs much more readily with nitrogen. The carbocation thus formed is a much better electron sink. Additional examples of this sink are

$$\begin{array}{cccc} H_3C & H_3C & H\oplus CH_3 & \\ C=0 & C=N\;\;+ & C=N & \triangle \\ H_3C & H_3C & H\;\;CH_3 & CH \\ & & & \|\;\;0 \end{array}$$

Example problem

Which of the following is the most reactive toward nucleophilic attack? Which is the next most reactive? Which is the least reactive?

$$\begin{array}{cccc} :\!\ddot{O}\!: & :\!NPh & \oplus\!:\!\ddot{O}H & \overset{\oplus}{H}NPh \\ \| & \| & \| & \| \\ Ph\text{–}C\text{–}H & Ph\text{–}C\text{–}H & Ph\text{–}C\text{–}H & Ph\text{–}C\text{–}H \end{array}$$

Answer: The most reactive is the protonated aldehyde. The protonated imine is next, for it is slightly more stable. The least reactive is the neutral imine because the anion formed upon nucleophilic attack is more basic than the anion formed upon attack of the neutral aldehyde.

Heteroatom–Carbon Triple Bonds

$$\ominus Nu:\!\curvearrowright\!\overset{\curvearrowleft}{\underset{R}{C}}\!\overset{Y}{\diagup} \;\;\rightarrow\;\; \underset{R}{\overset{Nu}{\diagdown}}C=\ddot{Y}\ominus$$

The most common representatives of this electron sink are the nitriles (Y equals N). Nitriles are much less reactive than ketones. The electronegativity of Y polarizes the triple bond so that carbon bears a partial plus charge. The two perpendicular double bonds can be treated separately, thus attack of a nucleophile is identical with that on the doubly bonded electron sink, C=Y. Some examples of this sink are Ph—C≡N, and CH$_3$—C≡O$^\oplus$.

Heterocumulenes

The heterocumulene derivatives, C=C=Y or Z=C=Y, are a minor variation of the polarized multiple-bond sink and undergo an identical nucleophilic attack at the C=Y bond. The only difference is that the intermediate is a resonance hybrid that protonates to form the most stable product. The product with the stronger bonds predominates.

$$\underset{\underset{:Nu^\ominus}{\diagdown}}{\overset{\diagup}{}}C=C=\overset{\curvearrowleft}{\ddot{Y}} \;\rightarrow\; \overset{\diagup}{\diagdown}C=C\!\underset{Nu}{-}\overset{\ominus}{\ddot{Y}} \;\leftrightarrow\; \overset{\ominus}{\overset{\diagup}{\diagdown}C}\!-\!\underset{Nu}{C}\!=\!Y \;\overset{p.t.}{\rightarrow}\; \overset{H}{\overset{\diagup}{\diagdown}}C\!-\!\underset{Nu}{C}\!=\!Y$$

Some examples of heterocumulene sinks are

$$\begin{array}{cccc} H & & & \\ C=C=0 & Ph\text{–}N=C=0 & 0=C=0 & R\text{–}N=C=N\text{–}R \\ H & & & \end{array}$$

Conjugate Acceptors

$$^{\ominus}Nu: \curvearrowright \underset{/}{\overset{\backslash}{C}} = \underset{\backslash}{\overset{/}{C}} \curvearrowright ^{ewg} \rightarrow \underset{/}{\overset{Nu}{\overset{\backslash}{C}}} - \underset{\backslash}{\overset{\ominus}{C}} ^{ewg}$$

The electron-withdrawing group polarizes the carbon–carbon double bond in a way similar to the C=Y electron sink and makes this sink a subset of that group (with Y equaling C—ewg). The electron-withdrawing group must be present for the double bond to accept electron density from a nucleophile. The attached electron-withdrawing group makes the C—ewg carbon more electronegative. **Trading C—ewg for Y is simply exchanging an electronegative carbon atom for an electronegative heteroatom. The better the electron-withdrawing group is, the better the electron sink will be.** More than one good electron-withdrawing group on a double bond, C=C(ewg)$_2$, gives a very reactive electron sink because the resulting anion is stabilized to a much greater extent.

A complication arises in that the nucleophile now has a choice; it can add in a conjugate fashion, or it can add to the electron withdrawing group directly. Section 8.6, Ambident Electrophiles, discusses this problem in depth. Examples of these sinks are

Triply Bonded and Allenic Conjugate Acceptors

$$^{\ominus}Nu: \curvearrowright \underset{}{-C \equiv C - ewg} \rightarrow \underset{/}{\overset{Nu}{\overset{\backslash}{C}}} = \overset{\ominus}{C} - ewg$$

$$^{\ominus}Nu: \underset{}{\overset{\backslash}{C} = C = \overset{/}{C}} ^{ewg} \rightarrow \underset{\backslash}{\overset{/}{C}} = \overset{Nu}{\underset{}{C}} - \overset{\ominus}{C} ^{ewg}$$

These two electron sinks are the least common of the six that make up the set of multiply bonded electron sinks with no attached leaving group. They behave very similarly to the conjugate acceptors previously discussed. The electronegative C—ewg group replaces the electronegative Y in accepting the electron flow from the nucleophile.

5.5 POLARIZED MULTIPLE BONDS WITH LEAVING GROUPS

Carboxyl Derivatives

$$^{\ominus}Nu: \curvearrowright \underset{L}{\overset{R}{\overset{\backslash}{C}}} = \overset{\curvearrowright}{Y} \rightarrow \pm \, Nu - \underset{L}{\overset{R}{\overset{\backslash}{\underset{\curvearrowleft}{C}}}} \overset{\curvearrowleft}{\overset{..}{Y}}^{\ominus} \rightarrow \underset{Nu}{\overset{R}{\overset{\backslash}{C}}} = Y \quad :L^{\ominus}$$

The most common representatives of this group are the carboxylic acid derivatives with Y as oxygen. To distinguish this group from polarized multiple bonds without leaving groups, the second step in the mechanism, the loss of the leaving group, is shown above. This is a two-step process: first nucleophile addition occurs, then the leaving group is lost. The rate-determining step is almost always the nucleophilic attack, and therefore the rate depends on the ability of this electron sink to attract and accept electron density from a nucleophile.

If the leaving group is a good electron donor it will delocalize and diminish the partial plus on the carbon of the carbonyl, decreasing its hardness and therefore decreasing its ability to attract a negatively charged nucleophile. Conjugation of a leaving group lone pair tends to make these systems more stable and less reactive. The barrier to reaction increases since this delocalization of the reactant is not possible in the tetrahedral intermediate. In an acyl chloride, the most reactive carboxyl derivative, this delocalization of the leaving group lone pair into the carbonyl is very minor because of poor $3p$–$2p$ overlap between the chlorine and the carbonyl carbon. **As L becomes a better donor, carboxyl derivatives become less reactive.**

$$ \underset{R}{\overset{:O:}{\underset{\diagdown}{\overset{\parallel}{C}}}}\diagup L \quad \leftrightarrow \quad \underset{R}{\overset{:\ddot{O}:^{\ominus}}{\underset{\diagdown}{\overset{|}{C}}}}\diagup L^{\oplus} $$

From a HOMO–LUMO viewpoint, the better donor L is, the more it raises the HOMO and LUMO of the system. Section 2.9 discussed the perturbation of a carbon–carbon pi bond by a donor; the carbon–oxygen pi bond is perturbed in the same manner. When the LUMO is raised in energy, it becomes less accessible. The interaction with the incoming nucleophile's HOMO is poorer because the raised LUMO is now farther away in energy.

Carboxylic acids are doubly poor sinks, for they can protonate the nucleophile (thus making it useless) and form the very unreactive carboxylate anion. Anions rarely attack anions because of the repulsion of like charges. **Only R—Li and LiAlH$_4$ are reactive enough to attack the carboxylate anion.** The less reactive nucleophiles will react only with the most reactive carboxyl derivatives. Using the donor trend, we can place ketones as more reactive than esters. Acyl halides, with the halide being both a poor donor and a good leaving group, are the most reactive carboxyl derivatives. A carboxyl derivative with a good leaving group is very reactive because the loss of the leaving group from the intermediate is often the driving force for the overall process. Acyl chlorides are considered more reactive than anhydrides for this reason. In general, the better the leaving group on the carboxyl derivative, the more reactive the it will be. The better donor group deactivation and the better leaving group activation of carboxyl derivatives do not often conflict, so either trend can be used to judge the reactivity of this electron sink.

$$ \overset{:O:}{\underset{}{\overset{\parallel}{R\text{-}C\text{-}\ddot{C}l:}}} \;>\; \overset{:O: \;\; :O:}{\underset{}{\overset{\parallel}{R\text{-}C\text{-}\ddot{O}\text{-}C\text{-}R}}} \;>>\; \overset{:O:}{\underset{}{\overset{\parallel}{R\text{-}C\text{-}\ddot{O}R}}} \;>\; \overset{:O:}{\underset{}{\overset{\parallel}{R\text{-}C\text{-}\ddot{N}R_2}}} \;>\; \overset{:O:}{\underset{}{\overset{\parallel}{R\text{-}C\text{-}\ddot{O}:^{\ominus}}}} $$

Best electron sink   Worst electron sink

One minor variation of this group is the carbonate derivatives, which have two leaving groups attached to the carbonyl instead of just one. These systems can be quite unreactive since the partial plus on the carbonyl is now diminished by two donor groups. Both leaving groups can be replaced by a nucleophile. If the leaving groups are different enough (for example, Cl and OEt), the monosubstitution product can be isolated.

$$ \underset{^{\ominus}Nu:}{\overset{:O:}{\overset{\parallel}{L\text{-}C\text{-}L}}} \rightarrow \underset{Nu}{\overset{^{\ominus}:\ddot{O}:}{\overset{|}{L\text{-}C\text{-}L}}} \rightarrow \underset{^{\ominus}Nu:}{\overset{:O: \;\; L:^{\ominus}}{\overset{\parallel}{L\text{-}C\text{-}Nu}}} \rightarrow \underset{Nu}{\overset{^{\ominus}:\ddot{O}: \;\; L:^{\ominus}}{\overset{|}{L\text{-}C\text{-}Nu}}} \rightarrow \overset{:O:}{\overset{\parallel}{Nu\text{-}C\text{-}Nu}} + 2L:^{\ominus} $$

Example problem

Which of the following is the most reactive toward nucleophilic attack? Which is the least reactive?

$$\overset{:\overset{\displaystyle ..}{O}:}{\underset{}{R-\overset{||}{C}-\overset{..}{\underset{..}{O}}R}} \qquad \overset{:\overset{..}{O}:\;\;:\overset{..}{O}:}{R-\overset{||}{C}-\overset{..}{\underset{..}{O}}-\overset{||}{C}-R} \qquad \overset{:\overset{..}{O}:}{R\overset{..}{\underset{..}{O}}-\overset{||}{C}-\overset{..}{\underset{..}{O}}R}$$

Answer: The most reactive bears the weakest donor. The anhydride is the most reactive. The ester is the next most reactive; least reactive is the carbonate.

Vinyl Leaving Groups

This less common group bears the same relationship to the carboxyl derivatives L–C=Y discussed in Section 5.5 as the conjugate acceptors C=C–ewg bear to the C=Y (Section 5.4) series. Again we have merely substituted C–ewg for the electronegative Y atom in the group. To illustrate how similar the reactivities are, if the ewg is C=Y, then the L–C=C–C=Y group is merely a vinylogous L–C=Y group. Aromatic rings that bear electron-withdrawing groups and a leaving group can be included in this classification. Again, more than one leaving group can be attached.

The electron-withdrawing group is absolutely necessary for the acceptance of electron density from a nucleophile. Simple olefinic halides, RCH=CH−X, are very unreactive because there is no place to put the electrons from the nucleophile.

Leaving Groups on Triple Bonds

Two very rare electron sinks, L−C≡Y and L−C≡C−ewg, the first represented by the acid bromide of HCN, Br−C≡N, react as expected for a multiply bonded electron sink with attached leaving group. The C−ewg again replaces Y.

5.6 SUMMARY OF GENERIC ELECTRON SINKS

Reactivity Comparison Between Various Classes

All the electron sinks are summarized in Table 5.1; however, the general reactivity trends are not as clearcut as they were for electron sources. Generally, look first for electron-deficient species, then for the others. The reactivity trend within each class is more important than to which class a species belongs. For example, **acyl chlorides,** the top end of the reactivity spectrum for polarized multiple bonds with leaving groups, **are more reactive than most alkyl halides; however, esters are less reactive than most alkyl halides. Acyl chlorides are more reactive than aldehydes. Nitriles tend to be about as reactive as amides.**

A method for classification of a given structure into its most appropriate generic class relies on your ability to recognize electron-deficient species, leaving groups, and polarized multiple bonds. A flow chart of the classification process for electron sinks can be found in the Appendix.

1. Carbocations and Lewis acids are usually highly reactive, and thus are found first.
2. Next look for any leaving groups. Is the leaving group bound to carbon, and if so, what is the hybridization of that carbon? Then sort into the two general classes: weak single bonds, and polarized multiple bonds with leaving groups.
3. Search for any polarized multiple bonds without leaving groups.
4. Finally, metal ions as sinks are sorted last since they tend to be the least reactive.

Table 5.1 The 18 Generic Electron Sinks

Generic class	Symbol	Examples
Electron-deficient species (5.2)		
Carbocations	$\geqslant C^{\oplus}$	$^{\oplus}C(CH_3)_3$
Lewis acids	$\geqslant A$	$BF_3, BH_3, AlCl_3$
Metal ions	$M^{2\oplus}$	$Hg(OAc)_2$
Weak single bonds (5.3)		
Acids	$H-L$ (or $H-A$)	$HCl, ArSO_3H, CH_3COOH$
Weak single bonds between heteroatoms	$Y-L$	$RS-SR, HO-OH, Br-Br,$ $HO-Cl$
Leaving groups bound to sp^3 carbon	$\geqslant C-L$	$CH_3CH_2-Br, (CH_3)_3C-Br,$ $(CH_3)_2CH-Cl, CH_3-I$ Ketals $R_2C(OCH_3)_2$ (2 poor Ls)
Polarized multiple bonds without leaving groups (5.4)		
Heteroatom-carbon multiple bonds	$C=Y$	Aldehydes: $RHC=O$ Ketones: $R_2C=O$ Imines: $R_2C=NR$
	$C\equiv Y$	Nitriles: $RC\equiv N$
Conjugate acceptors	$C=C-ewg$	Enones: $C=C-C=O$, Conjugated esters: $C=C-COOR$
	$C\equiv C-ewg$	$RC\equiv C-COOR$
Heterocumulenes	$C=C=Y$	Ketenes: $R_2C=C=O$
	$Z=C=Y$	Ph-N=C=O, RN=C=NR, O=C=O
	$C=C=C-ewg$	$(CH_3)_2C=C=CHCOOCH_3$
	$ewg-C=C=C-ewg$	ArCOCH=C=CHCOAr
Polarized multiple bonds with leaving groups (5.5)		
Carboxyl derivatives	$L-C=Y$	Acyl chlorides: RCOCl Anhydrides: RCOOCOR Esters: RCOOEt Carbonate derivatives: ClCOCl
Vinyl leaving groups	$L-C=C-ewg$	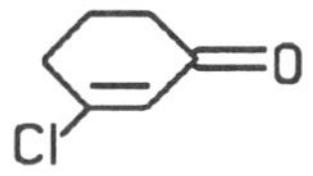
Leaving groups on triple bonds	$L-C\equiv Y$ $L-C\equiv C-ewg$	$Br-C\equiv N$ $Cl-C\equiv C-Cl$

ADDITIONAL EXERCISES

5.1 Give the generic class of each of the following sinks.

PhNCO	BF_3	HOBr	i-PrBr	Me_2CO
Ac_2O	CH_3Br	$AlCl_3$	$CH_2=CHCOCH_3$	CO_2
CH_3CN	$(CH_3)_2NCH_2^{\oplus}$	$AgNO_3$	CH_3COCl	$(CH_3)_3C^{\oplus}$

5.2 Give the generic class of each of the following sinks.

H_2SO_4	$PhSO_3CH_3$	CH_3SO_2Cl	$SOCl_2$	$CH_2=CHCN$
$COCl_2$	HOAc	H_2O_2	CH_3COOEt	$CH_3OSO_3CH_3$
PhCHO	EtOCOCl	Cl_2	$Cl-C\equiv N$	$H_2C=C=O$

5.3 Draw Lewis structures for all the species in problem 5.1 and designate which atom(s) on each would be attacked by the electron source in a reaction.

5.4 Draw Lewis structures for all the species in problem 5.2 and designate which atom(s) on each would be attacked by the electron source in a reaction.

5.5 Circle the best atom on each molecule for nucleophilic attack.

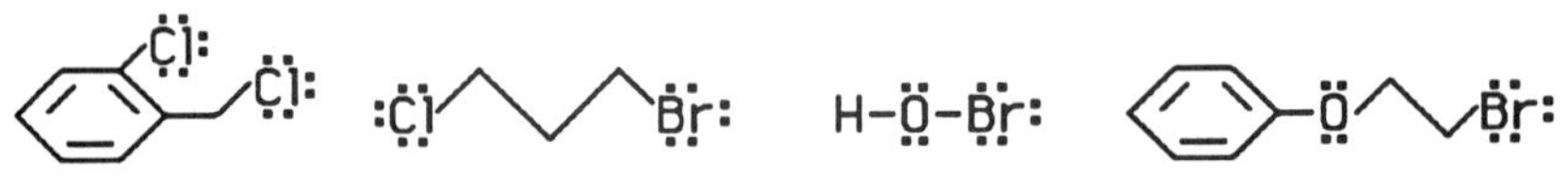

5.6 Trends: rank all species; use the numeral 1 to designate—

The most reactive toward nucleophilic attack

$$CH_3-OCH_3 \quad CH_3-Br \quad CH_3-NMe_2 \quad CH_3-I \quad CH_3-Cl$$

The most reactive conjugate acceptor

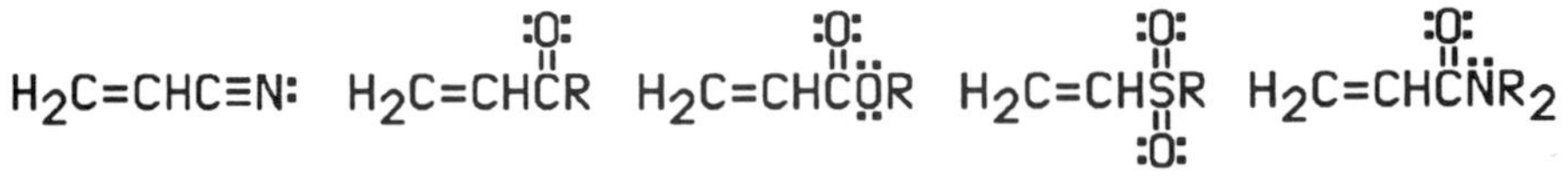

The most reactive toward nucleophilic attack

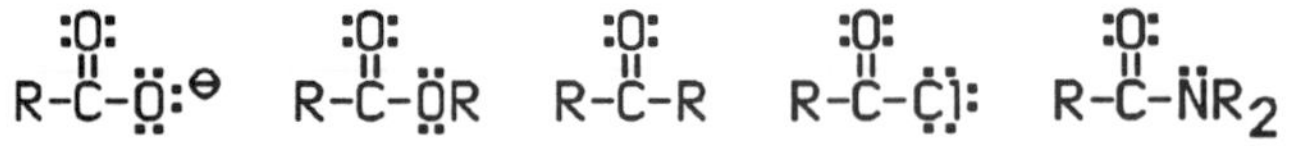

5.7 Explain in general terms why ketals, $R_2C(OR')_2$, fail to react (act as electron sinks) in basic media but react rapidly in acidic.

5.8 Sodium borohydride reduces ketones quickly but reacts very slowly to reduce esters. How quickly would you expect an aldehyde to react with sodium borohydride?

5.9 Circle the best atom on each molecule for nucleophilic attack. The comparison is between different classes of sinks.

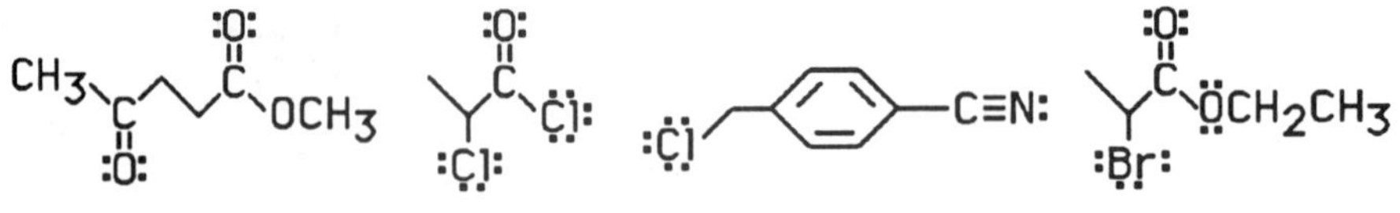

5.10 Thioesters, common in living systems, are more reactive than esters. Explain.

6

THE ELECTRON
FLOW
PATHWAYS

There are only four organic reaction types:

> Substitution – trade a nucleophile for a leaving group
> Addition – add a nucleophile or electrophile or both
> Elimination – lose a leaving group and usually a proton
> Rearrangement – produce a constitutional isomer

The first three can occur intramolecularly (within the same molecule) or intermolecularly (via a bimolecular collision), whereas rearrangement is only an intramolecular process. Termolecular (three molecule) collisions are rare and are considered a mechanistic possibility only when two of the three reacting molecules are loosely associated via hydrogen bonding or pi-complexation. The odds of a simultaneous collision of three independent molecules in the proper orientation for reaction are very, very slim.

6.1   THE DOZEN MOST COMMON PATHWAYS

Although there are thousands of different organic reactions, they occur by mechanisms that are a combination of relatively few pathways for electron flow. The following are 12 of the most common generic electron flow pathways. **These pathways should become a very important part of your mechanistic "vocabulary."** You will need to have an excellent command of these pathways to be able to combine them for the prediction of the mechanistic path that a reaction will take.

These electron flow pathways have been shown to be reasonable by many detailed mechanistic studies. Therefore if we confine our use of arrows to stepwise combinations of these pathways, taking into account each pathway's restrictions, we are pretty much assured that whatever we write will also be judged reasonable if not correct.

Beware of shortcuts that combine steps to save redrawing a structure. Any sequence of arrows on one structure implies that particular electron flow occurs in one step, which in turn demands that all the interacting orbitals be properly lined up. Any sequence of more than three arrows may have major $\Delta S^{\ddagger}$ problems. There are examples in published work of absolute nonsense written with arrows, so always be suspicious of any set of arrows that does not fit a known pathway.

All of the electron sources and sinks previously discussed can react by at least one of these pathways or a simple combination of them. We must learn what sort of functionality is required for each pathway and also notice the limitations on each pathway. Eventually we will be trying to decide among alternate routes, and the limitations are important so that we may narrow down the possibilities. Each of these pathways has four charge types since the source and/or sink can be charged or neutral: Nu with E, $Nu^{\ominus}$

with E, Nu with $E^\oplus$, and $Nu^\ominus$ with $E^\oplus$. Although any combination of E with Nu can be viewed as either an electrophilic attack or a nucleophilic attack, for consistency we will call it nucleophilic attack if the organic reactant is the sink and electrophilic attack if an inorganic reactant is the sink.

Because of microscopic reversibility, many of these pathways are merely the reverse of others. The transition states for each related pair of paths are similar (they need not be identical for the reaction conditions are often slightly different). A figure that illustrates the **approximate** orbital alignment and the transformation of the orbitals of the reactants into those of the product is given for each pair of paths.

We can group these pathways by the type of electron sink involved. There are several pathways for formation of a carbocation: a neutral species may lose a negatively charged piece (path D_N), a cationic species can lose a neutral piece (another charge type of path D_N), or a neutral species may bond with a positively charged one (path A_E). A carbocation can then rearrange (path 1,2R) to a carbocation of equal or greater stability. Finally, two pathways are available for quenching of carbocations: capture by a nucleophile (path A_N) or loss of a positively charged group (path D_E). In two pathways (S_N2 and E2) the electron flow from the electron source is concerted with the loss of the leaving group. Polarized multiple bond reactions comprise the last group of pathways (Ad_E3 and Ad_N).

Path p.t., Proton Transfer to and from an Anion or Lone Pair

The strongest acid possible in a solvent is the protonated solvent. Proton transfer can occur from any acidic to any basic groups or to and from the solvent. A common shortcut in writing mechanisms is to draw just the proton rather than the acid or the protonated solvent. Always check the proton transfer K_{eq} (Section 3.3).

Proton transfer is either a deprotonation of the reactant

or the microscopic reverse, protonation of the reactant.

The nonbonding lone pair electrons of the carbonyl group, C=O, are more available than the electrons of the pi bond; proton transfer occurs to and from the lone pairs rather than to and from the carbonyl pi bond.

Proton transfer can occur in any media depending on the charge type. **A solution that contains equal concentrations of a base and its conjugate acid has a pH that is equal to the pK_{abH} (Section 3.2).** The pH of the medium will be close to the pK_{abH} of the anion needed for reaction, because the anion must have a high enough concentration to react at a reasonable rate (Section 2.2). As an illustration, if a ketone enolate, pK_{abH} = 19.2, is needed for reaction, the media will definitely be basic.

Protons on heteroatoms (H−Y) are usually rather acidic because the heteroatom is electronegative, so deprotonation can occur with a rather weak base. An electronegative heteroatom, Y, can be replaced by an electronegative carbon atom, C−ewg.

Path limitations (Viewed as a Reactant Deprotonation)

Overlap: With H−C−ewg, the C−H bond and the adjacent p orbital of the pi system of the ewg must be roughly coplanar (Figure 6.1). There are no restrictions for H−Y, but there is a tendency for the proton transfer transition state to have a linear arrangement of b, H, and Y (the most common hydrogen bond is linear). Little angular dependence is expected since the base is overlapping with a spherical hydrogen 1s orbital.

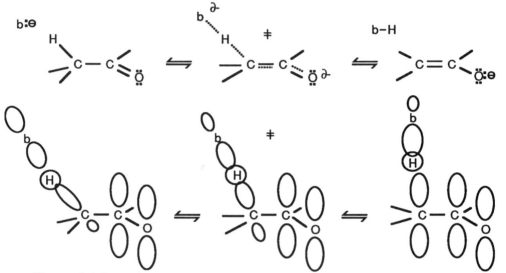

Figure 6.1 Loss of a proton adjacent to a carbonyl with the transition state shown in the center.

Base: strong enough to pull off proton (see pK_a chart).
Media: acidic, basic, or neutral. The pH is often near the pK_{abH} of the anion formed.
Solvent: polar to stabilize anion formed.
Steric: proton must be accessible to base.
Energetics: Use the proton transfer K_{eq} to get $\Delta G°$ (see Sections 2.1 and 3.3).

Kinetics: Many reactions are catalyzed by proton transfer. If the rate of the catalyzed reaction is determined by the reactant concentrations and just the pH, the [H⊕], the reaction is specific acid catalyzed. Specific-acid-catalyzed reactions do not show a rate change if the identity or concentration of the acid is changed as long as the pH remains the same. *Specific-acid catalysis* indicates that the proton-transfer equilibria are fast and not involved in the rate-determining step. For similar base-catalyzed reactions, the term *specific-base catalysis* is used.

Alternatively, if the presence of weaker proton donors, the [HA], contributes to the rate, the reaction is general acid catalyzed. General-acid-catalyzed reactions are dependent on the identity (pK_a) and concentration of the weaker acids present. *General-acid catalysis* indicates a reaction mechanism that has proton transfer occurring in the rate-determining step (for example, attack on a species hydrogen bonded to HA). For similar base-catalyzed reactions, the term *general-base catalysis* is used.

Path D$_N$, Ionization of a Leaving Group
(Dissociation, Nucleofugic)

$$\text{\ding{}C-L} \rightarrow \text{C}^{\oplus} \quad :\text{L}^{\ominus} \qquad or \qquad \overset{Y}{\underset{\|}{C}}\text{-L} \rightarrow \overset{Y}{\underset{\|}{C}}^{\oplus} \quad :\text{L}^{\ominus}$$

$$\text{\ding{}C-L}^{\oplus} \rightarrow \text{C}^{\oplus} \quad :\text{L} \qquad or \qquad \overset{Y}{\underset{\|}{C}}\text{-L}^{\oplus} \rightarrow \overset{Y}{\underset{\|}{C}}^{\oplus} \quad :\text{L}$$

Path limitations

Overlap: Leaving group departs along the axis of the bond (Figure 6.2). (Cleavage is a simple extension of a bond stretching vibration).

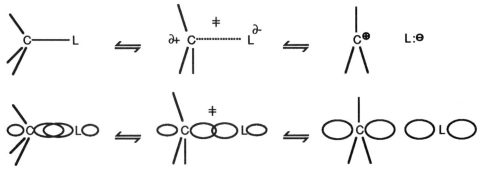

Figure 6.2 Ionization of a leaving group with the transition state in the center.

Cation: must be relatively stable.

Leaving group: For $\geqslant$C–L, the L must be good, often protonated. However, for Y=C–L or C=C–L, the L must be excellent or complexed with a Lewis acid to make up for the formation of the less stable cation.

Media: commonly acidic, occasionally neutral, rarely basic. The formation of unstable cations like the vinyl cation occurs only in strong acid.

Solvent: polar to stabilize cation.

Steric: Sterically crowded leaving groups tend to leave a little faster due to the strain released in going to the planar carbocation.

Energetics: often uphill since charge is created from a neutral species; however, see Section 8.3.

Path A$_N$, Trapping of an Electron-Deficient Species
(Association, Nucleophilic, the Reverse of the Previous Reaction)

$$\text{Nu:}^{\ominus}\overset{\frown}{}\text{C}^{\oplus} \rightarrow \text{Nu-C\ding{}} \qquad or \qquad \text{Nu:}^{\ominus}\overset{\frown}{}\overset{Y}{\underset{\|}{C}}^{\oplus} \rightarrow \text{Nu-}\overset{Y}{\underset{\|}{C}}$$

Included in this group would be any of the isoelectronic Lewis salt-forming reactions.

Path limitations

Overlap: same as path D$_N$; Nu must approach the empty p orbital along its axis so that good overlap occurs (Figure 6.2, but view the figure from right to left).

Media: commonly acidic, occasionally neutral, rarely basic.

Solvent: polar to stabilize cation.

Steric: relatively little problem because access to the cationic center is usually very good because the cation is flat.

Energetics: usually downhill since charge is neutralized.

Path A$_E$, Electrophile Addition to a Multiple Bond
(Association, Electrophilic)

An alkene is an average electron source and an aromatic compound is usually worse; therefore to get electrophilic addition to alkenes and aromatic compounds to occur one needs a good electron sink. Often a loose association of an electrophile with the pi electron cloud (called a pi-complex) occurs before the actual sigma bond formation step. The best electrophiles, cations, add easily. **The electrophile adds to give the most stable of the possible carbocation intermediates**

Path limitations

Overlap: Electrophile must approach, along the axis of the *p* orbital attacked, not from the side (Figure 6.3).

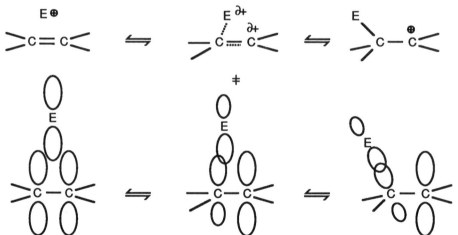

Figure 6.3 Electrophile addition to a double bond with the transition state shown in the center.

Cation formed: The most stable cation is formed. If the electrophile is the most stable of the possibilities, one expects the reverse reaction to be rapid and the equilibrium to favor reactants.

Media: most commonly acidic since good electrophiles are necessary for reaction. Occasionally neutral, rarely basic.

Solvent: often polar to stabilize carbocation.

Steric: will direct electrophile attack only if there is no difference in the two possible carbocations formed.

Energetics: usually uphill unless the electrophile is less stable than the carbocation.

The electrophile is commonly a proton. The reaction is usually uphill in energy unless the acid is very strong or the cation is very stable, and can easily reverse since most carbocations are very strong acids. As the carbocation becomes less stable, the equilibrium shifts toward reactants. This equilibrium can be shifted to form more of the carbocation by making the acid stronger. For protonation of a pi bond see Figure 6.4 (but view the figure from right to left).

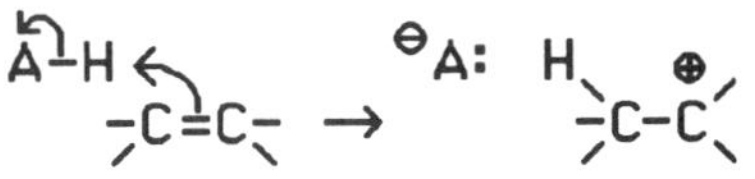

Path D_E, Electrophile Loss from a Cation To Form a Pi Bond
(Dissociation, Electrofugic, the Reverse of the Previous Reaction)

Path limitations

Overlap: C−E bond broken must be roughly coplanar with the empty *p* orbital of the carbocation so that a good pi bond is formed (Figure 6.3, right to left).
Media: acidic or neutral.
Electrophile lost (electrofuge): must be reasonably stable compared to carbocation.
Solvent: polar to stabilize cation.
Steric: little problem.
Energetics: downhill unless the electrophile is less stable than the carbocation.

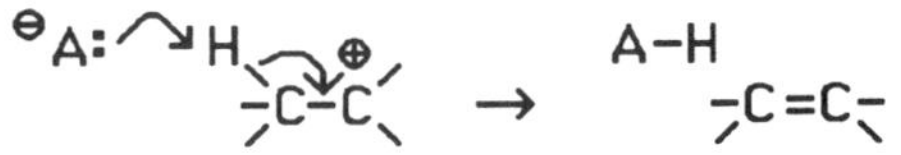

The electrophile lost is commonly a proton. This reaction is downhill in energy unless the A−H is a very strong acid. The orbital overlap is shown by Figure 6.4.

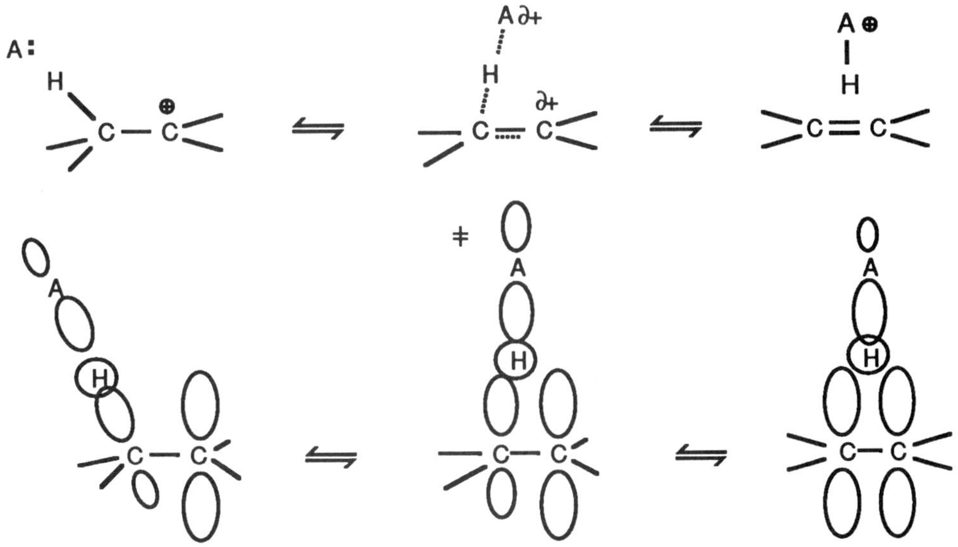

Figure 6.4 Loss of a proton from a carbocation with the transition state shown in the center.

Path 1,2R, 1,2 Rearrangement of a Carbocation

Rearrangement will occur whenever a full positive charge is formed on a carbon that has an adjacent group capable of shifting over to it to form a more stable cation. If the carbon bears only a partial plus, the tendency to rearrange is less. Rearrangement of alkyl groups to an anionic or radical center does not occur.

Path limitations

Overlap: The most important path restriction to rearrangements is that **orbital overlap must be maintained by the migrating group with both the atom it migrates from and the atom it migrates to.** At the transition state the migrating group orbital will overlap both atoms as shown in Figure 6.5. Usually a group migrates to the neighboring atom from the one it starts out on, a 1,2 shift.

Media: commonly acidic, occasionally neutral, rarely basic.

Migratory aptitudes: At the transition state the migrating group bears a partial positive charge. The groups that can best tolerate this partial plus migrate the best. In the rearrangements of carbocations, the order is $-H > -Ph > -CR_3 > -CHR_2 > -CH_2R > -CH_3$. In rigid systems, the group that migrates is the one that is able to achieve proper alignment with the p orbital of the carbocation.

Solvent: polar to stabilize cation.

Energetics: An energetically favorable rearrangement produces a more stable cation.

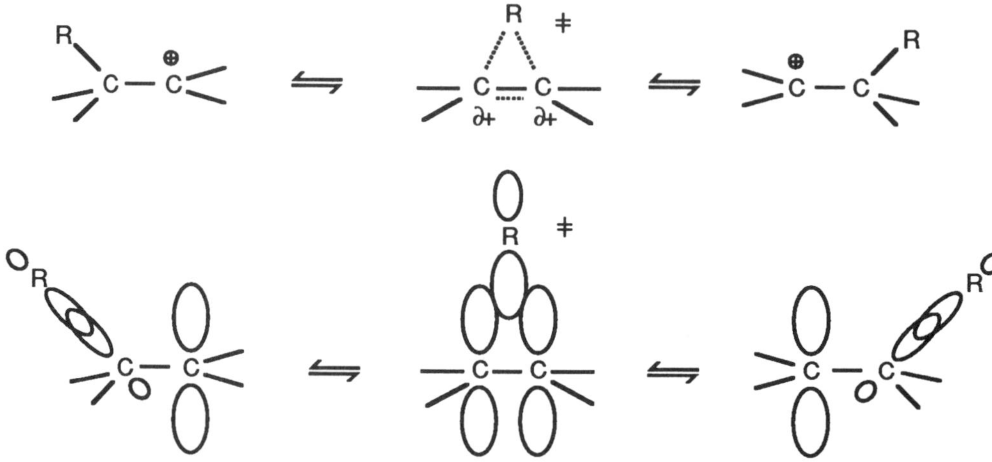

Figure 6.5 The 1,2 rearrangement of a carbocation. The transition state is in the center.

Path S$_N$2, The S$_N$2 Substitution

(Substitution, Nucleophilic, Bimolecular)

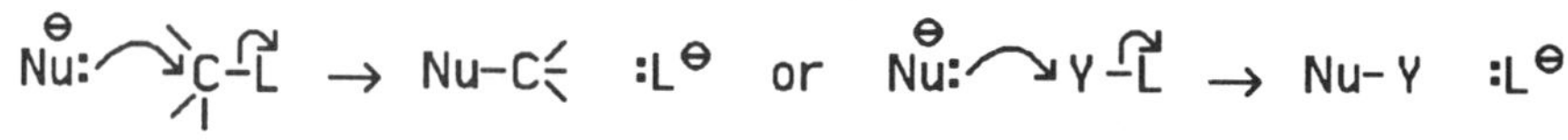

The leaving group is forced out by the nucleophile. The transition state is a five-coordinate carbon. The stereochemistry is affected, for **the tetrahedral configuration**

at the carbon atom is inverted. The transition state has a carbon p orbital partially bonded to the nucleophile on one side and the leaving group on the other (backside attack, see Figure 6.6). The 2 in the name S_N2 indicates that there are two reactant molecules involved in the rate-determining step (the nucleophile and the molecule attacked).

Path limitations

Overlap: Nu, carbon attacked, and L should be close to collinear; distortions are tolerated, for three-membered rings form easily.

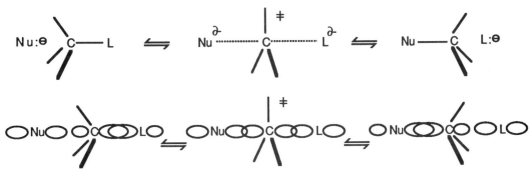

Figure 6.6 The S_N2 substitution with the transition state shown in the center.

Leaving group: good or better, occasionally fair.
Media: basic or neutral, occasionally acidic.
Solvent: polar or medium polarity.
Steric: Good access to carbon attacked is critical. Methyl, CH_3-L, and primary, CH_3CH_2-L, are best; secondary, $(CH_3)_2CH-L$, is possible if unhindered. **Tertiary**, $(CH_3)_3C-L$, **does not react by this pathway**. Neopentyl, $(CH_3)_3C-CH_2-L$, fails to react by this path because a methyl group always blocks the nucleophile's line of approach. **Aromatic, Ar−L, and vinylic, C=C−L, leaving groups do not react via the S_N2 process**.

Allylic, $H_2C=CH-CH_2-L$, benzylic, $PhCH_2-L$, and the related $:Z-CH_2-L$ react even more quickly than methyl. The overlap of the p orbital (of the double bond, aromatic, or lone pair) with the adjacent p orbital of the S_N2 transition state forms a pi-type bond that lowers the energy of the transition state and speeds up the reaction.

Energetics: Use the ΔpK_a rule (Section 3.9) to decide if the forward or reverse reaction is favored.

Path E2, The E2 Elimination
(Elimination, Bimolecular, Simultaneous Loss of a Proton and Leaving Group)

Path limitations (See Section 7.2j for more discussion)

Overlap: Both the C−H bond and the C−L bond must be roughly coplanar, as shown in Figure 6.7, or a twisted pi bond will be formed.

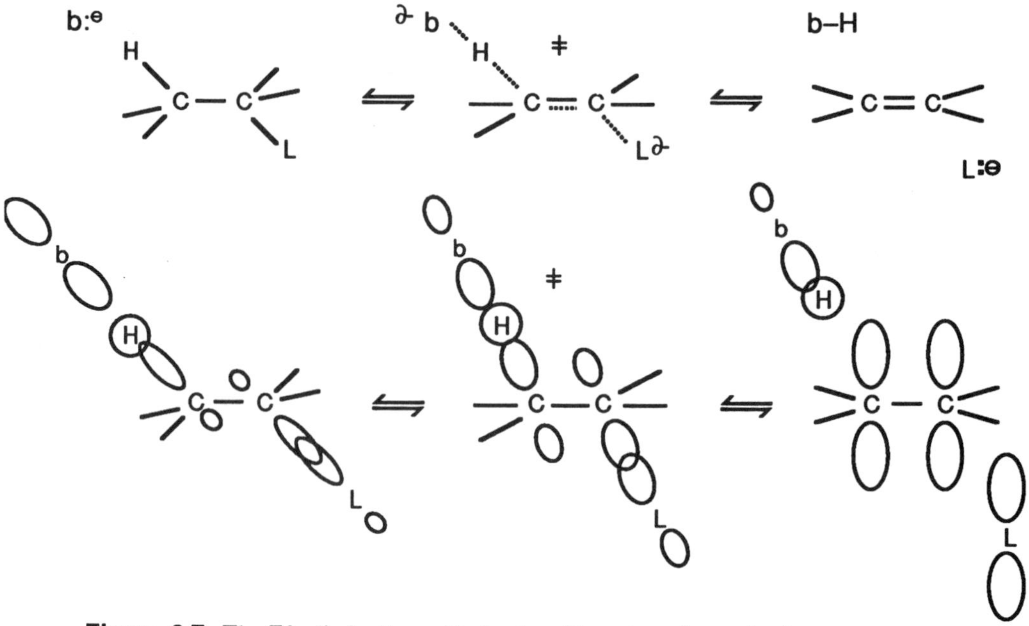

Figure 6.7 The E2 elimination with the transition state shown in the center.

Leaving group: fair or better.
Media: normally basic, occasionally neutral, rarely acidic.
Solvent: medium polarity.
Steric: Base must be able to approach the C—H bond easily. For this reason the *anti* elimination, H and L on opposite sides, is preferred over the *syn*, H and L on the same side. The leaving group crowds the incoming base in the syn.
Variations: A heteroatom can replace either carbon so that the elimination forms a carbon–heteroatom pi bond. For other variations see Section 6.4.

Path Ad$_E$3, The Ad$_E$3 Addition
(Addition, Electrophilic, Trimolecular, the Reverse of the Previous Reaction)

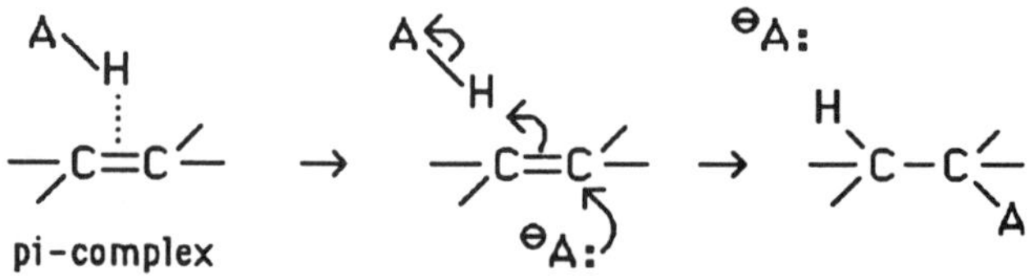

Pi complexation or hydrogen bonding serves further to polarize and activate the double bond toward an otherwise slow nucleophilic attack. (A dotted line indicates complexation only, does not change the electron count, and should not be confused with a solid line describing a bonding pair of electrons.)

Path limitations (See Sections 7.2k and 7.3 for more discussion)

Overlap: The H attacked, the pi bond, and the A$^\ominus$ must be roughly coplanar. Figure 6.7, viewed from right to left, with the substitution of A for both b and L, approximates the overlap requirements. The A$^\ominus$ attacks the largest partial plus of the pi-complex. Markovnikov's rule is followed even though there is no full positive charge formed.
Special: The H—A must be pi-complexed or hydrogen bonded to the pi bond.

Media: normally acidic, occasionally neutral.

Steric: The nucleophile, $A^{\ominus}$, must be able to approach the pi bond easily. For this reason *anti* addition is preferred over the *syn*. The H—A complexation crowds the incoming nucleophile in the latter.

Variation: The hetero AdE3 involves nucleophilic attack on a hydrogen-bonded complex between the heteroatom lone pair and a weak acid and is often called a general acid-catalyzed addition. Note that the stereoelectronic requirements are different since the hydrogen-bonded lone pair lies in the pi nodal plane, and the nucleophilic attack occurs perpendicular to this plane (Figure 6.8).

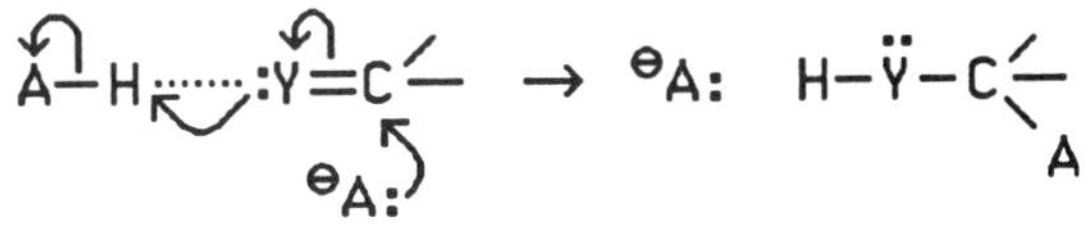

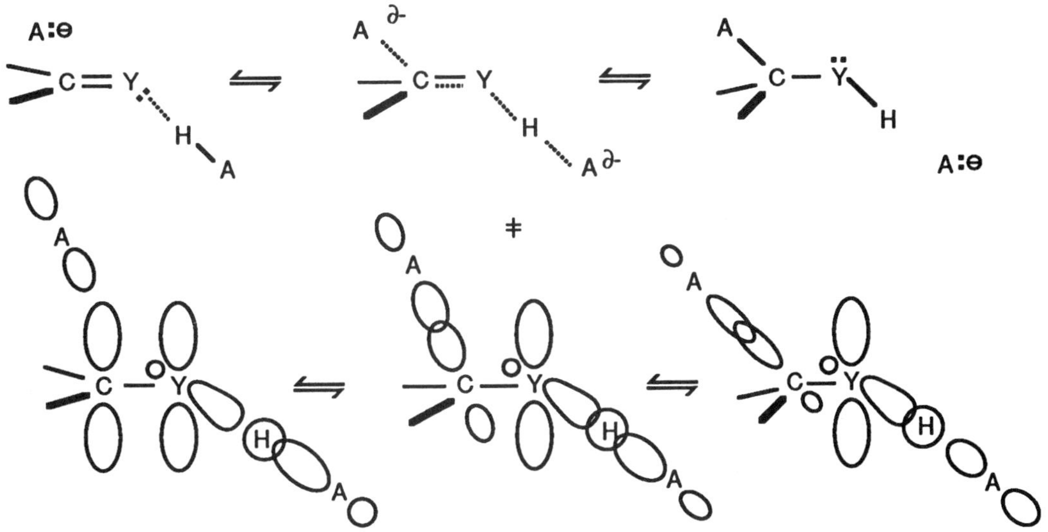

Figure 6.8 The hetero AdE3 addition with the transition state in the center.

The following set of arrows is a commonly drawn shortcut, and although it gets all the lines and dots in the right places, it ignores the fact that the hydrogen bond is usually to the lone pair and not to the less available pi bond.

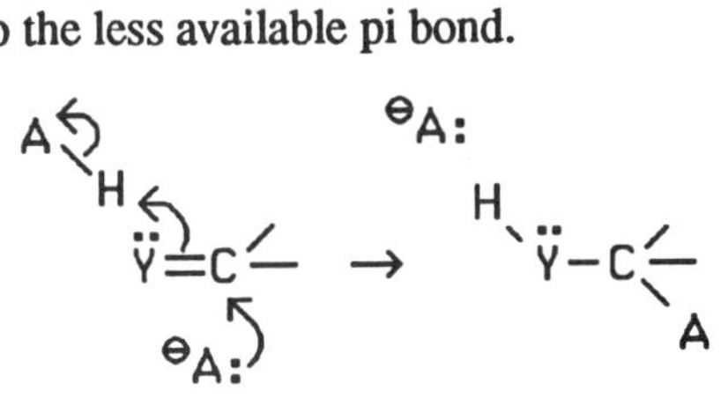

Path Ad$_N$, Nucleophilic Addition to a Polarized Multiple Bond
(Addition, Nucleophilic)

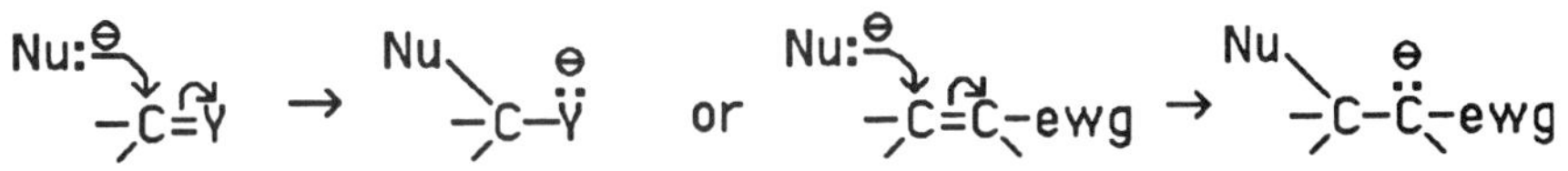

In this pathway electrons flow from the source to the multiply bonded carbon, break the pi bond, and produce a stable anion. As before, an electronegative heteroatom, Y, can be replaced by an electronegative carbon atom, C—ewg, and the electron flow does not change. When the electron sink is a carbon–carbon multiple bond with an electron-withdrawing group, the reaction is called *conjugate addition*. The electron-withdrawing group is essential, for without a means to stabilize the carbanion formed, electron flow cannot occur.

Path limitations (See Section 7.2k for more discussion)

Overlap: Nu must approach the ∂+ carbon attacked from behind and in the pi plane (Figure 6.9).

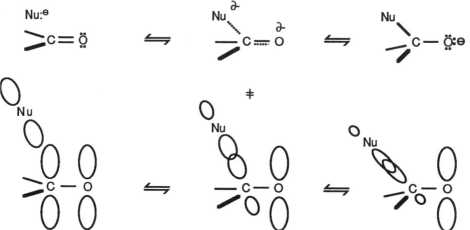

Figure 6.9 The addition to a polarized multiple bond with the transition state shown in the center.

Media: no limitation; however, if the nucleophile becomes protonated it no longer is a good Nu.

Solvent: often polar to stabilize the anion formed.

Steric: The approach of the Nu to the sink is not usually blocked, although rate differences are common. Example: Aldehydes are much more reactive than ketones toward Nu attack.

Energetics: Anion formed must usually be more stable than the Nu or the reaction will reverse by path E$_\beta$, the microscopic reverse of this path. However, in a protic solvent the initially formed anion may be rapidly and irreversibly protonated before the reverse reaction can occur. In acidic media, protonation can occur prior to nucleophilic attack (see the hetero Ad$_E$2 under path combinations).

Because for R—C≡Y, the two pi systems of the triple bond are perpendicular to each other, they behave independently. The pathway is the same as for double bonds with few path differences:

Overlap: not a problem due to the cylindrical shape of the triple bond.

Resultant anion: slightly more stable being sp^2 hybridized (for the same Y).

Steric: not a problem.

Path E$_\beta$, Beta Elimination from an Anion or Lone Pair
(The Reverse of the Previous Reaction)

Path limitations

Overlap: Anion *p* orbital and C—L bond must be roughly coplanar (Figure 6.9, right to left).

Anion: average.

Leaving group: average.

Media: no limitation. However, if the anion or lone pair becomes protonated, the electron pair needed to push out the leaving group is lost.

Solvent: polar to stabilize anion.

Energetics: use ΔpK_a rule (Section 3.9).

Path 6e, Concerted Six-Electron Pericyclic Reactions

The largest class of thermal concerted reactions involve six-electron cyclic transition states. Reactions include rearrangements and cycloadditions (two pieces forming a ring). Although the following reactions may all look different, they all involve six electrons (three arrows) going around in a circle (pericyclic).

Thermal rearrangements:

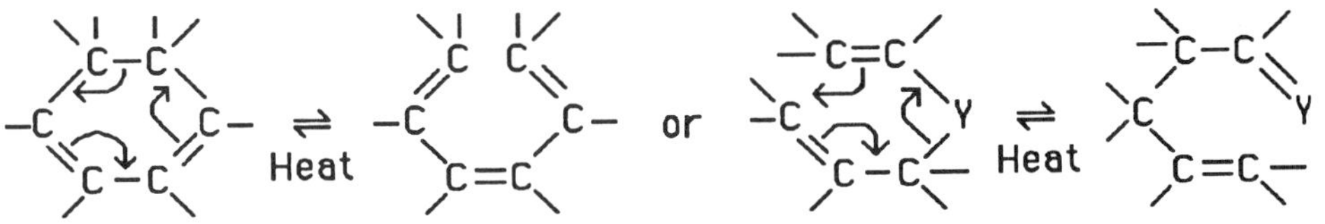

Thermal cycloadditions or cycloreversions (the reverse reaction):

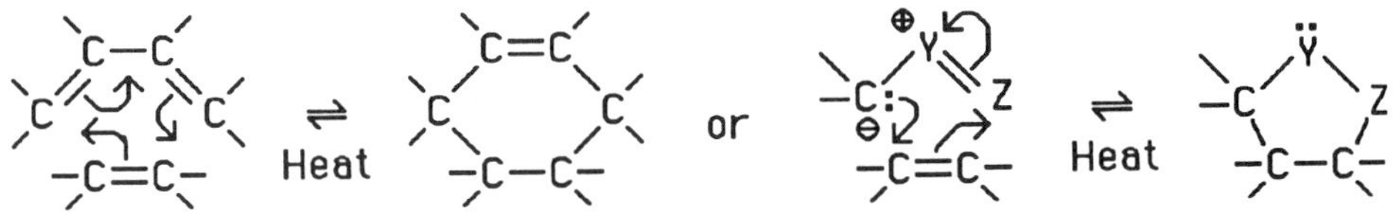

Metal-chelate-catalyzed additions:

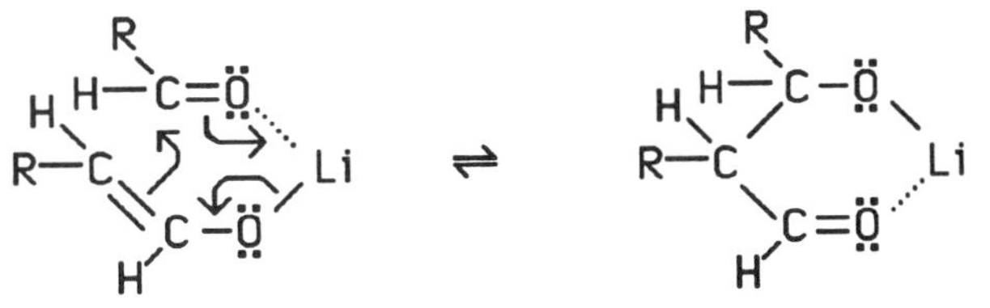

The transition state for these reactions has been described as aromatic since it has six electrons in orbitals that form a loop. In the neutral cases the direction that the arrows are drawn does not matter. The reverse reactions of all of these examples also occur. The major path limitation is to achieve the proper alignment for the cyclic transition state (see Section 7.7). There is a large diversity of compounds that react by this pathway; in many of the reactions, there are examples of reactions where a carbon atom has been replaced by a heteroatom or vice versa.

There are many reactions in which a metal ion is present and serves to hold the reacting partners together by complexation. Some of the metals for which this process is common are aluminum, magnesium, and lithium. In the following hydride transfer example, in addition to serving as a counterion for the $R_2HC\text{—}O^{\ominus}$, the aluminum acts as a Lewis acid catalyst complexing the carbonyl and making it more able to accept the electron flow.

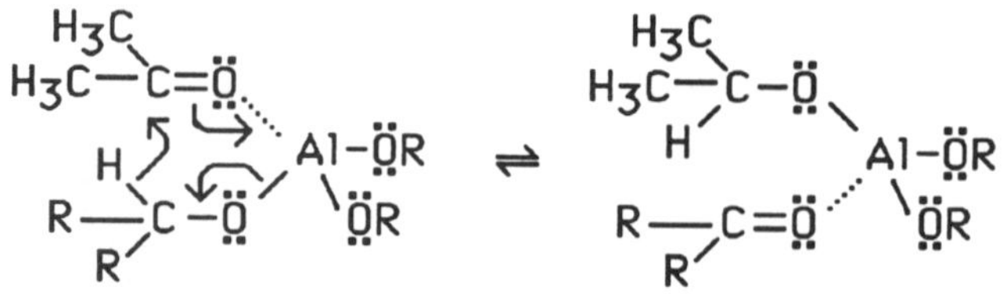

6.2 COMMON PATH COMBINATIONS

There are several pathway combinations that occur so frequently that they have been given a name. **These "phrases" in your mechanistic vocabulary are the next step toward your eventual construction of a grammatically correct mechanistic "sentence."**

S_N1 (Substitution, Nucleophilic, Unimolecular), $D_N + A_N$

The rate-determining ionization of a good leaving group, path D_N, produces a carbocation-leaving group ion pair that may or may not dissociate before the carbocation is trapped by a nucleophile, path A_N. Substrates that would produce a poor cation, CH_3—L, Ar—L, and C=C—L, do not react via the S_N1 process. The nucleophile attacks the carbocation or ion pair and gives the product in a fast second step. If the leaving group were on a chiral center, a racemic mixture can result because the nucleophile can attack the free carbocation from either the top or bottom face equally. Often, however, nucleophilic attack on the ion pair results in inversion, since the leaving group partially blocks one face. If the nucleophile is the solvent, this process is called solvolysis. (See Section 7.2f for more discussion of the various substitution types.)

E1   (Elimination, Unimolecular), $D_N + D_E$

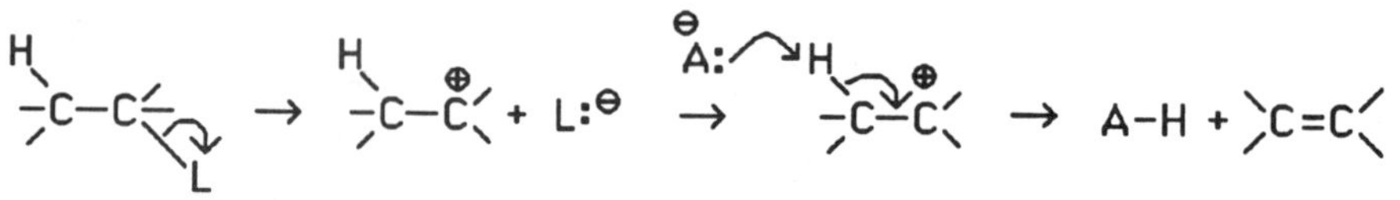

After rate-determining ionization of a good leaving group, path D_N, the cation is deprotonated by path D_E to produce a pi bond. As might be expected, the E1 process competes with the S_N1 process. **If $A^{\ominus}$ acts as a base, E1 occurs; if it acts as a nucleophile, S_N1 occurs.** Chapter 8.5 will discuss such decisions in detail. (See Section 7.2f for more discussion of the various elimination types.)

Ad_E2 (Addition, Electrophilic, Bimolecular), $A_E + A_N$

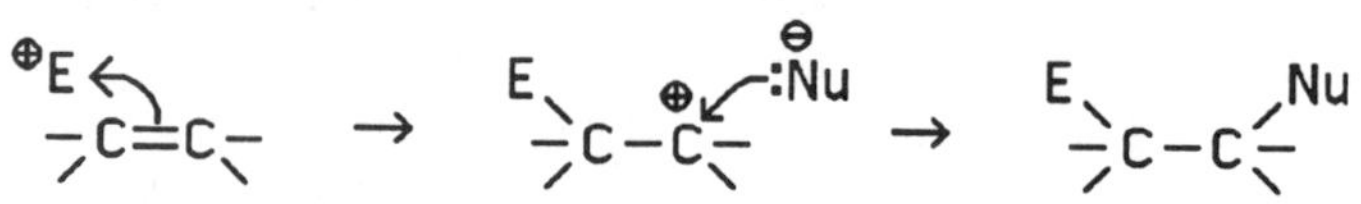

The Ad_E2 is the major addition pathway of alkenes and dienes. Electrophile addition to a pi bond or protonation of a pi bond, path A_E, produces the most stable cation, which is then trapped by a nucleophile, path A_N. (See Section 7.3 for more discussion of the various addition types.) If the electrophile is a proton, this reaction is the reverse of the E1 reaction. The reaction commonly produces a mixture of syn and anti addition.

Hetero Ad_E2, p.t. + Ad_N

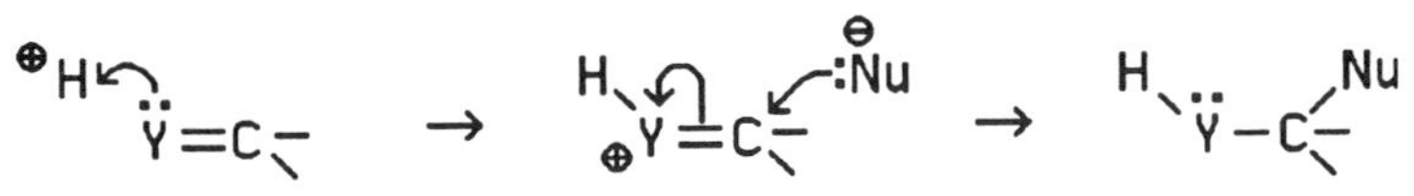

This reaction is a minor variant of the Ad_E2 in which the electrophile, usually a proton, attacks the lone pair rather then the less available heteroatom–carbon pi bond. The lone pair is protonated, path p.t., to give a highly polarized multiple bond that can be viewed as a stabilized carbocation. The nucleophilic attack on this carbocation could also be viewed as path A_N, trapping of a cation by a nucleophile, instead of path Ad_N.

Major   ↔   Minor

Lone-Pair-Assisted E1, E_β + p.t.

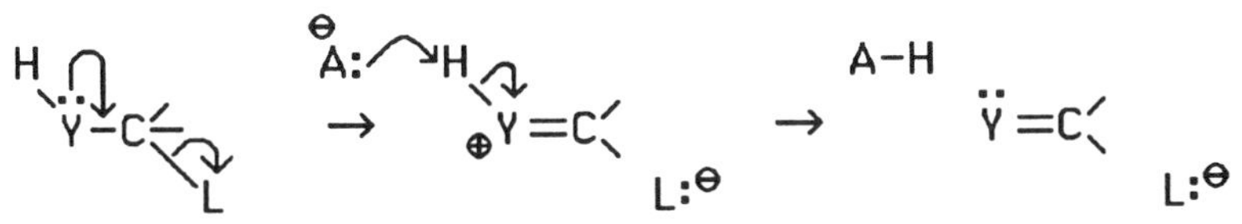

This reaction is the reverse of the hetero Ad_E2 reaction. The lone-pair-assisted E1 uses a properly aligned lone pair to expel the leaving group, path E_β. The resultant cation is then deprotonated, path p.t.

Electrophilic Aromatic Substitution, A_E + D_E

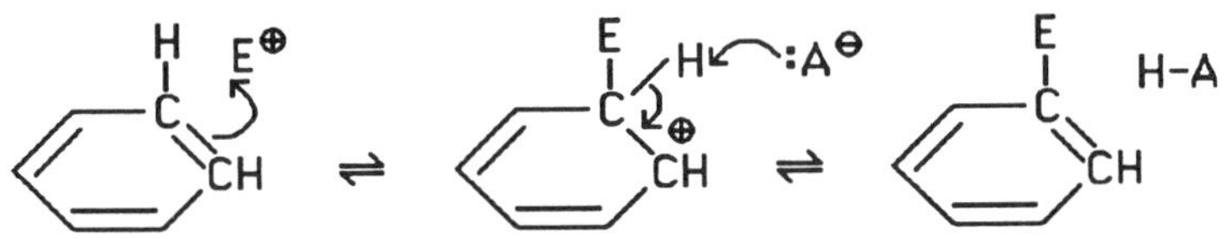

The electrophile adds to the pi bond of the aromatic ring, path A_E, followed by deprotonation of the cation formed, path D_E, restoring aromatic stabilization.

E1cB (Elimination, Unimolecular, Conjugate Base), p.t. + E_β

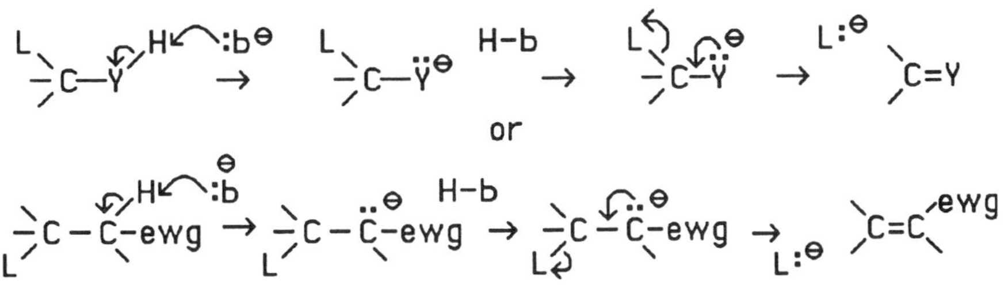

or

A stabilized anion forms first via path p.t., and then a fair leaving group departs in the slow step via the beta elimination from an anion path E_β. (See Section 7.2f for more discussion of the various elimination types.)

Ad$_N$2   (Addition, Nucleophilic, Bimolecular),   Ad$_N$ + p.t.

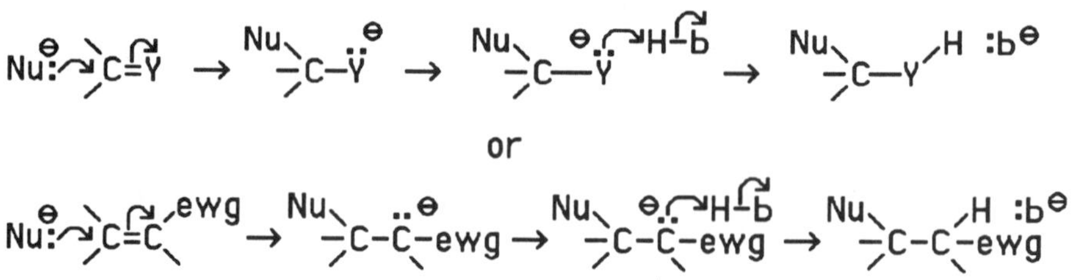

$$Nu: \!\!\curvearrowright\!\! C\!\!\equiv\!\!Y^{\ominus} \rightarrow \underset{}{Nu}\!\!-\!\!C\!\!-\!\!\ddot{Y}^{\ominus} \rightarrow \underset{}{Nu}\!\!-\!\!C\!\!-\!\!Y^{\ominus}\!\!\curvearrowright\!\!H\!\!-\!\!b \rightarrow \underset{}{Nu}\!\!-\!\!C\!\!-\!\!Y\!\!-\!\!H \; :b^{\ominus}$$

or

$$Nu:\!\!\curvearrowright\!\!C\!\!\equiv\!\!C\!\!-\!ewg \rightarrow \underset{}{Nu}\!\!-\!\!C\!\!-\!\!\ddot{C}^{\ominus}\!\!-ewg \rightarrow \underset{}{Nu}\!\!-\!\!C\!\!-\!\!\overset{\ominus}{C}\!\!-ewg\!\!\curvearrowright\!\!H\!\!-\!\!b \rightarrow \underset{}{Nu}\!\!-\!\!C\!\!-\!\!C\!\!-ewg\; H \; :b^{\ominus}$$

This reaction is the reverse of the E1cB elimination. A rate-determining addition of a nucleophile to the polarized multiple bond occurs by path Ad$_N$, producing a stabilized anion that is then protonated by path p.t.

Addition–Elimination,   Ad$_N$ + E$_\beta$

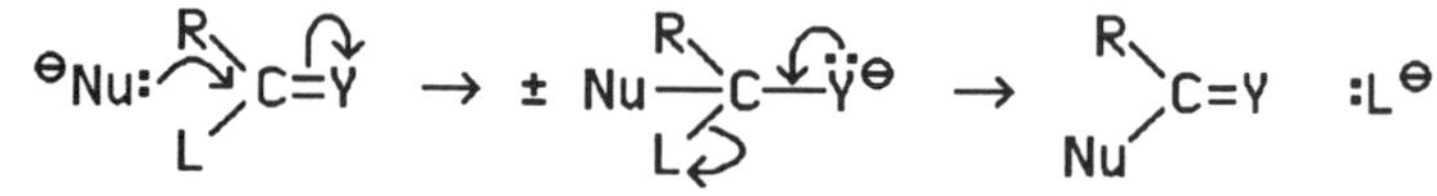

$$^{\ominus}Nu: \!\!\curvearrowright\!\! \overset{R}{\underset{L}{C}}\!\!=\!\!Y \rightarrow \pm\; Nu\!\!-\!\!\overset{R}{\underset{L}{C}}\!\!-\!\!\ddot{Y}^{\ominus} \rightarrow \overset{R}{\underset{Nu}{C}}\!\!=\!\!Y \quad :L^{\ominus}$$

This is the only route in basic media for replacement of a leaving group bound to a double bond. The most common substrates for this reaction are the carboxyl derivatives. A nucleophile adds first via path Ad$_N$, and then the leaving group departs via the beta elimination from an anion path E$_\beta$.

Tautomerization,   taut.

$$H\!\!-\!\!C\!\!-\!\!C\!\!=\!\!Z \rightleftharpoons C\!\!=\!\!C\!\!-\!\!Z\!\!-\!\!H$$

Tautomerization is the shift of a hydrogen from a carbon adjacent to a carbon–heteroatom double bond to the heteroatom itself (and the reverse process). It is an acid- or base-catalyzed equilibrium. Two examples are the keto/enol pair (Z = oxygen) and the imine/enamine pair (Z = nitrogen).

Acid catalysis (path p.t. followed by path D$_E$) goes via the lone-pair-stabilized carbocation.

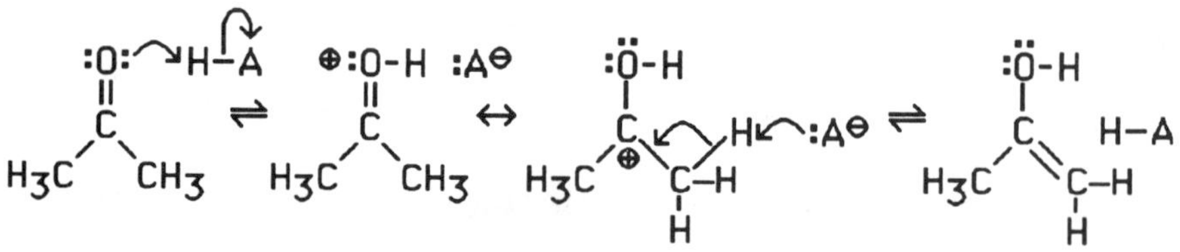

Keto   Carbocation   Enol

Base catalysis (path p.t. twice) goes via the enolate anion.

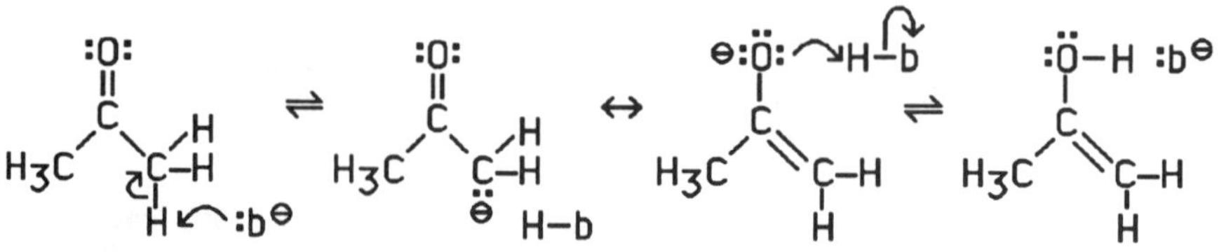

Keto   Enolate   Enol

Exercise

Using the previous structures as a guide, draw the arrows for the tautomerization of the enol form back to the keto form for both acidic and basic media. What pathways did you use?

Answer: The acidic route from the enol to the keto form is path A_E then p.t. Note that this backward route uses the reverse of each path.

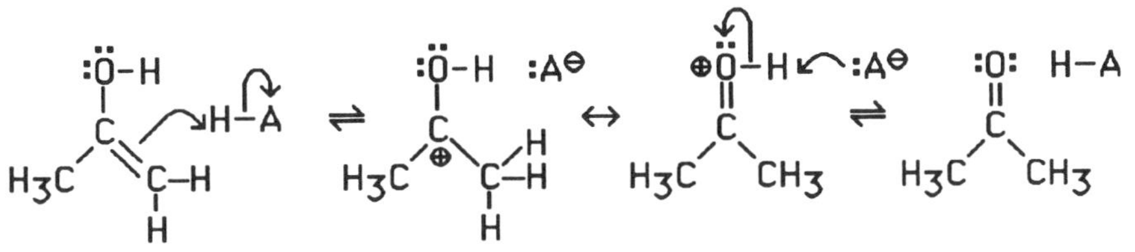

Keto   Carbocation   Enol

The basic route from the enol to the keto form is via proton transfer, path p.t. twice.

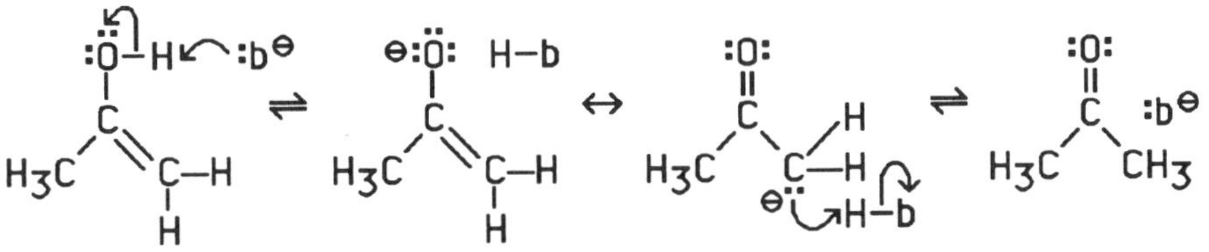

Keto   Enolate   Enol

6.3 EIGHT MINOR PATHWAYS

Path Ei, Thermal Internal *Syn* Elimination

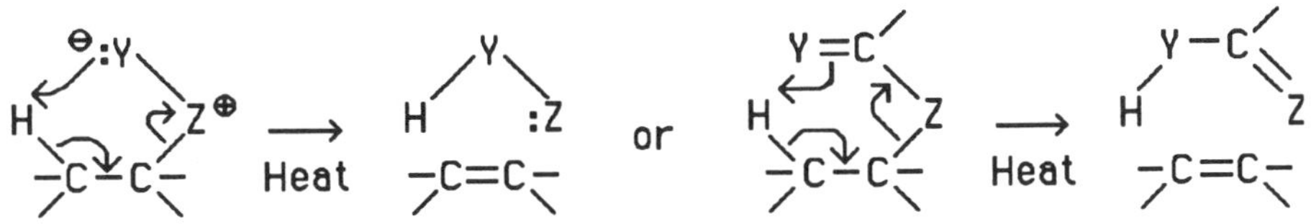

There are many possible reactants for this thermal internal *syn* elimination path. Five membered transition state examples include Y equals oxygen and Z is $^\oplus NR_2$ (amine oxides), SPh (sulfoxides), or SePh (selenoxides). Six-membered transition state examples include both Y and Z being oxygen (esters), or Y is sulfur and Z is oxygen (xanthates).

Path limitations

Overlap: The C–H bond and the C–Z bond involved in the elimination must be roughly coplanar for strong pi bond formation (Figure 6.10).

Anion: average.

Leaving group: average.

Media and solvent: no limitation; can occur in the gas phase.

Energetics: Poorer leaving groups require a higher temperature.

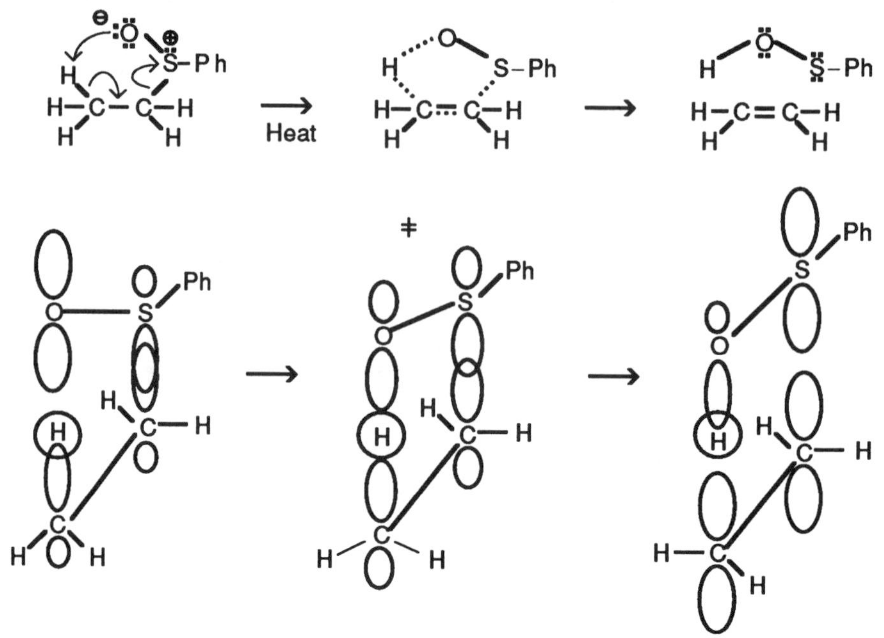

Figure 6.10 Thermal internal *syn* elimination with the transition state shown in the center.

An additional six-membered transition state example is decarboxylation. The overlap limitation for the new carbon–carbon pi bond requires the breaking C–C bond to align coplanar with the carbon *p* orbital of the carbonyl.

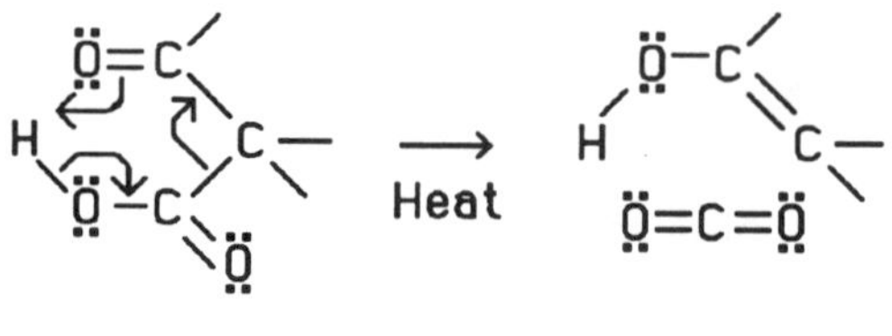

Path AdgB, General Base-Catalyzed Addition to a Polarized Multiple Bond

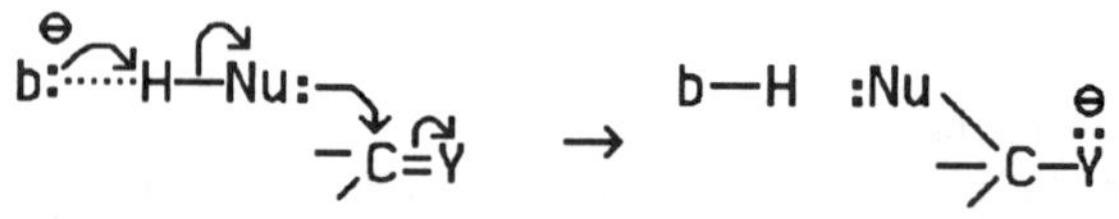

This pathway is very similar to path AdN with the difference that the nucleophile is poorer and is hydrogen bonded to a base when this pair collides with the polarized multiple bond. In this pathway the electron flow comes from the base to break the Nu–H bond, which in turn enhances the nucleophilicity of the nucleophile's lone pair. This lone pair attacks the multiply bonded carbon, breaks the pi bond, and produces a stable anion similar to path AdN.

Two processes are occurring to a varying extent: the hydrogen is being pulled off by the base, and the nucleophile lone pair is attacking the polarized multiple bond. If the nucleophile is especially poor, proton removal would need to be complete before a good enough nucleophile would be generated for the attack to proceed; this would then be path p.t. followed by path Ad_N. If the nucleophile is very good, proton removal would be unnecessary, and the reaction would go by path Ad_N. This path AdgB is used when there is a weak base present like $RCOO^\ominus$, pK_{abH} 4.8.

The following set of arrows is the frequently drawn shortcut; it ignores the role of the lone pair and incorrectly implies that the sigma bond is acting as a nucleophile. The four-arrow description is more rigorous and agrees better with what is known from enzymatic catalysis.

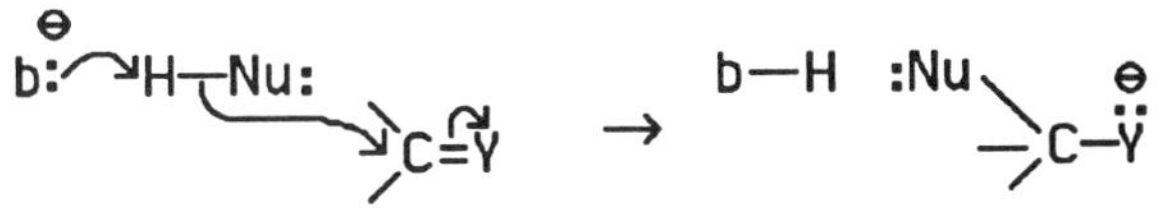

An electronegative heteroatom, Y, can be replaced by an electronegative carbon atom, C—ewg. The electron-withdrawing group is extremely necessary, for without a means to stabilize the carbanion formed, the electron flow cannot occur.

Path limitations

Overlap: Nu must approach the $\partial+$ carbon from behind and close to the pi plane (Figure 6.11).

Media: A weak base is needed; the medium is usually weakly basic to neutral.

Solvent: Protic and polar to stabilize the anion formed.

Steric: The approach of the Nu to the sink is usually not a problem.

Energetics: Anion formed must usually be more stable than the base or the reaction will reverse by path EgA.

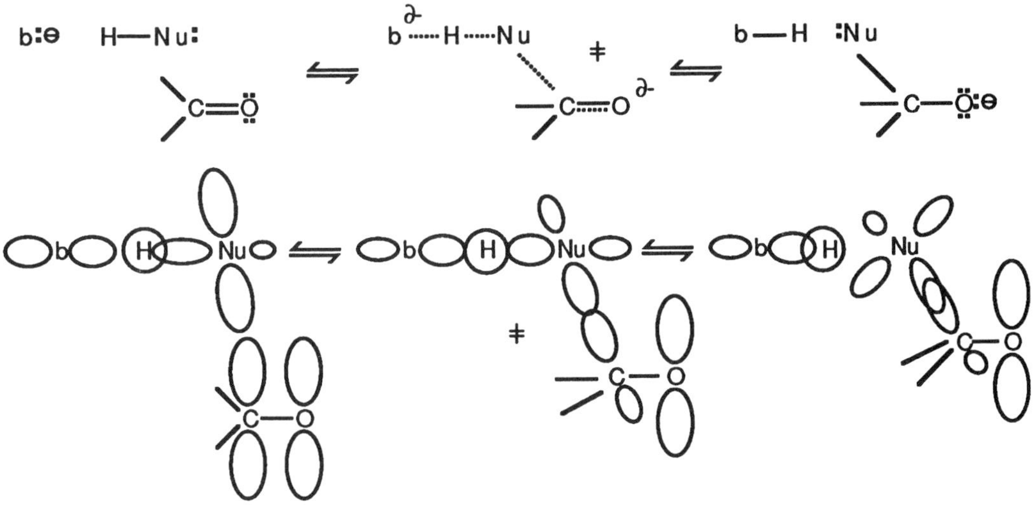

Figure 6.11 The general base-catalyzed addition to a polarized multiple bond with the transition state shown in the center.

Path EgA,   General Acid-Catalyzed Beta Elimination
(The Reverse of the Previous Reaction)

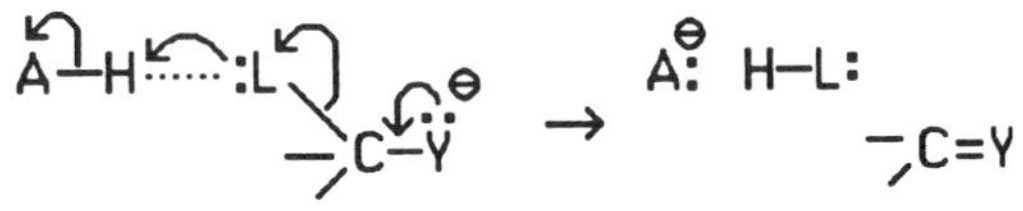

The difference between this and path E$_\beta$ is that the leaving group is poorer and must hydrogen bond to a weak acid (like RCOOH) to be good enough to depart. The hydrogen bond enhances the quality of the leaving group. Again, this path is the middle of three paths that involve different degrees of proton transfer. If the leaving group is very poor, it would need to be fully protonated before it would be good enough to depart; this would be path p.t. followed by path E$_\beta$. If the leaving group were good, then proton transfer would be unnecessary and the reaction would proceed by path E$_\beta$.

To save drawing an extra arrow, chemists will commonly draw the following set of arrows, a shortcut that does get all the lines and dots in the right places but has some problems that are more obvious in this direction than they were in path AdgB, the microscopic reverse.

The problems with the three-arrow shortcut are that it ignores the necessary role of hydrogen bonding to the lone pair of the leaving group, that it protonates the unavailable sigma bond electron pair, and that it does not quite agree with intramolecular and enzymatic studies. As in path AdgB, this process is more rigorously drawn with four arrows to emphasize the role of the lone pair.

Path limitations

Overlap: Anion p orbital and C—L bond must be roughly coplanar (see Figure 6.11, right to left).
Anion: average.
Leaving group: average, but must have a lone pair to hydrogen bond to the weak acid.
Media: A weak acid is needed; the medium is usually weakly acidic to neutral.
Solvent: Protic and polar to stabilize anion.
Energetics: Use ΔpK_a rule (Section 3.9).

Summary of Polarized Multiple Bond Reactions

Figure 6.12 shows the top view of an energy surface that relates all three **basic media** mechanisms for addition of a nucleophile to a polarized multiple bond. The reactants are in the upper left corner, and the products are in the lower right corner. If the base is weak and the nucleophile is good, the nucleophile adds (path Ad$_N$) and later gets deprotonated (path p.t.). If the nucleophile is poor or the base very strong, deprotonation occurs first (path p.t.) followed by addition to the polarized multiple bond (path Ad$_N$). If the base is weak, and the nucleophile is mediocre, and a weak base is present, then the general base-catalyzed process (path AdgB) occurs.

There are three **acidic media** mechanisms that can occur depending on the acidity of the media: acid catalysis, protonation followed by nucleophilic attack (path p.t. then Ad$_N$) would occur in strong acid with weak nucleophiles; protonation at the same time as nucleophilic attack (path Ad$_E$3) would occur in weak acids; and finally, nucleophilic attack followed by protonation (path Ad$_N$ then p.t.) would occur with good nucleophiles in weakly acidic media. Figure 6.13 shows the top view of an energy surface that relates all three mechanisms.

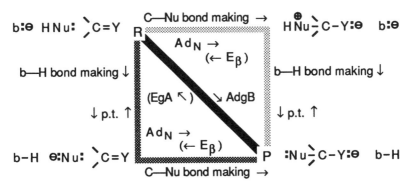

Figure 6.12 Polarized multiple bond addition/elimination mechanisms in basic media. The reverse processes are given in parenthesis.

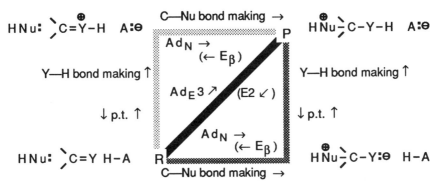

Figure 6.13 Polarized multiple bond addition/elimination mechanisms in acidic media. The reverse processes are given in parenthesis.

These next paths require rather unusual starting materials or some other aspect that makes them relatively uncommon. One-electron processes, which are slightly more common, are covered separately in Chapter 10.

Path NuL, Nu−L Additions
(Three-membered ring formation)

This pathway shows some variation on the timing of when the leaving group falls off. Occasionally the addition may be concerted as shown (with bromine or peracids, for example), but more often the leaving group loss and the nucleophilic attack on the pi bond are separate steps. The leaving group may fall off before attack, or if the multiple bond is capable of stabilizing an anion, as C=Y is, then the leaving group may fall off after nucleophilic attack (for a thorough discussion see Section 7.5).

Path limitations (Concerted)

Overlap: Nu−L must approach the pi bond from above or below and not from the side. The Nu, the L, and the double-bond carbons should be close to coplanar, as shown in Figure 6.14. The nucleophile and leaving group orbitals are perpendicular or close to it, and thus can react independently. Although a synchronous example is shown, almost all examples are not synchronous or even concerted.

Leaving group: good to fair, may leave before addition occurs.

Steric: usually no problem.

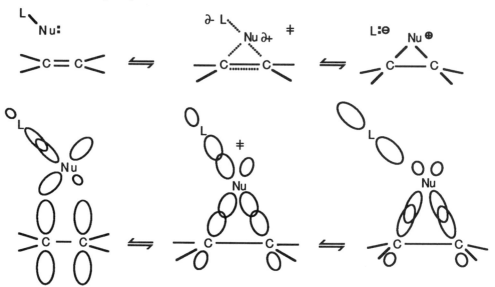

Figure 6.14 The addition of a Nu−L species to form a three-membered ring with the transition state shown in the center.

Path 1,2RL, 1,2 Rearrangement with Loss of Leaving Group

Often rearrangement occurs as shown with a path reminiscent of the generic S_N2, a backside displacement of the leaving group by the migrating group. There are some rearrangements where it appears that the leaving group departs first, as in the S_N1 rather than the S_N2, so that the electron sink the group migrates to is the carbocation (we would then call it path D_N followed by 1,2R). Sometimes the leaving group can be replaced by a polarized multiple bond as the electron sink, shown in the following reaction. In essence, rearrangement can proceed with any of our three general classes of electron sinks: an empty p orbital, a weak single bond to a leaving group, or a polarized multiple bond.

Path limitations

Overlap: as shown by Figure 6.15, the migrating bond and the leaving group bond must lie in the same plane for good overlap.

Leaving group: excellent to fair, with excellent leaving groups the pushing $Y^{\ominus}$ group may not be needed and can be replaced by R. Heteroatom Z can be replaced by a carbon.

Solvent: usually polar.

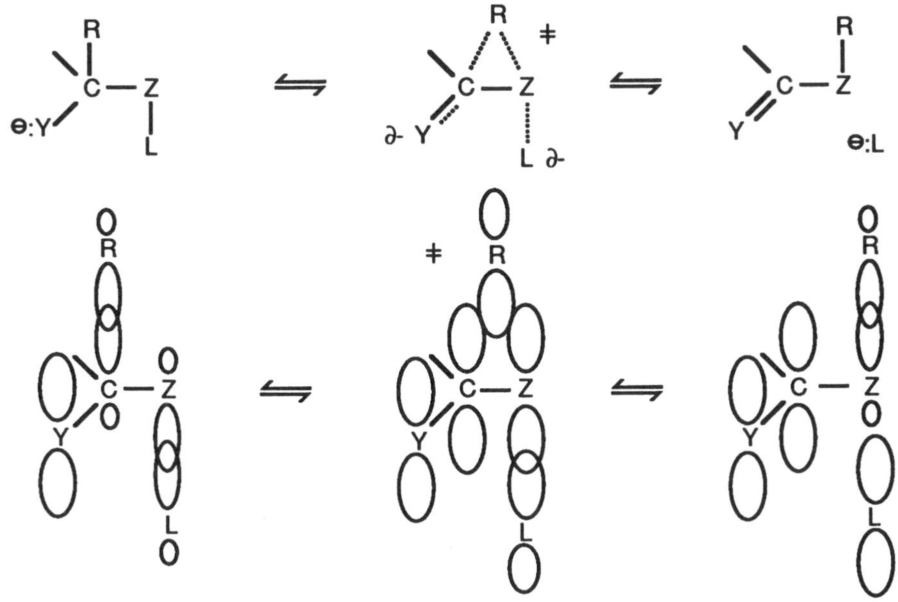

Figure 6.15 Rearrangement with loss of a leaving group with the transition state shown in the center.

Path 4e, Four-Center, Four-Electron

Four-center, four-electron processes do not occur thermally. The transition states for these unfavorable reactions have been described as having antiaromatic destabilization because they have four electrons in a normal closed loop. There are **three exceptions** that go by this path; all have some unusual orbital arrangement to allow them to bypass the problem of antiaromatic destabilization of their transition states.

The addition of boranes to pi bonds occurs by this pathway (also the less common but isoelectronic trivalent aluminum hydrides, such as diisobutylaluminum hydride, $[(CH_3)_2CHCH_2]_2AlH$, add by this pathway). The overlap path limitation is that all four atoms must be coplanar (Figure 6.16). A pi-complex of the empty p orbital on boron with the double bond forms first, then collapses by this four-electron process to the product. Boron is the electrophile and hydrogen is the nucleophile. Because the nucleophile orbital and electrophile orbital are perpendicular (therefore do not interact), the loop is not closed. The direction of electrophilic addition is determined by the formation of the most stable partial plus.

Alternatively, the two bond formation steps may not be synchronous but occur so close together in time that the intermediate has an insignificant lifetime. Electron density from the double bond flows into the empty orbital on boron, creating an electron-deficient center on a carbon to which the now electron-rich boron can donate a hydride. The addition is *syn*; both boron and hydrogen add to the same face of the double bond.

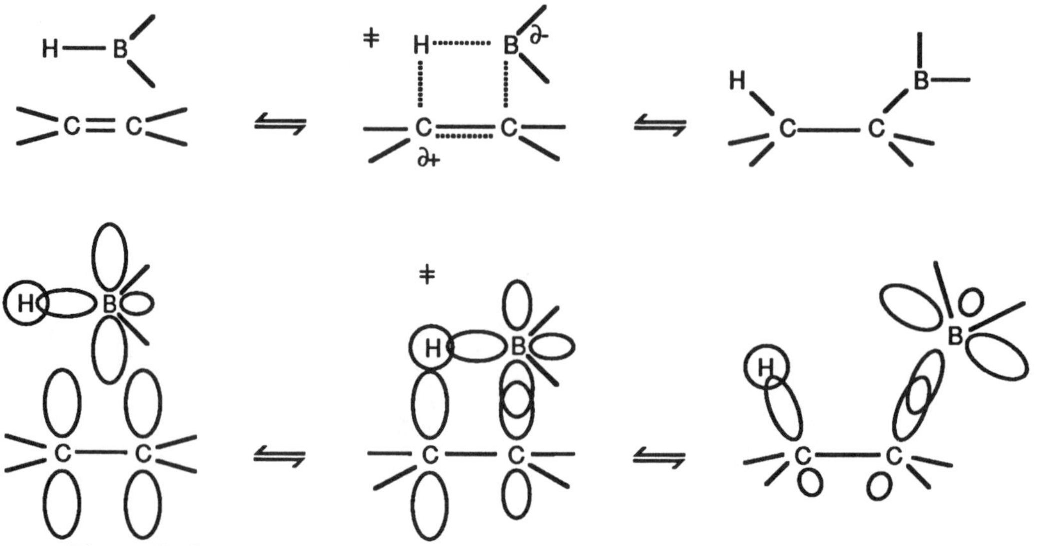

Figure 6.16 The addition of a borane to a double bond with the transition state shown in the center.

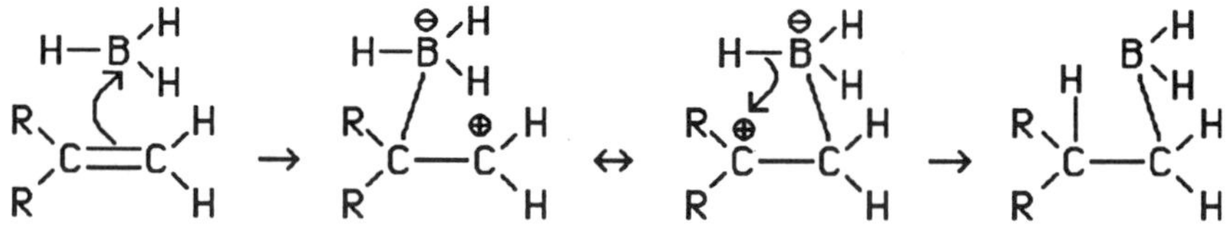

A second exception occurs with the elimination of phosphine oxides from oxaphosphetanes. The pentacoordinate phosphorus serves as the electron source in an internal elimination of the oxygen bound to phosphorus.

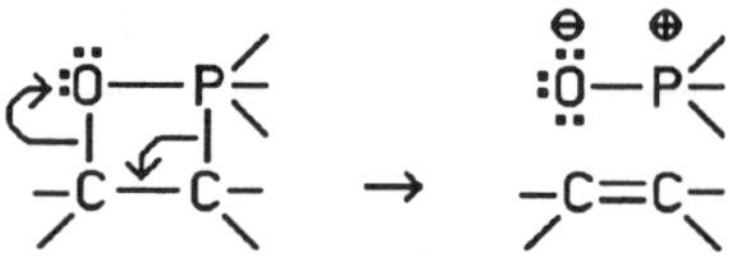

The last exception has an unusual twist, literally. The transition state is a strange loop with a half-twist, a Möbius loop. In contrast to a normal loop, Möbius loops are predicted to be stable with $4n$ electrons in them. Given enough heat, the cyclobutene sigma bond twists open to form a diene.

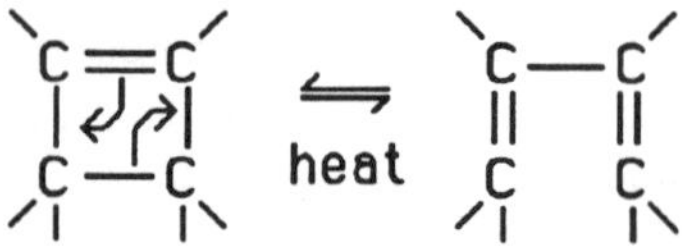

Four-electron electrocyclic reaction from a HOMO–LUMO perspective
(A supplementary, more advanced explanation)

HOMO–LUMO principles (Section 2.7) allow us to predict whether the ends rotate in the same direction (conrotatory) or in opposite directions (disrotatory)(Figure 6.17). If the process is favorable, as it is in this case when the rotation is in the same direction, then the σ-LUMO will overlap with the π-HOMO in a bonding manner to give a bonding molecular orbital of the product. Stereochemical labels on the cyclobutene confirm that this electrocyclic ring opening is conrotatory.

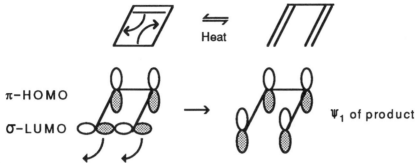

π–HOMO

σ–LUMO

Ψ₁ of product

Figure 6.17 The thermal conrotatory ring opening of cyclobutene.

Path pent., Substitution via a Pentacovalent Intermediate
(For Atoms Capable of Pentacoordinate Bonding Only, Not for Carbon)

$$^{\ominus}Nu\colon \curvearrowright P-L \; \rightleftharpoons \; \left[Nu-\overset{\ominus}{\underset{\curvearrowleft}{P}}\overset{|}{L} \right] \rightleftharpoons \; Nu-P\overset{<}{} \;+\; \colon L^{\ominus}$$

A nucleophile adds forming a trigonal bipyramidal pentacovalent intermediate that then ejects the leaving group in the microscopic reverse of the nucleophilic addition reaction. The pentacovalent intermediate is often short-lived, and this is shown by enclosing it in brackets. The pentacovalent intermediate bonds that are colinear are called axial (or apical) bonds, whereas the three bonds that lie in the plane perpendicular are called equatorial. Three common elements that can react by this pathway are silicon, phosphorus, and sulfur. Hydrolysis of phosphate esters like DNA occurs by this path.

Path limitations

Overlap: Attacking Nu forms an axial bond; therefore by microscopic reversibility, the departing L must leave from an axial position. The orbital overlap is similar to that shown for the S_N2 (Figure 6.6), but the structure in the center of the figure is now an intermediate rather than a transition state.

Leaving group: good to fair. If the leaving group is very good, the intermediate would be expected to have a short lifetime and may become just a transition state (S_N2).

Media: acidic, basic, or neutral.

Solvent: commonly polar or medium polarity.

Steric: access usually not a problem since bond lengths are longer.

Special: Pentacoordinate intermediates are possible only for second-row or higher elements, not for first-row elements.

Path H⊖ t., Hydride Transfer to a Cationic Center

$$R\underset{}{\overset{}{\searrow}} \overset{}{\underset{}{-C}}\!\overset{\frown}{-H} \;\searrow\! \overset{|}{\underset{|}{C}}{}^{\oplus} \;\rightleftharpoons\; \overset{R}{\underset{}{\overset{|}{C}}}{}^{\oplus} \quad H-C\overset{<}{}$$

Hydride is never a free species floating around in solution. A carbocation can abstract a hydride to form a more stable carbocation.

Path limitations

Overlap: Only that the carbocation must approach within bonding distance to the hydride transferred. This pathway is related to path 1,2R, rearrangement of a carbocation, in which a group migrates with its pair of bonding electrons to an adjacent carbocation. In this minor path, a hydrogen with its bonding electron pair is passed to a nonadjacent carbocation; the empty *p* orbital of the carbocation must get close enough to allow the hydride to be partially bonded to both atoms at the transition state, as in Figure 6.18.

Carbocation formed: good.

Media: acidic, occasionally neutral.

Solvent: commonly polar or medium polarity.

Steric: not a problem.

Energetics: An energetically favorable transfer forms a more stable carbocation.

Related path: If the carbocation is replaced by a partial plus center, then a pushing $Y^\ominus$ is required on the hydride donor, and the reaction tends to go by path 6e (see hydride transfer example).

Biochemical example: Oxidations by $NAD^\oplus$ commonly go by this path (See Section 7.8f).

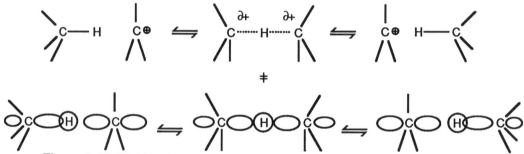

Figure 6.18 Hydride abstraction by a cation with the transition state in the center.

Cross-Checks for Suspected Additional Minor Paths

Any suspected new electron flow path should be put through a series of tests before it is accepted as a new pathway: first, check to see that it is not just a combination of or a variation on already known paths (see Section 6.4); second, check the orbital alignment with molecular models — any orbitals that will become double bonds in the product must be able to get close to coplanar in the starting material; third, check the product for stability especially if it is charged; fourth, use molecular models to look for steric and strain problems; fifth, check that the electronics fit theory, HOMO–LUMO, HSAB, stereoelectronic effects, etc.; sixth, check the energetics of the process with a ΔH calculation or the ΔpK_a rule; lastly, be suspicious anyway, it's healthy.

6.4 VARIATIONS ON A THEME

The pathways discussed should be considered as generic; the route for electron flow is important, but the identity of the atoms in each path can easily change. The variety of the acids, bases, electrophiles, nucleophiles, and leaving groups may sometimes make an otherwise familiar reaction look unrecognizable. Therefore, it is very important that you be able to recognize and classify groups into their respective generic classes.

Besides having several charge types, the E2 pathway has many variants, a few of which will be discussed in this section as illustrations of how the identity of the atoms can change while the electron flow stays the same. As was shown earlier, a heteroatom can be replaced for carbon and not affect the process at all:

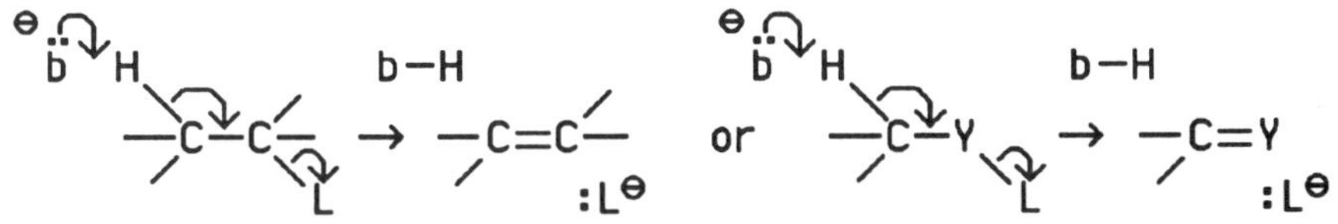

The dehalogenation reaction is a variant of the E2 where the H and L are now halogens, commonly bromine, and the base has been replaced by iodide or metallic zinc. The orbital alignment restrictions are the same as for the E2 reaction:

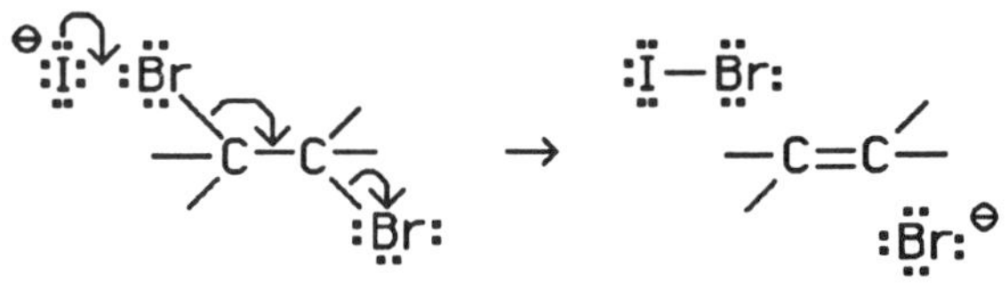

The fragmentation reaction is an E2 elimination with another electron source replacing the base acting on a C—H bond. All orbitals that form double bonds must be coplanar and usually must be arranged as shown. The source can also be $COO^{\ominus}$, in which Y is oxygen, and then the fragmentation would be called a decarboxylation and produce CO_2.

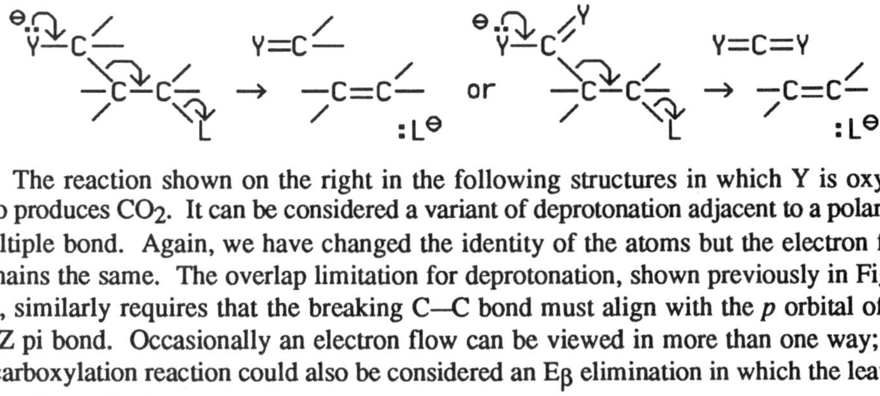

The reaction shown on the right in the following structures in which Y is oxygen also produces CO_2. It can be considered a variant of deprotonation adjacent to a polarized multiple bond. Again, we have changed the identity of the atoms but the electron flow remains the same. The overlap limitation for deprotonation, shown previously in Figure 6.1, similarly requires that the breaking C—C bond must align with the p orbital of the C=Z pi bond. Occasionally an electron flow can be viewed in more than one way; this decarboxylation reaction could also be considered an E_β elimination in which the leaving group is an allylic system.

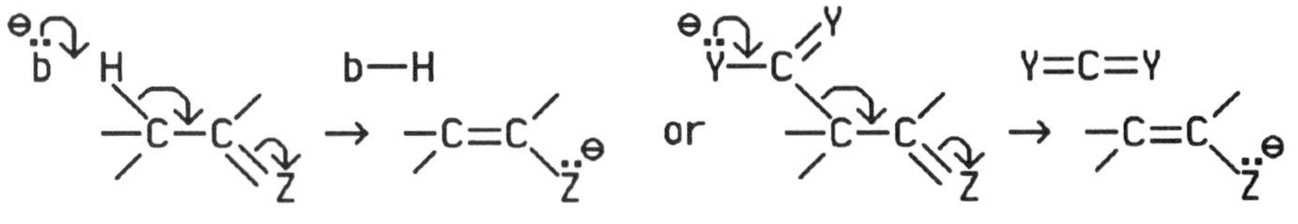

As this next example illustrates for another pathway, additional conjugation may not change the system significantly (the principle of vinylogy, Section 1.7). The S_N2' reaction is just a vinylogous S_N2 reaction:

However, the overlap restrictions now include two more atoms and may be therefore more difficult to achieve. For the S_N2' the Nu, p orbital of carbon attacked, and C—L bond should be coplanar (often *syn*). This variant is found mostly in rigid systems where the C=C—C—L is locked into the proper alignment. Freely rotating systems go by the normal S_N2 reaction unless the S_N2 site is sterically blocked, and the S_N2' site is open. Triply bonded systems C≡C—C—L (propargyl leaving groups) do not have as critical an alignment problem; one of the two pi bonds will be close to the correct position.

Another topic that is appropriate under the variations-on-a-theme rubric is the extent of proton transfer between groups in protic solvents. This topic has been explored for polarized multiple bonds, as illustrated previously by Figures 6.10 and 6.11. The general acid/base-catalyzed pair of paths, AdgB and EgA, represent a relatively well explored middle ground of partial proton transfer. Similar to path EgA, catalysis by hydrogen-bonded species may be occurring any time a reasonably basic leaving group is lost in a protic solvent. Likewise the reverse reaction similar to path AdgB, the enhancement of nucleophilicity by hydrogen bonding, may occur in a protic solvent whenever the nucleophile bears a reasonably acidic hydrogen.

Enzymes can have several catalytic groups in the cavity of their active site. When the reactant fits into that active site, these catalytic groups may function together in a push–pull catalysis. For example, the top view of the energy surface for enolization is shown in Figure 6.19. The reactants are in the lower left corner and the products are in the upper right. The acid-catalyzed process forms the O—H bond before the C—H bond breaks (up, then right), whereas the base catalyzed process breaks the C—H bond before the O—H bond forms (right, then up). The push–pull catalysis is the diagonal route in which the C—H bond breaks as the O—H bond forms.

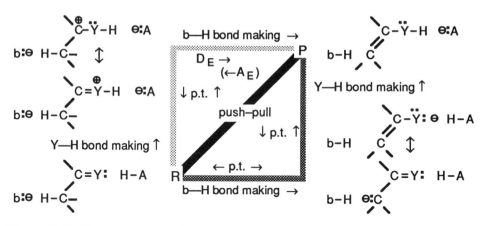

Figure 6.19 The energy surface for enolization. The lightest path is acid catalyzed, the gray path is base catalyzed, and the black path is the push–pull-catalyzed path. Reverse paths are in parentheses.

This section should close with a reminder about charge types. Almost all of the pathways have several charge types. Nucleophiles can be anionic or neutral; electrophiles can be cationic or neutral, and leaving groups can depart as anions or neutrals. Reactions catalyzed by a Brønsted acid can often be similarly catalyzed by a Lewis acid or by a positively charged metal ion.

6.5 COMMON ERRORS

It is a debatable pedagogic point whether or not to show students common errors since they may remember that they "saw it somewhere in the text" and not that it was incorrect. However, the pitfalls of using arrows are so numerous that something needs to be said. Also, the student will be exposed to these errors in print anyway since these mistakes are common in published works now that using arrows to explain reactions has become fashionable. Novices who use arrows only to get all the bonds in the right places tend to ignore the continuity of the electron flow and the path restrictions.

The incorrect example will appear on the left with the corrected version to the right of it, followed by an explanation on why the left one is incorrect. The errors are grouped by type, starting with the most simple and ending with the more subtle.

Lewis Structure Errors

It is impossible to flow electrons correctly if you cannot keep track of them accurately. The structure on the left has exceeded the octet for nitrogen. Since nitrogen is a first-row element, it cannot expand its valence shell.

Ligand-Rich Versus Electron-Deficient Errors

Groups like $NH_4^{\oplus}$ and $H_3O^{\oplus}$ are cationic but not electron-deficient because both nitrogen and oxygen have a complete octet. For example, the nitrogen in $NH_4^{\oplus}$ has four groups bound to it, completing its octet and giving it a formal charge of +1. Any attack by a nucleophile on it would form a fifth bond to nitrogen and exceed its octet. First-row elements, like C, N, and O, cannot exceed their octet.

Proton Transfer Errors

In this case, the arrows are not at fault, but rather the process that they represent is incorrect. Often beginners will forget to check the K_{eq} for proton transfer and use a base to remove a proton that is not acidic. It should be assumed that the hydrogen is not acidic if you cannot find the appropriate or a related compound on the pK_a chart. For example, a hydrogen on a carbon atom that bears only an electron donor group is not acidic. As an illustration, you will not find diethyl ether, $CH_3CH_2OCH_2CH_3$, on the pK_a chart; none of its hydrogens are acidic. The carbonyl hydrogen of an aldehyde, RCHO, also is not acidic. If you cannot find the conjugate acid of a proposed leaving group (or that of a closely related species) on the chart, then suspect that that group may not be a leaving group. It is very important that you know how to use the pK_a chart.

Off-the-Path Errors

Until you have the principles of mechanistic organic chemistry thoroughly mastered, it is best to restrict your mechanistic proposals to simple combinations of the electron flow pathways. You may see a shortcut that with several arrows would allow you to transform the lines and dots of the Lewis structure of the reactant into the lines and dots of the product, but that is not the point of it. What you are trying to do with arrows is to guess what is actually going on in the reaction, and for that you should use processes that are actually known to exist. The pathways are a very powerful mechanistic vocabulary, and there are very few mechanistic processes that cannot be expressed as a simple combination of them.

Energetics Errors

If you fail to check the energetics of a step, you may propose a step that is not energetically reasonable. You are trying to propose the most reasonable **lowest energy** route to product. To do that you must always monitor the energetics of each step. Even if only one step has an impossible barrier, the route is not correct. Steps that form highly unstable intermediates are instantly suspect. Cross-check all intermediates with the appropriate stability trend from Chapter 3.

Charge Errors

Typical organic solvents have difficulty stabilizing adjacent like charges. Therefore, avoid forming adjacent like charges unless the reaction solvent is water. **Very few organic intermediates have multiple charges.**

Charge must be conserved. The total charge on both sides of the transformation arrow must be the same. Charge is part of your electron count and must balance through each step.

Selectivity Errors

Make sure that you look at all the reactive groups in the reaction mixture. A reaction is almost always the combination of the best electron source with the best electron sink. If two different electron sinks, for example, are present under the same reaction conditions, the more reactive of the two will react first. If the reaction conditions are sufficiently vigorous, both may react.

Charge Balance Errors Combined with Line Structure

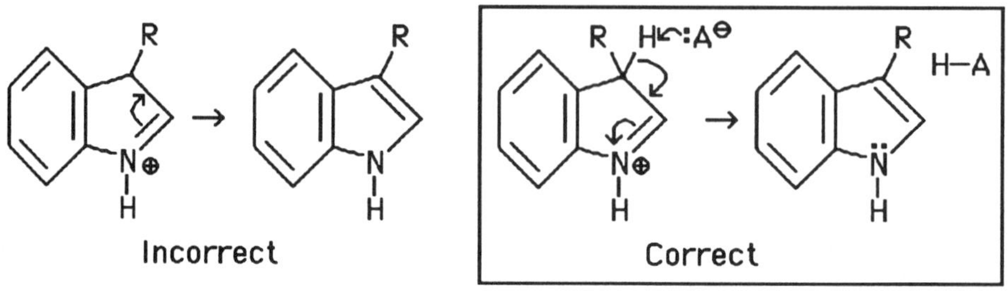

The loss of a formal charge may be difficult to notice in reactions that drop off a proton because the proton is commonly not drawn as one of the products. Since the proton that was lost was not drawn in line structure on the left, it was simply forgotten. The arrow made a double bond in the right place but flowed the electrons the wrong way: away from the positive center and not toward it. Besides the loss of the formal charge, the lone pair on nitrogen was lost; it was assumed in the product but the incorrect arrow did not form it. The right side shows the correct electron flow using path D_E, deprotonation of a carbocation to form a pi bond.

Electron Flow Backwards Error

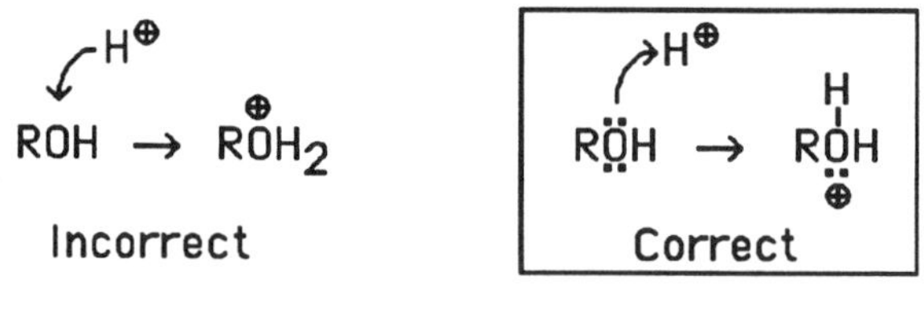

A proton has no electrons at all, so an arrow cannot start from it. This arrow error is seen in some biochemistry texts to mean "put the proton there." Always use arrows only for electron movement, nothing else. Arrows will always head toward positive centers and away from negative ones. The correct path is p.t., protonation of a lone pair.

Electron Flow Continuity Errors

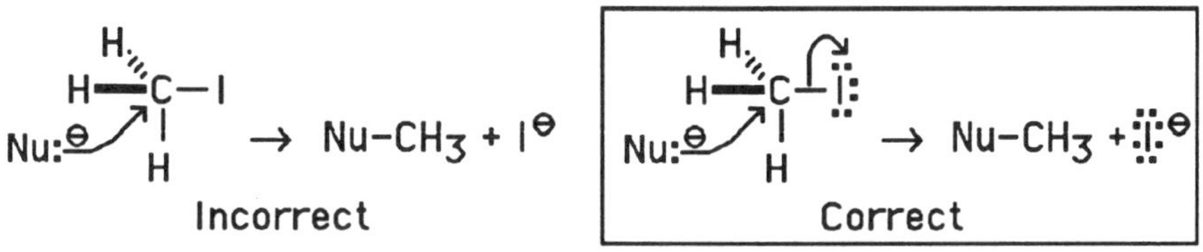

To be fair, some authors just draw the first arrow of an electron flow and expect the reader to supply the rest; it is best to see the entire flow drawn out. The correct path is the S_N2 substitution.

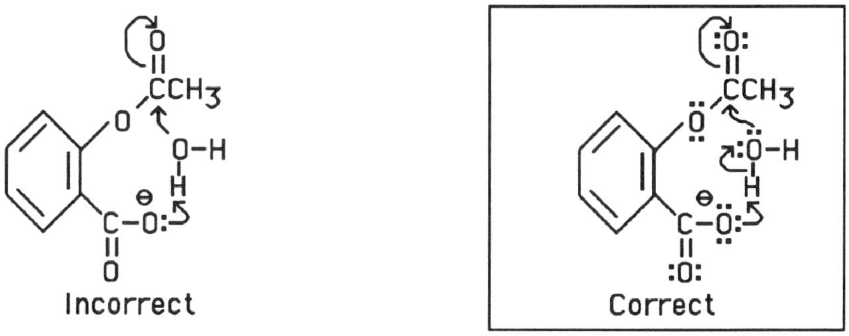

Occasionally an arrow in the middle of the flow is forgotten, quite often the breaking of a C—H or O—H bond. In the left example, the origin of the middle arrow is unclear; it appears to come from a lone pair position, but the lone pairs have been omitted. The correct path is AdgB, general base-catalyzed addition to a polarized multiple bond.

Curved Arrows That Miss Their Target

Beginners are often not careful in their drawing of arrows and may start an arrow from the wrong bond or put the head in between or on the wrong atoms. Since it is easy to spot, we won't give an example here. Always draw out all the electron pairs on any group that you are going to draw arrows near. Start the first arrow from a bond or electron pair, not from the formal charge.

Errors from Shortcuts to Save Arrows or Compress Steps

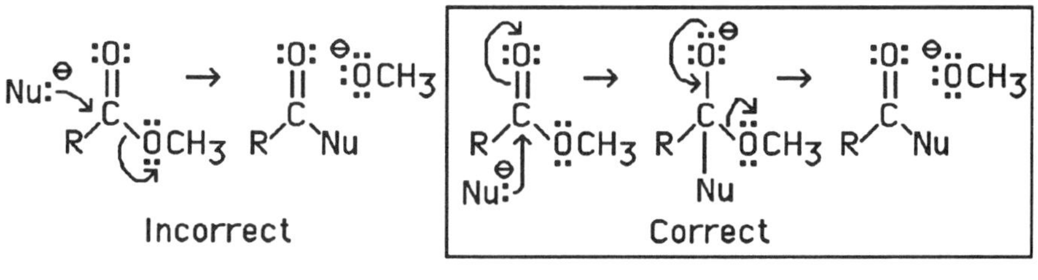

Although the direct nucleophilic displacement, the S_N2, of leaving groups bound to a sp^2 center is a mechanism that has been searched for, with one or two rare exceptions there is little evidence in the literature that it has ever been found. The correct mechanism is path Ad$_N$, addition to a polarized multiple bond, followed by path E$_\beta$, beta elimination from an anion.

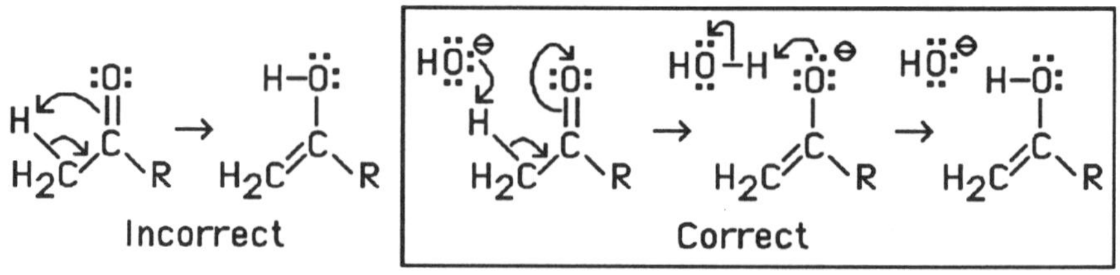

This shortcut certainly saves writing many structures and arrows, but there is good experimental evidence that this tautomerization definitely needs to be catalyzed by acid or base and will not proceed on its own as drawn on the left. As drawn on the left, it is a four-center, four-electron process, that with very few exceptions (minor path 4e) does not occur thermally. The correct process is either base (shown on the right) or acid catalyzed. For a more thorough discussion of this tautomerization process see Section 6.2.

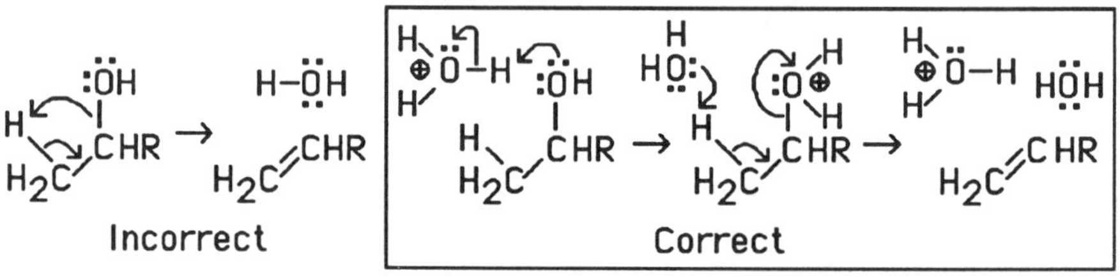

Alcohols do not do uncatalyzed eliminations of water at reasonable temperatures. The carbon–oxygen bond is not a base, and the carbon–hydrogen bond not an acid. Again, the process drawn on the left is a four-center, four-electron process, which with very few exceptions (minor path 4e) does not occur thermally. The most common route for water elimination is by acid catalysis, as shown on the right: path p.t., protonation of a lone pair, followed by the E2 elimination. If the carbocation is reasonably stable, the reaction may proceed via the E1 process (Section 6.2).

Errors That Ignore the Polarization of a Bond

These errors are nearly impossible to spot from the arrows alone since the error is not the arrows themselves but that they run counter to the polarization of a bond. The valid resonance forms of a molecule show us the internal polarization of its bonds.

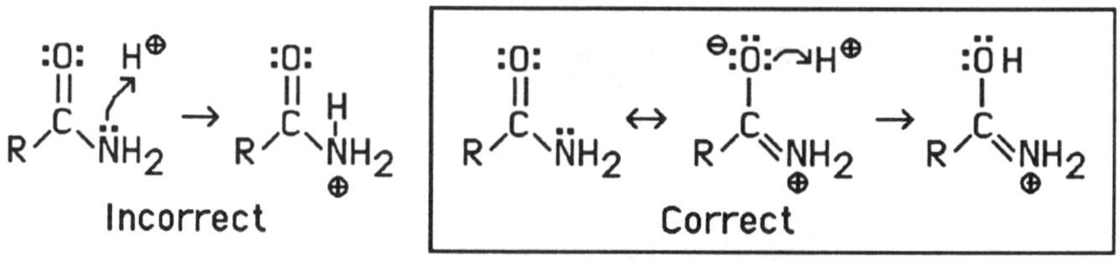

In the above example, the arrows are drawn on the neutral resonance form, and so the polarization of the amide group was forgotten. The resonance forms show that the partially negative oxygen and not the partially positive nitrogen will attack the electrophile as shown on the right (path p.t., protonation of a lone pair). Another common student mistake is to forget Markovnikov's rule when adding an electrophile to a pi bond.

Mixed-Media Errors

	Acidic	or	Basic
Incorrect		Correct	

Sometimes students will optimize the source and the sink and forget that those two species cannot exist in the same media. Since the protonated ester pK_a is -6.5 and the pK_{abH} of hydroxide is 15.7, they span over 22 pK_a units. Therefore any media that would be acidic enough to have some concentration of the protonated ester would have essentially no concentration of hydroxide ion. The right examples show two correct alternatives: acidic media with the protonated ester and neutral water as the nucleophile or basic media with the neutral ester and hydroxide as the nucleophile (path Ad_N, addition to a polarized multiple bond).

Another common mixed-media error of students analyzing two-step reactions is to involve species from both steps at the same time. Many organometallics, for example, have acidic water as a second step, the workup. The first step is in strongly basic media and the second step acidic; in your mechanisms, keep the steps separate. Since the pK_{abH} of most organometallics is very high, they are strong bases and react quickly and sometimes violently with acidic water to produce the alkane.

Orbital Alignment Errors

The topic of how much angular "slop" there is in a transition state is currently a debated topic. We can get agreement that perpendicular orbitals do not interact. However, most calculations of the actual shape of orbitals show them to be much fatter than they tend to be drawn in beginning textbooks. It now appears that a 10° deviation from the optimum has a negligible effect.

The most important place to watch orbital overlap is in the formation of pi bonds. Pi bonds that have about a 30° twist can be made but are reactive. Since substitution competes with elimination, it does not take very much of a twist to tilt the balance away from elimination. There are many examples to show that changing the leaving group from approximately coplanar with the hydrogen to 60° out of alignment shuts off the E2 elimination process entirely. Be especially careful of proposing eliminations in rigid systems.

Incorrect	Correct

The decarboxylation on the left, written as a variant of the E2 (Section 6.4) contains a serious overlap error. The error is that the source and sink are about 60° out of being coplanar, clearly beyond the distortion limits of the E2 elimination (Figure 6.20).

Using the pathways in this chapter, we can propose a reasonable alternative elimination, the E1 process (Section 6.2), that produces the same product. If the leaving

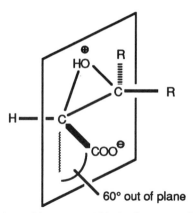

Figure 6.20 An illustration of improper orbital alignment. The C, C, and O of the three-membered ring define the plane. The carboxylate sticks out 60° from the plane of the ring.

group departs first, the carbocation formed can rotate about the single bond until the carbon–carbon bond of the source is coplanar with the empty *p* orbital. Then the loss of carbon dioxide can occur easily to form the pi bond of the enol product.

Chemical structures drawn flat on a page may not reflect accurately the three-dimensional nature of the compound drawn. If you have any suspicion that a reaction step has poor orbital alignment, or is physically impossible (inaccessibility of a reactive site, for example), make the molecular model.

6.6 MAJOR PATH SUMMARY

p.t., Proton Transfer to and from an Anion or Lone Pair

A_N, Trapping of an Electron-Deficient Species

$$\overset{\ominus}{Nu:} \curvearrowright \overset{|}{\underset{|}{C}}\overset{\oplus}{} \rightarrow Nu-C\overset{\diagdown}{\diagup} \quad or \quad \overset{\ominus}{Nu:} \curvearrowright \overset{\overset{Y}{\|}}{\underset{|}{C}}\overset{\oplus}{} \rightarrow Nu-\overset{\overset{Y}{\|}}{C}\diagdown$$

A_E, Electrophile Addition to a Multiple Bond

$$\overset{\oplus}{E}\curvearrowleft \quad \underset{|}{\overset{|}{C}}=\underset{|}{\overset{|}{C}} \quad \rightarrow \quad \underset{|}{\overset{E}{C}}-\underset{|}{\overset{\oplus}{C}}$$

D_E, Electrophile Loss from a Cation To Form a Pi Bond

$$E\curvearrowright \underset{|}{\overset{|}{C}}\overset{\oplus}{\underset{|}{C}}\diagup \quad \rightarrow \quad \overset{E^{\oplus}}{} \quad \diagdown C=C\diagup$$

1,2R, Rearrangement of a Carbocation

$$R\curvearrowright \underset{|}{\overset{|}{C}}-\underset{|}{\overset{\oplus}{C}}\diagup \quad \rightarrow \quad \diagdown \overset{\oplus}{C}-\underset{|}{\overset{R}{C}}\diagdown$$

S_N2, The S_N2 Substitution

$$\overset{\ominus}{Nu:}\curvearrowright \underset{|}{\overset{|}{C}}\overset{\frown}{L} \rightarrow Nu-C\overset{\diagdown}{\diagup} \ :L^{\ominus} \quad or \quad \overset{\ominus}{Nu:}\curvearrowright Y\overset{\frown}{L} \rightarrow Nu-Y \ :L^{\ominus}$$

E2, The E2 Elimination

$$\overset{\ominus}{b:}\curvearrowright \overset{H}{\underset{|}{\overset{|}{C}}}\curvearrowright \underset{\underset{L}{\curvearrowright}}{\overset{|}{C}}\diagup \rightarrow \overset{b\diagdown H}{\underset{|}{\overset{|}{C}}}=C\diagup \underset{L:^{\ominus}}{}$$

Ad_E3, The Ad_E3 Addition

$$\overset{A-H}{\underset{|}{\overset{|}{C}}}\cdots \overset{|}{\underset{|}{C}}\diagup \rightarrow \overset{A\curvearrowleft H}{\underset{\ominus A:\curvearrowright}{\overset{|}{C}}=C\diagup} \rightarrow \overset{\ominus A:}{\overset{H}{\underset{A}{\overset{|}{C}}}}-C\diagup$$
$$pi\text{-}complex$$

Ad_N, Nucleophilic Addition to a Polarized Multiple Bond

$$\overset{Nu:^{\ominus}}{\underset{|}{\overset{|}{C}}}\overset{\frown}{Y} \rightarrow \overset{Nu}{\underset{|}{\overset{|}{C}}}-\overset{\ominus}{\underset{\cdot\cdot}{Y}} \quad or \quad \overset{Nu:^{\ominus}}{\underset{|}{\overset{|}{C}}}\overset{\frown}{C}-ewg \rightarrow \overset{Nu}{\underset{|}{\overset{|}{C}}}-\overset{\ominus}{\underset{\cdot\cdot}{C}}-ewg$$

E_β, Beta Elimination from an Anion or Lone Pair

$$L\curvearrowright \underset{|}{\overset{|}{C}}\overset{\ominus}{\underset{\cdot\cdot}{Y}} \rightarrow \overset{L:^{\ominus}}{} \underset{|}{\overset{|}{C}}=Y \quad or \quad L\curvearrowright \underset{|}{\overset{|}{C}}\overset{\ominus}{\underset{\cdot\cdot}{C}}-ewg \rightarrow \overset{L:^{\ominus}}{} \underset{|}{\overset{|}{C}}=C-ewg$$

Path 6e, Concerted Six-Electron Pericyclic Reactions

Thermal rearrangements:

$$\rightleftharpoons \underset{Heat}{} \qquad or \qquad \rightleftharpoons \underset{Heat}{}$$

Thermal cycloadditions or cycloreversions:

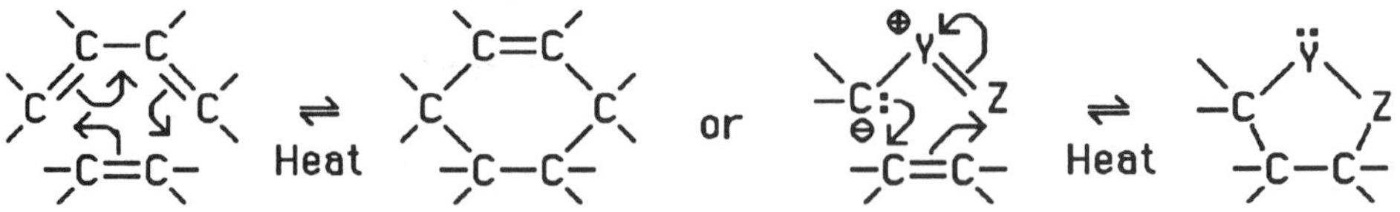

Metal chelate catalyzed additions:

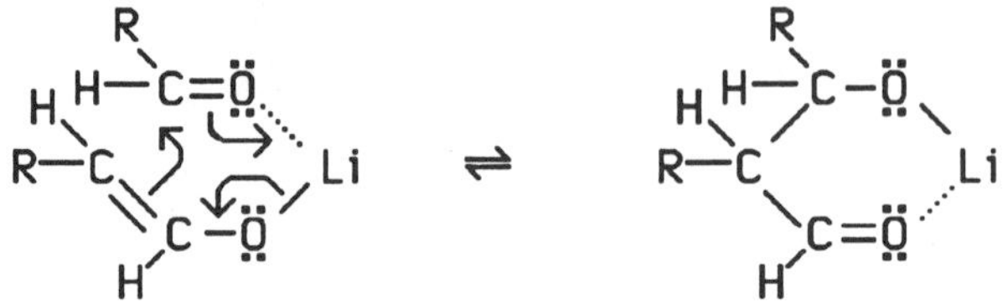

ADDITIONAL EXERCISES

6.1 Go back to the worked examples in Chapter 1, Section 1.3, and place the name of the path over each transformation arrow.

6.2 Go back to Chapter 1, exercises 1.5 and 1.6, and place the name of the path over each transformation arrow.

6.3 Draw a general example of the process indicated.
   The S_N2 substitution
   E_β, Beta elimination from an anion
   Ei, Thermal internal syn elimination
   E2
   p.t., Proton transfer to and from an anion or lone pair
   Ad_E3
   D_N, Ionization of a leaving group
   S_N1
   Ad_N2
   A_N, Trapping of a cation by a nucleophile
   E1
   A_E, Electrophile addition to a multiple bond
   Ad_E2
   1,2R, Rearrangement of a carbocation

6.4 Draw a general example of the process indicated.
   E1cB
   Ad_N, Nucleophilic addition to a polarized multiple bond
   Addition–Elimination
   6e, Thermal, six-electron pericyclic reactions
   Nu—L additions
   Electrophilic aromatic substitution
   Hetero Ad_E2
   Tautomerization
   D_E, Electrophile loss from a cation to form a pi bond
   Lone-pair-assisted E1

6.4 (continued) Draw a general example of the process indicated.
EgA, General acid-catalyzed beta elimination
pent., Substitution via a pentacovalent intermediate
H⊖ t., Hydride transfer to a cationic center
AdgB, General base-catalyzed addition to a polarized multiple bond

6.5 Select from the lists in problems 6.3 and 6.4 the dozen most common pathways.

6.6 Select from the lists in problems 6.3 and 6.4 the ten common path combinations.

6.7 Select from the lists in problems 6.3 and 6.4 the four acidic media additions of an electrophile to a multiple bond.

6.8 Select from the lists in problems 6.3 and 6.4 the three basic media nucleophilic additions to a multiple bond.

6.9 Select from the lists in problems 6.3 and 6.4 the two nucleophilic substitutions of a group bound to a sp^3 hybridized carbon.

6.10 Select from the lists in problems 6.3 and 6.4 the two substitutions of a group bound to a sp^2 hybridized carbon.

6.11 Select from the lists in problems 6.3 and 6.4 the five eliminations of an H and a leaving group from a neutral molecule.

6.12 State the overlap limitation of the E2 process and for the rearrangement of a carbocation.

6.13 From biochemical studies, a general acid catalyst is most effective if its pK_a is close to the pH of the medium. Explain this by examining the extremes. What would happen if the acid were very acidic, a much lower pK_a; then consider if the acid were not very acidic, a much higher pK_a?

6.14 Classify the electron sink and the electron source for the following reaction. Classify the reaction into one of the four general types. Using the sink as a guide, list the possible paths. Using the path restrictions, pick the only path that fits for this reaction.

$$CH_3CH_2{-}\overset{..}{\underset{..}{I}}{:} \ + \ Bu_3P{:} \ \rightarrow \ Bu_3\overset{\oplus}{P}{-}CH_2CH_3 \ + \ {:}\overset{..}{\underset{..}{I}}{:}^{\ominus}$$

6.15 Many reactions give product mixtures. Give the product of an S_N2 and also the product of an S_N2' on the following compound. Use Nu to symbolize the nucleophile.

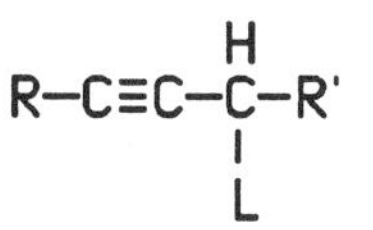

$$R{-}C{\equiv}C{-}\overset{\displaystyle H}{\underset{\displaystyle L}{\overset{|}{\underset{|}{C}}}}{-}R'$$

7

Interaction of Electron Sources and Sinks

7.1 SOURCE AND SINK MATRIX

A Correlation Matrix Is the Best Way To Display All the Simple Interactions Between Two Groups

7.2 DETAILED MATRIX-CELL DISCUSSIONS

7.2a Lone Pair Sources Reacting with Acids
7.2b Metal Hydrides Reacting with Acids
7.2c Organometallics Reacting with Acids
7.2d Allylic Sources Reacting with Acids
7.2e Bases Reacting with Acids
7.2f Lone Pair Sources Reacting with Y–L or ⩾C–L
7.2g Hydrides Reacting with Y–L or ⩾C–L
7.2h Organometallics Reacting with Y–L or ⩾C–L
7.2i Allylic Sources Reacting with Y–L or ⩾C–L
7.2j Bases Reacting with Y–L or ⩾C–L, Eliminations
7.2k Lone Pair Sources Reacting with C=Y and R–C≡N
7.2l Hydrides Reacting with C=Y and R–C≡N
7.2m Organometallics Reacting with C=Y and R–C≡N
7.2n Allylic Sources Reacting with C=Y and R–C≡N
7.2o Bases Reacting with C=Y and R–C≡N, Enolization
7.2p Conjugate Addition by Lone Pair Sources to C=C–ewg
7.2q Conjugate Addition by Metal Hydrides to C=C–ewg
7.2r Conjugate Addition by Organometallics to C=C–ewg
7.2s Conjugate Addition by Allylic Sources to C=C–ewg
7.2t Bases Reacting with C=C–ewg, Extended Enolates
7.2u Lone Pair Sources Reacting with L–C=O
7.2v Hydrides Reacting with L–C=O
7.2w Organometallics Reacting with L–C=O
7.2x Allylic Sources Reacting with L–C=O, Acylation
7.2y Bases Reacting with L–C=O, Enolization

7.3 ELECTROPHILIC ADDITIONS

7.3a Lone Pairs as Electron Sources
7.3b Alkenes, Dienes, and Alkynes as Electron Sources
7.3c Electrophilic Aromatic Substitution

7.4 REARRANGEMENTS WITH LOSS OF A LEAVING GROUP

Good Orbital Alignment Is Crucial

The flow of electron density, symbolized by a set of arrows, comes from the generic electron sources discussed in Chapter 4, via the pathways of Chapter 6, and ends in one of the generic electron sinks covered in Chapter 5. In this chapter we link the sources and sinks with their appropriate pathways and discuss specific examples of each process.

7.1 SOURCE AND SINK MATRIX

A correlation matrix is the best way to display all the simple interactions between two groups so that no interactions are left out accidently. Before we draw a matrix for any given set of data, we must first recognize the essential elements that make up the system. The elements must be classified then into generic groupings, and the entire list of generic groups must be tested for completeness. Finally the interaction of the generic groupings can be displayed with a matrix.

Almost all of the reactions in organic chemistry can be described as simple bimolecular collisions between 1 of the 12 generic electron sources (Section 4.7) and 1 of the 18 generic electron sinks (Section 5.6). Many reaction mechanisms are multistep processes where the product of the initial collision of source and sink then collides with another reactant and continues on.

The easiest way to represent all the possible combinations of electron sources and sinks is with a matrix, each cell of the matrix corresponding to a specific combination of source and sink. A giant matrix containing all the electron sources and sinks can be drawn, but it would be 12 x 18 and thereby contain 216 individual cells. This large matrix contains many cells in which no reaction occurs and many cells that have been little researched.

For simplification, Table 7.1, a much smaller matrix that contains representatives of the more common sources and sinks, will be used in this section and discussed in Section 7.2. Given within each cell of the matrix is the subsection in Section 7.2 that covers the appropriate pathways for electron flow for that particular combination of source and sink.

Any important combinations not included in the simplified matrix are covered in Sections 7.3 to 7.8.

Table 7.1 Correlation Matrix for the More Common Electron Sources and Sinks: An Index to Section 7.2.

Electron Sinks	Common Electron Sources				
	Lone pair (Z:)	Hydride ($MH_4^\ominus$)	Organometallic (R–M)	Allylic (:Z–C=C)	Base acting on an adjacent CH
H–A	7.2a	7.2b	7.2c	7.2d	7.2e
Y–L	7.2f	7.2g	7.2h	7.2i	7.2j
⋛C–L	7.2f	7.2g	7.2h	7.2i	7.2j
C=Y	7.2k	7.2l	7.2m	7.2n	7.2o
C≡N	7.2k	7.2l	7.2m	7.2n	7.2o
C=C–ewg	7.2p	7.2q	7.2r	7.2s	7.2t
L–C=Y	7.2u	7.2u	7.2w	7.2x	7.2y

7.2   DETAILED MATRIX-CELL DISCUSSIONS

7.2a   Lone Pair Sources Reacting with Acids

Acids react by simple protonation of the electron source via proton transfer (see Section 3.3 for the calculation of the K_{eq}).

$$(CH_3)_3C\ddot{O}H \quad \curvearrowright H—\ddot{Br}: \xrightarrow{p.t.} (CH_3)_3C\overset{H}{\underset{|\oplus}{\ddot{O}}}H \; + \; {}^{\ominus}:\ddot{Br}:$$

7.2b   Metal Hydrides Reacting with Acids

Aluminum hydrides react violently with protic solvents and acids (path p.t.) to produce hydrogen gas and can be used only in *aprotic* neutral or basic media. The borohydrides, $BH_4^\ominus$, are weaker sources and they react much more slowly with protic solvents than do the aluminum hydrides. The trivalent aluminum or boron species is now electron deficient and can react with a lone pair donor to form a tetravalent Lewis acid–base salt (path A_N) that can donate hydride again. This repeats until all hydrogen–metal bonds have been used.

$$\overset{H}{\underset{H}{H-\overset{\ominus}{Al}-H}} \curvearrowright H—\ddot{O}: \; Li^\oplus \xrightarrow{p.t.} \overset{H}{\underset{H}{H-Al}} \; H_2 \; {}^{\ominus}:\ddot{O}-H \; Li^\oplus \xrightarrow{A_N} \overset{H}{\underset{H}{H-\overset{\ominus}{Al}-\ddot{O}H}} \; Li^\oplus$$

$$\overset{H}{\underset{H}{H\ddot{O}-\overset{\ominus}{Al}-H}} \curvearrowright H—\ddot{O}: \xrightarrow{p.t.} \overset{H}{\underset{H}{H\ddot{O}-Al}} \; H_2 \; {}^{\ominus}:\ddot{O}-H \xrightarrow{Repeat} \to \to H_2 \; + \; \overset{:\ddot{O}H}{\underset{:\ddot{O}:}{H\ddot{O}-Al}} \; {}^{\ominus}:\ddot{O}-H$$

7.2c   Organometallics Reacting with Acids

Organometallics can act as bases and get protonated by any acidic hydrogen (path p.t.). If the acidic hydrogen is a C–H bond, a new and more stable organometallic is slowly produced. Reaction with an acidic O–H bond is fast.

$$Li-CH_3 \leftrightarrow Li^\oplus \; {}^\ominus CH_3 + \curvearrowright H—\ddot{O}-CH_3 \xrightarrow{p.t.} Li^\oplus + CH_4 + {}^\ominus:\ddot{O}-CH_3$$

7.2d   Allylic Sources Reacting with Acids

Allylic sources are ambident nucleophiles (Sections 4.4 and 8.4) and can therefore attack an electrophile at either of two sites. Since proton transfer is commonly reversible and rapid, an equilibrium mixture is quickly achieved. When the source is anionic, the protonation occurs most rapidly on the heteroatom (hard–hard, path p.t.) but can also occur on carbon. Equilibration to the more stable product occurs by proton transfer (Section 6.2, tautomerization). A ΔH calculation (Section 2.1) will verify the carbon protonated species is the more stable product.

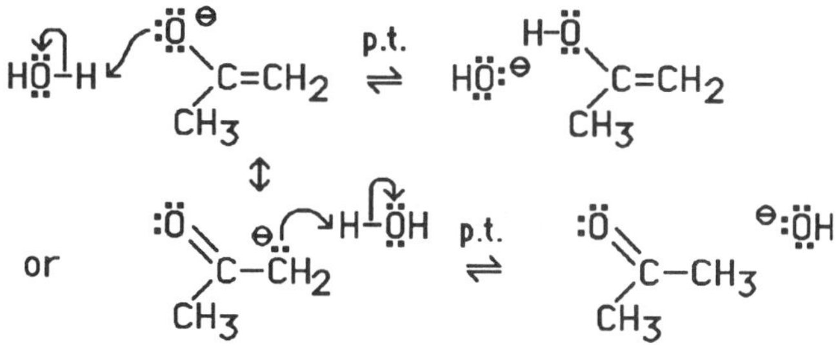

When the source is neutral, protonation can occur to produce the heteroatom protonated species or the stable heterosubstituted carbocation. Protonation on the more basic heteroatom lone pair is the lower energy process, but the heteroatom protonated species has little it can do but revert to reactants.

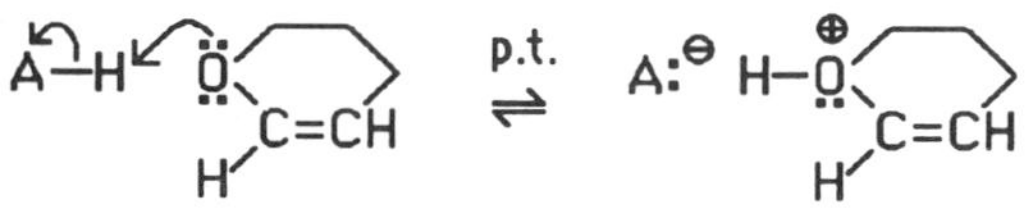

However, protonation on carbon creates a heterosubstituted carbocation that can easily be captured (path A_N) by solvent or another nucleophile.

Ad_E2 example (with follow-up proton transfer to solvent):

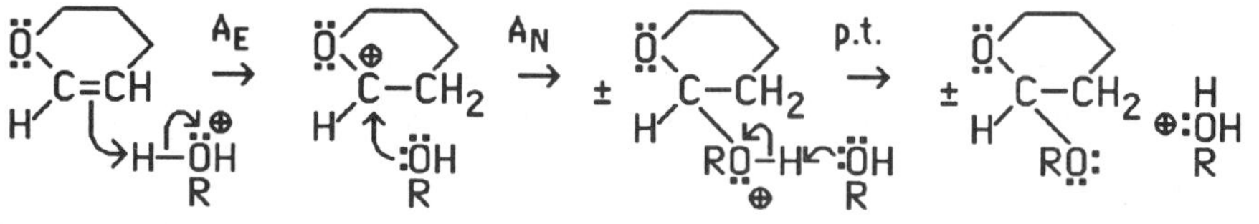

7.2e   Bases Reacting with Acids

This simple neutralization reaction always forms the weaker base.

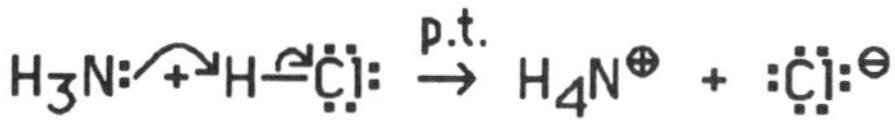

$$H_3N: \curvearrowright H \!-\! \ddot{C}l: \xrightarrow{\text{p.t.}} H_4N^{\oplus} + :\ddot{C}l:^{\ominus}$$

7.2f   Lone Pair Sources Reacting with Y–L or $\geqslant$C–L

The path decision between substitution and elimination is considered in depth in Section 8.5.

Sink Y–L

Since Y usually represents an electronegative atom, $Y^{\oplus}$ is a poor cation. The path therefore is the generic S_N2. The ΔpK_a rule is helpful in determining the position of equilibrium in the displacement of a leaving group by a nucleophile.

Example (with follow-up proton transfer):

$$R_3N: \overset{\frown}{\ddot{O}} \overset{\ominus}{\ddot{O}} H \overset{S_N2}{\rightarrow} R_3\overset{\oplus}{N}-\ddot{O}-H \overset{\curvearrowleft}{\ddot{O}}-H \overset{p.t.}{\rightarrow} R_3\overset{\oplus}{N}-\ddot{O}:^{\ominus} + H\ddot{O}H$$

Two reasons to mention phosphorus and sulfur halides as sinks separately are that although sulfur probably goes by the S_N2, most likely the substitution at phosphorus goes via the pentacovalent intermediate (path pent.). Secondly, if the electron source is an oxygen, a follow-up reaction often occurs in which O—Y serves as a leaving group. Overall, an OH group is converted to a better leaving group, then is eliminated to the alkene (see Section 7.2j) or displaced by a nucleophile, commonly the leaving group on the Y—L.

Example — no follow-up reaction:

$$R-\ddot{O}:^{\ominus} + \overset{:\ddot{O}:}{\underset{Ar}{\overset{\parallel}{S}}}-\ddot{C}l: \overset{S_N2}{\rightarrow} R-\ddot{O}-\overset{:\ddot{O}:}{\underset{:\ddot{O}:}{\overset{\parallel}{S}}}-Ar + :\ddot{C}l:^{\ominus}$$

Follow-up substitution, covered in the next subsection:

$$RCH_2\ddot{O}:^{\ominus} \overset{\frown}{} PBr_3 \overset{pent.}{\rightarrow} \left[RCH_2\ddot{O}-\overset{\ominus}{P}Br_2 \right] \overset{pent.}{\rightarrow} RCH_2\ddot{O}-PBr_2 \overset{S_N2}{\rightarrow} RCH_2 + :\ddot{O}-PBr_2$$

Sink ⩾ C—L, alkylation

There are two pathways that differ only in the timing of when the nucleophile attacks and when the leaving group falls off: **Generic S_N1** — leaving group departs first (rate-determining) by path D_N, producing a reasonably stable cation; the cation is then trapped by the nucleophile, path A_N. **Generic S_N2** — Simultaneous attack of nucleophile and loss of the leaving group. It is best to consider a S_N1/S_N2 spectrum where there is competition between the rate of ionization of the leaving group and the rate of nucleophilic attack on the substrate or partially ionized substrate. See Section 8.3 for a discussion of the decision of when to expect the ionization of the leaving group. Table 7.2 outlines the characteristics of the extremes of the spectrum.

TABLE 7.2 The S_N1/S_N2 Spectrum Extremes

Variable	S_N1 (Path D_N then A_N)	S_N2 (Concerted)
Carbocation	Must be good because it is an intermediate (tertiary or better)	Not formed, therefore carbocation stability is not a concern
Nucleophile	Weaker Nu tolerated since the sink is good (often neutral)	Good Nu required (commonly charged)
Site of attack	Often hindered	Must be open
Stereochemistry	Racemization or inversion	Inversion
Media	Often acidic	Often basic
Leaving group	Excellent	Average

Exceptions: Vinylic, C=C—L, or aryl, Ar—L, leaving groups do not substitute by S_N2 and are very slow by S_N1 since the carbocation formed is so unstable. $(CH_3)_3CCH_2$—L is too hindered to react by S_N2 and rearranges under S_N1 conditions.
S_N2 example (poor cation):

$$Ph\overset{:O:}{\overset{\parallel}{C}}-\ddot{O}:^{\ominus} + CH_3-\ddot{I}: \overset{S_N2}{\rightarrow} Ph\overset{:O:}{\overset{\parallel}{C}}-\ddot{O}-CH_3 + :\ddot{I}:^{\ominus}$$

S_N1 example (hindered site) with follow-up proton transfer:

$$Me_3C\text{--}\overset{..}{\underset{..}{Br}}: \overset{D_N}{\to} Me_3\overset{\oplus}{C} \quad :\overset{..}{\underset{..}{Br}}:^\ominus \overset{A_N}{\to} Me_3C\text{--}\overset{\oplus}{\underset{..}{O}}\text{--}H \overset{p.t.}{\to} Me_3C\overset{..}{\underset{..}{O}}H + H_3\overset{..}{O}{}^{\oplus}$$

$$\overset{}{\underset{:\overset{..}{O}H_2}{}} \qquad\qquad\qquad \overset{}{\underset{H \curvearrowleft :\overset{..}{O}H_2}{}}$$

The substitution energy surface

Although the actual energy surface for the simplest substitution reaction is much more complex than the simplified surfaces drawn in the following figures, we can use these surfaces to understand how the changes in reaction conditions would affect the reaction path.

The S_N1/S_N2 spectrum can easily be visualized as an energy surface, as in Figure 7.1, where the front horizontal axis is the degree of C−Nu bond making, the axis going back to front is the degree of C−L bond breaking, and the vertical axis is increasing energy. Reactants are in the back right corner at point R. The back left corner corresponds to only C−Nu bond making without C−L bond breaking, thus forming a pentacovalent intermediate that is a high-energy point since carbon cannot expand its valence shell. The front right corner corresponds to only C−L bond breaking and therefore is the carbocation; the front left corner is the substitution product P. The lowest energy path on the surface is the dark line from R to P.

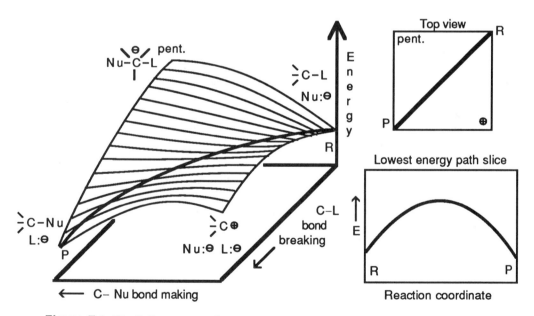

Figure 7.1 The S_N2 energy surface.

It is now possible to understand the S_N1/S_N2 spectrum trends by watching how the substitution surface folds and/or tilts in response to a change. As shown in Figure 7.1, if the carbocation is poor, its corner will be high in energy; since the pentacovalent corner is also high in energy, the surface folds down the middle. The reaction will go by the S_N2 process in which C−Nu bond making and C−L bond breaking occur to an equal extent, and the lowest energy path will follow the *diagonal* through the "pass" between the two "mountains" created by the high-energy carbocation and high-energy pentacovalent intermediate.

If the carbocation is rather stable, the corner on the diagram corresponding to the carbocation will be lowered in energy, and now the lowest energy path will go via the carbocation: the S_N1 substitution (Figure 7.2). The energy barrier for the loss of the leaving group is usually higher than that of the trapping of the cation by the nucleophile.

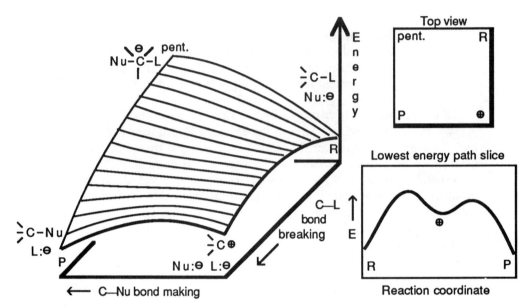

Figure 7.2 The S_N1 energy surface.

Steric hindrance raises the pentacovalent intermediate to an even higher energy, since it is the most crowded, with five groups around the central atom. The carbocation with only three groups around the central atom is least affected by steric hindrance. Therefore, increasing steric hindrance raises the pentacovalent intermediate greatly and the carbocation little, resulting in a tilting of the surface toward the formation of the carbocation.

If the leaving group is better, C—L bond breaking is easier, and therefore the entire front edge is lowered because all points along that edge have the C—L bond broken. The effect of a better leaving group lowers the carbocation corner, favoring S_N1, and lowers the product. This makes the overall reaction more favorable.

A better nucleophile makes C—Nu bond making easier, and therefore lowers the entire left edge, since all points along that edge have the C—Nu bond made. Therefore, the effect of a better nucleophile is to lower the pentacovalent corner somewhat, favoring S_N2, and to lower the product, again making the overall reaction more favorable.

The only way to lower the pentacovalent intermediate corner far enough so that the reaction proceeds through that intermediate is to have the ability to stabilize that intermediate, which second-row and higher elements can do (Figure 7.3). Substitution at phosphorus commonly proceeds through a pentacovalent intermediate, path pent., as shown in the following example.

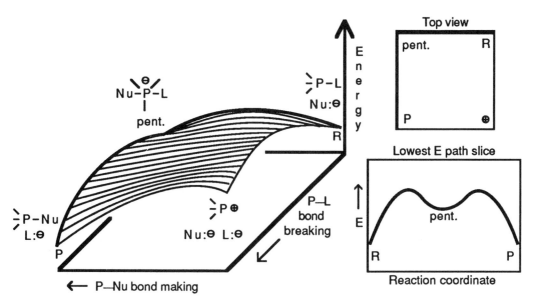

Figure 7.3 The energy surface for substitution via a pentacovalent intermediate.

7.2g Metal Hydrides Reacting with Y—L or $\geqslant$C—L

The substitution reaction occurs in neutral or basic media via the S_N2 pathway to replace the leaving group with hydrogen. Often the metal cation can form a complex with the leaving group to aid in its leaving.

Y—L example:

$$
\overset{\ominus}{\underset{\underset{H}{|}}{\overset{\overset{H}{|}}{H-B}}}{-}H \quad \overset{\curvearrowright}{}\overset{..}{Se}\overset{..}{\underset{Ph}{-}}\overset{..}{Se}Ph \quad Na^{\oplus} \quad \overset{S_N2}{\rightarrow} \quad \overset{\overset{H}{|}}{\underset{\underset{H}{|}}{H-B}} \quad + \quad H\overset{..}{Se}Ph \quad + \quad {}^{\ominus}{:}\overset{..}{Se}Ph \quad Na^{\oplus}
$$

$\geqslant$C—L example:

$$
\overset{\ominus}{\underset{\underset{H}{|}\ RCH_2}{\overset{\overset{H}{|}}{H-Al}}}{-}H \quad \overset{\curvearrowright}{}CH_2{-}\overset{..}{\underset{..}{Br}}{:} \ Li^{\oplus} \quad \overset{S_N2}{\rightarrow} \quad \overset{\overset{H}{|}}{\underset{\underset{H}{|}\ RCH_2}{H-Al}}CH_2 \quad + \quad {:}\overset{..}{\underset{..}{Br}}{:}^{\ominus} \ Li^{\oplus}
$$

7.2h Organometallics Reacting with Y—L or $\geqslant$C—L

Since organometallics react with acids (Section 7.2c), they can be used only in basic or neutral media and in aprotic solvents. The most common reaction is substitution via S_N2.

Y—L example:

$$
BrMgPh \longleftrightarrow {:}\overset{..}{\underset{..}{Br}}{:}^{\ominus} \ Mg^{2+} \ Ph{:}^{\ominus}\overset{\curvearrowright}{}{:}\overset{..}{\underset{..}{I}}{-}\overset{..}{\underset{..}{I}}{:} \quad \overset{S_N2}{\rightarrow} \quad Ph{-}\overset{..}{\underset{..}{I}}{:} \ + \ {:}\overset{..}{\underset{..}{Br}}{:}^{\ominus} \ Mg^{2+} \ {:}\overset{..}{\underset{..}{I}}{:}^{\ominus}
$$

$\geqslant$C—L example:

$$
{:}\overset{..}{\underset{..}{Br}}{:}^{\ominus} \ Mg^{2+} \ \underset{CH_3CH_2CH_2}{PhH_2C{:}^{\ominus}}\overset{\curvearrowright}{}CH_2{-}\overset{\overset{..}{\underset{..}{O}}{:}}{\underset{\underset{..}{\overset{..}{O}}{:}}{\overset{||}{\underset{||}{S}}}}{-}Ar \quad \overset{S_N2}{\rightarrow} \quad \underset{CH_3CH_2CH_2}{PhH_2C{-}CH_2} \ + \ {:}\overset{..}{\underset{..}{Br}}{:}^{\ominus} \ Mg^{2+} \ {}^{\ominus}{:}\overset{\overset{..}{\underset{..}{O}}{:}}{\underset{\underset{..}{\overset{..}{O}}{:}}{\overset{||}{\underset{||}{S}}}}{-}Ar
$$

7.2l Allylic Sources Reacting with Y—L or $\geqslant$C—L

An allylic electron source can be considered to be a vinylogous lone pair. Sometimes the lone pair itself is the nucleophile and can undergo reactions similar to the lone pair sources in the first column of the matrix. However, most often the nucleophile is the end carbon atom of the pi bond. The heteroatom lone pair is much harder than the pi bond, and the decision of which to use as the nucleophile is based on the hardness of the electrophile (HSAB principle) and will be discussed at greater length in Section 8.4.

Y—L as sink

The reaction is almost exclusively by the S_N2 pathway because $Y^\oplus$ is usually a poor cation. However the substitution at phosphorus and silicon usually goes via the pentacovalent intermediate (path pent.). The following examples show the dual reactivity of the same allylic source, the enolate ion. Z as Nu (hard E, Si—Cl bond polarized due to a large electronegativity difference):

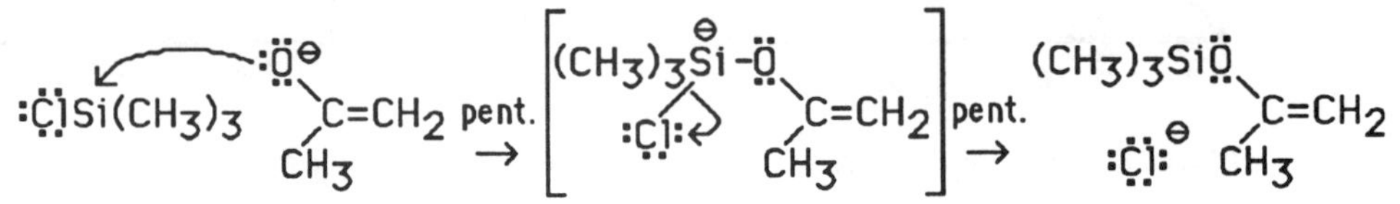

C as Nu (soft E, Br—Br bond not polarized at all):

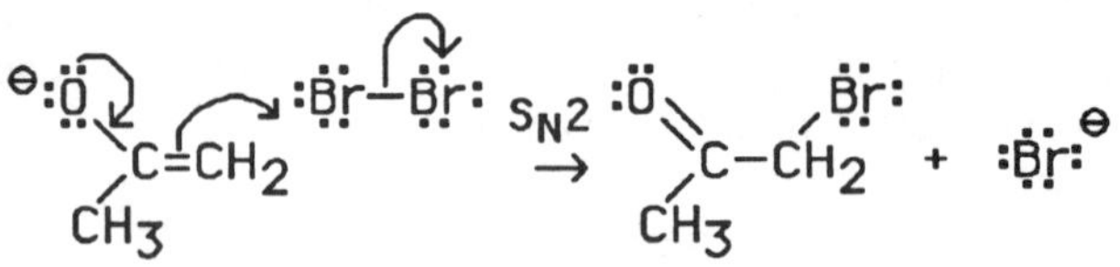

$\geqslant$ C—L as sink, alkylation

Enamines, N—C=C, usually alkylate on carbon with this sink, and produce an iminium ion, $^\oplus$N=C—C—R, which is normally hydrolyzed to the ketone in an acidic water workup (the reverse of imine formation, Section 7.2k).
Enamine alkylation example:

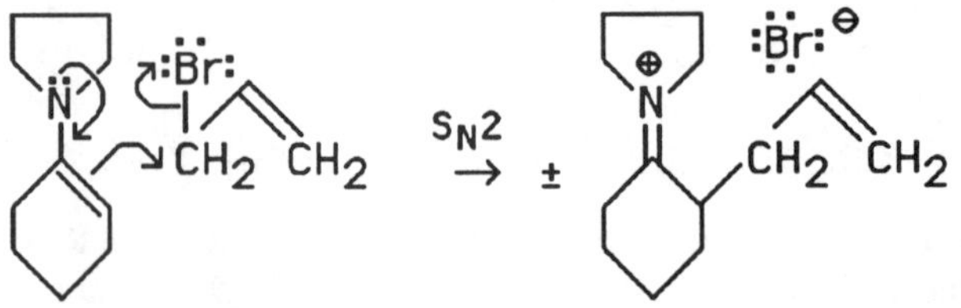

However, with simple enolates, $^\ominus$O—C=C, alkylation can be a messy reaction. The alkylation product and the enolate can usually proton transfer, producing new electron sources and leading to undesired products. The best way to run this reaction is to add the source into an excess of sink and hope that the alkylation proceeds faster than proton transfer.

Vinylogous allylic electron sources

One can speed up alkylation relative to proton transfer by stabilizing the enolate anion with an additional ewg, making it softer and less basic. For example, the acidic

CH_2 of malonates, $CH_2(COOEt)_2$, or acetoacetates, CH_3COCH_2COOEt, deprotonates easily and makes an excellent nucleophile. The ester can later be hydrolyzed to the acid (Section 7.2u) and removed by decarboxylation via path Ei. This ester is a detachable ewg that makes the enolate less reactive, more selective.

Acetoacetate example:

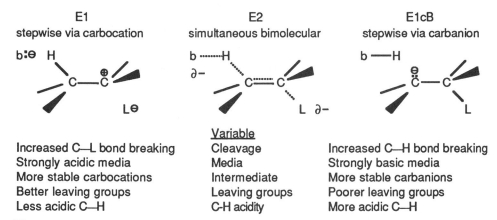

7.2j Bases Reacting with Y—L or ⩾C—L, Eliminations

There are three pathways that differ only in the timing of when the proton is pulled off and when the leaving group falls off. **Generic E1** — The leaving group departs first (rate-determining) by path D_N, producing a reasonably stable cation; the proton is then lost by path D_E, forming the alkene. **Generic E2** — Simultaneous loss of proton and leaving group. **Generic E1cB** — Proton transfer forms an anion stabilized by an electronegative atom, and then the leaving group departs (rate-determining) by path E_β. Path selection between substitution and elimination is considered in depth in Section 8.5.

Sink Y—L, oxidation

Y is an electronegative heteroatom, and thus it forms a poor cation, so, the E1 path is out. The bases used in this reaction are usually weak, ruling out the E1cB, leaving the E2 or the Ei as the major path. For some examples of this reaction see Section 8.5, NuL Reacting with ⩾C—L, and also see Section 7.8e, Chromium Species as Oxidants.

Sink ⩾C—L alkene formation — the E1/E2/E1cB spectrum

The factors that decide which elimination path occurs are listed in Figure 7.4. For the E1 and the E1cB, solvation of the ionic intermediate is very important. The E1 and E1cB should be considered the two extremes of a spectrum of elimination mechanisms in which the E2 is in the center.

E1	E2	E1cB
stepwise via carbocation	simultaneous bimolecular	stepwise via carbanion

	Variable	
Increased C—L bond breaking	Cleavage	Increased C—H bond breaking
Strongly acidic media	Media	Strongly basic media
More stable carbocations	Intermediate	More stable carbanions
Better leaving groups	Leaving groups	Poorer leaving groups
Less acidic C—H	C-H acidity	More acidic C—H

Figure 7.4 The E1/E2/E1cB spectrum.

Mechanisms occur that do not cleanly fit the label of E1, E2, or E1cB. If the reaction is partway between E2 and E1, it is called E1-like, for the C—L bond is broken to a greater extent than the C—H bond, and a partial plus builds up on the leaving group carbon; this adjacent partial plus makes the C—H bond more acidic and therefore more easily broken. If the reaction is partway between E2 and E1cB, it is called E1cB-like, for the C—H bond is broken to a greater extent than the C-L bond, and a partial minus builds up on the carbon adjacent to the leaving group; this adjacent partial minus aids in pushing off the leaving group.

The different elimination paths often produce different alkene constitutional isomers as products (*regiochemistry*). For the E2 reaction the C—H bond and the C—L bond must lie in the same plane, preferably *anti* to each other for steric reasons. The protons, shown in bold, on either carbon adjacent to the leaving group can be in proper alignment for reaction; thus different products are produced (the *cis* product is also formed).

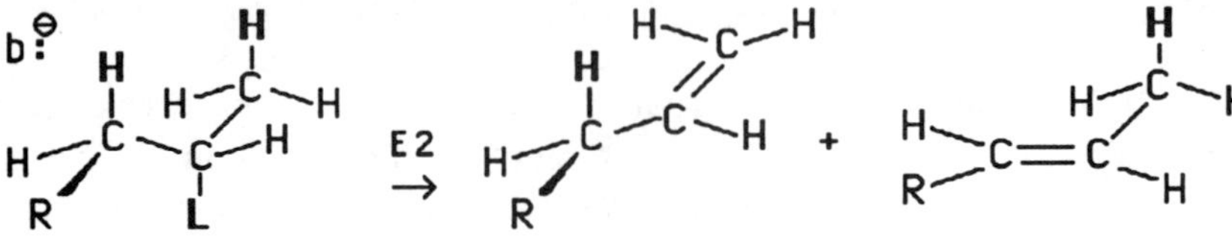

The regiochemistry of the elimination depends on the type of elimination process that occurs. **The E1 process favors the formation of the more substituted alkene** because reversible protonation of the double bond occurs and creates an equilibrium mixture that favors the more stable product. **The E2 regiochemistry is controlled by the need to minimize steric interactions in the transition state**; the size of the base is important because one proton may be more accessible than another. **The E1cB regiochemistry is determined by the loss of the most acidic proton.**

In addition to the formation of different regioisomers, elimination reactions can produce different stereoisomers, for example, *cis* and *trans* alkenes. Since the *trans* isomer is usually of lower energy because of steric reasons, it usually predominates over the *cis* isomer in the product mixture. The same factors that determined the regiochemistry also influence the stereochemistry.

E1 example (good L and cation, forms most substituted alkene):

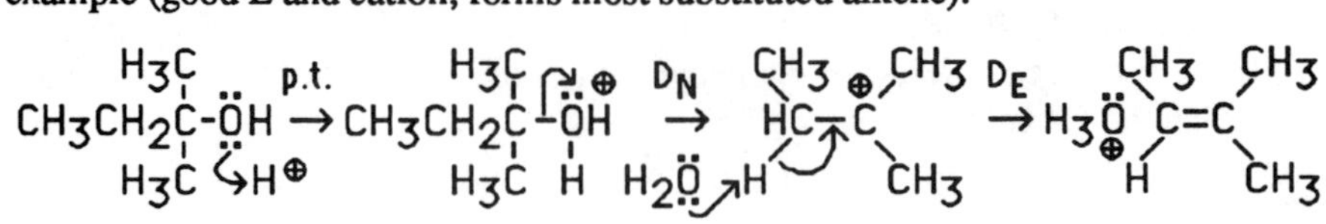

E2 example (concerted):

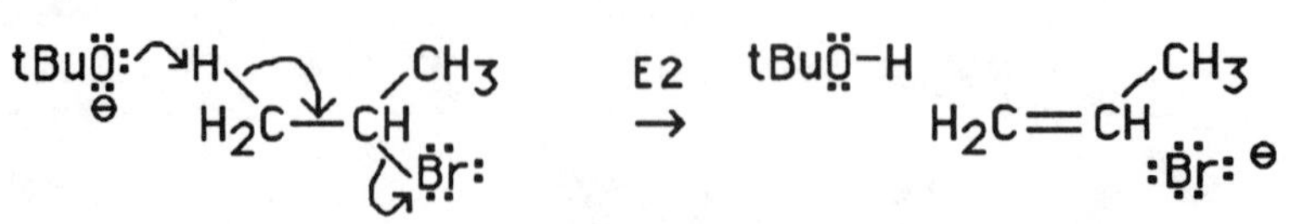

E1cB example (ewg makes C—H acidic; anion needed to kick out poor L):

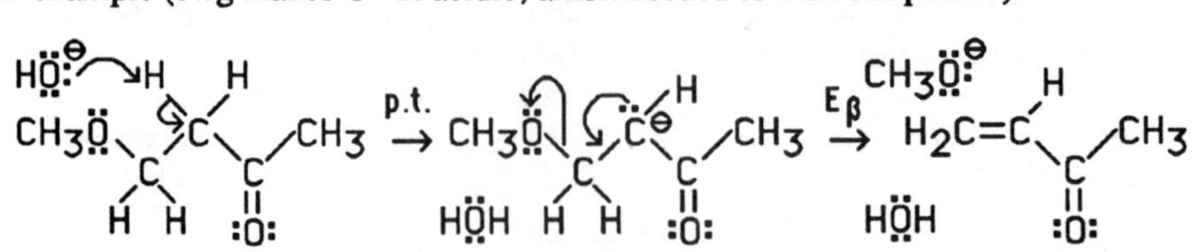

Heteroatom Variations

Anionic eliminations of the following type will also be considered as belonging to the E1cB class of eliminations. Since protons on heteroatoms are usually rather acidic due to the electronegativity of the heteroatom, the first step, deprotonation, can occur with a rather weak base. The second step is beta elimination from an anion, E_β, the reverse of nucleophile addition to a polarized multiple bond.

Hetero E1cB example (see Figure 7.10 for an energy diagram):

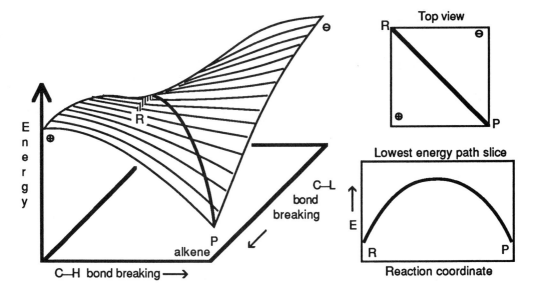

Likewise a lone pair on a heteroatom can aid the loss of the leaving group in the E1 process. The following example shows first the protonation of the leaving group followed by the lone-pair-assisted E1, path E_β followed by proton transfer.

A hetero E2 example was discussed in Section 6.4, Variations on a Theme.

The elimination energy surface

The elimination spectrum can be visualized as an energy surface (Figure 7.5), where the front horizontal axis is the degree of C—H bond breaking, the axis going from back to front is the degree of C—L bond breaking, and the vertical axis is energy. Reactants are in the back left corner at point R; the back right corner corresponds to only C—H bond breaking and thus is the carbanion; the front left corner corresponds to only C—L bond breaking and therefore is the carbocation; the front right corner, P, is the product alkene. The lowest-energy path is the dark line from R to P (shaded if behind the surface).

Figure 7.5 The E2 energy surface.

If both carbocation and carbanion are poor, the two corners corresponding to them will be high in energy. The reaction will go by the E2 process in which C—H bond breaking and C—L bond breaking occur to an equal extent, and the lowest energy path will follow the diagonal through the saddle resulting from the high-energy carbocation and carbanion. If the carbocation is rather stable, the corner on the diagram corresponding to the carbocation will be lowered in energy, and now the lowest energy path will go via the carbocation, the E1 elimination (Figure 7.6).

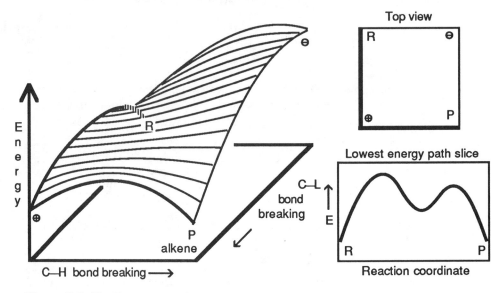

Figure 7.6 The E1 energy surface.

If, on the other hand, the carbanion is reasonably stable and the carbocation is not, the corner on the diagram corresponding to the carbanion will be lowered in energy instead, and now the lowest energy path will go via the carbanion, thus the E1cB elimination (Figure 7.7).

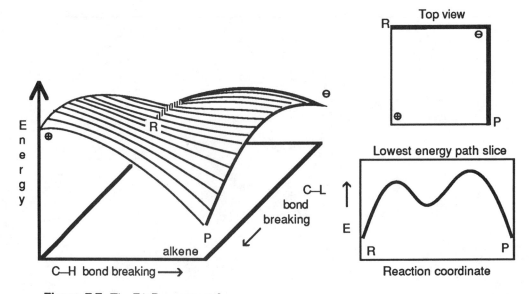

Figure 7.7 The E1cB energy surface.

Finally, if the carbocation is more stable than the carbanion but neither are very good, the two "mountains" on the diagram may be of different heights and will have a "pass'" that is off to one side, not exactly on the diagonal (Figure 7.8). In this case, the lowest energy path to product takes a curve through that pass. More C—L bond breaking than C—H bond breaking occurs in going to the transition state "pass," a partial plus builds up on the leaving group carbon at the transition state, and the reaction is called E1-like. Had the carbocation been slightly more stable, the lowest energy path would have fallen into the carbocation energy well, and the process would be E1, not E1-like.

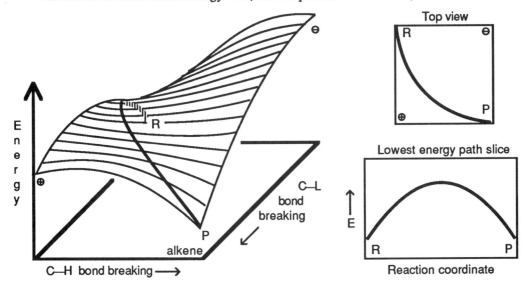

Figure 7.8 The E1-like energy surface.

In a similar manner, if the carbanion is more stable than the carbocation, the "pass" will be on the carbanion side and more C—H bond breaking than C—L bond breaking occurs in going to the transition state. A partial minus charge builds up on the carbon adjacent to the leaving group carbon at the transition state, helping to push out the leaving group, and the reaction is E1cB-like.

Also for the E1/E2/E1cB spectrum, similar to what was seen in the Sn1/Sn2 spectrum, a better leaving group will lower the front edge, making the overall process more favorable and tilting the surface toward the E1 path. If the H is more acidic or the base stronger (making deprotonation easier), the right-hand edge will be lowered, again making the elimination more favorable but this time tilting the surface toward the E1cB.

7.2k Lone Pair Sources Reacting with C=Y and R—C≡N

C=Y as sinks

These reactions are very common (see Figures 6.12 and 6.13 for the interrelationship of the available paths in acidic and in basic media). **In basic media the attack of the nucleophile on a polarized multiple bond forms an anion that in the workup of the reaction is usually protonated (path combination Ad$_N$2).**

In protic solvents if the nucleophilic attack forms a stronger base, a following irreversible proton transfer step may make the overall reaction favorable. This very favorable step is the driving force for the reaction. A look at two energy diagrams for this reaction that differ only in the proton transfer step is instructive. The overall $\Delta G°$ value is known from experiment to be -2 kcal/mol (-8.4 kJ/mol). By estimating the pK_{abH} of the intermediate at about 13, we can get a value for the proton transfer K_{eq}, which gives an approximate $\Delta G°$ value of - 5 kcal/mol (-21 kJ/mol) for the proton transfer. By subtraction, we get the $\Delta G°$ for the formation of the intermediate to be about +3 kcal/mol (+12.6 kJ/mol), and can construct an energy diagram, shown in Figure 7.9.

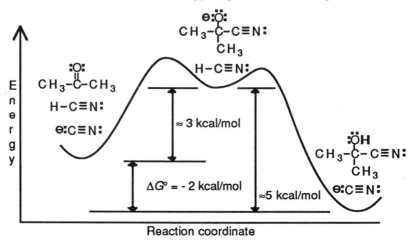

Figure 7.9 The energy diagram for cyanohydrin formation.

If we change the acid, the first part of the diagram remains the same since the acid is just a spectator in the first step. With a weaker acid the proton transfer step is not favorable, and the equilibrium favors starting materials (Figure 7.10).

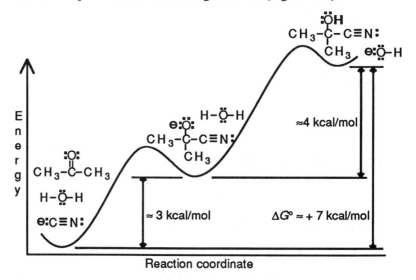

Figure 7.10 The energy diagram for cyanohydrin cleavage.

In acidic media, polarized multiple bonds often undergo acid catalysis, and **a common mode of addition is the Ad_E2.** Loss of the proton from the nucleophile to solvent regenerates the neutral compound. Two common examples of this easily reversible reaction are the formation of *hydrates* (NuH is H_2O) and *hemiketals* (NuH is ROH). Usually this reaction favors reactants.

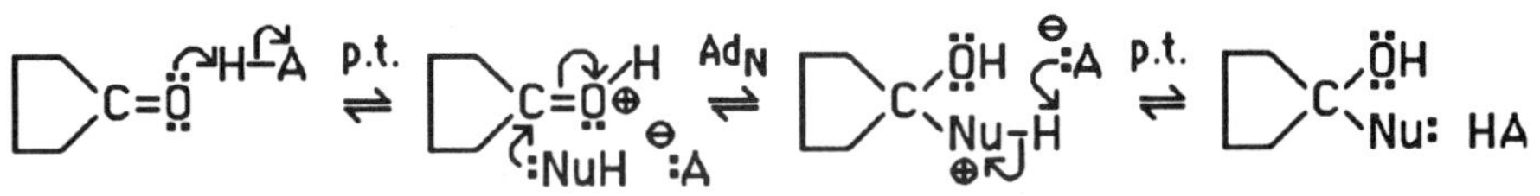

If the following equilibria are **driven by the removal of water**, a follow-up reaction can occur in acid in which the OH group gets protonated (path p.t.), then is lost as water (path D_N), and the carbocation formed either is trapped or loses a proton. The identity of the nucleophile is used to determine the route that the reaction will take.

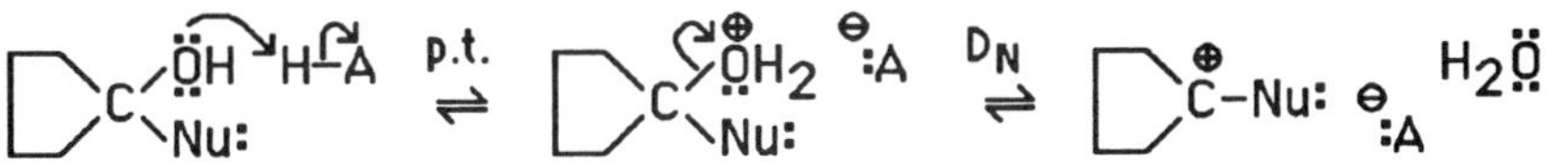

The carbocation can add a second nucleophile (path A_N followed by path p.t.).
This is common for alcohol sources (NuH is ROH), in acidic media, with removal of water to form *ketals*.

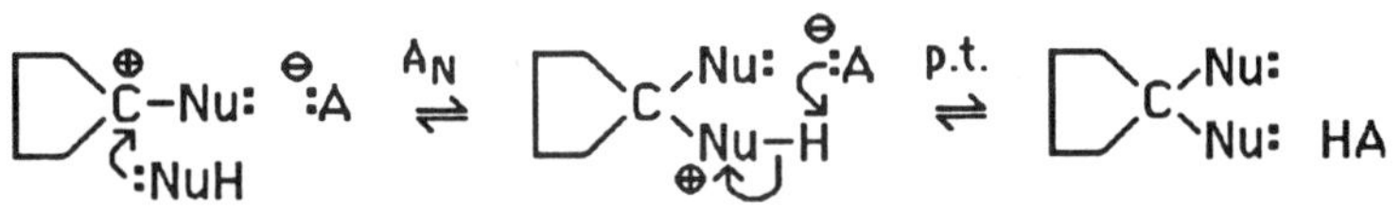

Alternatively the C—H bond next to the carbocation can be deprotonated (path D_E). This is common for secondary amine sources (NuH is R_2NH), in acidic media, with removal of water to form *enamines*.

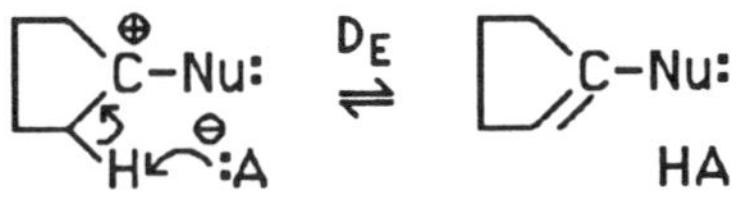

However, if there were two H's on the original nucleophile (for example RNH_2), we could now remove that second H on the nucleophile to yield a stronger heteroatom–carbon double bond (path D_E). This is common for primary amine sources (NuH_2 is RNH_2), in acidic media, with removal of water to form *imines*. If the NuH_2 is $HONH_2$ the reaction forms an *oxime*.

R—C≡Y as sinks

The two pi bonds in a triple bond are perpendicular to each other, so they act independently. In the following example after attack of the Nu by path Ad_N, a follow-up protonation occurs from the protic solvent, then tautomerization to the more stable amide.

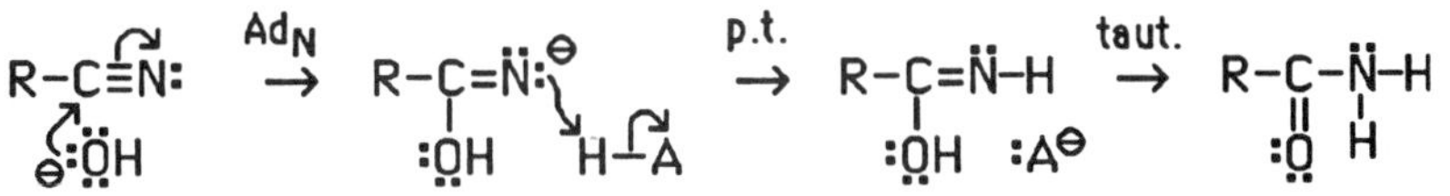

7.2l   Metal Hydrides Reacting with C=Y and R-C≡N

C=Y as sinks

Hydride is added to the multiple bond (path Ad$_N$) as expected. Usually the metal cation complexes with the Y heteroatom; nucleophilic attack on the polarized multiple bond will then yield a product that is stabilized by ion pairing.

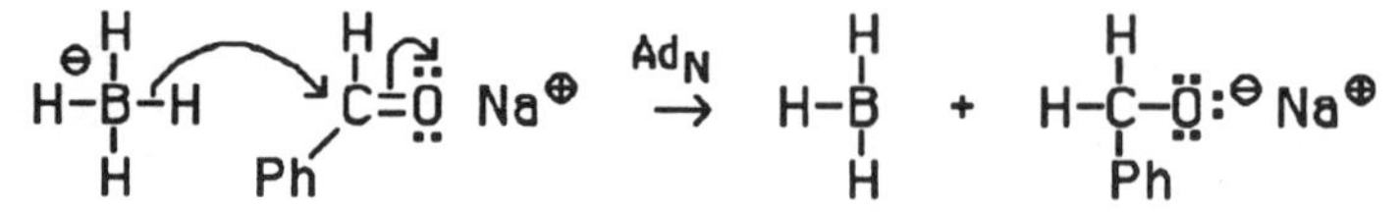

R—C≡N as sinks

With LiAlH$_4$ a second attack on the nitrile occurs. The product of the first addition can form a Lewis salt, which (before or after acting as a hydride source) catalyzes the second addition. The product, after the acidic water in the workup protonates the nitrogen and dissociates it from the metal, is the amine. This double addition can be prevented by the use of a less reactive aluminum hydride or by the use of diisobutylaluminum hydride, [(CH$_3$)$_2$CHCH$_2$]$_2$AlH.

Double addition example (path Ad$_N$, then A$_N$, followed by the Lewis salt serving as a hydride source, then Ad$_N$ again):

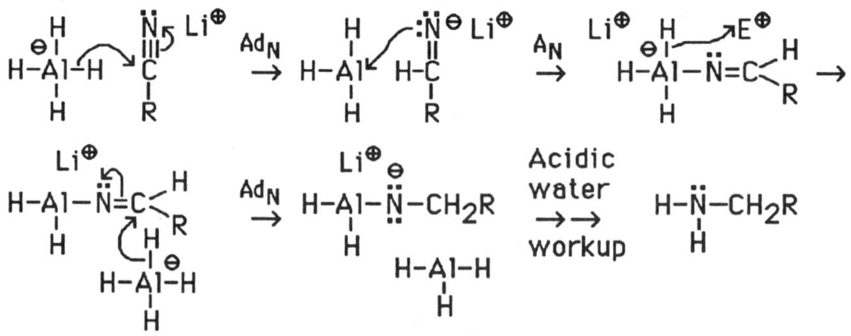

7.2m   Organometallics Reacting with C=Y and R—C≡N

Organometallics react with this sink by addition to the multiple bond (path Ad$_N$). The more covalent, less reactive organometallics, like R$_2$Cd, react very slowly with almost all of these sinks, whereas organomagnesiums, RMgX, and organolithiums react quickly. Complexation of the metal ion to the Y heteroatom catalyzes this reaction. Organometallics react much faster as nucleophiles than as bases (carbon–acid, carbon–base proton transfer is slow).

C=Y example:

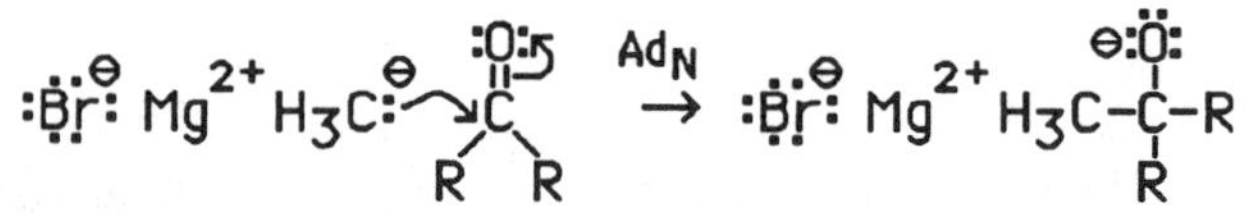

C≡Y example:

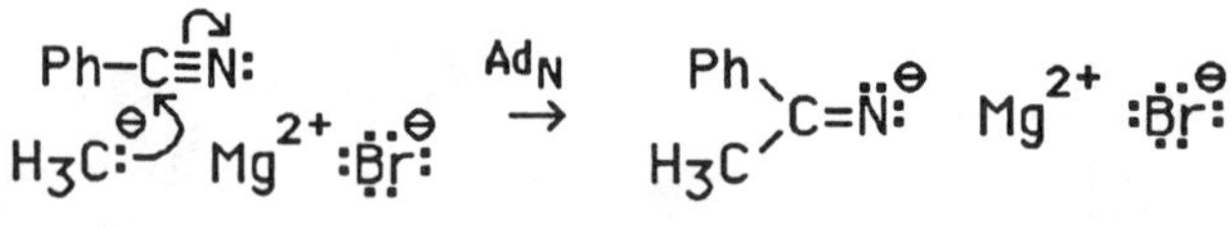

7.2n Allylic Sources Reacting with C=Y and R-C≡N

C=Y as sink, aldol reaction

Under equilibrium conditions (thermodynamic control), the allylic source adds to the polarized multiple bond (path Ad_N). The source can serve as a base and deprotonate the sink, creating a mixture of sources and sinks thus a messy statistical mixture of products. Clean products result if the source is just the deprotonated sink or if the sink has no acidic protons. With ketones, the equilibrium of the attack step favors the starting materials, and therefore the reaction goes to completion only if driven by elimination. In the following example, the source is the deprotonated sink.

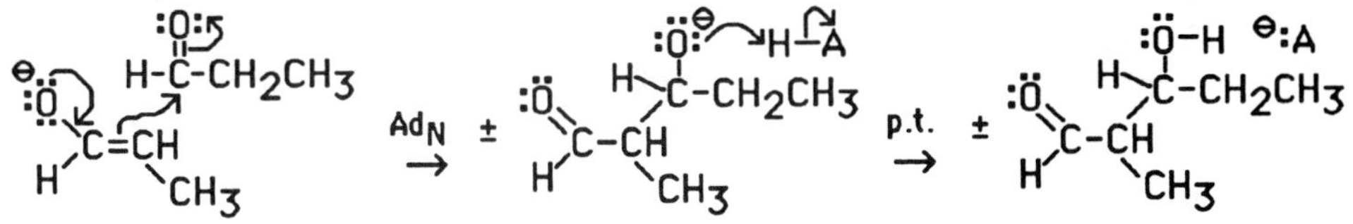

If the reaction is heated, elimination usually follows via the E1cB path combination.

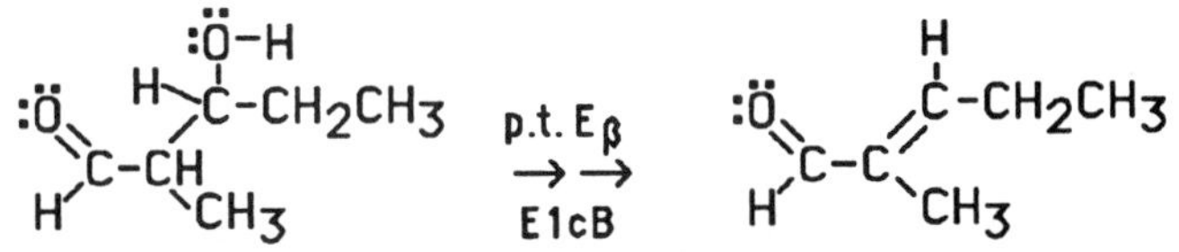

Non-enolizable sink example (with follow-up elimination):

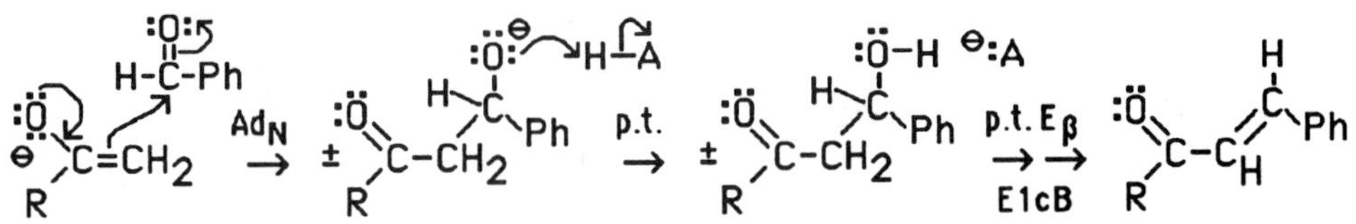

Under kinetic control the aldol reaction is very stereospecific. The lithium enolate is generated in an aprotic solvent, and then the carbonyl compound is added. The reaction proceeds via the metal-chelated path 6e. The minimization of steric effects in the chair transition state and the stereochemistry of the enolate determine the stereochemistry of the product.

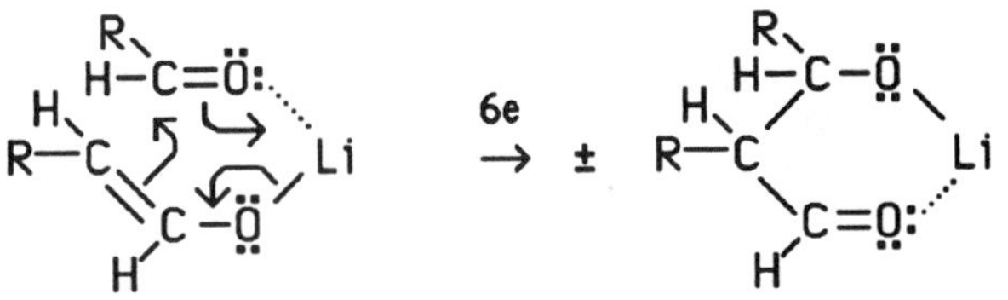

R—C≡Y as sinks

The following example is an intramolecular reaction similar to the aldol reaction. A deprotonated nitrile acts as a nucleophile and adds via an Ad_N path to another nitrile.

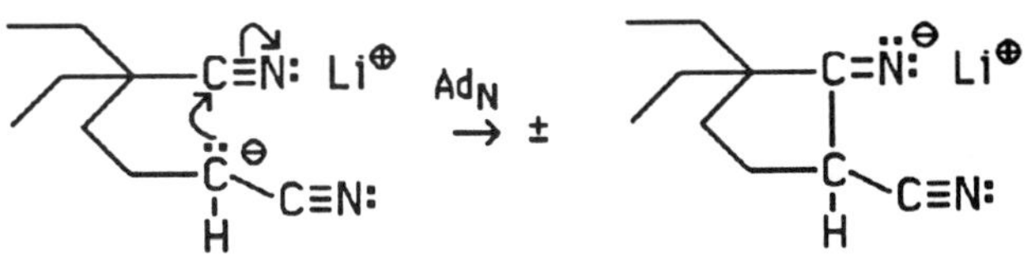

7.2o   Bases Reacting with C=Y and R—C≡N, Enolization

The polarized multiple bond makes adjacent CH bonds acidic; deprotonation generates an allylic electron source.

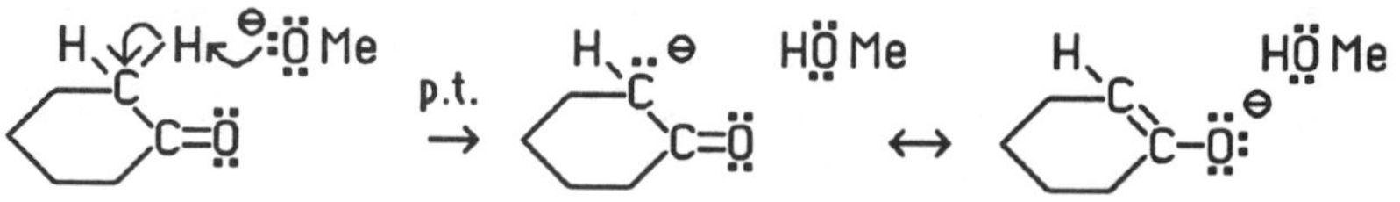

7.2p Conjugate Addition by Lone Pair Sources to C=C—ewg

When an electronegative carbon atom replaces the electronegative Y atom, the AdN process is called *conjugate addition*. Competition between direct addition to the ewg and conjugate addition is discussed in Section 8.6 (the conjugate addition site is softer). This reaction goes by the AdN2, the reverse of the E1cB elimination reaction. (See Section 7.3 and Figure 7.11 for the complete interrelationships of addition and elimination mechanisms.) A rapid, irreversible proton transfer from the alcohol solvent is the driving force for this example.

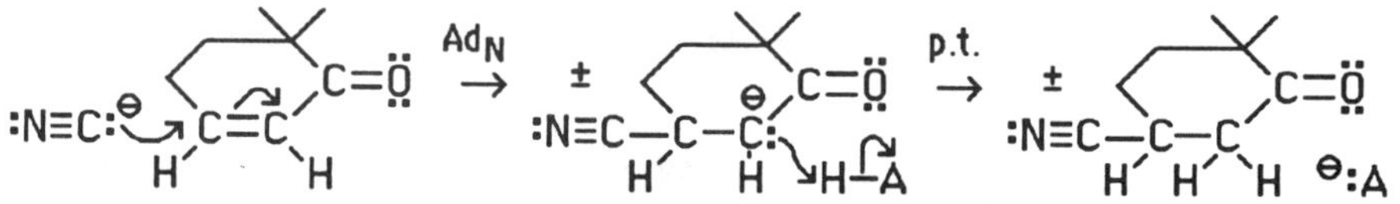

7.2q   Conjugate Addition by Metal Hydrides to C=C—ewg

Aluminum hydride, $AlH_4^\ominus$, is relatively hard; therefore conjugate addition is usually only a minor side reaction. The hydride adds to the electron-withdrawing group and not the double bond. Borohydride, $BH_4^\ominus$, since it is softer, has a much greater tendency toward conjugate addition but is often difficult to predict (see Section 8.6). Commonly conjugate addition is followed by a second attack on the electron-withdrawing group to produce a doubly reduced product.

7.2r   Conjugate Addition by Organometallics to C=C—ewg

Since the $\partial+$ at the conjugate position is less, the carbon atom is softer, and softer organometallics tend to favor conjugate attack (see Section 8.6).
Example (the cuprate is a soft organometallic):

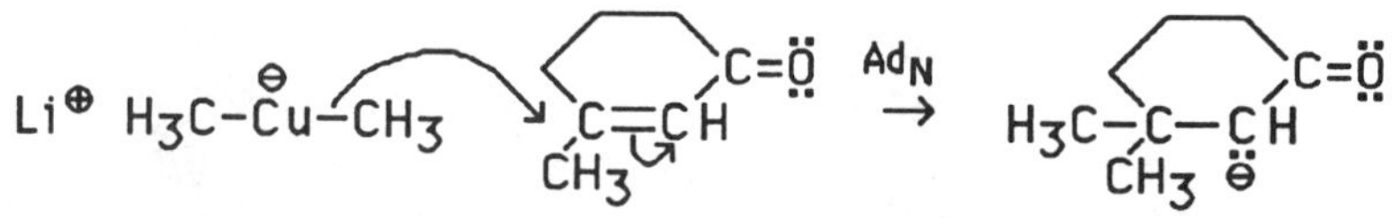

7.2s   Conjugate Addition by Allylic Sources to C=C—ewg

The reaction is AdN2 addition to the polarized multiple bond (AdN, then p.t.). This reaction is easy to predict because the most stable product is formed under these equilibrium conditions. HSAB theory predicts that the combination of the two softest sites is favored.

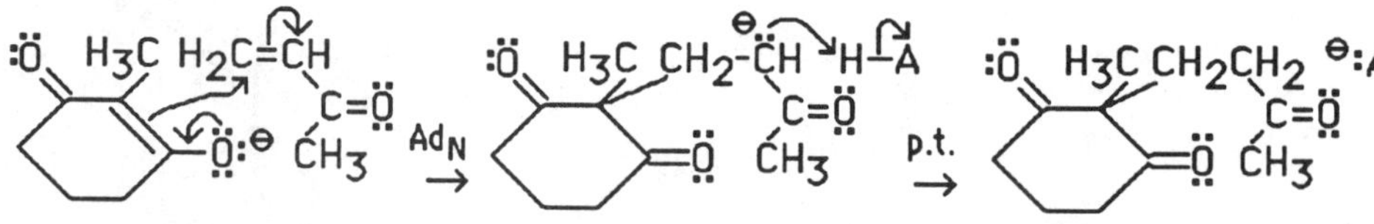

7.2t Bases Reacting with C=C−ewg, Extended Enolates

These deprotonations generate the vinylogous allylic sources. Deprotonation can also occur on the CH₃ next to the carbonyl in the following example. The choice of which hydrogen to remove is covered in Section 8.8, Kinetic vs. Thermodynamic Enolate Formation.

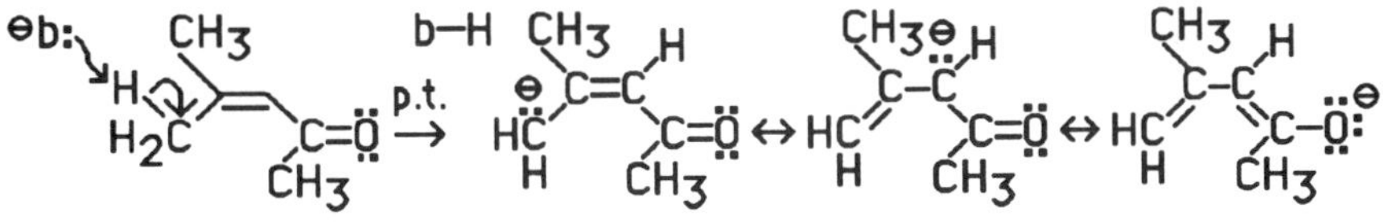

7.2u Lone Pair Sources Reacting with L−C=O

The most common representatives of the L−C=Y class of electron sinks are the carboxyl derivatives with Y equal to oxygen. **In basic media there is only one pathway: the addition–elimination path, path Ad$_N$ + E$_\beta$** (see Section 6.2). The leaving group should be a more stable anion than the nucleophile, or the reaction will reverse at the tetrahedral intermediate. A follow-up reaction of a second addition to the polarized multiple bond occasionally occurs. With lone pair sources a second addition is rare because the nucleophile is usually a relatively stable species; the second tetrahedral intermediate tends to kick it back out (see Section 8.2).

Basic media addition–elimination example (path Ad$_N$ + E$_\beta$):

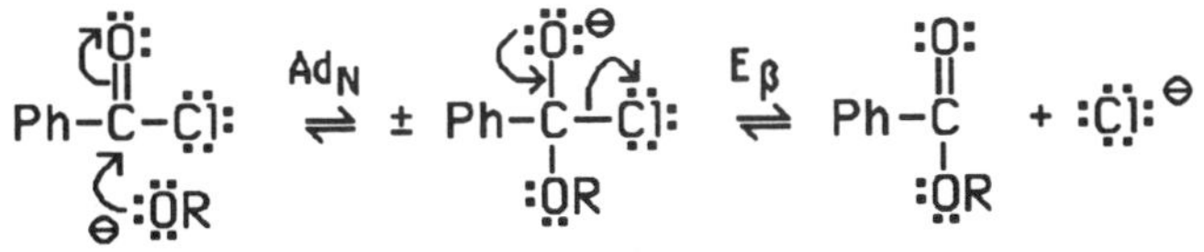

In acidic media, the carbonyl lone pair can get protonated to produce a much better electron sink. This better electron sink can be attacked by weaker nucleophiles. Then proton transfer to the leaving group makes it a better leaving group. The reaction is reversible; the position of equilibrium is determined by mass balance.

Acidic media example (path p.t., Ad$_N$, p.t., p.t., E$_\beta$, and p.t.):

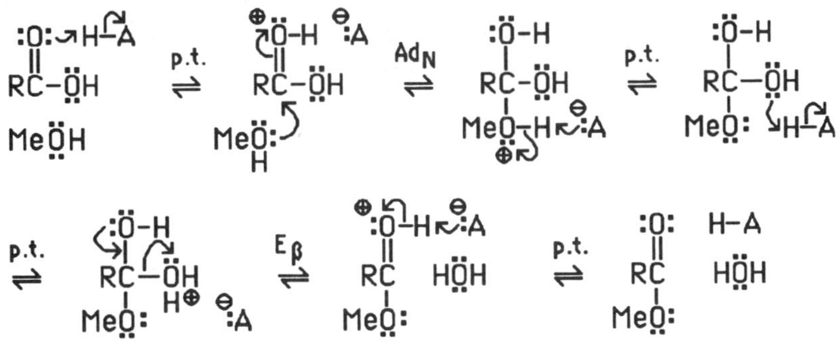

7.2v Metal Hydrides Reacting with L−C=O

The product of the initial reduction is most often more reactive than the starting material, therefore a second addition is very common (Ad$_N$, then E$_\beta$, then Ad$_N$). An Ad$_N$ + E$_\beta$ product can be obtained with acyl halides and one equivalent of a less reactive metal hydride at low temperature. Borohydrides selectively react with aldehydes and ketones in the presence of less reactive esters and amides.

Second-addition example (path Ad_N then E_β, Lewis salt formation with the aluminium species to create the nucleophilic aluminate, followed by a second Ad_N):

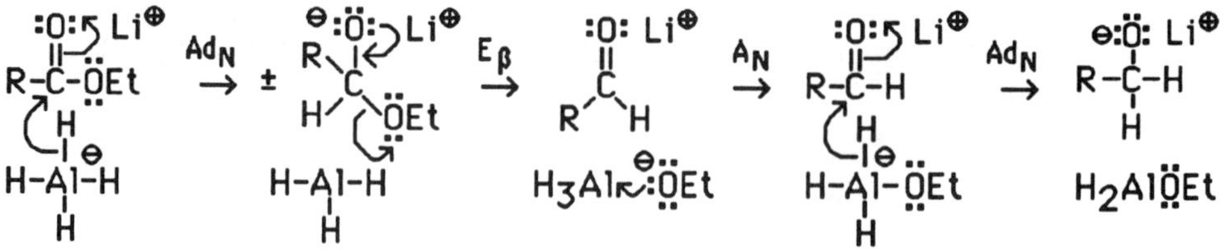

If the L is poor as in amides, Lewis salt formation allows an aluminum oxide to be kicked out instead of the amine (path Ad_N, then A_N, followed by the Lewis salt serving as a hydride source, then E_β, and then Ad_N).

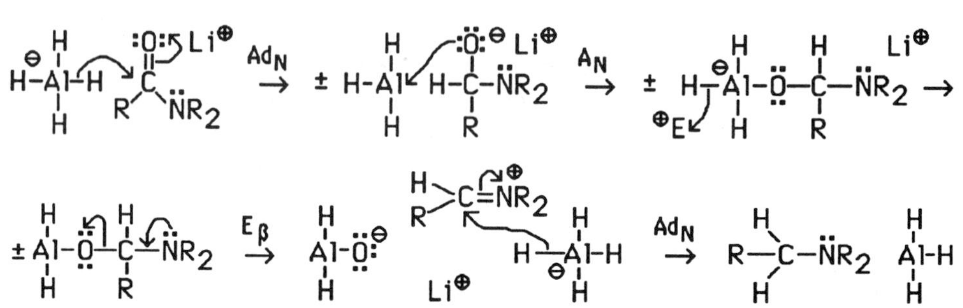

7.2w   Organometallics Reacting with L—C=O

Organometallics follow the addition–elimination pathway ($Ad_N + E_\beta$) characteristic of this sink. However, the products of the addition–elimination path often undergo a second attack by the organometallic because the $Ad_N + E_\beta$ product is often more reactive than the original substrate The reactivity ranking of the electron sinks and sources is important (see Section 8.2).

Second-addition example (path Ad_N, then E_β, then Ad_N again):

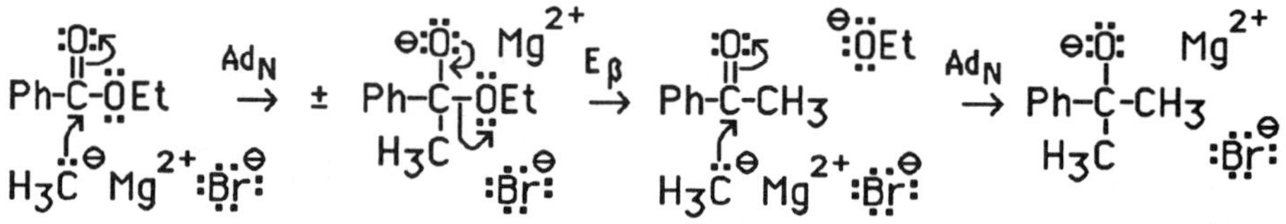

Mono-addition example (path Ad_N, then E_β):

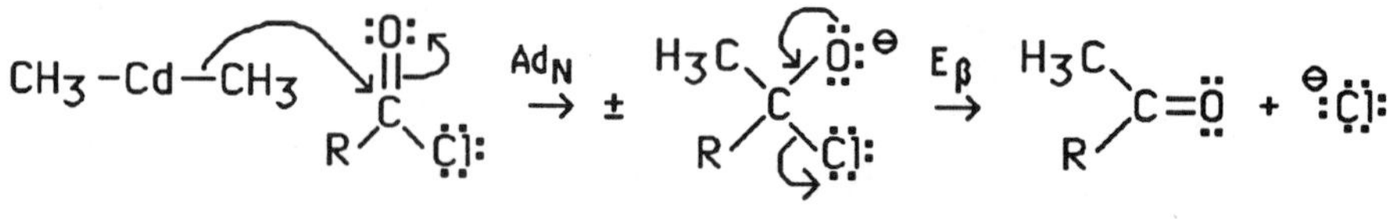

7.2x   Allylic Sources Reacting with L—C=O, Acylation

Carboxyl derivatives such as esters are attacked almost exclusively by the carbon of the allylic source via path Ad_N, then E_β (Section 8.4). As in Section 7.2n, the enolate source should be the deprotonated sink or the sink should not be enolizable. A final proton transfer to give a stabilized anion is frequently the driving force for the reaction. Example:

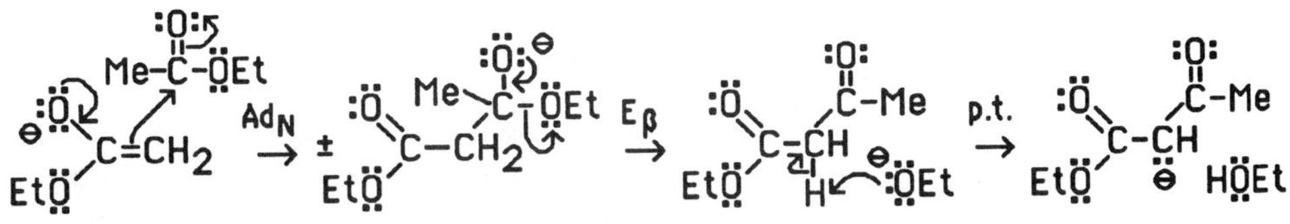

7.2y Bases Reacting with L—C=O, Enolization

These enolates are usually generated by deprotonation below 0° C. At higher temperatures the enolate ejects the leaving group via path E_β to form a reactive heterocumulene (usually an undesirable side reaction).
Enolization example:

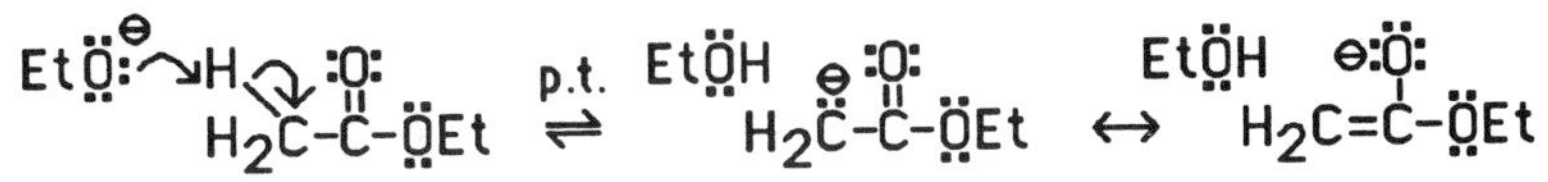

7.3   Electrophilic Additions

7.3a   Lone Pairs as Electron Sources

Empty _P_ as sink

The trapping of a cation by a nucleophile, A_N, can be considered the electrophilic addition of a cation to a lone pair.

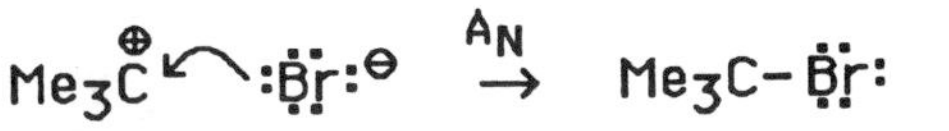

The electrophilic addition of a Lewis acid to a lone pair to form a Lewis salt is an isoelectronic reaction.
Example, borate ester hydrolysis:

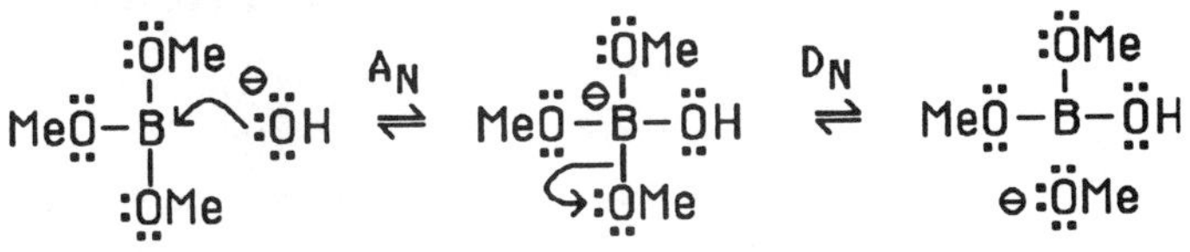

7.3b   Alkenes, Dienes, and Alkynes as Electron Sources

Empty _P_ as sink

Since the product of the carbocation addition to an alkene via path A_E is also a carbocation that can rearrange and/or attack another alkene molecule (polymerization), unwanted product mixtures can result.

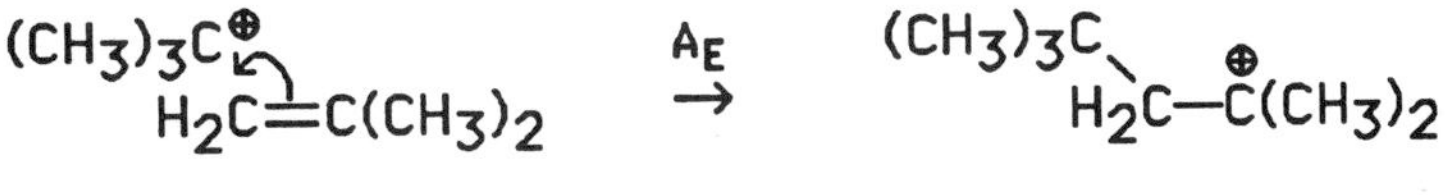

Borane, $BH_3 \rightleftharpoons B_2H_6$, isoelectronic with carbocations, adds exclusively *syn* to multiple bonds. Initially a pi-complex is believed to form, followed by rapid hydride donation to the largest ∂^+ atom (4e path). The reaction repeats to produce trialkylboranes, R_3B.

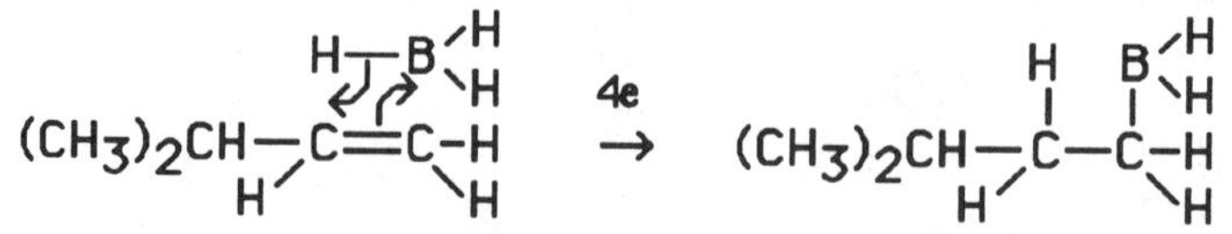

H – A as sink – electrophilic addition of acids to alkenes

There are two mechanisms by which acids add to alkenes, the Ad_E2 and the Ad_E3 processes. The reaction is run in the dark to minimize free radical side reactions. In the Ad_E2, the proton adds by pathway A_E to produce the more stable carbocation intermediate; in a second step the cation is trapped by a nucleophile, path A_N. The reverse reaction is the E1 elimination pathway.
Ad_E2 example:

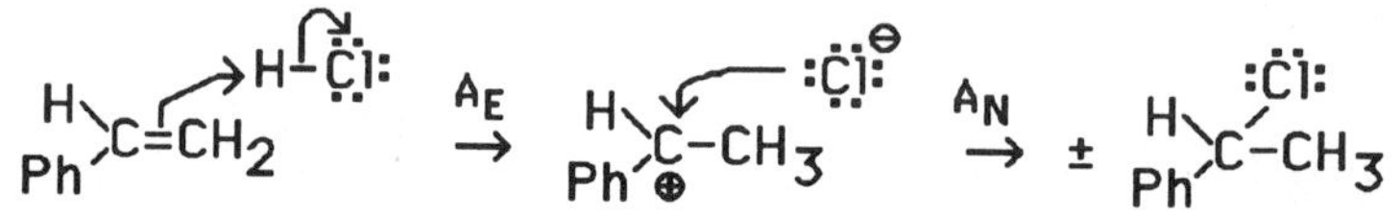

The Ad_E3 occurs when the cation is not as stable. Instead of proton transfer, a pi-complex forms, which polarizes and activates the double bond. The nucleophile attacks the largest partial plus of the complex and gives the same Markovnikov orientation of product as the Ad_E2 process. The pi-complex usually blocks one face of the double bond, therefore the nucleophile adds to the opposite face, giving overall *anti* addition as the predominant process. The reverse reaction is the E2 elimination pathway.
Ad_E3 Example:

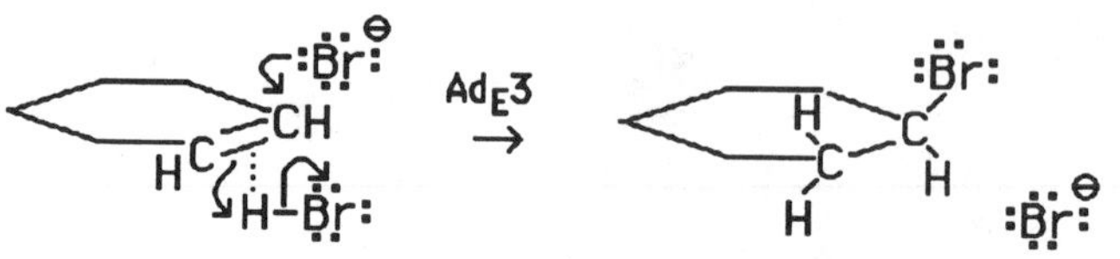

The interrelationship of the addition and elimination processes

Addition and elimination reactions, being complementary processes, occur on essentially the same energy surface. Conditions that lower the alkene corner favor elimination, whereas conditions that raise it or lower the addition product corner favor addition. A top view of the general addition/elimination energy surface is shown in Figure 7.11.

The same factors that folded and tilted the surface, discussed in the elimination section, have the same effect when the surface is tilted to favor addition. If the cation and anion are relatively unstable, the surface folds and the lowest energy path is the Ad_E3 process. If an ewg stabilizes the anion, the Ad_N2 process is the lowest energy route. If, however, the cation is stabilized instead, the Ad_E2 is energetically the best.

Electrophilic addition of H – A to alkynes

The two perpendicular pi bonds act independently of one another. Alkynes commonly undergo Markovnikov *anti* addition of acids by the Ad_E3 process because the

Ad$_E$2 would go via a less stable vinyl cation intermediate. Slightly more vigorous conditions will favor the double-addition product.

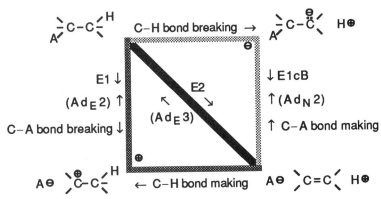

Figure 7.11 The top view of the addition/elimination surface. The addition reactions are in parenthesis.

Electrophilic addition can be followed by tautomerization.

Electrophilic addition of H−A to dienes C=C−C=C

An electrophile will attack a diene to produce the reasonably stable allylic cation. The partial charge on each end of the allylic system will be different if the substitution at each end of the allylic unit is different. Since the attraction of unlike charges contributes greatly to the bringing of nucleophile and electrophile together (hard–hard), **the larger the partial plus is on an atom, the greater a negatively charged nucleophile is attracted to it.**

For the simplest diene, 1,3-butadiene, electrophile attack creates an allylic cation where the greatest partial plus is on the end of the allylic system bearing the methyl donor group.

If the rate of nucleophile attack governs the product distribution (the nucleophilic attack is not reversible at the given temperature), the major product will be the 1,2-

addition product that results from Nu attack at the greatest partial plus (kinetic control, Section 2.2). However, if the addition of the nucleophile is reversible (higher temperature or the Nu is a good L), then the 1,4-addition product will be formed because the more substituted double bond is the more stable product (thermodynamic control, Section 2.2). Suspect kinetic control if the reaction temperature is significantly below 0°C.
Diene example (Ad$_E$2):

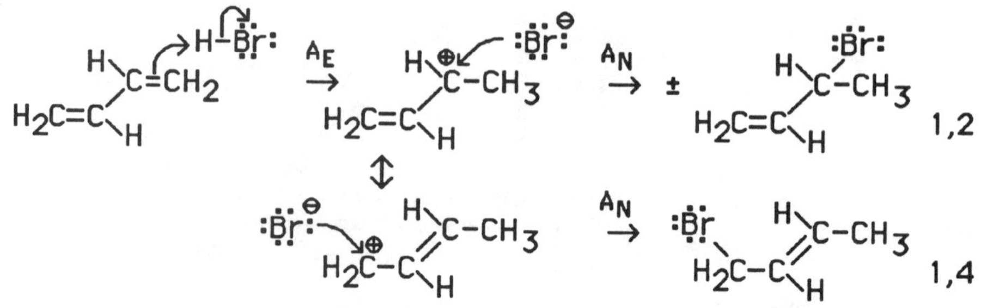

Y—L as sink — electrophilic addition of Y—L to alkenes

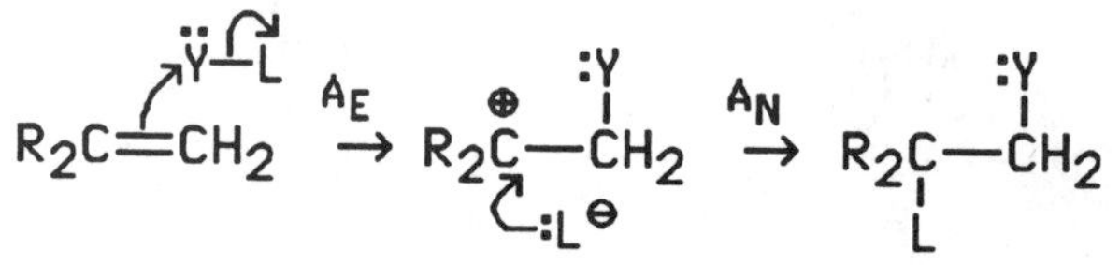

This reaction and the electrophilic addition of acids compose almost all of the ionic reactions of alkenes as electron sources. When Y bears a lone pair, the intermediate carbocation can be further stabilized by sharing the lone pair electrons with the cationic center in a bridge to form a three-membered ring, resonance form B in the following example. Although we have drawn the Ad$_E$2 process (A$_E$, followed by A$_N$) in this section, it would often be just as correct to use the Nu—L path to get to the bridged ion directly followed by path A$_N$ (see Nu—L, Section 7.5). An example is bromination, in which the leaving group traps the cation (reaction run in non-nucleophilic solvent).

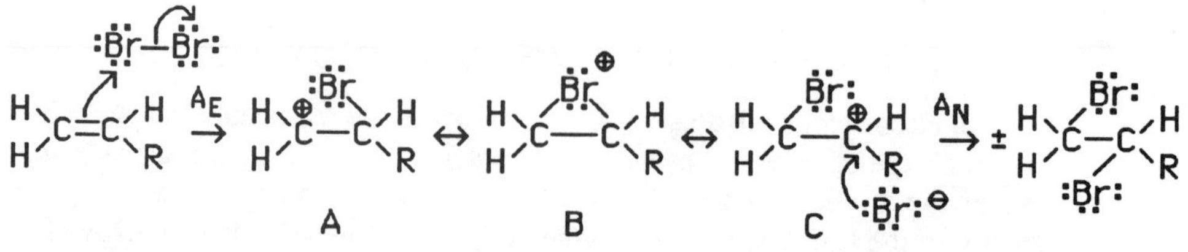

Bridging gives all atoms an octet, and some energy is released in forming a new bond, but this is at the expense of creating ring strain and putting the positive charge partially on the heteroatom. Highly electronegative elements like fluorine and oxygen are relatively poor at bridging. Less electronegative elements have a greater tendency to share their lone pairs and bridge. Stabilization of the carbocation by electron-donor groups favors the open form. Bridging is favored if the cation is otherwise poorly stabilized.

The second step in the electrophilic addition of Y—L to alkenes is the attack of a nucleophile on the cationic intermediate by the original leaving group or the solvent. The bridging Y blocks one face, so the nucleophile comes in from the opposite face, giving overall *anti* addition. The site of attack will depend on which resonance form, A or C, contributes more to the overall hybrid. The nucleophile will attack the carbon with the greater partial plus.

Solvent trap example (same reaction run in water solvent):

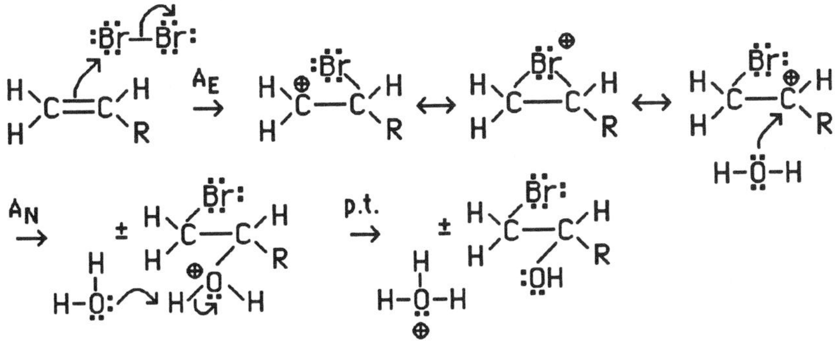

Electrophilic addition of Y—L to alkynes

Alkynes undergo the same set of reactions as alkenes but are slightly less reactive because the intermediates involved are less stable. For the Ad_E2 process, the vinyl cation intermediate formed is less stable than the alkyl carbocations formed when electrophiles attack alkenes. Addition to the vinyl cation produces a mixture of *syn* and *anti* addition. Stabilizing the vinyl cation by bridging is less favorable since the bridged ion is more strained and may have some antiaromatic character.

The Ad_E3 process competes with the Ad_E2 and can make the overall addition *anti*. The Ad_E2 process is seen when there is some additional stabilization of the vinyl cation. Double additions are frequent. Often the product of an electrophilic addition is less reactive than the original alkyne, and thus the reaction can be stopped before a second addition occurs if run with a equimolar amount of electrophile and alkyne.

Ad_E3 example (note the *anti* addition):

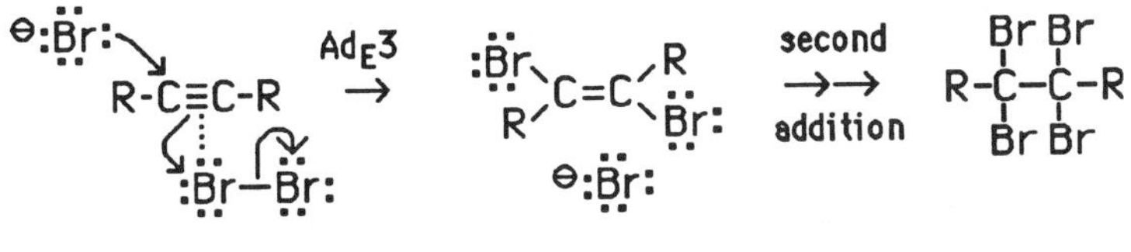

Electrophilic addition of Y—L to dienes

Since the allylic cation formed upon addition of an electrophile to an alkene is relatively stable, dienes add Y—L by the Ad_E2 process. Both 1,2 and 1,4 addition can occur. At higher temperatures, the more stable 1,4 product predominates.

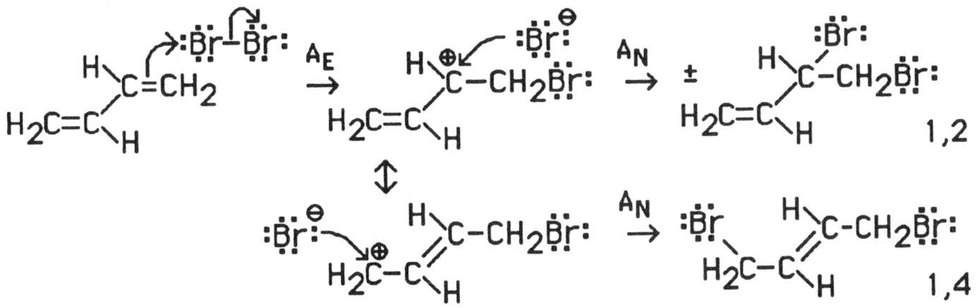

7.3c   Electrophilic Aromatic Substitution

There is only one pathway by which the aromatic electron source reacts. After the electrophile attacks (path A_E), the intermediate delocalized carbocation, the *sigma complex*, loses an electrophile, commonly a proton (path D_E) to restore the aromaticity to the ring.

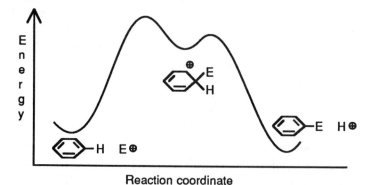

Usually the attacking electrophile is more reactive than a proton, and therefore the rate-determining step is electrophile attack. Proton loss from the sigma complex will be an easier process than loss of a reactive electrophile. Figure 7.12 shows a typical energy diagram for an electrophilic aromatic substitution.

Figure 7.12   A common energy diagram for electrophilic aromatic substitution.

Groups present on the ring before the electrophile attacks not only dictate the reactivity of the ring but also the position of attack. An electron-donor group attached to the ring can stabilize the sigma complex if the electrophile attacks either ortho or para to the group. In one resonance form of either sigma complex, the positive charge is adjacent to the electron donor. By stabilizing the sigma complex, the electron donors both increase the reactivity of the ring (by increasing the electron availability) and direct the incoming electrophile ortho or para. If the attached donor group is sterically larger than methyl, para attack will dominate.

Ortho attack:

Para attack:

If the group on the ring is an electron-withdrawing group, the reactivity of the ring drops far enough so that the reaction often fails. The least deactivated positions on the ring are meta to the electron-withdrawing group. The incoming electrophile is directed meta, going via the least destabilized sigma complex. In no resonance form of the sigma complex is the positive charge adjacent to the electron withdrawing group.

Meta attack:

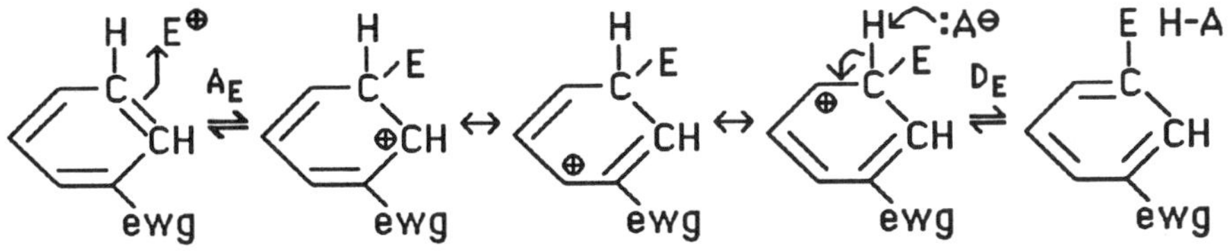

Occasionally the electrophile will attack the ring carbon that bears the group itself (ipso attack), but this site is definitely more hindered. If the group can form a cation equal to or more stable than the incoming electrophile, the group can depart from the sigma complex as a cation (path D_E) and thus be replaced by the electrophile.

Ipso attack:

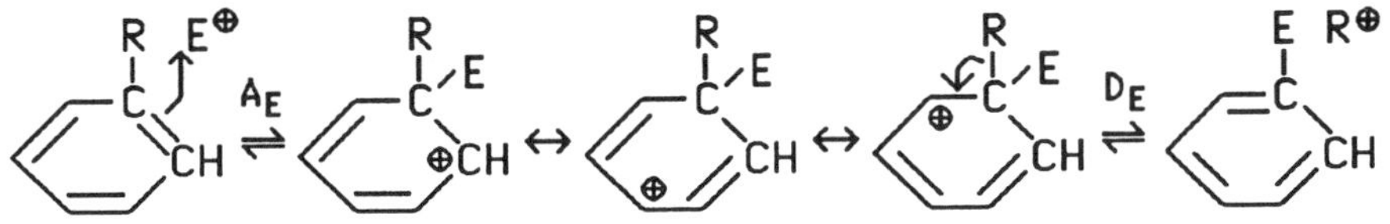

If more than one group is on the ring, the one that can stabilize the sigma complex the most will be the one that directs electrophile attack. The reaction proceeds through the most stable of the possible sigma complexes. For example, if two donors are on the ring, the electrophile attacks ortho or para to the best donor.

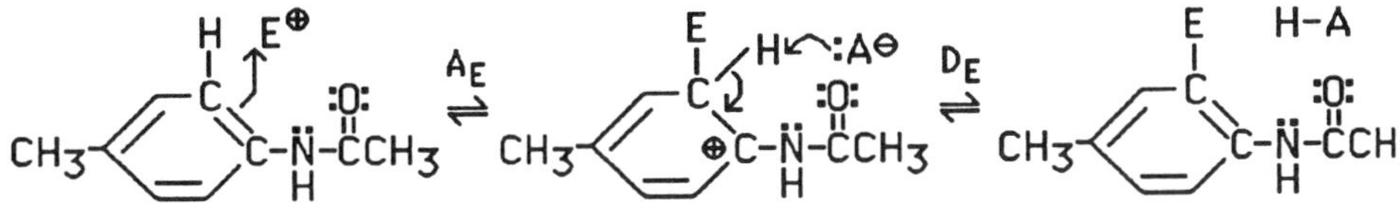

When two donors are meta to each other, attack at the position between them is extremely difficult because of steric hindrance.

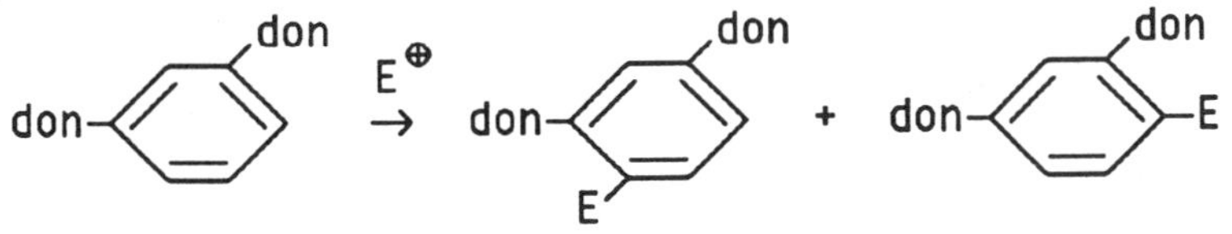

Other aromatics as electron sources

The stability of the sigma complex is the guiding principle when one attempts to determine the site of electrophilic attack on any aromatic ring. The more approximately equal energy resonance structures the carbocation has, generally the more stable it is. In the reactions of heteroaromatics, resonance forms for the sigma complex which require a heteroatom to have an incomplete octet are especially poor. Pyridine undergoes electrophilic attack meta to the nitrogen.

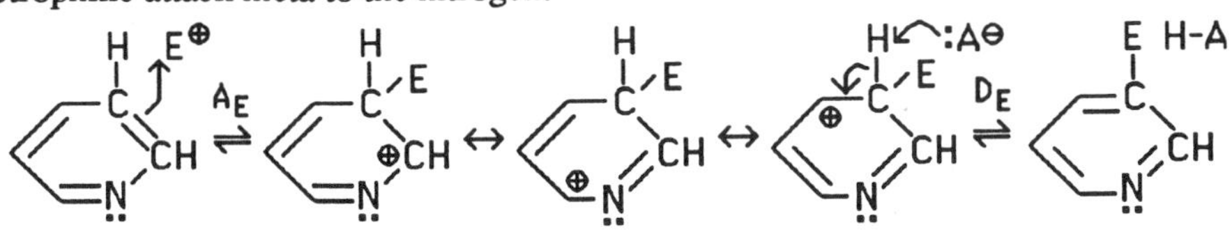

No attack at the ortho or para positions occurs.

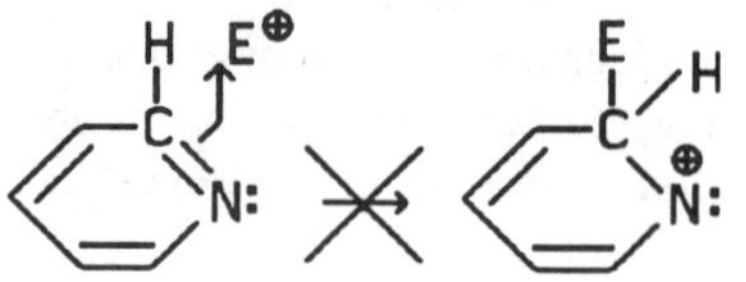

However, the attack of the electrophile on the nitrogen lone pair of pyridine is a competing reaction. Remember that the nitrogen lone pair of pyridine is perpendicular to the pi system of the aromatic ring and therefore cannot be used to stabilize any sigma complex.

Five-membered ring heteroaromatics undergo electrophilic aromatic substitution at the position ortho to the heteroatom. The most stabilized sigma complex dominates. Furan example:

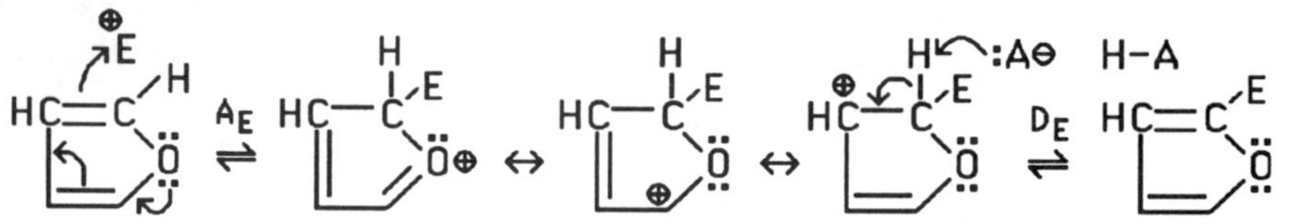

Naphthalene undergoes electrophilic aromatic substitution at the position next to the second ring. The sigma complex leading to the product that is obtained has more resonance structures indicating a more delocalized charge than that of the alternative.

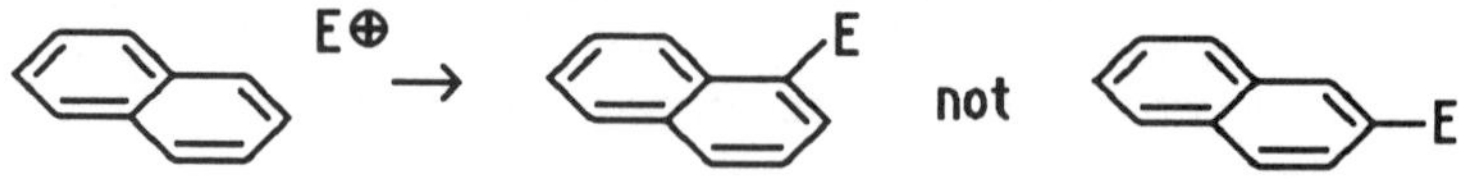

Aromatics reacting with H − A

This reaction is noticeable only if ipso attack triggers rearrangement or loss of a group on the ring.
Desulfonation example:

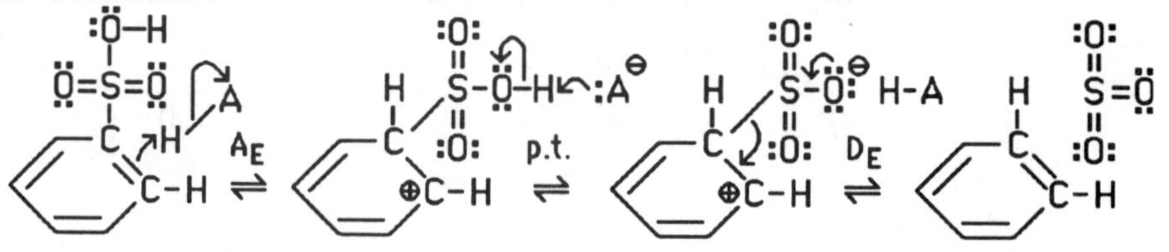

Aromatics reacting with Y − L as sinks

Most of the reactions fall into this class. Many of the electrophiles are cationic, and their generation and use is similar to an S_N1 process. Others merely require that a Lewis acid catalyze the departure of the leaving group as in the following chlorination reaction.

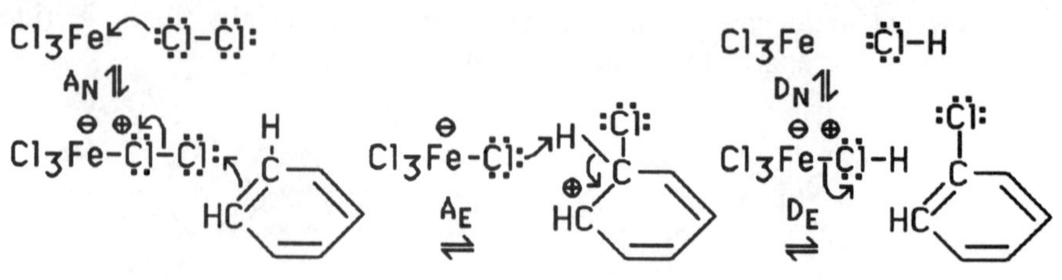

Friedel–Crafts alkylation – aromatic rings with simple carbocations as sinks

This reaction has several limitations and problems. The reaction fails for deactivated rings. Second, if it can, the carbocation almost always rearranges to a more stable carbocation. Third, once the electrophile is attached, it activates the ring for further attack so sequential electrophile additions to give multiple substituted aromatics are possible. The electrophiles are generated in a variety of ways.

$$R\text{–}OH + H^{\oplus} \overset{p.t.}{\rightleftharpoons} R\text{–}\overset{\oplus}{O}H_2 \overset{D_N}{\rightleftharpoons} R^{\oplus} + H_2O$$

$$R\text{–}X + AlX_3 \overset{A_N}{\rightleftharpoons} R\text{–}\overset{\oplus}{X}\text{–}\overset{\ominus}{AlX_3} \overset{D_N}{\rightleftharpoons} R^{\oplus} + AlX_4{}^{\ominus}$$

$$\rangle C{=}C\langle + H^{\oplus} \overset{A_E}{\rightleftharpoons} \rangle \overset{\oplus}{C}\text{–}C\langle H$$

Alkylation example:

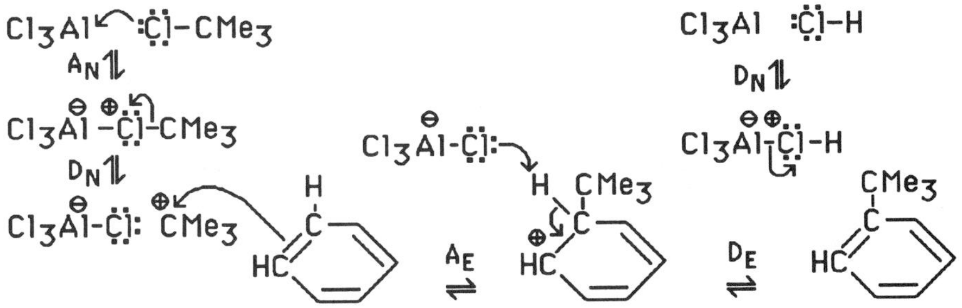

Friedel–Crafts – aromatic rings with lone-pair-stabilized carbocations as sinks

The electrophile is most often a carbonyl that is either protonated or complexed to a strong Lewis acid. The product can enter into a second Friedel–Crafts alkylation so that often the final product is a result of two attacks. The synthesis of the insecticide DDT is a good example.

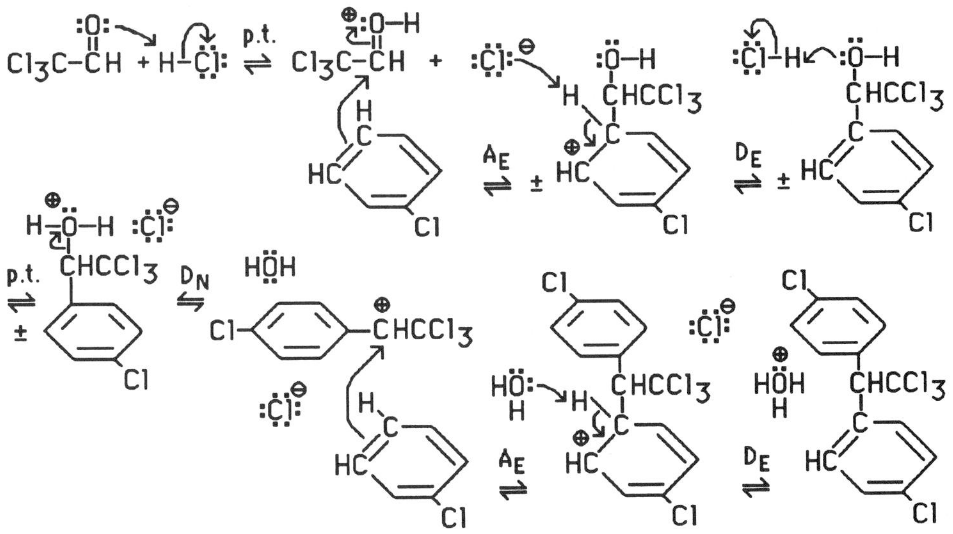

Friedel–Crafts acylation – aromatic rings with lone-pair-stabilized vinyl carbocations as sinks

The electrophile is the acylium ion, $R-C\equiv O^{\oplus}$, generated by Lewis acid catalyzed ionization of a leaving group (path D_N) from acyl halides or acid anhydrides. The reaction fails for deactivated rings (Ar−ewg, meta directors). After the electrophile adds it deactivates the ring toward further attack. No rearrangement of the electrophile occurs. Acylation example:

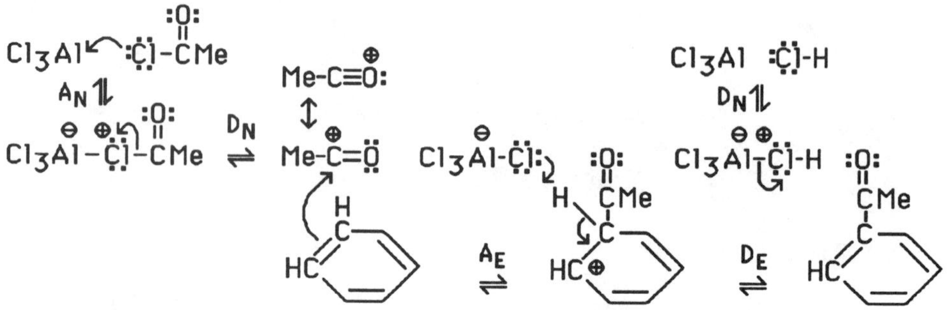

7.4   REARRANGEMENTS WITH LOSS OF A LEAVING GROUP

The most important path restriction on rearrangements is that orbital overlap must be maintained by the migrating group with both the atom it migrates from and the atom it migrates to. At the transition state the migrating group will overlap both atoms. Often rearrangement can occur by a path (1,2RL) reminiscent of the generic S_N2. Good orbital alignment is crucial; the group that is best lined up to displace the leaving group from the backside is the one that preferentially migrates. Occasionally the leaving group will depart first (path D_N), and then rearrangement occurs (path 1,2R) in a path resembling the S_N1 process. Examples of migration to oxygen and migration from boron will be given in Section 7.5, Nu−L pathways.

Migration to carbon (with acid-catalyzed loss of the leaving group):

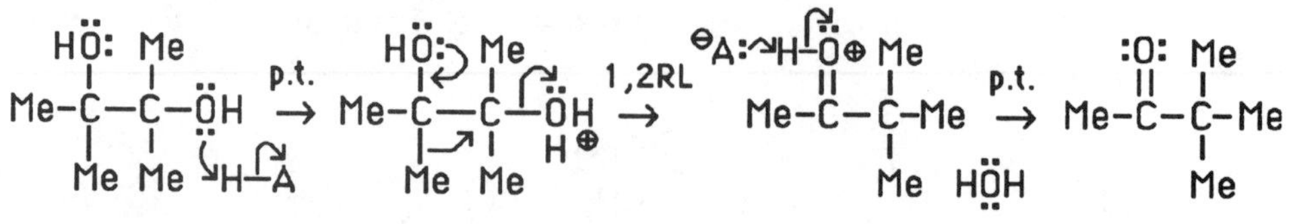

Vinyl migrating and leaving group (followed by Nu attack on the heterocumulene and then decarboxylation):

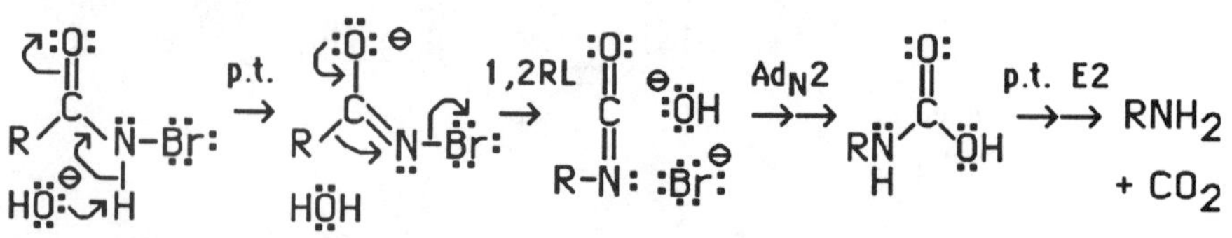

Migration to nitrogen (in this example, the leaving group returns as a nucleophile to trap the heterosubstituted vinyl cation):

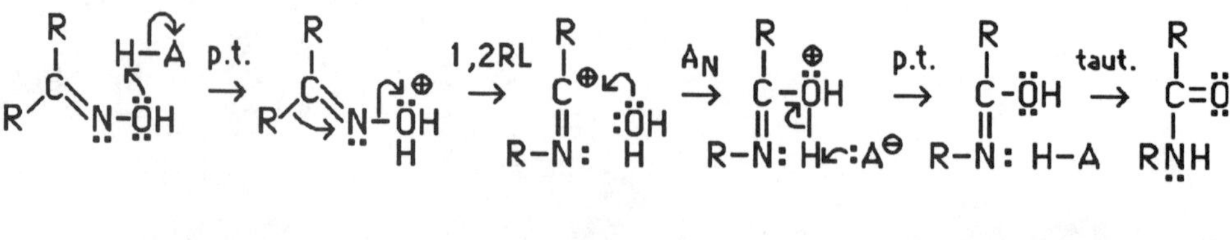

7.5 Nu—L REACTIONS

The Nu—L class of reagents can be recognized by a nucleophilic center to which a leaving group is directly bonded. **This unusual arrangement in which the same atom can serve both as a sink and a source is characteristic of the Nu—L class,** and thus their reactions are reasonably similar, most often three-membered ring formation or addition followed by rearrangement. The following list gives several examples of the Nu-L class.

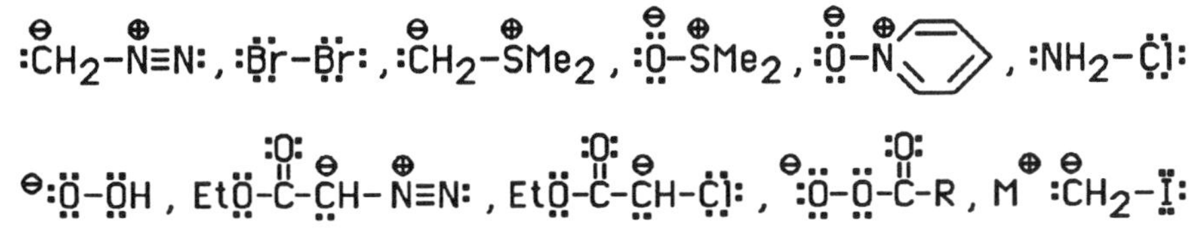

$$:CH_2-N\equiv N: \;,\; :\ddot{B}r-\ddot{B}r: \;,\; :CH_2-SMe_2 \;,\; :\ddot{O}-SMe_2 \;,\; :\ddot{O}-N\langle\rangle \;,\; :NH_2-\ddot{C}l:$$

$$^{\ominus}:\ddot{O}-\ddot{O}H \;,\; Et\ddot{O}-\overset{O}{\overset{\|}{C}}-\ddot{C}H-N\equiv N: \;,\; Et\ddot{O}-\overset{O}{\overset{\|}{C}}-\ddot{C}H-\ddot{C}l: \;,\; ^{\ominus}:\ddot{O}-\ddot{O}-\overset{O}{\overset{\|}{C}}-R \;,\; M^{\oplus}\,{}^{\ominus}:CH_2-\ddot{I}:$$

Nu—L Reacting with Trialkylboranes

Trialkylboranes, R_3B, react with Nu—L reagents to form Lewis acid–base salts that rearrange by migration of the alkyl group to the Nu with loss of the leaving group (path 1,2RL). The process repeats until all the alkyl groups have reacted. The borate that is formed can be hydrolyzed to give the free RNuH (see Section 7.3a). This is a very general and useful set of reactions.

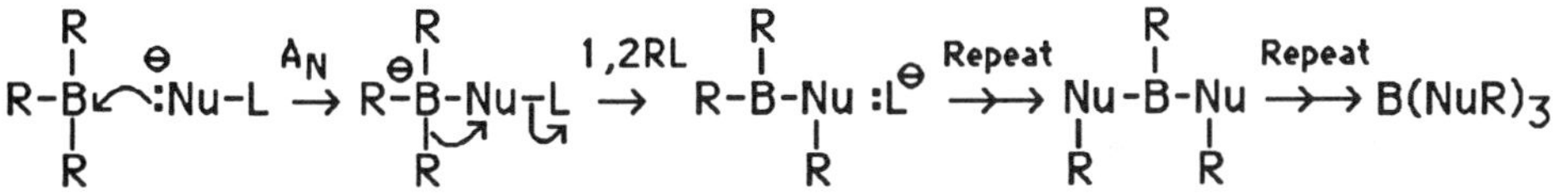

Nu—L Reacting with Acids

The Nu portion of the Nu—L can be protonated with acids, producing a reactive species with which the conjugate base of the acid may react.

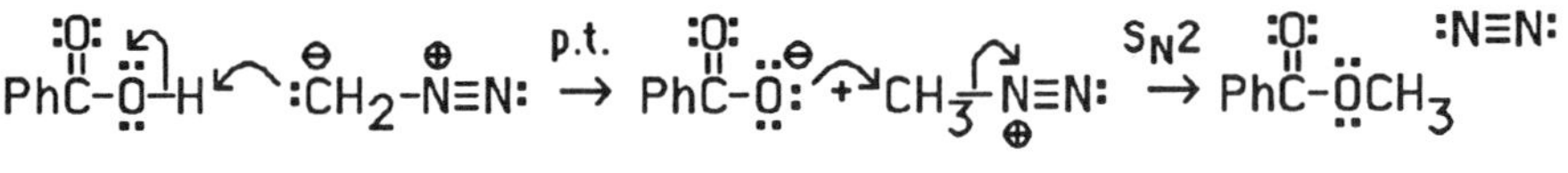

Nu—L Reacting with ⩾C—L

The Nu—L can substitute the leaving group on carbon and can be followed by an elimination reaction.

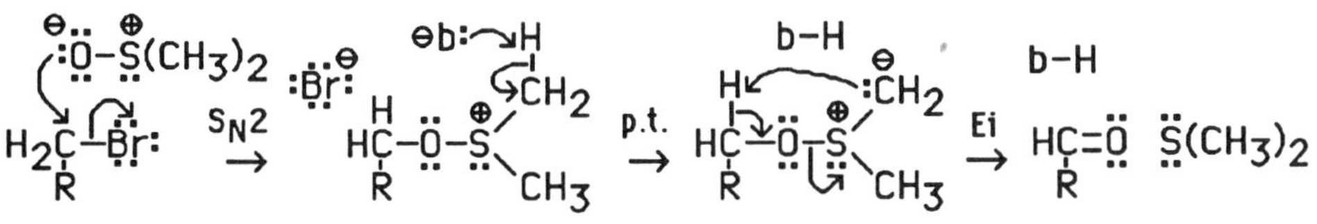

Nu—L Reacting with C=C

Leaving group drops off first: carbenes

Carbenes bear a resemblance to carbocations in that there is an empty *p* orbital that can behave as an electron sink. However, a full orbital that can serve as an electron source is on the same atom. Trihalomethyl anion loses halide, forming the reactive dichlorocarbene, a neutral, electron-deficient, electrophilic intermediate. Stabilization in dichlorocarbene results from the interaction of the full lone pair orbitals of chlorine with the empty p orbital of the carbene. If the donors on the carbene are good enough, the carbene becomes nucleophilic (Section 2.9). With few exceptions, carbenes react stereospecifically with double bonds to produce three-membered rings (Figure 7.13).

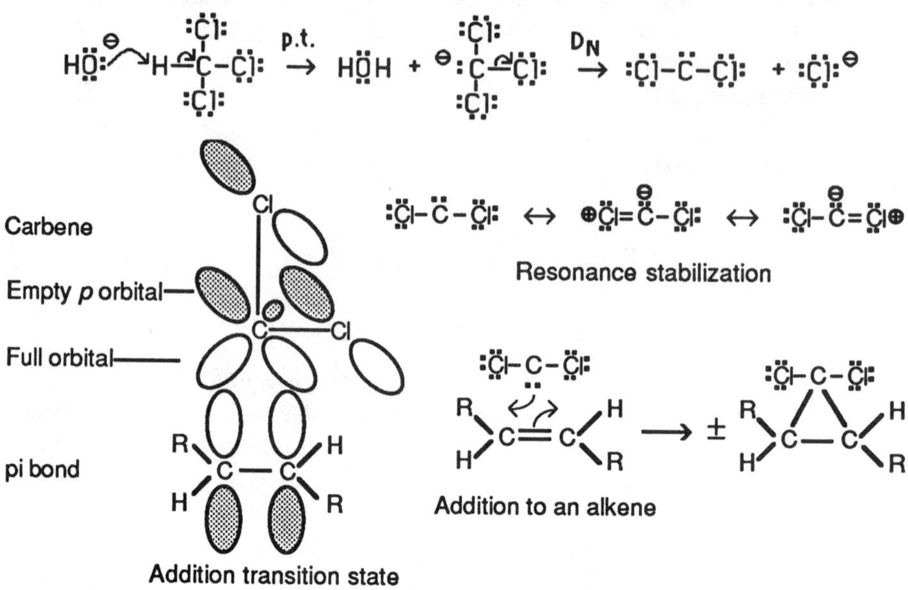

Figure 7.13 Carbene addition.

Leaving group loss concerted with attack (path NuL)

These reactions are characteristically stereospecific. The formation of a bromonium ion from electrophilic attack of bromine on a pi bond was the first example of this path (Section 7.3b).

The organometallics of this subset are called carbenoids because of their similar reactivity to carbenes. Carbenoids are generated by the reaction of a halide with a metal by halogen–metal exchange. The presumed sequence is shown in the following example. The structure of the organometallic is currently under debate.

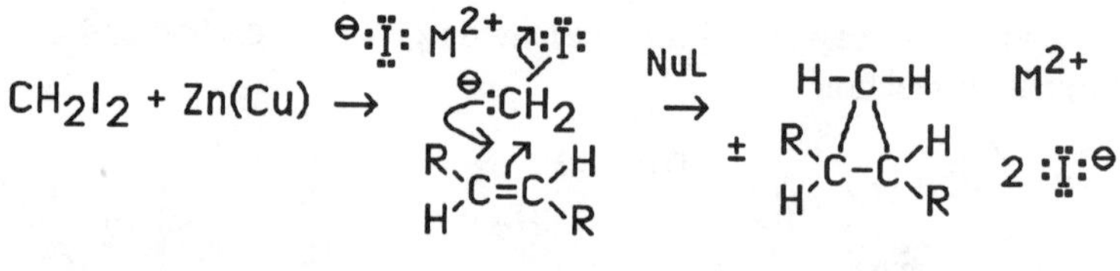

Although they are not considered carbenoids, peracids are mechanistically similar in their additions to alkenes. The intramolecular hydrogen bond (shown by two resonance forms) partially breaks the O—H bond, making it a better nucleophile.

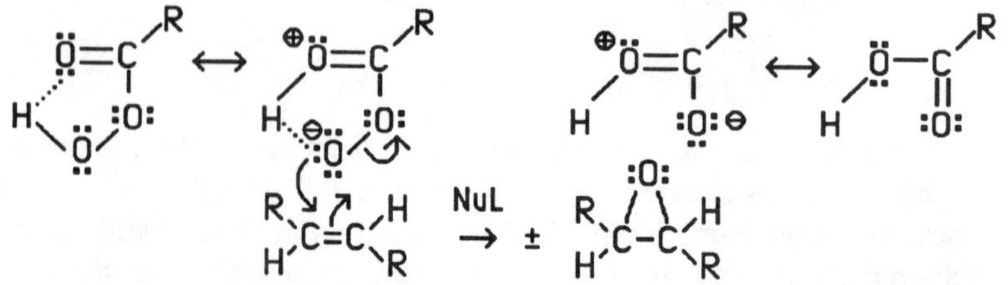

Nu—L Reacting with C=Y

Leaving group drops off after attack

This mechanism implies a new intermediate, one that stabilizes the anion formed upon nucleophilic attack. The Nu—L species acts first as an electron source, then in a second step functions as an electron sink. The most common members of this subset, the ylides, have adjacent plus and minus charges, $^{\ominus}$Nu—L$^{\oplus}$. This mechanism is a combination of the addition to a polarized multiple bond (path Ad_N) followed by an internal S_N2.

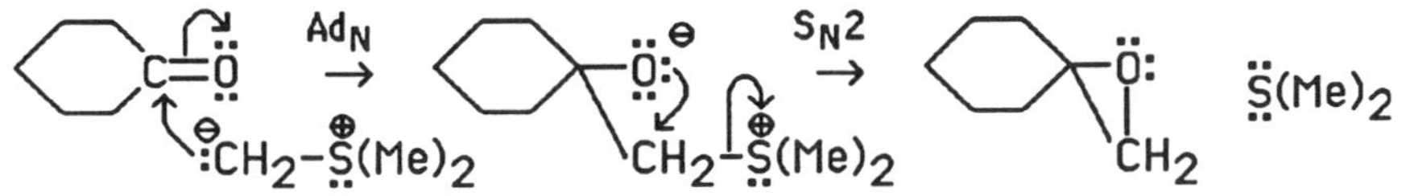

Phosphorous ylides are almost the only representative of a mechanistic variant. Rather than being displaced in an internal S_N2 reaction, the phosphorus bonds to oxygen and then undergoes an elimination (path 4e) as the very stable phosphine oxide.

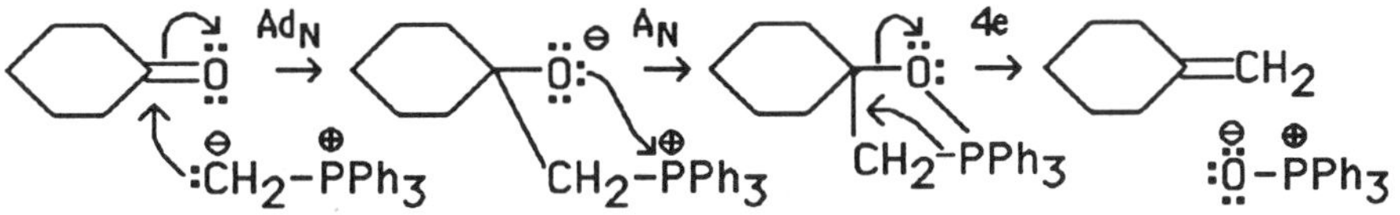

The Nu—L species can attack the polarized multiple bond (path Ad_N), then rearrange (path 1,2RL) with loss of the leaving group, discussed in the last section.

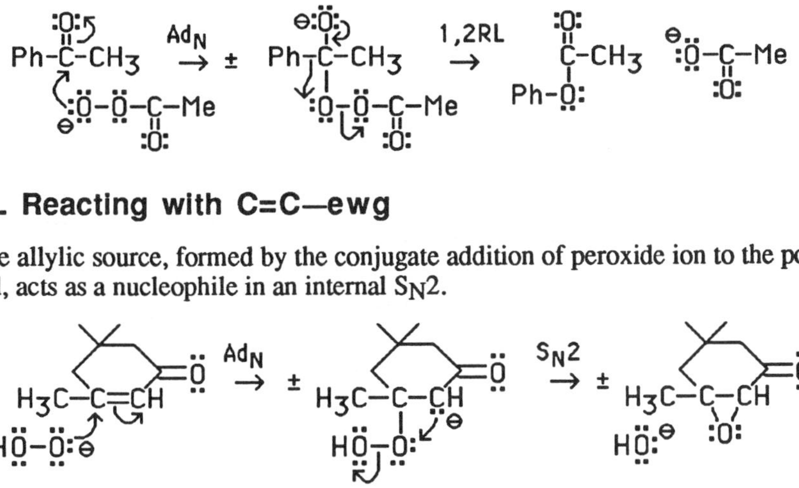

Nu—L Reacting with C=C—ewg

The allylic source, formed by the conjugate addition of peroxide ion to the polarized pi bond, acts as a nucleophile in an internal S_N2.

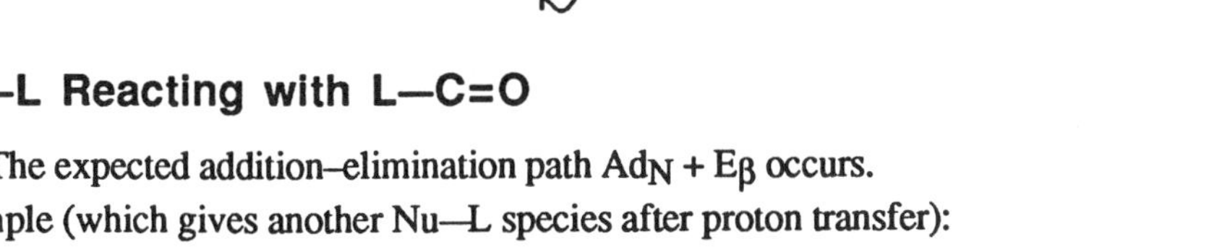

Nu—L Reacting with L—C=O

The expected addition–elimination path $Ad_N + E_\beta$ occurs.

Example (which gives another Nu—L species after proton transfer):

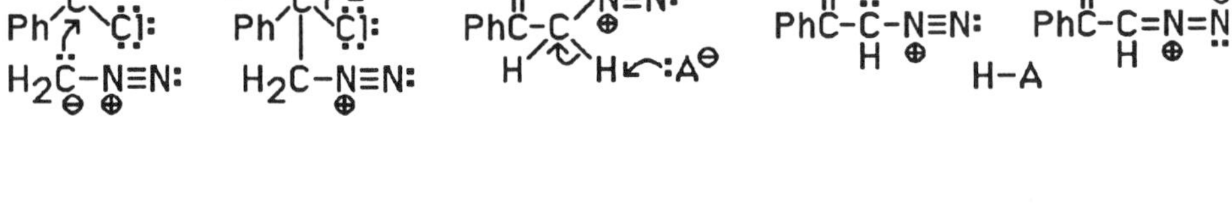

Azide ion addition–elimination is followed by a rearrangement with loss of nitrogen (path 1,2RL) in the following example.

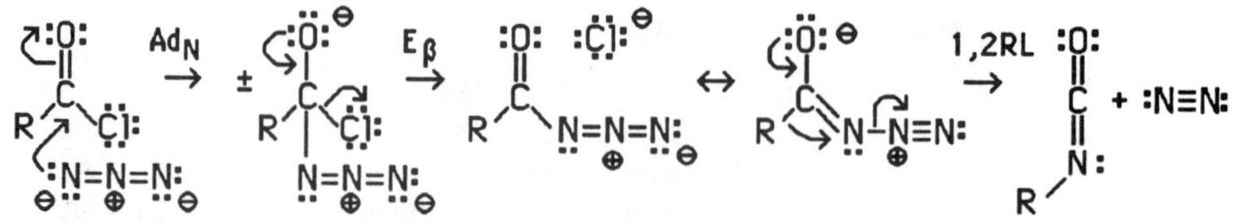

7.6   FRAGMENTATIONS

Reverse-aldol

A fragmentation reaction is one that breaks a molecule apart, usually into at least two pieces, by breaking a carbon–carbon bond. A fragmentation can be considered to be a type of elimination, similar to the E1cB or the E2 (see Section 6.4), in which a weakened carbon–carbon single bond is broken. A fragmentation has the same path restrictions as the elimination it resembles. The reverse of the aldol reaction (Section 7.2n) is an E1cB type of fragmentation in which the leaving group is an enolate:

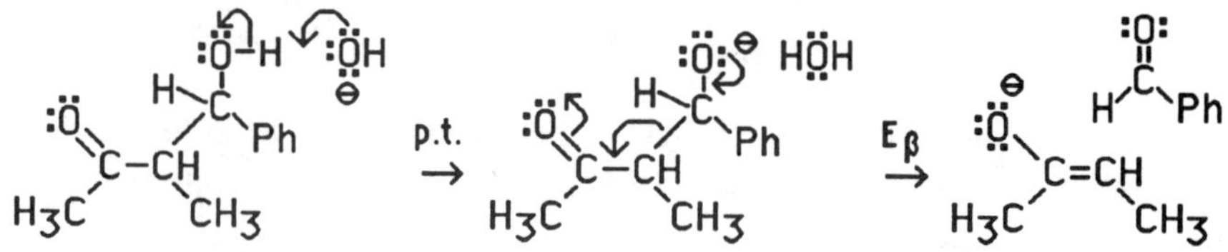

Thiamine-Catalyzed Decarboxylation of Pyruvate

Thiamine catalyzes the decarboxylation: of pyruvate, $CH_3COCOO^\ominus$, to acetaldehyde, CH_3CHO, in a mechanistically similar process. The fragmentation step is an E_β path in which the leaving group is an enamine (discussed in Section 6.4, Variations on a Theme). Thiamine is an elegantly optimized catalyst that adds to pyruvate, accepts the electron flow from the $COO^\ominus$, flows the electrons back out to pick up a proton, then detaches to start the process all over again. Without the thiamine, the decarboxylation does not occur, for the reaction would have to produce the highly unstable acyl anion.

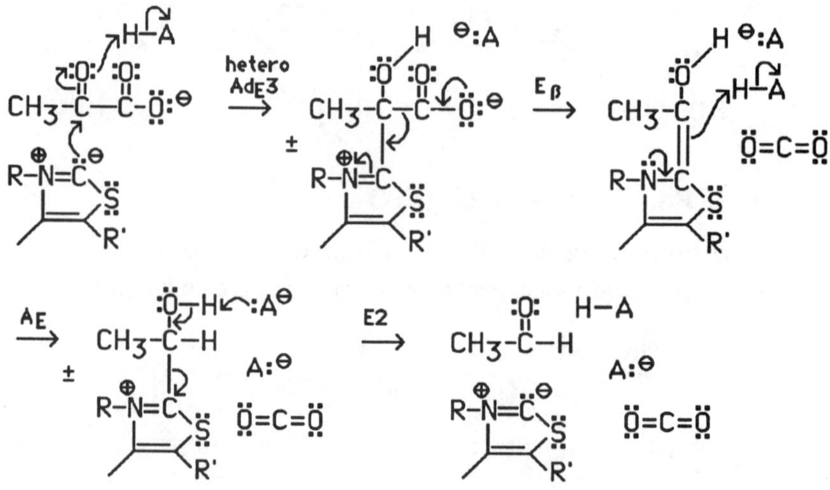

7.7 CONCERTED SIX-ELECTRON PERICYCLIC REACTIONS

4 + 2 Cycloadditions

The Diels–Alder reaction is a six-electron cycloaddition reaction between a diene (contributing four pi electrons to the transition state) and dienophile (contributing two pi electrons). It is an extremely useful cycloaddition because it goes in high yield and with predictable stereochemistry. The stereochemistry of both the starting pieces is preserved in the product; for example, if two groups on the dienophile are *cis*, they remain *cis* in the product. The net reaction breaks two pi bonds and makes two sigma bonds and therefore is about 40 kcal/mol (167 kJ/mol) exothermic. The six-membered transition state for the reaction resembles a folded cyclohexane (a boat conformation). Figure 7.14 shows the progress of the reaction with line structures and with orbitals. Notice how the *p* orbitals of the pi bonds rehybridize into the sigma bonds of the product.

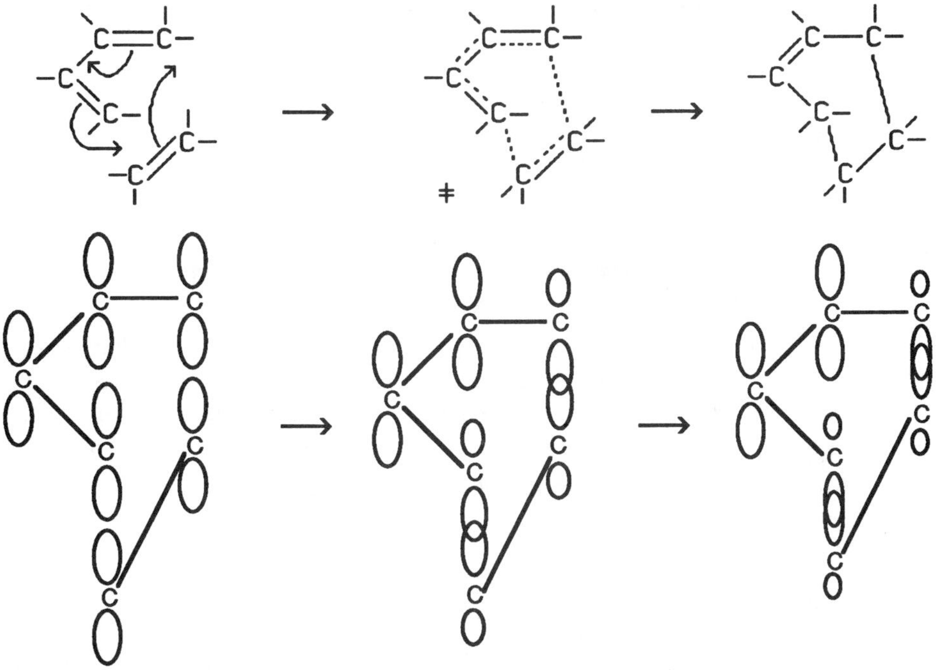

Figure 7.14 The Diels–Alder reaction with the transition state in the center.

The Diels–Alder reaction has much higher yields if an electron-withdrawing group is conjugated with the dienophile pi bond. Surprisingly, the product is the one with the electron-withdrawing group underneath, *endo*, rather than *exo*, sticking out away. Favorable orbital interactions between the electron-withdrawing group and the diene have been invoked to rationalize this endo stereochemistry.

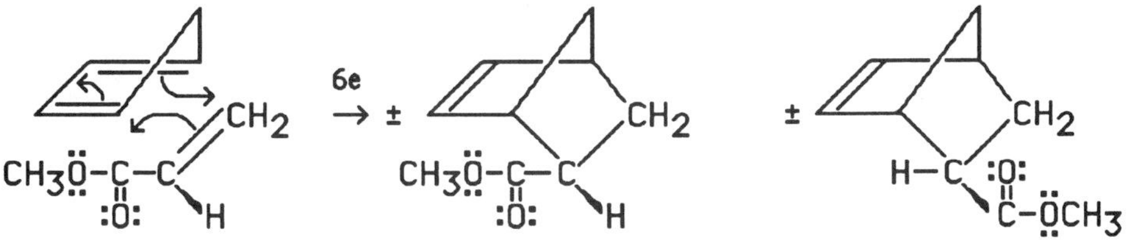

90% endo   10% exo

The five-membered transition state 4 + 2 cycloaddition is less common than the six. The four-pi electron piece is called a 1,3-dipole because it reacts as if it has a minus and a plus separated by one atom, $^{\oplus}$A-B-C$^{\ominus}$. The orbital arrangement for the 1,3-dipolar addition is similar to that for the Diels–Alder:

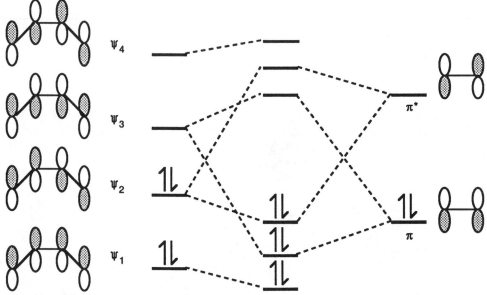

Cycloreversions are the reverse of cycloadditions and occur by the same mechanism. The ΔG for the cycloreversion process becomes favorable (negative) with increased temperature because the $-T\Delta S$ term dominates over the ΔH term. The reverse Diels–Alder of dicyclopentadiene into two molecules of cyclopentadiene is a common example:

Heat to 170°C
→
←
Cool to 25°C

4 + 2 cycloadditions from a HOMO–LUMO perspective
(A supplementary, more advanced explanation)

The soft–soft interaction of filled with empty orbitals is the major attractive force in the transition state because there is little hard–hard interaction. In an unsubstituted system two soft–soft interactions stabilize the transition state: the filled Ψ_2 molecular orbital of the diene interacting with the unfilled π^* of the dienophile, also the empty Ψ_3 of the diene interacting with the filled π of the dienophile. Figure 7.15 is the complete interaction diagram.

Figure 7.15 The Diels–Alder interaction diagram.

However, for the simplest system, 1,3-butadiene reacting with ethene, the HOMO–LUMO interactions are average, for the interacting orbitals are reasonably far apart in energy. The interaction can be increased (Section 2.9), usually by placing an ewg on the dienophile (lowering the LUMO, π^*), and sometimes by placing a donor on the diene (raising the HOMO, Ψ_2). In addition, these substituents polarize the pi bond they are attached to and introduce a hard–hard factor into the reaction. With substituents on the diene and dienophile, different constitutional isomers are possible. The regiochemistry is determined by the best overlap of the HOMO–LUMO pair closest in energy. The following structures show the regiochemistry for the two most common combinations (the controlling interaction is the unfilled π^* of the dienophile with the filled Ψ_2 of the diene).

If the donor were on the dienophile and the ewg on the diene (reversed polarity), the controlling interaction would be the filled π of the dienophile with the empty Ψ_3 of the diene. The following structures show the regiochemistry for the reversed polarity Diels–Alder.

If the dienophile has a substituent that extends its pi system, the best transition state for the reaction maximizes overlap between the two pi systems. To achieve this greater degree of interaction the dienophile must place the substituent underneath the diene, endo. The Diels-Alder of two dienes has such a transition state with an additional interaction between the unfilled diene Ψ_3 MO and the filled diene Ψ_2 MO (Figure 7.16).

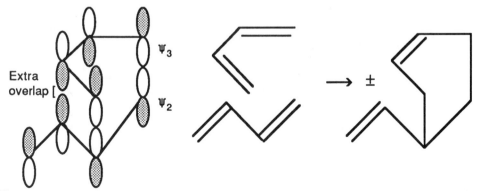

Figure 7.16 The Diels–Alder reaction of two dienes with the extra overlap of endo addition.

Six-Electron Cyclic Rearrangements

Two common six-membered transition state rearrangements of 1,5-dienes have either oxygen or carbon as Y. Both are reversible and favor the more stable product as expected. The transition state for the reaction commonly resembles the chair conformation of cyclohexane (Figure 7.17).

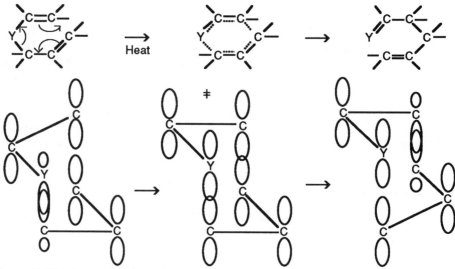

Figure 7.17 The concerted six electron cyclic rearrangement of a 1,5-diene.

The six-electron electrocyclic reaction bears a similarity to the other six-electron cyclic rearrangements but has a different transition state. The sigma bond of a cyclohexadiene twists open to form a hexatriene.

Six-electron electrocyclic reaction from a HOMO–LUMO perspective
(A supplementary, more advanced explanation)

As was seen for the four electron electrocyclic ring opening (path 4e), the direction that the ends rotate when the sigma bond twists open is important (Figure 7.18). In the six-electron case, the ends rotate in opposite directions, *disrotatory*. The ring twists open so that the σ LUMO overlaps with the π HOMO in a bonding manner to produce a bonding molecular orbital of the product.

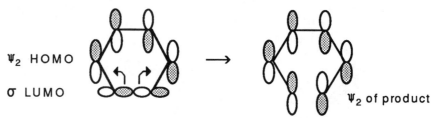

Ψ_2 HOMO

σ LUMO

Ψ_2 of product

Figure 7.18 The disrotatory opening of cyclohexadiene.

7.8   MISCELLANEOUS REACTIONS

7.8a Vinyl Leaving Groups as Electron Sinks

These electron sinks are much less common. If the ewg is a carbonyl and the leaving group is chloride, this electron sink is a vinylogous acid chloride.

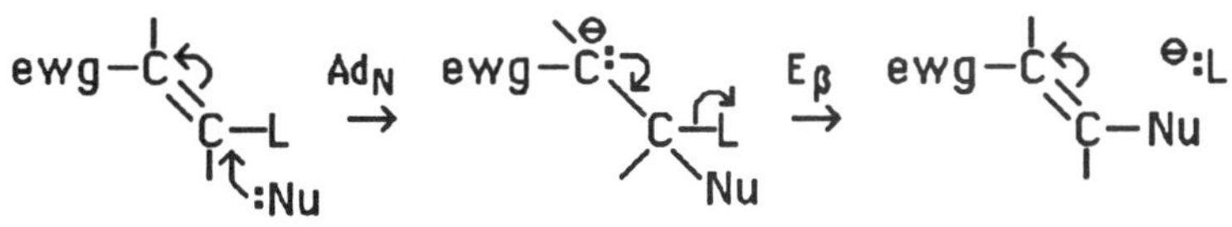

If the double bond is part of an aromatic ring, this process is called nucleophilic aromatic substitution. There are three routes available for this very poor electron sink, Ar—L . If the leaving group is good — for example, chloride — but there are no electron-withdrawing groups on the ring, temperatures over 300°C will be required to get reaction. Nitro groups and other good electron-withdrawing groups on the ring lower the temperature for reaction by stabilizing the intermediate carbanion.

$Ad_N + E_\beta$ example:

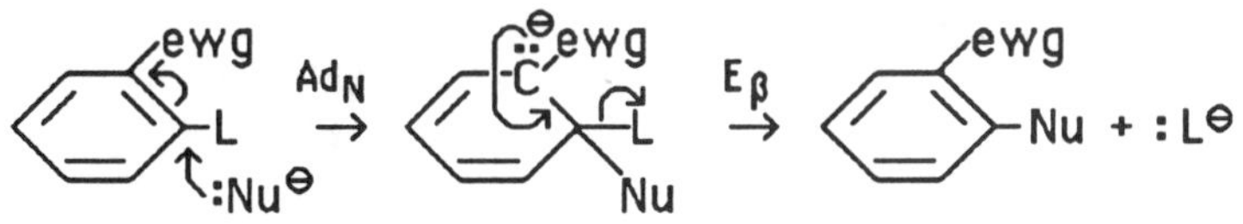

An unusual reaction occurs with aromatic halides and strong bases; the substitution reaction can go via elimination to the highly reactive benzyne followed by addition, path E2 then Ad_N2.

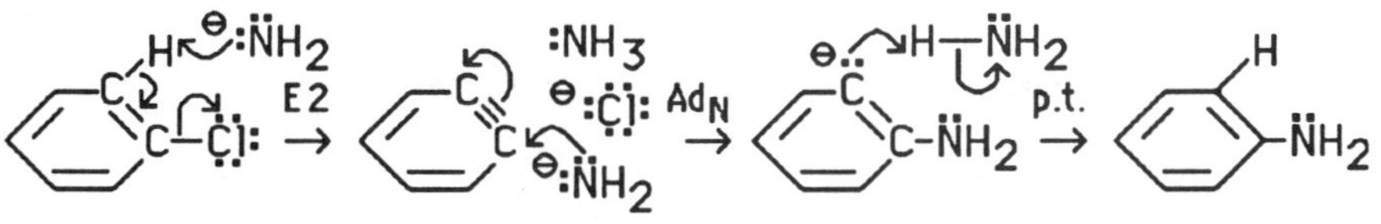

If the leaving group is excellent — namely, N≡N in diazonium ions — an S_N1-like path becomes available. The many mechanisms for substitution on these diazonium ions are complex, some involve electron transfer, radical intermediates, or copper catalysis.

7.8b Carbonate Derivatives as Electron Sinks

These are represented by the following structures and can undergo (depending on the quality of the leaving group) single, double, and triple additions by nucleophiles via the addition–elimination path. The tendency toward single-addition is highest with haloformates (center of the following structures) because one leaving group is much better than the other.

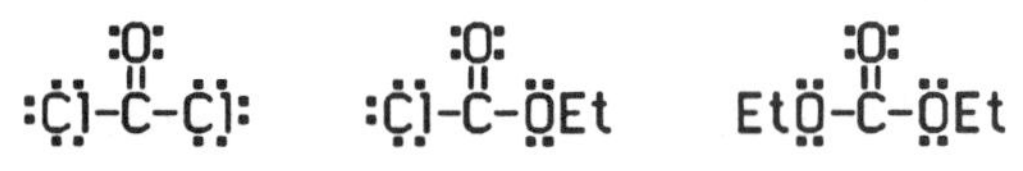

Carbonate derivatives with one group as OH and the other an L are not stable with respect to an elimination reaction to produce carbon dioxide.

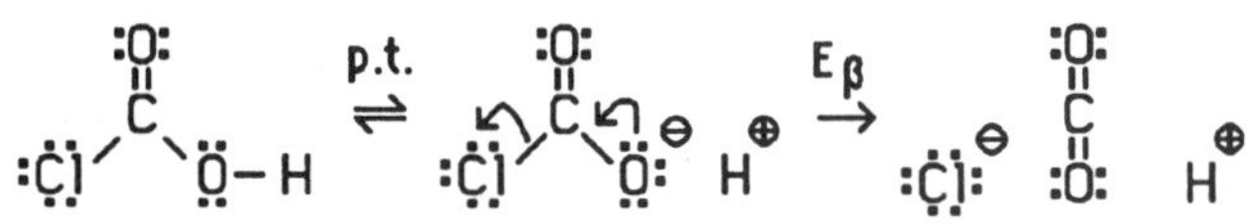

7.8c Heterocumulenes as Electron Sinks

Heterocumulenes are attacked at the central carbon. In acidic media, proton transfer may occur first. In basic media, the resultant anion is protonated in the workup.

Lone pair source example:

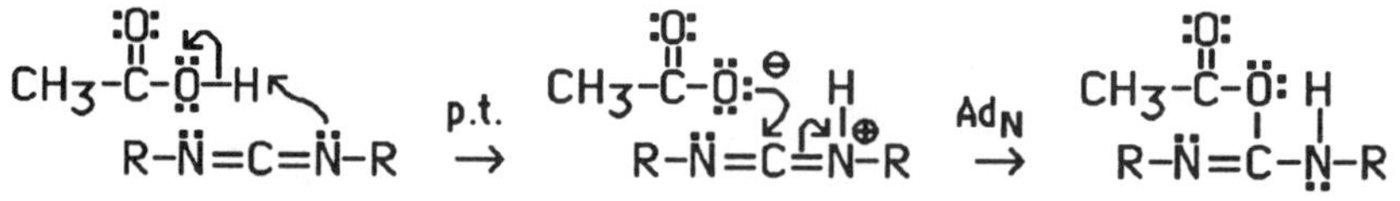

Organometallic example (acid water workup protonates the carboxylate anion):

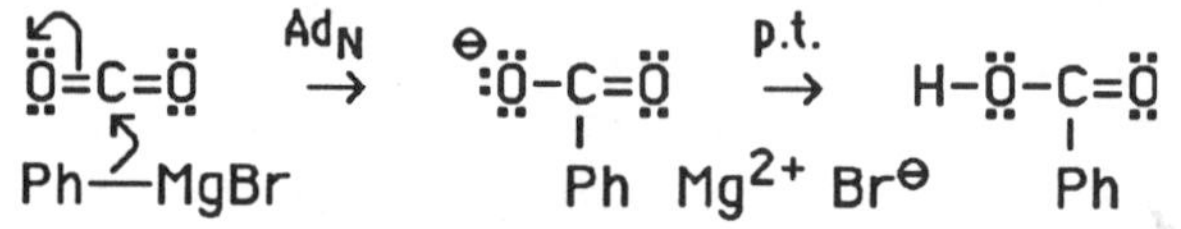

7.8d   Metal Ions as Electron Sinks

A good example of this reaction is the oxymercuration reaction. The nucleophile is commonly the solvent, usually water or an alcohol. The metal can be removed in a subsequent step. Many enzymes make use of metal electrophiles to activate substrates for attack by nucleophiles.

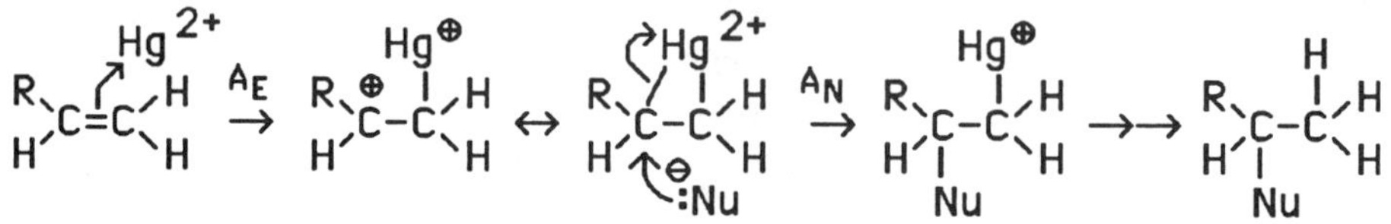

7.8e   Chromium Species as Oxidants

Chromium trioxide and chromic acid, H_2CrO_4, are commonly used to oxidize alcohols. The alcohol adds to the chromium species to give a chromate ester, which undergoes elimination by an E2 process.

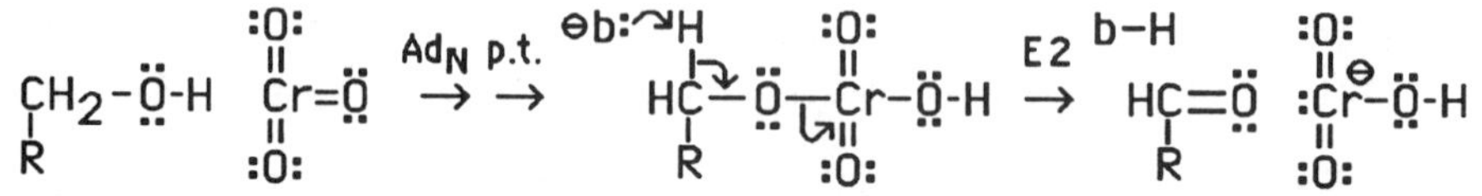

If water is present, the initially formed aldehyde adds water to form a hydrate, which is further oxidized to the carboxylic acid.

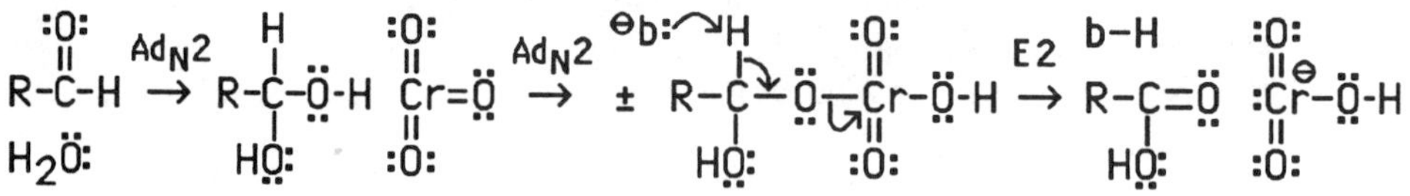

7.8f   NADH Reduction of Acetaldehyde

NADH (nicotinamide adenine dinucleotide) is a biochemical source of hydride. In the following example NADH reduces acetaldehyde to ethanol via minor pathway $H^\ominus$ t., hydride transfer to a cationic center. A zinc ion helps to polarize the acetaldehyde carbonyl to make it a better electron sink. The formation of the aromatic pyridinium ring in the $NAD^\oplus$ product helps balance the energetics of this easily reversible reaction.

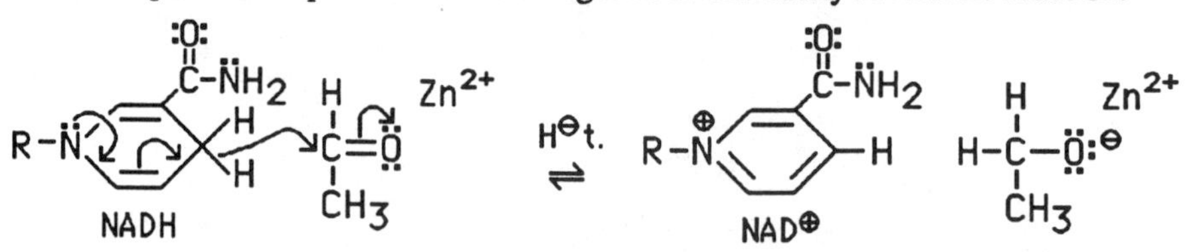

7.9 PRODUCT MATRIX SUMMARY

The product matrix (Table 7.3) serves as a summary of most of Chapter 7. It is the source and sink matrix that acted as an index to Section 7.2 expanded to include most of the common reactions. The product of the reaction of generic source and sink and an occasional comment appear in the cells of the product matrix. It cannot be stressed too much that **everything depends on your ability to recognize the generic class of the sources and sinks** that are present in the reaction mixture. Many times a species has a dual reactivity and therefore may fit into more than one generic class. A common example of this is that most anions can behave as a nucleophile or as a base. Chapter 8 discusses the common major decisions.

Table 7.3 Product Matrix

Sink	Z:⊖ as Nu	MH₄⊖	R–M	C=C–Z:⊖	⟩C=C⟨	ArH	B:⊖ and adj. C–H
⟩C⊕	trap ⟩C–Z	rare[a] ⟩C–H	rare[a] ⟩C–R	add.[b] ⟩C–Z–C=C	add.[c] ⟩C–C–C⊕	Ar subs. ⟩C–Ar	D_E ⟩C=C⟨
H–A	p.t. H–Z	p.t. H₂↑	p.t. R–H	p.t.[b] HC–C=Z	add. H–C–C–A	p.t. H–Ar	p.t. H–B
Y–L	subs. Z–Y	subs. H–Y	subs. R–Y	subs.[b] Y–C–C=Z	add. Y–C–C–L	Ar subs.[d] Y–Ar	elim. ⟩C=Y
⟩C–L	subs. ⟩C–Z	subs. ⟩C–H	subs. ⟩C–R	subs.[b] ⟩C–C–C=Z	no reaction[e]	Ar subs.[d] ⟩C–Ar	elim. ⟩C=C⟨
⟩C=Y	add. ⟩C(Z)(Y⊖)	add. ⟩C(H)(Y⊖)	add. ⟩C(R)(Y⊖)	add.[b] ⟩C(Y⊖)C–C=Z	no reaction[e]	Ar subs.[c,d] ⟩C(Ar)(YH)	p.t. ⟩C=C–Y⊖
–C≡Y	add. ⟩C=Y⊖ Z	add.[c] ⟩C=Y⊖ H	add. ⟩C=Y⊖ R	add.[b] ⟩C=Y C–C=Z	no reaction[e]	Ar subs.[d] ⟩C=YH Ar	p.t. ⟩C=C–Y⊖
⟩C=C–ewg	1,4-add.[b] ⟩C–C–ewg⊖ Z	1,4-add.[b] ⟩C–C–ewg⊖ H	1,4-add.[b] ⟩C–C–ewg⊖ R	add.[b] ⟩C–C–ewg C–C=Z	no reaction[e]	Ar subs.[b,d] ⟩C–C–ewg Ar–H	p.t. ⟩C–C–ewg⊖ C=C
⟩C=Y / L	add.-elim. ⟩C=Y Z	add.-elim.[c] ⟩C=Y H	add.-elim. ⟩C=Y R	add.-elim.[b] ⟩C=Y C–C=Z	no reaction[e]	Ar subs.[d] ⟩C=Y Ar	p.t. ⟨C=Y C⊖

[a] Media exclusion. [b] Ambident, Chapter 8. [c] Multiple add possible. [d] With acid catalyst.
[e] Reacts in the presence of a strong acid catalyst; often gives polymers or mixtures.

ADDITIONAL EXERCISES

7.1 Trends: rank all species using the numeral 1 to designate
the most reactive for S_N1

the most reactive for S_N2

7.2 State what controls the regiochemistry of the E1, the E2, and the E1cB.

7.3 Using the energy surfaces in Section 7.2f, predict how increasing the solvent
polarity should affect the competition between S_N1 and S_N2.

7.4 Draw the substitution product and decide whether the reaction is S_N1 or S_N2.

(a)

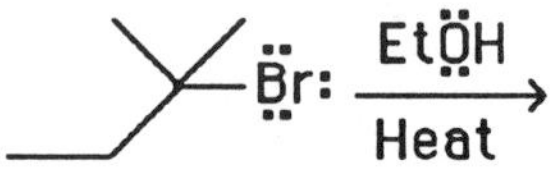

(b)

7.5 Draw the elimination product and decide whether the reaction is E1, E2, or E1cB.

(a)

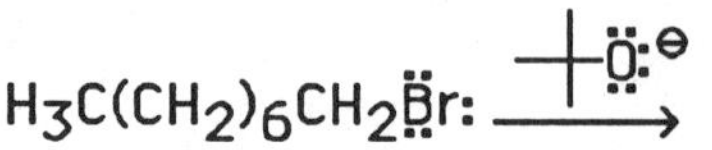

(b)

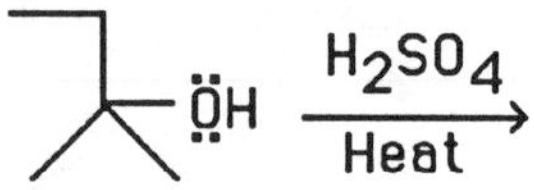

(c)

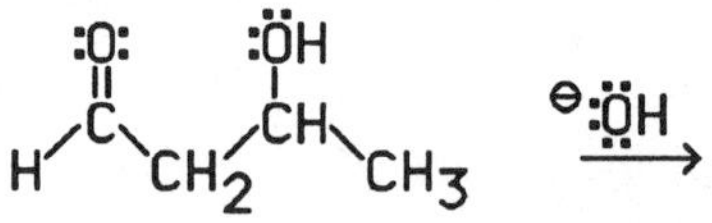

(For problems 7.6 to 7.10, refer to the Hints to Problems from Chapters 7, 8, and 9 if
you need assistance.)

7.6 Classify each reactant into its generic class of source and sink, then use the
appropriate electron flow pathway to rationalize the product of these **one-step** reactions.
These mechanisms are simple in that there is only one source and one sink present in the
reaction mixture. Find the source and the sink, then combine them with the appropriate
pathways discussed in this chapter.

(a)

$$(CH_3)_3C\text{-}\ddot{B}r\text{:} + Et\ddot{O}\text{:}^\ominus \rightarrow Et\ddot{O}H + (CH_3)_2C=CH_2 + \text{:}\ddot{B}r\text{:}^\ominus$$

7.6 (continued)

(b)

$$PhC\equiv C:^{\ominus} + CH_3CH_2\ddot{O}-\overset{\overset{:O:}{\|}}{\underset{:O:}{S}}-\langle\!\!\rangle- \rightarrow PhC\equiv C-CH_2CH_3 + {}^{\ominus}:\ddot{O}-\overset{\overset{:O:}{\|}}{\underset{:O:}{S}}-\langle\!\!\rangle-$$

(c)

7.7 Classify each reactant into its generic class of source and sink, then use the appropriate electron flow pathway to rationalize the product of these **two-step** reactions. There is only one source and one sink present in the reaction mixture. Find the source and the sink, then combine them with the appropriate pathways. Draw the product of the first step, then examine the product of the overall reaction to determine which bonds must be made or broken in the second step.

(a)

$$Ph_2CH-\ddot{\overset{\cdot}{C}l}: + {}^{\ominus}:\overset{\oplus}{N}=\overset{\oplus}{N}=\ddot{N}:^{\ominus} \rightarrow PH_2CH-\overset{\oplus}{N}=\overset{\oplus}{N}=\ddot{N}:^{\ominus} + :\ddot{\overset{\cdot}{C}l}:^{\ominus}$$

(b)

$$Ph-\overset{\overset{:O:}{\|}}{C}-H + {}^{\ominus}:CH_2-\overset{\oplus}{\ddot{S}}(CH_3)_2 \rightarrow \pm Ph-\overset{\overset{:O:}{|}}{\underset{H}{C}}-CH_2 + \ddot{S}(CH_3)_2$$

(c)

$$(Et\ddot{O}-\overset{\overset{:O:}{\|}}{C}-)_2\overset{\ominus}{C}H + H_2C=\overset{\overset{:O:}{\|}}{\underset{Ph}{C}}-C-\ddot{O}Et \overset{Et\ddot{O}H}{\rightarrow} \pm (Et\ddot{O}-\overset{\overset{:O:}{\|}}{C}-)_2CHCH_2\overset{\overset{H}{|}}{\underset{Ph}{C}}-\overset{\overset{:O:}{\|}}{C}-\ddot{O}Et$$

7.8 Classify each reactant into its generic class of source and sink, then use the appropriate electron flow pathway to rationalize the product of these **two-step** reactions. The first step is a favorable proton transfer that generates the electron source for the reaction; there is only one electron sink.

(a)

$$Et\ddot{O}-\overset{\overset{:O:}{\|}}{C}-CH_2-\overset{\overset{:O:}{\|}}{C}-\ddot{O}Et \xrightarrow[\text{2) } CH_3(CH_2)_3\ddot{\overset{\cdot}{B}r}:]{\text{1) }{}^{\ominus}:\ddot{O}-Et} Et\ddot{O}-\overset{\overset{:O:}{\|}}{C}-CH-\overset{\overset{:O:}{\|}}{C}-\ddot{O}Et \quad :\ddot{\overset{\cdot}{B}r}:^{\ominus} \quad \overset{H\ddot{O}Et}{CH_2CH_2CH_2CH_3}$$

(b)

$$CH_3-\overset{\overset{:\ddot{O}-H}{|}}{\underset{H}{C}}-C\equiv N: + {}^{\ominus}:\ddot{O}-H \rightarrow CH_3-\overset{\overset{:O:}{\|}}{C}-H + {}^{\ominus}:C\equiv N: + H-\ddot{O}-H$$

(c)

$$CH_3(CH_2)_{10}CH_2\ddot{O}H + H\ddot{\overset{\cdot}{B}r}: \rightleftharpoons CH_3(CH_2)_{10}CH_2\ddot{\overset{\cdot}{B}r}: + H_2\ddot{O}$$

7.9 Classify each reactant into its generic class of source and sink, then use the appropriate electron flow pathway to predict the product of these **one-step** reactions.

(a)

$$PhMgBr + CO_2 \rightarrow$$

7.9 (continued)

(b)

$$Ph_3P: \; + \; CH_3-\ddot{I}: \; \rightarrow$$

(c)

$$(C_6H_{11})_2BH \; + \; CH_2=CHCH_2CH_3 \; \rightarrow$$

7.10 Classify each reactant into its generic class of source and sink, then use the appropriate electron flow pathway to predict the product of these **two-step** reactions. Use a common path combination.

(a)

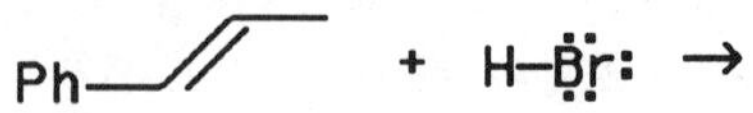

$$Ph-\!\!/\!\!/ \quad + \quad H-\ddot{B}r: \; \rightarrow$$

(b)

$$Ph-\overset{\overset{\displaystyle :O:}{\|}}{C}-\ddot{O}CH_3 \; + \; {}^{\ominus}:\ddot{O}-C(CH_3)_3 \quad \rightarrow$$

(c)

$$(CH_3)_3C^{\oplus} \; + \; \langle\!\!\!\bigcirc\!\!\!\rangle \; \rightarrow$$

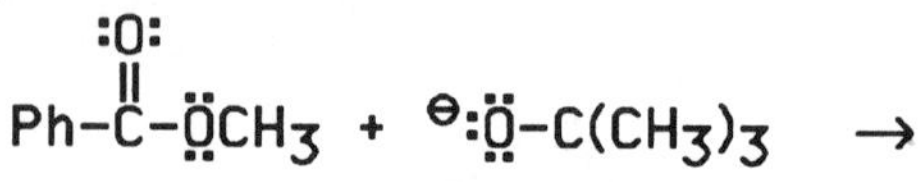

8

DECISIONS,
DECISIONS

8.1 DECISION POINT RECOGNITION

The most important question that you must continually ask yourself when examining any step in a process that has known and unknown branch points is, What else can happen? This may seem to add complexity, but obvious routes often do not work and therefore lead to frustration and the the question, Where did I go wrong? The answer is usually, You forgot to watch for alternative routes and proceeded with the first idea that occurred to you.

Draw out every step of the process and look for alternative routes from each step; try to find at least one alternative for each step. For hints check resonance forms, proton transfer, alternative sources and sinks, less likely combinations, and higher-energy species. Specifically look for the more obvious decision points covered in the rest of this chapter.

If you find that you habitually miss a decision point or two, then a good idea would be to make up a checklist of reminders until new habits form. Refer to this checklist whenever you are working a problem. Continue to personalize this checklist by adding reminders to watch for common errors that you find yourself repeating.

8.2 MULTIPLE ADDITIONS

Nucleophilic Addition to L—C=O

The products of an addition–elimination path very often undergo a second attack by the nucleophile ($Ad_N + E_\beta$ then a second Ad_N) because the product of the first addition–elimination is also a reactive electron sink. If the addition–elimination product is more reactive than the original starting material, a second addition almost certainly will occur. A ranking of the reactivity of electron sinks and sources is very important for path selection. For each reaction, the useful limit must be found for each reactivity spectrum. How far down the reactivity trend for the electron source can we go and still have the reaction with a particular electron sink occur at a useful rate? There are five ways that the reaction may stop after the initial $Ad_N + E_\beta$:

1. The Nu is such a weak base that the second addition easily reverses. Heteroatom lone pair sources almost never add twice.

2. The initial electron sink must be more reactive than the product of the addition, and the electron source, although reactive enough to add once, is not reactive enough to make a second addition to the less reactive compound. Organometallics in the reactivity range of organocadmiums, organozincs, and organocoppers (metal electronegativities are 1.69, 1.65, 1.90, respectively) add quickly to acyl chlorides but slowly to the ketone products.

3. Exceptionally large electron sources can be sterically too big to add a second time.

4. The leaving group is so poor (in amides, for example) that it cannot be ejected from the tetrahedral intermediate.

5. One equivalent or less of the electron source was added to the sink at a low enough temperature (usually -78°C) so that the loss of the leaving group from the tetrahedral intermediate by path E_β does not occur to a significant extent before the electron source is used up. Tetrahydrofuran is the preferred solvent.
Example:

Additions of Nucleophiles to Nitriles

With few exceptions, organometallics add only once to nitriles (Section 7.2 m). Only if there is a very strong Lewis acid present to activate the anionic intermediate of the first nucleophilic attack will a second addition take place. An example of this second addition is the reduction of nitriles with LiAlH$_4$, which proceeds all the way to the amine (Section 7.2 l). The AlH$_3$ formed from the initial reduction acts as a Lewis acid to catalyze the second addition. This second addition can be prevented by using a less reactive aluminum hydride or diisobutylaluminum hydride, [(CH$_3$)$_2$CHCH$_2$]$_2$AlH. (See Section 9.5, example 4 for an explanation of the reactions in the acidic water workup.) Example :

$$Ph-C\equiv N: \quad \overset{4e}{\rightarrow} \quad Ph-C=N: \quad H_3O^{\oplus} \quad :O: \\ H-Al\overset{R}{\underset{R}{\diagdown}} \qquad\qquad H \ Al\overset{R}{\underset{R}{\diagdown}} \quad \rightarrow \rightarrow \quad Ph-\overset{\|}{C}-H$$

8.3 IONIZATION OF LEAVING GROUPS

The ionization of a leaving group depends primarily on three factors, the ability of the solvent to stabilize the charges formed, the stability of the carbocation, and the quality of the leaving group. Table 8.1 lists the general trends:

$$R \overset{\frown}{-} L \overset{D_N}{\rightarrow} R^{\oplus} + :L^{\ominus}$$

Table 8.1 The Ionization of Leaving Groups

Carbocation	pK_{aHL}	Solvent	Ionize?
Primary	Any	Any	Rarely
Secondary	Above 0	Any	Rarely
	Below 0	Polar	Slow
Tertiary	Any	Nonpolar	Rarely
	7 or higher	Polar	Rarely
	0 to 6	Polar	Medium
	Below 0	Polar	Fast

To summarize, definitely expect ionization of a leaving group to occur with good leaving groups, in polar solvents, with carbocations of equal or greater stability than tertiary. Do not expect ionization of a leaving group to occur in a nonpolar solvent.

8.4 AMBIDENT NUCLEOPHILES

$$\overset{\ominus}{Z}\overset{E^{\oplus}}{\diagdown}{\underset{\diagup}{C}=C\overset{\diagdown}{\diagup}} \rightarrow E-Z\underset{\diagup}{\overset{\diagdown}{C}}=C\overset{\diagdown}{\diagup} \quad or \quad \overset{\ominus}{Z}\overset{E^{\oplus}}{\diagup}{\underset{\diagup}{C}=C\overset{\diagdown}{\diagup}} \rightarrow Z\underset{\diagup}{\overset{\diagdown}{C}}-C\overset{E}{\diagup}$$

All allylic sources have the capability to act as a nucleophile on either ∂- end of the source. Since the two ends of the allylic source commonly differ in electronegativity, the charge on each end differs. The two ends commonly differ in polarizability also. Since most reactions have both a hard–hard and a soft–soft component, it is important to determine which is the dominant effect. For soft electrophiles, the soft–soft component is most important; therefore the atom with the greatest polarizability will be the best nucleophile.

For hard electrophiles, the hard–hard component is most important, therefore the atom with the largest partial minus will be the best nucleophile (charge control). However, this end of the allylic source is often sterically blocked from being a good nucleophile by ion pairing (for anionic allylic sources) or hydrogen bonding by a hydroxylic solvent. Highly polar aprotic solvents that break up ion pairing but cannot hydrogen bond to the allylic source leave the end with the largest partial minus free to serve as a nucleophile; the products of charge control are seen with hard electrophiles in these solvents.

Alkylation of Enols, Enolates, and Enamines with $\geqslant$C—L

The heteroatom is much harder than the carbon atom and bears more of the charge because of its greater electronegativity. The decision whether to use the heteroatom or the carbon atom as the nucleophile is based on the hardness of the electrophile. Hard electrophiles will attack the heteroatom (hard with hard). Since carbocations are harder electrophiles (greater charge) than $\geqslant$C—L, the tendency to alkylate on the heteroatom will increase as the reaction mechanism shifts from S_N2 toward S_N1. With $\geqslant$C—L, as the leaving group gets more electronegative, the partial plus on carbon increases, and the electrophile gets harder; therefore the amount of alkylation on the heteroatom lone pair increases — for example, oxonium ions, $(CH_3)_3O^\oplus$, sulfate esters, $CH_3OSO_2OCH_3$, and sulfonate esters, $ArSO_2OCH_3$, tend to alkylate on the heteroatom. Trimethylsilyl chloride, $(CH_3)_3Si$—Cl, having a highly polarized silicon–chlorine sigma bond, alkylates on the heteroatom.

Enolate general example:

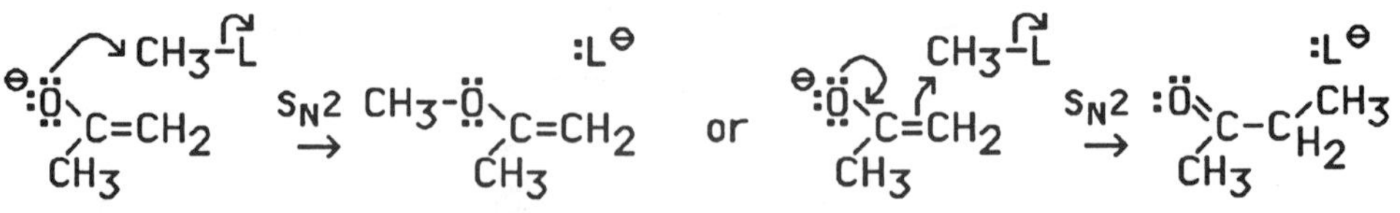

Alkylation on oxygen   Alkylation on carbon

A special case and not really considered an enolate, phenoxide usually alkylates on oxygen because alkylation on carbon would interrupt the aromaticity of the ring:

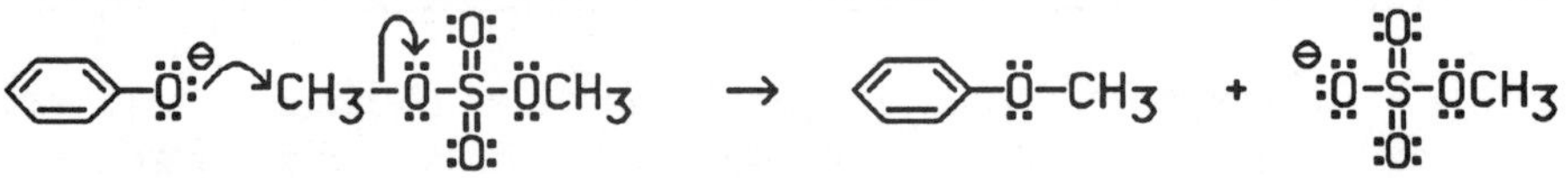

With anionic allylic sources, highly polar aprotic solvents increase the amount of lone pair alkylation because poor solvent stabilization of the anion leaves the heteroatom end less hindered by solvent. Conversely, groups bound to the heteroatom increase the steric hindrance about it, and therefore decrease the tendency toward heteroatom alkylation. For example, enamines (R_2N—C=C), tend to alkylate on carbon rather than nitrogen (as shown in Section 7.2i).

Since alkylation is usually not reversible, the products are the result of kinetic control. However, iminium ions, CH_2=$N(CH_3)_2^\oplus$, are stabilized electrophiles and can reversibly add to enols to produce the thermodynamically more stable C alkylated product.

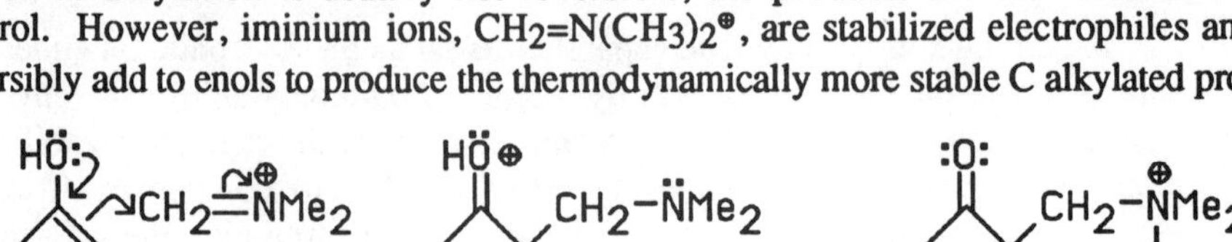

In the following list are the extremes for the alkylation reaction of allylic sources, however product mixtures occur rather often.

Alkylation on Z (less common)   Alkylation on C (much more common)
Hard electrophiles   Soft electrophiles
Highly polar aprotic solvent   Ether or alcohol solvent
Heteroatom accessible   Carbon atom accessible
Irreversible reactions   Reversible reactions

Acylation of Enols, Enolates, and Enamines with O=C—L

Acylation on oxygen   Acylation on carbon

For the most common conditions, **basic media**, acylation almost always goes on carbon. A ΔH calculation predicts the C acylated compound to be thermodynamically more stable than the Z acylated (mainly because of the greater C=Z bond strength in the C acylated product). Acylation on the heteroatom produces a product that can be attacked by another molecule of the allylic source to produce the C acylated compound. The product of any equilibrium is the thermodynamic C acylated product.

The basic media exceptions can be easily understood if we invoke HSAB theory and realize that the kinetic and thermodynamic products are different. As L becomes a poorer donor, the partial plus on the acyl carbon increases making it harder. Acylation on the heteroatom of the allylic source is fast for acyl halides and anhydrides where the acyl carbon is harder (greater partial plus) than the acyl carbon of esters. If the reaction is under kinetic control (allylic source added to an excess of acyl halide or anhydride), the Z acylated product is formed; however, if equilibration occurs (excess of allylic source), the product will be the C acylated, thermodynamic product.

In **acidic media**, the electron sink is most often the carbocation produced from protonating the acylating agent and therefore the sink is very hard. Attack by the Z end (harder end) of the allylic source is very fast. For enols, the Z acylated kinetic product can be isolated. Since the Z acylated enol is itself an allylic source (but weaker), it can be forced by more vigorous conditions to equilibrate to the more stable C acylated product. For enamines, the Z acylated enamine is a good acylating agent; any excess of enamine will attack it and equilibrate it to the more stable C acylated product.

In summary, acylation on Z is achieved with hard electrophiles (acyl halides) under kinetic control; acylation on C is achieved with any electrophile under thermodynamic control. As with alkylation, C acylation is more common.

Amides and Amidates

In amides the oxygen is the better nucleophile because it is partially minus, and the nitrogen is partially plus. The partial plus on the nitrogen repels any like charged electrophile. However, in the amidate anion both nitrogen and oxygen share the negative charge, and therefore the softer nitrogen is commonly the better nucleophile.

Amide resonance forms   Amidate resonance forms

8.5   SUBSTITUTION VS. ELIMINATION

Nucleophilicity vs. basicity is perhaps the most difficult decision to make because almost any anion can serve as either a base (elimination) or a nucleophile (substitution):

$$A:^{\ominus} \ + \ H_2C-C{\leq}L \ \rightarrow \ A-H \ + \ {>}C{=}C{<} \ + \ ^{\ominus}{:}L$$

$$\text{or } A:^{\ominus} \ + \ H_2C-C{\leq}L \ \rightarrow \ H_2C-C{\leq}A \ + \ ^{\ominus}{:}L$$

Again HSAB theory is useful. In general, the C—H bond is significantly harder than the C—L bond. Therefore the softer anions will tend toward substitution, and the harder anions will tend toward elimination.

However, the situation is not that simple, for the accessibility of the substitution site and the size of the anion play a major role also. If the anion is hindered, it is a poor nucleophile. If the site is so hindered that the nucleophile cannot attack it, the balance tilts toward elimination. **The substitution/elimination decision becomes a function of three major variables: nucleophilicity, basicity, and steric hindrance.**

The substitution energy surface can be placed adjacent to the elimination surface since they share the C—L bond-breaking axis. Now the factors that tilted each of the surfaces can be used to understand the competition between substitution and elimination.

In acidic media the S_N1 competes with the E1 process; the first step of both processes is to lose the leaving group to form the cation. Carbocations are excellent electron sinks and tend to react quickly with low selectivity. Equilibrium thermodynamics is therefore the best way to bias this surface (Figure 8.1).

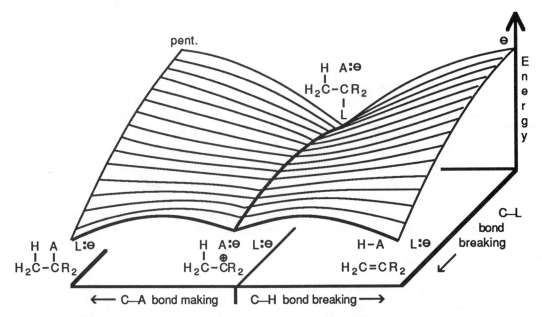

Figure 8.1 The $S_N1/E1$ energy surface.

The C—Nu corner can be raised by making the Nu poor and also a good leaving group so that it falls off again returning to the carbocation. The alkene corner is now the lowest point on the surface and is favored by any equilibrium. An example of this is the elimination of alcohols by concentrated phosphoric acid.

An attempt to bias this surface toward substitution presents a slight problem. The C—Nu corner cannot be lowered very much by making the nucleophile better because the acidic media that the ionization reaction prefers will protonate any good nucleophiles. The best way to increase substitution is to use a low basicity nucleophile and to shift the equilibrium by mass balance: make the nucleophile the solvent if possible (solvolysis). An example of this is the solvolysis of *tert*-butyl bromide in water to produce *tert*-butyl alcohol.

There normally is competition in basic media between the S_N2 and E2. In contrast to the previous surface, Figure 8.2 shows that the split in paths occurs much earlier because the carbocation is much higher in energy and not an intermediate. The choice of which path to take now depends heavily on kinetics because the system is not usually reversible.

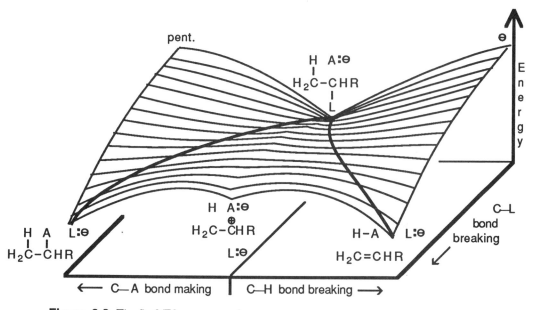

Figure 8.2 The S_N2/E2 energy surface.

To bias this surface toward elimination, one makes the S_N2 a higher-energy route by using hard, highly hindered strong bases, which are poor nucleophiles. Strong bases make the deprotonation energetically more favorable thus lower the energy of the right edge. Poor nucleophiles make the substitution reaction more difficult thus raise the energy of the left edge of the diagram. The entire surface then tilts toward elimination. Two highly hindered strong bases with low nucleophilicities, $(CH_3)_3CO^{\ominus}$ $K^{\oplus}$ and $(i\text{-}Pr)_2N^{\ominus}$ $Li^{\oplus}$, are used to optimize elimination.

Conversely, good, unhindered, soft nucleophiles of low basicity bias the surface toward substitution by tilting it exactly the opposite way. Poor bases raise the energy of the right edge; good nucleophiles lower the energy of the left edge; the entire surface tilts toward substitution.

As was mentioned earlier, the groups around the C—L site make a large difference in whether substitution or elimination occurs. As the C—L site gets more hindered, the left side is raised in energy and the surface then tilts toward elimination.

To summarize, substitution versus elimination is a multivariable decision that we can break down into major and minor variables. The three major variables are C—L site hindrance, nucleophilicity, and basicity; the minor variables are reaction temperature (higher temperature favors elimination), and the electronegativity of the leaving group (more electronegative leaving groups make the C—L carbon harder). A major but less common variable is that electron-withdrawing groups can make the C—H so acidic that elimination dominates.

We need to consider each different type of C—L site as a function of the other two major variables, nucleophilicity and basicity. A three-dimensional correlation matrix is the best way to do this (Figure 8.3).

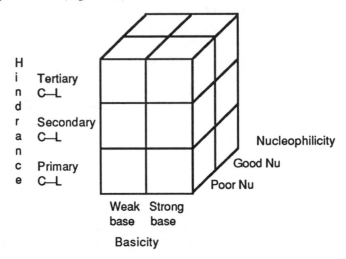

Figure 8.3 A three-dimensional correlation matrix for substitution versus elimination.

We divide each axis into convenient portions corresponding to the most common cases (although we actually have a continuum in three dimensions). To make the individual cells in this matrix easier to view, we will consider each vertical layer separately, and show them as individual two-dimensional matrices in Figure 8.4.

Examples

	Weak base	Strong base	
Good nucleophile	I$^\ominus$ RS$^\ominus$	HO$^\ominus$ EtO$^\ominus$	Good nucleophile
Poor nucleophile	H$_2$O EtOH	t–BuO$^\ominus$ R$_2$N$^\ominus$	Poor nucleophile

Unhindered primary

	Weak base	Strong base
Good nucleophile	Subst.	Subst.
Poor nucleophile	Subst.	Mixture

Secondary

	Weak base	Strong base	
Good nucleophile	Subst.	Mixture	Good nucleophile
Poor nucleophile	Subst.	Elim.	Poor nucleophile

Tertiary

	Weak base	Strong base
Good nucleophile	Mixture	Elim.
Poor nucleophile	Subst.	Elim.

Figure 8.4 Substitution/elimination matrices. The matrix in the upper left gives some common examples of each class of reagents.

Notice how the mixture quadrant in the primary matrix moves around to next quadrant in the secondary matrix, then again in the tertiary matrix, each time leaving behind an elimination quadrant.

For each mixture quadrant more information is required to be able to make a decision on substitution vs. elimination. The major variables have balanced out, and therefore the minor variables can tilt the balance. In the primary mixture quadrant, if the leaving group is a halide, elimination is the major product. If, however, the leaving group is a sulfonate, its electronegativity increases the partial plus on carbon (making it harder); substitution is then the major product. In the secondary mixture quadrant, elimination is common for any reaction that is heated. Elimination has a slightly higher energy barrier than substitution and is therefore favored by higher temperatures. In the tertiary mixture quadrant, a fine tuning of basicity is important, for even mild bases like acetate cause elimination to occur.

For the weak-base, poor-nucleophile quadrant, substitution occurs by the S_N2 path for primary and secondary substrates but because the electron source is poor by definition, the reaction can be very slow if the leaving group is not very good. Tertiary substrates, in the weak-base, poor-nucleophile quadrant, substitute by the S_N1 mechanism.

Even though a group is somewhat removed from the site, it can still hinder the approach of a nucleophile. Although $R-CH_2-L$ is a general description of a primary site, the size of R is very important because it can change which hindrance matrix best applies. The matrix for hindered primary leaving groups, isobutyl for example, $(CH_3)_2CH-CH_2-L$, is similar to the matrix for secondary. The matrix for hindered secondary is similar to that of tertiary. Vinyl leaving groups, $H_2C=CH-L$, do not substitute under normal conditions and with a strong base eliminate to the alkyne. For CH_3-L systems, substitution occurs in all quadrants because elimination is not a structural possibility.

We treat as an exception our less common major variable, the presence of an electron-withdrawing group. Leaving groups beta to an electron withdrawing group, $ewg-CH_2CH_2-L$, commonly eliminate because the increased acidity of the H makes the $C-H$ easier to break thereby favoring elimination.

All multivariable decisions are messy, and if this presentation appears too clean to hold universally, you are correct in your suspicion. We are predicting the predominant product in a product mixture; rarely is the predicted product the only product. There are examples where a major variable joined with the other minor variables (and an experimental technique or two) can outweigh the other two major variables. Alcohol dehydration is such a case. Alcohols are heated in concentrated sulfuric or phosphoric acid, and the alkene is distilled out of the reaction mixture into another flask. The elimination is forced: the media is highly acidic and dehydrating; sulfate and phosphate are weak bases, poor nucleophiles, and good leaving groups; heating not only favors elimination but also helps remove the product, thus displacing any equilibrium.

8.6 AMBIENT ELECTROPHILES

Conjugated Ketone Systems – Enones

Enone   1,4 Attack   or   1,2 Attack

A conjugated system with two $\partial+$ sites can be attacked at either site. The thermodynamic product will be the most stable of the possible products. A ΔH calculation shows that the thermodynamic product is the **1,4 or conjugate addition product**, primarily because of the greater bond strength of the C=O bond than C=C bond. Any equilibrium would produce the conjugate addition product.

Because the effect of an electronegative group diminishes with distance, the carbonyl carbon in this system will have the greatest partial plus and therefore will be harder and attract a hard, negatively charged nucleophile best; the **1,2 product or normal addition product** will be the kinetic product for hard nucleophiles. Soft nucleophiles therefore prefer conjugate attack. Finally, if one site is very sterically hindered, attack at the more open site will dominate.

In summary, highly reactive, harder organometallics add irreversibly to an enone and produce the 1,2 product. The softer, more stable delocalized anions add reversibly and therefore produce the 1, 4 product. Organomagnesium reagents give 1,2 addition except when the carbonyl of the enone is hindered (by a phenyl or larger group). Several nucleophiles and their predominant site of attack on an enone are presented in Table 8.2.

Table 8.2 The Mode of Attack of Various Nucleophiles on Enones

Nucleophile	Favored Attack	Rationale
R-Li	1,2	Very hard organometallic
R-Mg-X	1,2 (usually)	Hard organometallic
R_2CuLi	1,4	Soft organometallic
R-Mg-X/CuX	1,4	Makes soft organometallic by transmetallation
Enolates	1,4	Soft Nu, reversible addition
Lone pair Nu	1,4	Stable Nu, reversible addition
$LiAlH_4$	1,2 (mainly)	Hard hydride source
$NaBH_4$	1,4 (mainly)	Softer hydride source

Miscellaneous Ambident Electrophiles

As the carbonyl becomes more reactive, 1,2 attack will become more predominant; in contrast to what was found for enones, borohydride attacks 1,2 on conjugated aldehydes. All organomagnesium reagents add 1,2 on conjugated aldehydes. With lone pair nucleophiles (often very hard), conjugated acyl halides undergo 1,2 attack (hard carbonyl, often irreversible 1,2 addition) whereas conjugated esters undergo 1,4 attack (softer carbonyl, more reversible 1,2 addition). Only 1,2 attack occurs on Ph—ewg because conjugate attack would disrupt the aromaticity of the aromatic ring.

8.7   INTERMOLECULAR VS. INTRAMOLECULAR

One of the determining factors in reaction rates is how frequently the reaction partners collide. If collisions are more frequent, the reaction rate will be faster. Increasing the concentration of a reactant invariably increases the collision frequency and therefore the reaction rate. If the nucleophile and electrophile are part of the same molecule, they may collide much more often than is possible in even the most concentrated solutions. Therefore intramolecular reaction rates can easily exceed intermolecular ones. The energy of the cyclic transition state for an intramolecular process depends on the rigidity and size of the loop of atoms in the cyclic transition state and on the orbital alignment restrictions of the process involved.

For the S_N2 reaction shown in the following example, the determining factor for ring closure in a freely rotating chain is its size. The formation of very small rings requires the bending of sigma bonds, creating significant ring strain. In the formation of large rings (rings greater than eight atoms), the ends that must be brought together to form the ring may collide so infrequently that bimolecular reactions can easily compete. The optimum ring size for closure is a five-membered ring; six is not as good; seven starts to get too large. A three-membered ring forms easily, for although there is much ring strain, the number of effective collisions is great since the ends are so close. Four-membered ring formation is poor because the ring strain has not dropped off appreciably, but the number of effective collisions has.

Example:

$$HO:^{\ominus} \; + \; ^{\ominus}:O-(CH_2)_nCH_2-L \quad \xrightarrow{S_N2} \quad ^{\ominus}:O-(CH_2)_nCH_2-O-H \; + \; (CH_2)_n\!\!\diagup\!\!\overset{O}{\diagdown}\!\!CH_2$$

Intermolecular S_N2 or   Intramolecular S_N2

Although there is some variation among systems, the general order is 5 > 6 > 3 > 7 > 4 > 8 to 10-membered transition states. In more concentrated solutions intermolecular reactions may be competitive with four-membered ring formation. In summary, **any intramolecular process with a 5-, 6-, 3-, or 7-membered transition state is almost always faster than the corresponding intermolecular process.**

In forming a ring on a substrate that has some conformational rigidity, you must be sure that the ends can reach each other. In fact, any intramolecular process should be checked with molecular models to see if the reacting partners can interact as needed for the reaction.

8.8 KINETIC VS. THERMODYNAMIC ENOLATE FORMATION

$$H_3C-\overset{\overset{\displaystyle :O:}{\|}}{C}-CH_2R \quad \xrightarrow[\text{Base}]{\text{p.t.}} \quad H_2C=\overset{\overset{\displaystyle :\overset{..}{O}:^{\ominus}}{|}}{C}-CH_2R \quad \text{and/or} \quad H_3C-\overset{\overset{\displaystyle :\overset{..}{O}:^{\ominus}}{|}}{C}=CHR$$

Kinetic   Thermodynamic

An unsymmetrical ketone can deprotonate to form two possible enolates, and the choice of which enolate forms depends on whether the deprotonation was carried out under kinetic or thermodynamic control. The thermodynamic enolate is the most stable of the possible enolates; in general the more substituted enolate is the more stable (reflecting the double-bond stabilities). The kinetic enolate is the result of deprotonating the most accessible acidic hydrogen. Not only are the three hydrogens on a methyl more accessible than a hydrogen on a more substituted center, but also one of them is bound to be close to the proper orbital alignment needed for deprotonation.

Deprotonation of the ketone must be fast and complete for kinetic control of enolate formation. No equilibration of the enolates can be allowed to occur. Optimum conditions for kinetic control of deprotonation are to add the ketone slowly to an excess of very strong base (usually i-Pr$_2$NLi, the anion of diisopropyl amine, $pK_{abH} = 36$) in an aprotic solvent (such as dry tetrahydrofuran or dimethoxyethane). Since the K_{eq} for deprotonation of a ketone with this base is $10^{(36 - 19.2)} = 10^{+16.8}$, the reaction is complete and irreversible.

Any equilibrium will produce the thermodynamically most stable enolate. The most stable enolate will have the greatest charge delocalization. In the following example, the thermodynamically favored enolate is conjugated; the kinetically favored enolate is not. Common conditions for thermodynamic control are to use average bases (like sodium ethoxide or potassium *tert*-butoxide, pK_{abH}'s 16 to 19) in alcohol solvents. Proton transfer equilibria rapidly occur among base, solvent, ketone, and enolate. Sodium hydride or potassium hydride in an ether solvent are thermodynamic reaction conditions that allow equilibration between the ketone and the enolate.

Example:

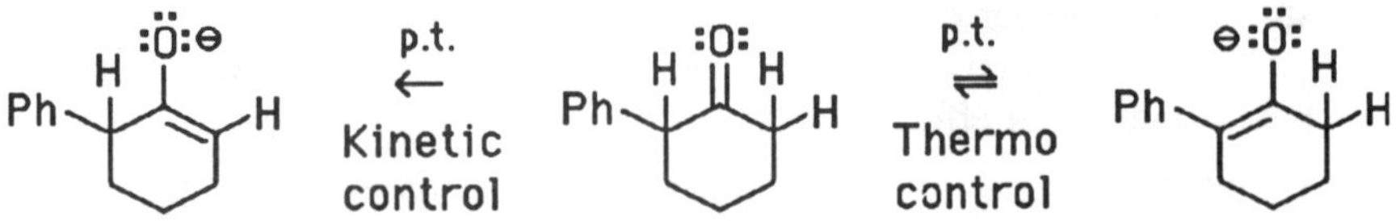

ADDITIONAL EXERCISES

(Refer to the Hints to Problems from Chapters 7, 8, and 9 if you need assistance.)

8.1 For each of the following reactions decide whether substitution or elimination predominates.

	Substrate	Reagent	Temperature
			(assume 25°C if not given)
(a)	CH₃CH₂CH₂CH₂OH	HBr	
(b)	CH₃CH₂CH₂CH₂OTs	*t*-BuO⊖	
(c)	CH₃CH₂CH₂CH₂Br	*t*-BuO ⊖	
(d)	(CH₃)₂CHCH₂Br	I⊖	
(e)	(CH₃)₂CHBr	EtOH	
(f)	(CH₃)₂CHBr	EtO⊖	55°C
(g)	(CH₃)₂CHCH₂CH₂Cl	[(CH₃)₂CH]₂N⊖	
(h)	(CH₃)₂CHBr	CH₃COO⊖	
(i)	(CH₃)₂CHBr	EtS⊖	
(j)	(CH₃)₂CHCHBrCH₃	CH₃COO⊖	55°C
(k)	CH₃CH₂CBr(CH₃)₂	CH₃OH	
(l)	(CH₃CH₂)₃CCl	CH₃O⊖	
(m)	(CH₃)₃CBr	CH₃COO⊖	

8.2 For each of the following reactions a decision discussed in this chapter is required. Choose which product would be preferred.

(a)

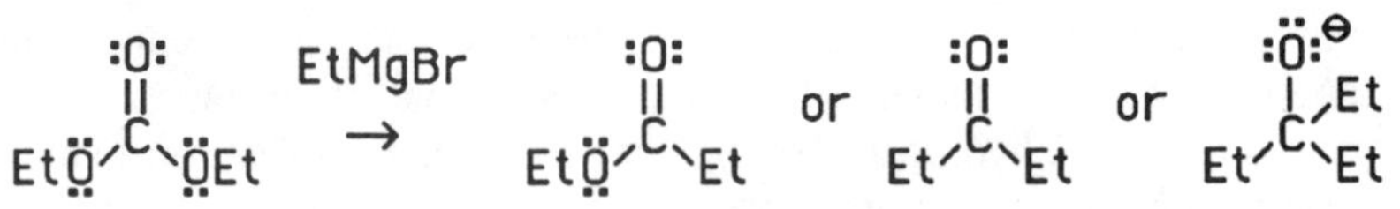

(b)

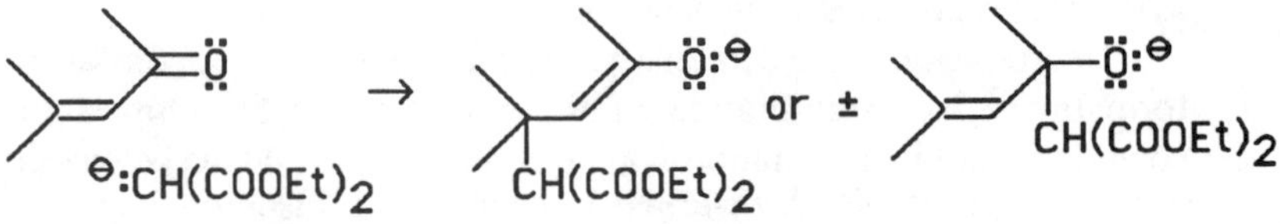

(c)

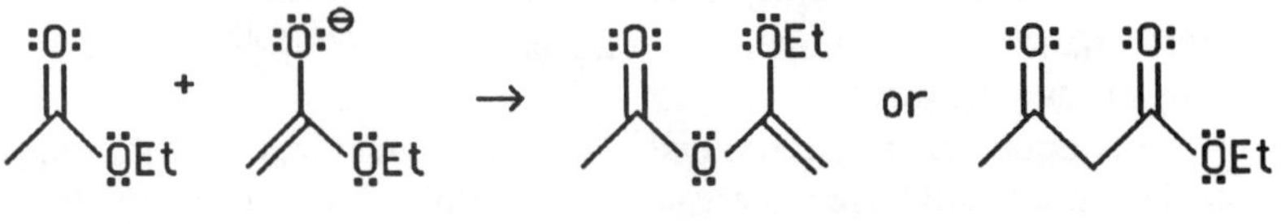

8.2 (continued)

(d)

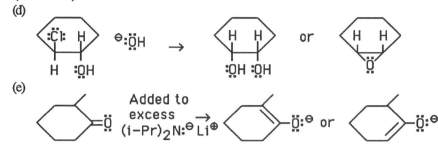

(e)

8.3 Provide a mechanism for the formation of the preferred products in problem 8.2. Combine the source and sink with the appropriate pathway.

8.4 Each of the following reactions has only one electron source and sink. Give the product of the following reactions; a decision discussed in this chapter is required.

(a)

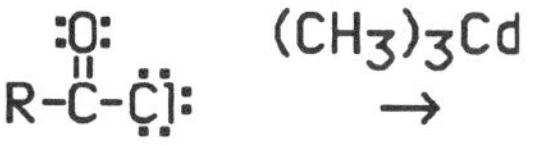

$$
\text{R-}\overset{\displaystyle :\overset{..}{O}:}{\underset{}{\overset{\|}{C}}}\text{-}\overset{..}{\underset{..}{C}}\text{l}: \qquad (CH_3)_3Cd \;\rightarrow
$$

(b)

$$
\text{Ph-}\overset{\displaystyle :\overset{..}{O}:}{\underset{}{\overset{\|}{C}}}\text{-}\overset{..}{\underset{..}{O}}\text{Et} \qquad \text{Ph}-MgBr \;\rightarrow
$$

(c)

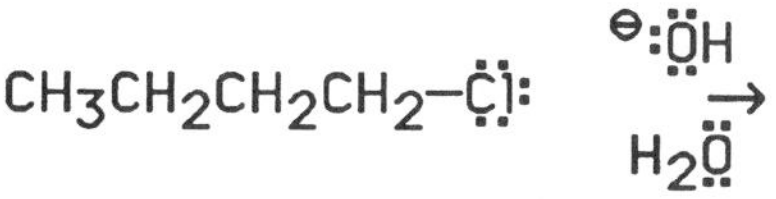

$$
CH_3CH_2CH_2CH_2\text{-}\overset{..}{\underset{..}{C}}\text{l}: \qquad \overset{\ominus :\overset{..}{O}H}{\underset{H_2\overset{..}{\underset{..}{O}}}{\rightarrow}}
$$

(d)

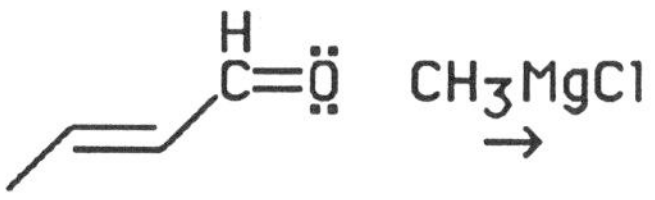

$$
\overset{H}{\underset{}{}}\text{C=}\overset{..}{\underset{..}{O}} \qquad CH_3MgCl \;\rightarrow
$$

(e)

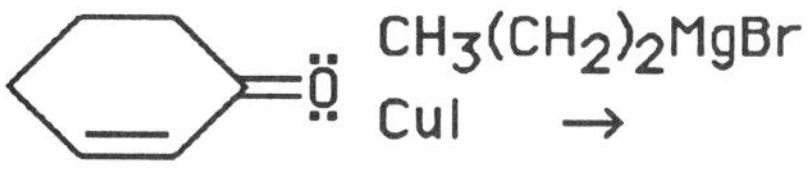

$$
\text{=}\overset{..}{\underset{..}{O}} \qquad \overset{CH_3(CH_2)_2MgBr}{\underset{CuI}{}} \;\rightarrow
$$

(f)

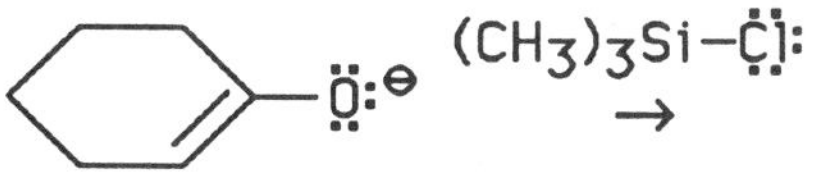

$$
\text{-}\overset{..}{\underset{..}{O}}\text{:}^{\ominus} \qquad (CH_3)_3Si\text{-}\overset{..}{\underset{..}{C}}\text{l}: \;\rightarrow
$$

(g)

$$
\text{-}\overset{..}{\underset{..}{O}}\text{-}\overset{\displaystyle :O:}{\underset{:O:}{\overset{\|}{S}}}\text{-}\!\!\!\!\text{-}CH_3 \qquad +\overset{..}{\underset{..}{O}}\text{:}^{\ominus} \;\rightarrow
$$

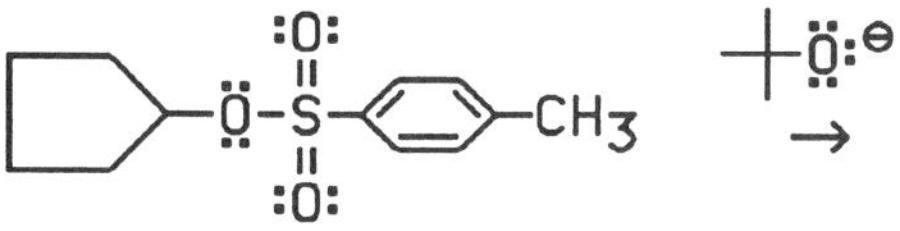

9

CHOOSING THE MOST PROBABLE PATH

9.1 PROBLEM-SOLVING IN GENERAL

Study the Material before Attempting the Problems; Establish an Informational Hierarchy while You Study; Assemble a "Toolbox" of Commonly Needed Items for Problem Solving; Read the Problem Carefully – Understand the Problem; Gather Together All Applicable Information about the Problem before Starting It; Classify into Generic Groupings; Recognize Possible Intermediate Goals; Always Write Down Any Possibility That You Consider; Have a Systematic Method to Your Answer Search; Always Look for Alternatives and Write Them Down – Generate All Paths and then Select the Best; Recognize the Generic Form of Each Step; Work Carefully and Cross-Check the Work as You Go; Don't Skip Steps – Look for Any Hidden Decision Points; Make a Scratch Sheet into an Idea Map; Beware of Memorization; Watch for Bad Habits; Stay on the Pathways; Don't Force the Answer! If Stuck, Don't Just Stare at the Page, Draw Something on It! Recognize the Limits Placed by the Reaction Conditions; Beware of Limits That You Place on the Problem; If Still Stuck, Retrace the Route, Examining All Other Possibilities at Each Decision Point; When Done, Always Go Back and Check the Answer; Practice, Practice, Practice; What Can Be Learned From the Problem and the Methods You Used To Solve It That Would Be Applicable to Other Problems?

9.2 GENERAL MECHANISTIC CROSS-CHECKS

Electron Flow Pathway Check, Completeness Check, Media Check, Energetics and Stability Check, Charge and Typo Check

9.3 THE PATH SELECTION PROCESS

Understand the System – Look Around and Gather Helpful Information
Find Possible Routes – Find the Paths, Explore Short Distances
Evaluate and Cross-Check – Pick the Lowest Energy Route, Cross-Check
Repeat the Process

9.4 REACTION MECHANISM STRATEGIES

Helpful Information Can be Derived from the Product; Make Sure the Chemical Equation Balances; Identify the Carbons That Belong to the Starting Materials within the Product; Decide What the Original Nucleophile–Electrophile Pairs Might Have Been; Consider Any Reasonable Proton Transfer; *Generate All the Possible Paths, Then Select the Best;* If It Does Not Make Sense, It's Wrong

9.5 WORKED MECHANISM EXAMPLES

(1) Alkyne Hydrobromination
(2) Aldol Condensation
(3) Ketal Bromination
(4) Organometallic on a Nitrile
(5) Glucose to Fructose

9.6   WORKED PRODUCT PREDICTION EXAMPLES
     (1) The Reaction of a Carboxylate and 4-Bromophenacyl Bromide
     (2) The Reaction of an Ester and Ethoxide
     (3) The Reaction of a Ketal and Acidic Water
     (4) The Reaction of an Amide, Basic Water, and Heat
     (5) The Reaction of a Carboxylic Acid and Thionyl Chloride
9.7   METHODS FOR TESTING MECHANISMS
     Initial Studies; Catalysis; Inhibition; Isolation or Detection of Intermediates;
     Labels; Stereochemistry and Chirality; Kinetics; Solvent Effects; Substituent
     Effects; Primary Deuterium Isotope Effects; Barrier Data

9.1   Problem-Solving in General

The following hints are intended to help you avoid the most common "traps" or conceptual blocks encountered in solving a problem in organic chemistry or actually in any field. Specific decisions were covered in Chapter 8 and procedures applicable to various problem types are covered later in this chapter. All of these procedures are extra work when they are compared to "just having the answer pop into your head," but they make the problem-solving process more systematic and more reliable; they make it easier to retrace your thinking when checking the answer and simpler to "debug" when things go wrong.

The more complex and difficult a problem is, the less likely you will be able to see the answer from the start. Therefore you must begin work on the problem without any clear idea of your goal. If so far you have always seen the answer from the start, working on a complex problem is, in a way, like flying on instruments through a cloud bank when you are used to visual navigation. Very often students will just stare at the problem and wait for the answer to hit them. With difficult problems this can be a very long wait indeed. These procedures allow you to pursue the answer actively rather than wait for it to come to you.

Study the Material before Attempting the Problems

Study with a pencil to underline important ideas; make margin notes about things that do not make sense; write down questions to ask the instructor; make a written summary of what you have read. Understand the trends, vocabulary, and principles that are needed to solve the problem. Target your study to concentrate on weak areas as indicated by exams and homework.

Establish an Informational Hierarchy while You Study

You tend to remember the things that you use the most and those that are most recent. The only way you can raise the mental awareness of the important concepts over the more recent insignificant fact is to **review the most important material more often**. As any course of study progresses, the incoming facts and important concepts must be put in their place in the hierarchy. You manage an information overload by allowing the little fact to be forgotten after it has served its purpose of being an example for a more important principle. The older, important concepts must be continually be raised above the newer facts by additional review (even though you feel you know the concepts very well; you are just reinforcing their importance by keeping them active). Continuously review the classification of the sources and sinks and the electron flow pathways.

Assemble a "Toolbox" of Commonly Needed Items for Problem Solving

As a start toward this, the Appendix gathers together important tools: a pK_a chart, a bond-strength table, a list of sources and sinks, the electron flow pathways, trends, rules of thumb, and other useful items. Continue to customize your toolbox with any valuable tools that you need.

Read the Problem Carefully — Understand the Problem

Be sure to know what is asked. Can you rephrase the question in your own words? Make a model or draw a figure. Can you break the problem into separate, smaller, more easily solved units? Can you give a rough estimate of the overall process?

Gather Together All Applicable Information about the Problem before Starting It

Look to see that there is enough information to solve the problem. Search for relationships between the data and what is wanted. Write down any chemical equations that relate to the problem. What are the assumptions and limitations that come with the problem? What principles was the problem designed to illustrate?

Classify into Generic Groupings

Back off from the problem and get an overview of it. Have you seen a problem of the same generic class before? Often changes in the hydrocarbon "grease" around a reactive site will disguise the similarities to reactions that you have seen before: don't slip on the grease! What are the *real* similarities and differences? Decide what parts of that related problem are useful. Have you seen a problem with a similar start or finish point?

Recognize Possible Intermediate Goals

For a mechanism question, number the atoms in the starting material and try to find those atoms in the product. Find out what bonds need to be broken or made. On what atoms do you need to generate nucleophiles or electrophiles? What are all the possible starting points? Are there any possible last steps? Middle steps?

Always Write Down Any Possibility That You Consider

Work on paper, not "in your head." The biggest problem-solving trap is to shoot down a possibility "in the air" without writing it down first. **Writing down something that you are considering forces you to take more time to think about it** and produces a written record that it has been checked out.

Have a Systematic Method to Your Answer Search

Random trial and error does not work on complex problems. A systematic search for the answer may seem like a long and tedious process, but it is much quicker and more reliable than random shots in the dark.

Always Look for Alternatives and Write Them Down — Generate All Paths and then Select the Best

Pick the best path and check it out, but always keep in mind the other alternatives in case the first one does not work out. In writing down the alternatives, you may get other ideas. Don't waste a lot of time at an apparent dead end before going back to check out other possibilities.

Recognize the Generic Form of Each Step

Classify each step or group of steps as a substitution, elimination, addition, or rearrangement. Then check for other alternative forms of the same process that may fit the reaction conditions better. For example, your first impulse in a particular problem step might be to write some sort of elimination because it seems to get you closer to the answer. If you recognize that what you did was a type of elimination, you would know that eliminations could go by either an E1, E2, E1cB, or Ei mechanism. You can then pick the most appropriate one, rather than remaining with the first one that occurred to you.

Work Carefully and Cross-Check the Work as You Go

Simple errors can lead to some rather absurd predictions. Arrow pushing is a form of electron bookkeeping; therefore the arrows and Lewis structures must correlate and be accurately drawn for the results to make sense. The total charge on both sides of the transformation arrow should be the same. Pay attention to detail.

Don't Skip Steps — Look for Any Hidden Decision Points

One common trap is to skip a step rather than take the time to write out the intermediate. If there is another route from the unwritten intermediate, the decision point will be missed and difficult to locate on a cross-check. Be continually on the lookout for the major decisions discussed in Chapter 8. The most common and also the most commonly missed decision is the nucleophile versus base decision.

Make a Scratch Sheet into an Idea Map

Keep your scratch sheets neat. Be organized on the scratch paper so you can trace your thoughts and be able to go back to other ideas and check thinking **without getting lost**. Remember, you are exploring unknown territory; draw yourself a map as you go.

Beware of Memorization

Memorization is never a satisfactory or reliable substitute for understanding what you are doing. Beware of mindlessly "turning the crank" on a problem; it might not be connected to anything at all.

Watch for Bad Habits

Do not stop at the first answer you find but continue to look for alternative better solutions. Make up a list of your own problem-solving bad habits and refer to it when you work a problem. For example, if you usually forget to check for nucleophilicity vs. basicity, write yourself a reminder. Know the common "potholes" on the route to your answer.

Stay on the Pathways

Although it is tempting to get the problem over with by using a burst of arrows and as few steps as possible, our object is to write a **reasonable** hypothesis, not necessarily the shortest one. **Construct your mechanistic "sentences" with known, tested "words," the electron flow pathways.** Mechanistic steps that you invent as a beginner may not be reasonable.

Don't Force the Answer!

Be patient. The correct answer will seem to have a natural flow to it; nothing will have to be forced to fit. Some dead ends can seem frustratingly close to the final answer, and there is a tendency, born out of this frustration, to "hammer home" the "last step" even though you know it is not right. Put the hammer away when working a jigsaw puzzle.

If Stuck, Don't Just Stare at the Page, Draw Something on It!

Reread the problem, then draw resonance forms, three-dimensional diagrams, other equilibria, less probable possible paths, anything to search actively for that crucial hint that is needed. **Blank paper gives no hints.**

Recognize the Limits Placed by the Reaction Conditions

Is the reaction in acidic or basic media, in protic or aprotic solvents?   Classify the electron sink, then sort out those pathways that do not use that sink to help you reduce the number of paths you need consider. Is the reaction site accessible or is it blocked by large groups? Are there constraints on orbital alignment that need to be met?

Beware of Limits That You Place on the Problem

Sometimes we have a tendency to make an initial guess as to the region in which the solution will be found, and later when the solution is not found within those limits to forget that those limits were not imposed by the problem. Without being aware of it, we can artificially confine our search to an area that does not contain the answer.

If Still Stuck, Retrace the Route, Examining All Other Possibilities at Each Decision Point

Force yourself to consider other alternatives. One tends to be reluctant to cross-check a hard decision, especially if it was a hassle to make. When cross-checking decisions, make sure that you **go back all the way to the first decision** at the start. Check all decisions, especially those that seem obvious.

When Done, Always Go Back and Check the Answer

Can you see the answer now at a glance? Is there a second way to get the same result? Does it make sense? Use the law of microscopic reversibility to check alternatives on reversible steps. Is the assumed causality correct?

Practice, Practice, Practice

Only by working many problems and many different types of problems can you get good at problem-solving. Reading the material is not sufficient; you need to be able to **use** what you've learned. You need to work actively with the material to build a network of interrelated concepts. Working problems allows you to find your weak areas before an exam reveals them.

What Can Be Learned From the Problem and the Methods You Used To Solve It That Would Be Applicable to Other Problems?

Learn not only from your mistakes but also from your successes!

9.2   GENERAL MECHANISTIC CROSS-CHECKS

For a mechanism to be considered reasonable it must fit into the body of knowledge that has accumulated over the years. The following are some points that need to get checked against any mechanistic proposal in order to judge whether or not the mechanism is reasonable. Section 6.5 gives examples of errors that these checks are designed to catch.

Electron Flow Pathway Check

The most important check is to make sure that each mechanistic step is a valid electron flow path. There is a temptation, arising out of frustration, to rearrange the lines and dots of reactant into the lines and dots of product with a mechanistically meaningless barrage of arrows just to get the problem over with. **It is important that you assemble your mechanistic hypothesis from known steps, the 20 electron flow pathways.** Check that the electron flow starts at a lone pair or bond that is a good electron source. Make sure that the flow continues without interruption and ends at a good electron sink. Did you consider alternate pathways that achieve the same overall transformation and then pick the most appropriate?

Completeness Check

When you generate the possible paths make sure that you have not left out any important alternatives, then pick the most probable and check that decision. A common problem-solving error is to race off with the first possibility that looks reasonable, failing to check all the possibilities. **This completeness check is especially important for the first step where the initial direction that you set out to explore is determined.** If your first step is in error, you may be reluctant, several steps later, to go back to the beginning and reconsider. Rather than retreat to the start, the typical student is more likely, as frustration builds, to force an incorrect answer just to be done with the problem.

Media Check

An important cross-check is the media restriction. Paths that form reactive cations almost exclusively occur in acidic media. Likewise paths that form reactive anions are the domain of basic media. No medium can be both a strong acid and a strong base; it would neutralize itself. **The reactive species in equilibrium-controlled reactions have a limited range of acidities.** For example, in neutral water the hydronium ion concentration and the hydroxide ion concentration are both 10^{-7} mol/L. Their relative concentrations are defined by $K_w = [\text{H}^{\oplus}][\text{OH}^{\ominus}] = 10^{-14}$. Their pK_a values span 17.4 pK_a units. Would a reaction mechanism proposal be reasonable if it required both hydronium and hydroxide ions? At what point does the span of pK_a values become unreasonable because the needed ions have so low a concentration that the reaction rate essentially stops?

To attempt to answer these questions, let's return to the example given in Section 3.3, the deprotonation of an ester (pK_a 24) by ethoxide ion (pK_{abH} 16); the pK_a span is 8. This small concentration of the ester enolate can then enter into a reaction with the original ester (see the product prediction example two in Section 9.6 for a discussion of the entire surface). If we try to run this reaction catalyzed by phenoxide ion (pK_{abH} 10) instead of ethoxide, the reaction fails; no product is isolated, only unreacted starting materials. There are two ways that we could interpret this result: first that the base was too weak, and no appreciable concentration of ester enolate was formed. A second interpretation is that there no longer was sufficient driving force for the reaction since the anion of product is now a stronger base (only by 0.7 pK_a units) than phenoxide; the reaction is no longer forming a weaker base. The pK_a span with phenoxide ion was 14 pK_a units.

Therefore as a first approximation, **a pKa span of 8 units is acceptable, but a span of 14 pKa units is probably not.** In your mechanisms, make sure that the span of pK_a's of any species that are in equilibrium do not stray too far from the acceptable range. If several steps occur in the same medium, check the pK_a span of all species within those steps. In a way, we are just restating the ΔpK_a rule (which set a limit of eight for a single step) and applying it to all species in equilibrium throughout the reaction. Certainly the pK_a span at which the reaction rate becomes so slow that it is unacceptable is open to debate.

Exceptions are expected to arise when the driving force for a reaction is an irreversible proton transfer at the diffusion controlled limit in water or a similar protic solvent. With a maximum bimolecular rate constant of 10^{10} L/mol-sec and a solvent concentration of 55.5 mol/L, a reactive intermediate could conceivably have a concentration as low as 10^{-16} mol/L, and the reaction still proceed at a reasonable rate at room temperature.

Energetics and Stability Check

We have three ways to check the energetics of a process. For proton transfer reactions we can calculate a K_{eq}. A K_{eq} greater than 1 indicates that products are favored and that the $\Delta G°$ is negative. If neutral compounds are transformed into neutrals, we can calculate the ΔH for the reaction from the bond strengths. If the reactants or products are ionic or neutral, we can use the ΔpK_a rule. The most reliable of these three is the K_{eq}, next probably the ΔH, and least the ΔpK_a rule. A common error is to forget to check the proton transfer K_{eq}. Use the trends to gauge the stability of any intermediates formed in the reaction. Common errors include the failure to judge the stability of carbocations, carbanions, and leaving groups.

Charge and Typo Check

Make sure that the Lewis structures and their formal charges are correct. An error on a formal charge can be especially dangerous since the attraction of opposite charges is used to predict reactive sites. The charge must balance on both sides of the reaction arrow. Do not accumulate charges on your intermediates as they proceed through the reaction. Avoid generating any species with adjacent like charges. Look for line-structure errors, particularly for the "vaporization" of an H that was not drawn in the line structure.

9.3   THE PATH-SELECTION PROCESS

One can view the reaction process as a wandering over an energy surface via the lowest energy path toward the lowest accessible point on the surface. The reacting partners have a limited amount of energy; only certain transition states and intermediates can be achieved and will end up at the **lowest accessible** point on the surface, which may not necessarily be the lowest point on the surface.

The most useful mental models approximate some physical reality. Since we will be exploring a energy surface with peaks, passes, and valleys unknown to us, the best mental model would be a mountainous wilderness, but one luckily crossed with trails. Imagine being dropped off in the middle of this wilderness at a trail junction, assured that one of the many trails leads down and out. We know that the desired route will be the lowest energy route. We certainly do not want to panic and get off the trails into real unknown territory or to guess randomly at a direction and wander off. We must carefully and logically explore the nearby wilderness drawing a map as we go.

Armed with the tools that we have learned in the previous chapters, we can predict where the lowest energy path may go. The **electron flow pathways are the trails to guide us** in this wilderness of an unfamiliar energy surface. Some trails may dead end in high valleys that have no reasonable exit other than the one we came in. Others may lead us to a low point (side product) on the energy surface but not the lowest accessible point (major product). The reacting partners are also exploring these dead ends, and very few reactions produce only one product .

What do we need in order to succeed? We need the ability to draw a good map (if you don't want to repeat some paths or wander around in circles), the ability to recognize whether we are going uphill or down (energy awareness), the ability to recognize all the paths and when they branch (you can't stay on the trails if you can't recognize them), and finally the ability to cross-check what you have done (was the hike to a certain point a silly thing to do or is it on the right track?)

The exploration process is really quite simple. The hard part is to stick with it when things are not going well and you are feeling lost; it's easy to get discouraged and do something rash just to be done with the struggle. The other time that it is hard to stick with it is when things are going so well that the end is in sight, and it feels that it might be fine to abandon the trails and just make a straight run for the end. Both arise out of the natural tendency to rely on insight because the alternative, rigorous logical analysis, is a lot of hard work.

So what is the search process? First understand where you are. Observe your surroundings; take inventory; classify what you have to work with. Second, start drawing your map by marking the start point down and note all the paths that lead from it. Find a downhill path or at best the least uphill path. Follow that path to the next point and mark that on your map. Again observe your surroundings. Now cross-check that going to this new point is indeed a reasonable thing to have done (here is where you make use of all those trends). Cross-check that the energy of the route is reasonable (with the ΔpK_a rule or K_{eq} calculation or ΔH calculation — whichever is appropriate). Now look for all the new paths leading off this new point, and start the process again. Always keep in mind your starting point and do not climb too high before returning to a lower point to explore other routes.

How do you go about learning this method of exploration? First learn the needed skills (Chapters 1—8), then try them out on easy problems with the help of an experienced guide. On these guided tours we will explore the entire surface, including all its dead ends, so that you feel you know the territory. We will gradually explore more difficult surfaces until you feel confident enough to venture out on your own. Let's look at each step of the exploration process in more detail.

Understand the System

Write down the Lewis structure of the reactants, complete with formal charges, and draw any major resonance forms. Look for leaving groups, polarized single and multiple bonds, acids and bases. Classify into generic sources and sinks and then rank them. **The** reaction usually occurs between the best source and sink. Above all, note if the media is acidic or basic. In basic media, find the best base, then locate any acidic hydrogens within range (not much more than 8 pK_a units above the pK_{abH} of the base). In acidic media, identify the best sites for protonation. Likewise, do not create a species that is much more than 8 pK_a units more acidic than your acid.

Find Possible Routes

Generate all the possible paths. The number of trails are few since only those pathways that have the same sources and sinks as the reacting partners need be considered. The media will also restrict paths. Do not use acid media paths when the medium is definitely basic and vice versa. Draw the Lewis structure of the product. Never discard a route without first taking a small amount of time to write down the result (this will force you to consider it more carefully). Again, use scratch paper to draw a map of this new energy surface that you are exploring so that you can keep track of where you have been, note branches in the path that you have passed but not explored (and may want to come back to), and write down all the possibilities branching from the spot that you are currently considering. Drawing such a map may seem like it would slow down the process of reaching the goal, but actually it speeds it up because much less time is spent lost, wandering aimlessly, taking random shots at the answer.

Evaluate and Cross-Check

So that we do not have to explore every path, but only the most probable ones, the trends allow us to gauge the relative height of the passes and high valleys that we have to choose among. Especially cross-check the media restriction and the stability of intermediates. We can often use either the $\Delta p K_a$ rule on our alternatives or the calculation of the ΔH or the K_{eq} to make sure the energetics of the process are reasonable. **Generate and select** the best path. As with any problem for which you cannot see the route to the answer at the start, you must be very careful at the beginning to select the best route, not just the first one you see. Double-check any major decision.

Repeat The Process

With mechanism problems, the final product is given, so the end point is unambiguous. The problem-solving thought process resembles a deliberate search from the starting point which eventually triggers an insight coming from the goal. The methods of deliberate searching can be learned by a novice; insight comes by being familiar with the territory and will get better with practice.

With product-prediction problems, there will always be some doubt as to whether you are at the lowest accessible point on the surface or are in the energy well of a side product. Hopefully, the end should be obvious because you will definitely be downhill from your starting point, and it will seem that all alternatives for several steps around are definitely uphill. Again, knowing when you are done also improves with practice.

9.4   REACTION MECHANISM STRATEGIES

Mechanism problems are easier than product-prediction problems because helpful information can be derived from the product. Given the starting and ending points you need only supply the lowest energy path between them. You may not be able to see the answer to a long mechanism problem at the start; just let a logical analysis carry you to the answer. Don't panic and try to force a quick (and incorrect) answer.

Commonly, organic equations are not balanced; so make sure the chemical equation balances. This may give you a hint in that it may identify another product; for example, water may have been lost. If there is a small piece lost in the reaction, try to find out where it came from.

Next identify the carbons that belong to the starting materials within the product (number the carbons if you have trouble) so that you can tell what bonds have been made or broken. If the starting material structure just will not fit into the product structure (taking into consideration any small pieces lost), rearrangement may have occurred.

Look at each new bond and then at the polarization of the corresponding sites in the starting materials so that you can decide what the original nucleophile–electrophile pairs might have been for those bonds. If more than one new bond has been made, consider all possible sequences of bond formation, not just the first one that occurs to you. Look at adjacent functionality to get a hint as to whether it would stabilize an electrophile's or nucleophile's partial or full charge. Ask how proton transfer might have generated the needed nucleophiles and electrophiles. Consider any reasonable proton transfer and always check the K_{eq} value (at best $\geq 10^{-8}$).

Use the general trends to guide the decision process. At every decision point, generate all the possible paths, then select the best one on the basis of trends. For example, if there is a choice of making two possible cations, rank their stability and pick the route that goes through the most stable cation. If nothing looks reasonable, return to any decision point where the trends indicated that an alternative was also feasible. Use "hindsight" (knowing the final product) as a guide through any difficult decisions.

Finally, do not combine steps to save drawing out the structure again. Keep to the pathways to avoid the trap of merely rearranging the lines and dots of the structures on the page. Be as rigorous in your own practice as you would be on an exam; you are trying to develop good intellectual habits. The prime directive is this:

If it does not make sense, it's wrong.

9.5 Worked Mechanism Examples

The problems at the end of Chapters 7 and 8 were short and simple mechanism and product-prediction problems. Chapter 7 problems had only one source and sink and no major decisions to be made, and those in Chapter 8 were the same except one major decision was required. We are now going to cover problems of gradually increasing complexity: more than one source and sink, longer problems, more decisions, and more alternate routes. For each example, cover each step and try to provide the expected information; compare yours with what is given, then move on to the next portion.

1. Alkyne Hydrobromination

Give a reasonable mechanism for the following reaction.

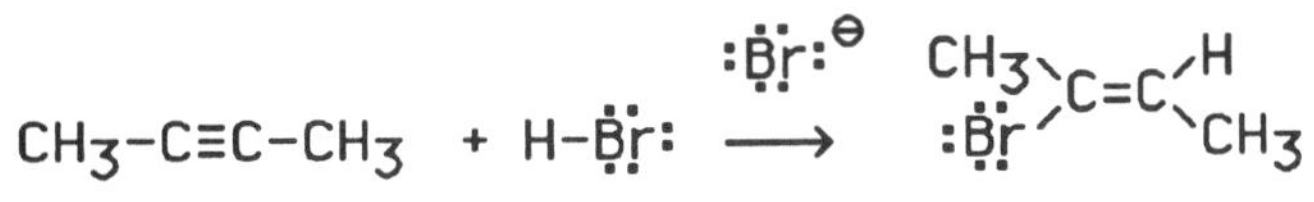

$$CH_3-C\equiv C-CH_3 \ + \ H-\overset{\cdot\cdot}{\underset{\cdot\cdot}{Br}}: \ \longrightarrow \ \overset{:\overset{\cdot\cdot}{\underset{\cdot\cdot}{Br}}:^{\ominus}}{\underset{:\overset{\cdot\cdot}{\underset{\cdot\cdot}{Br}}:}{}} \ \overset{CH_3}{\underset{}{}}C=C\overset{H}{\underset{CH_3}{}}$$

Understand the system

Balanced? Yes. Generic process? An addition has occurred. Medium? Strongly acidic, HBr pK_a is -9. Sources? Triple bond and the bromide lone pair. Sinks? The H—Br bond is easily broken. Leaving groups? Bromide. Stereochemistry of addition is *anti* since the H and the Br end up on opposite sides of the double bond.

Find possible routes

We are at a well-known trail junction, the starting point of the addition surface. We must judge the possible routes as to how appropriate they are to the sources, sinks, and the media. There are three possible routes that we must decide among: the AdN2 addition (path AdN followed by p.t.),

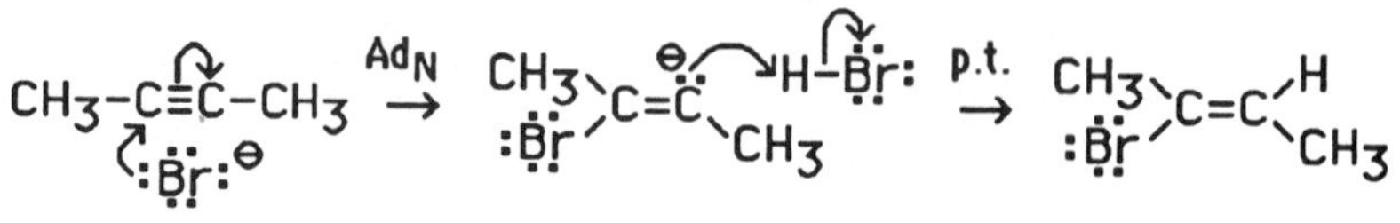

$$CH_3-C{\equiv}C-CH_3 \xrightarrow{Ad_N} CH_3\!\!\diagdown_{C=C}\diagup^{H-\ddot{B}r:}_{CH_3} \quad \ddot{:}\ddot{Br}: \xrightarrow{p.t.} CH_3\!\!\diagdown_{C=C}\diagup^{H}_{CH_3}$$

the AdE2 addition (path AE followed by AN),

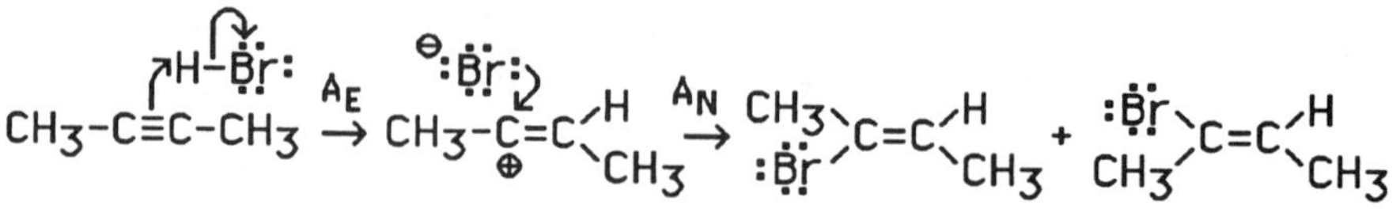

$$CH_3-C{\equiv}C-CH_3 \xrightarrow{A_E} CH_3-\underset{\oplus}{C}{=}C\diagup^{H}_{CH_3} \xrightarrow{A_N} CH_3\!\!\diagdown_{C=C}\diagup^{H}_{CH_3} + \ddot{:}\ddot{Br}\!\!\diagdown_{C=C}\diagup^{H}_{CH_3}$$

and the AdE3 addition.

$$CH_3-C{\equiv}C-CH_3 \xrightarrow{Ad_E3} CH_3\!\!\diagdown_{C=C}\diagup^{H}_{CH_3} \quad :\ddot{Br}:^{\ominus}$$

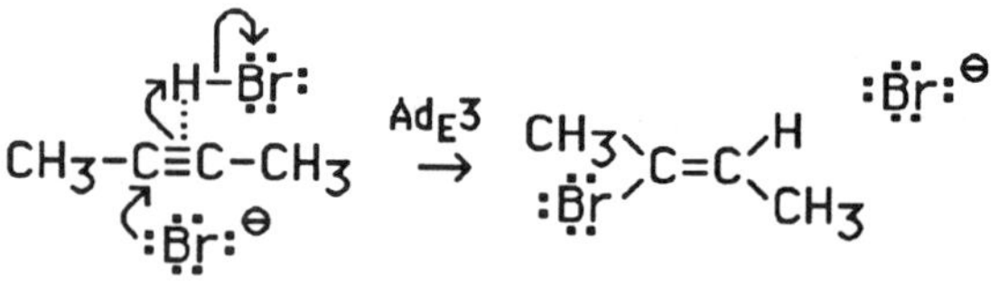

Evaluate and cross-check

The AdN2 is not appropriate for acidic media because it produces a highly basic anion, pK_{abH} 44. The ΔpK_a rule would likewise throw out the AdN2 since bromide at pK_{abH} -9 would never be expected to form a product anion of pK_{abH} 44. This route is so uphill that it would never occur. In this example to illustrate the sorting process, we showed all the possibilities, but in later mechanism examples we will consider only those routes that are contenders for the lowest energy route.

The AdE2 and AdE3 processes are appropriate for the acidic medium. The AdE2 goes via an unstable secondary vinyl carbocation. A second problem is that that carbocation is expected to be linear and could therefore trap bromide from either the top or the bottom face to produce a mixture of stereoisomers.

The AdE3 avoids the unstable carbocation and rationalizes the stereochemistry of the product. The attack is *anti*, the bromide attacks the bottom face since the top attack is blocked by the pi complex of the alkyne with HBr. In this case the addition surface has folded so that the AdE3 is the lowest energy route proceeding up through a pass between the two high corners of an unstable anion and an unstable cation. For an example of this surface, see the E2 surface (the reverse reaction) Figure 7.5. The AdE3 route explains the stereochemistry and is reasonable for the alkyne substrate and the strongly acidic medium.

2. Aldol Condensation

Give a reasonable mechanism for the following reaction.

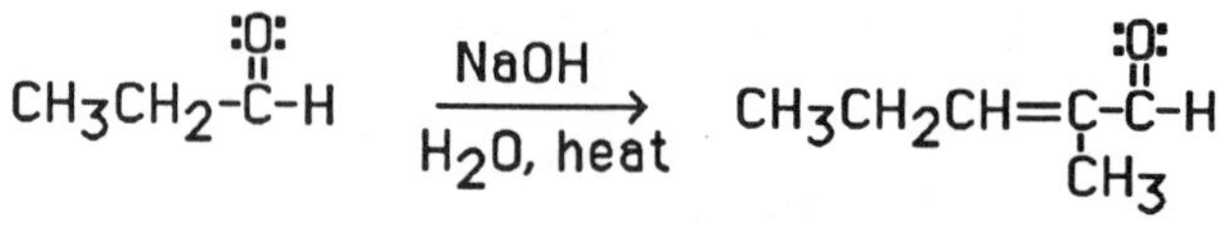

$$CH_3CH_2-\overset{\overset{\textstyle :\!\ddot{O}:}{\|}}{C}-H \xrightarrow[\text{H}_2\text{O, heat}]{\text{NaOH}} CH_3CH_2CH{=}\underset{\underset{\textstyle CH_3}{|}}{C}-\overset{\overset{\textstyle :\!\ddot{O}:}{\|}}{C}-H$$

Understand the system

Balanced? No, the left-hand side requires 2 mol of aldehyde, and the right-hand side needs a water molecule to balance. Find the pieces of the two reactants within the product. Note that the reaction has occurred between the aldehyde carbonyl of one reactant and the carbon next to the aldehyde on the other.

$$CH_3CH_2CH{=}C{-}C{-}H$$
$$\overset{\displaystyle :O:}{\underset{CH_3}{\|}}$$

Generic process? An addition and an elimination have occurred. Medium? Definitely basic, predominant anion is hydroxide, pK_{abH} 15.7, whose pK_a would give a useful proton transfer K_{eq} up to about pK_a 24. Sources? The carbonyl lone pair, water lone pair, and hydroxide anion. Best source? Hydroxide anion, a lone pair source that can behave as a nucleophile or as a base. Sinks? Polarized multiple bond, the aldehyde carbonyl. Acidic H's? Water and the CH_2 next to the aldehyde, pK_a 16.7, within range of hydroxide. Leaving groups? None. Resonance forms?

$$CH_3{-}CH_2{-}\overset{:O:}{\overset{\|}{C}H} \quad \longleftrightarrow \quad CH_3{-}CH_2{-}\overset{:\ddot{O}:^{\ominus}}{\underset{\oplus}{C}H}$$

Find possible routes

There are just two that fit the medium, sources and sinks: deprotonation to form an anion (path p.t.),

$$CH_3{-}CH{-}\overset{:O:}{\overset{\|}{C}H} \quad \underset{}{\overset{p.t.}{\rightleftharpoons}} \quad CH_3{-}\overset{\ominus}{C}H{-}\overset{:O:}{\overset{\|}{C}H}$$
$$\underset{H\,\curvearrowleft\,^{\ominus}:\ddot{O}H}{} \qquad\qquad \underset{H\ddot{O}H}{}$$

or addition to a polarized multiple bond (path Ad_N).

$$CH_3{-}CH_2{-}\overset{\curvearrowleft:O:}{\overset{\|}{C}H} \quad \overset{Ad_N}{\rightleftharpoons} \quad CH_3{-}CH_2{-}\overset{:\ddot{O}:^{\ominus}}{\underset{:\ddot{O}H}{C}H}$$
$$\underset{\curvearrowleft^{\ominus}:\ddot{O}H}{}$$

Evaluate and cross-check

Since the pK_a for both products can only be estimated, and they both come out to be around pK_{abH} 16 to 17 versus 14 to 15, they are too close in energy to distinguish with our crude ΔpK_a rule. We should explore both routes. A use of "product hindsight" shows that no additional hydroxyl groups are attached, thus making this nucleophilic attack path less probable. If this were a product prediction problem and not a mechanism problem we would have to explore both routes.

Understand the system

We need to record only that which has changed from our original observations. We now have a new source, the delocalized anion just formed. Resonance forms:

$$CH_3{-}\overset{\ominus}{\overset{}{C}H}{-}\overset{:O:}{\overset{\|}{C}H} \quad \longleftrightarrow \quad CH_3{-}CH{=}\overset{:\ddot{O}:^{\ominus}}{C}H$$

Find possible routes

Using the new source, we find that there are just two paths that fit the medium, sources and sinks: protonation of an anion, p.t., and addition to a polarized multiple bond, Ad_N. There are two sites that can serve as a base or as a nucleophile on this ambident allylic source. We have four choices to evaluate: proton transfer to oxygen,

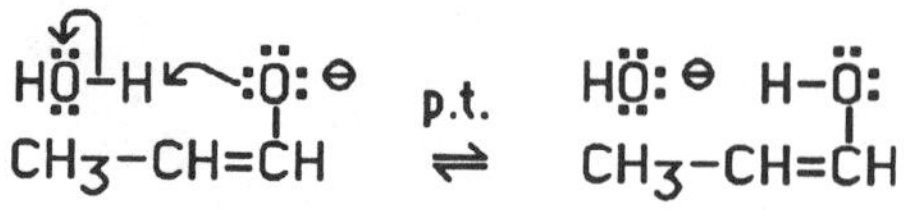

$$HO-H \quad :O:^{\ominus} \quad \xrightarrow{\text{p.t.}} \quad HO:^{\ominus} \quad H-O:$$
$$CH_3-CH=CH \qquad \rightleftharpoons \qquad CH_3-CH=CH$$

or to carbon,

$$HO-H \quad {}^{\ominus} \quad :O: \quad \xrightarrow{\text{p.t.}} \quad HO:^{\ominus} \quad H \quad :O:$$
$$CH_3-CH-CH \qquad \rightleftharpoons \qquad CH_3-CH-CH$$

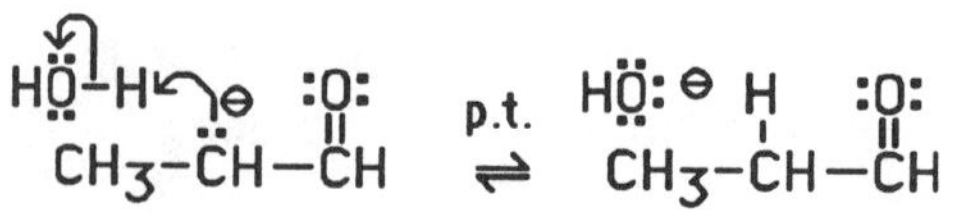

or addition to a polarized multiple bond, path Ad_N, by carbon,

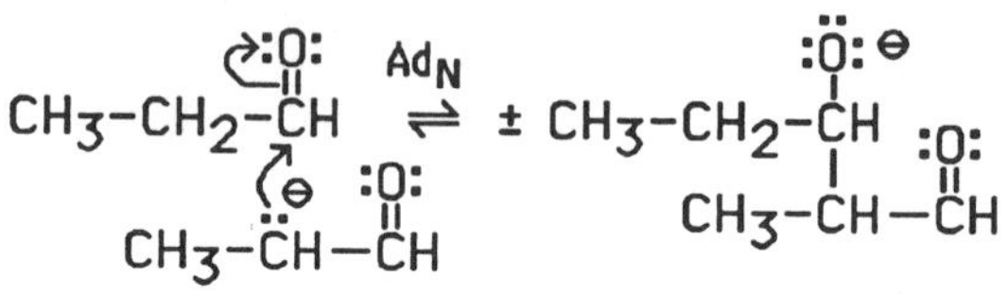

$$:O: \quad Ad_N \qquad :O:^{\ominus}$$
$$CH_3-CH_2-CH \quad \rightleftharpoons \quad \pm \; CH_3-CH_2-CH \quad :O:$$
$$^{\ominus} \quad :O: \qquad\qquad CH_3-CH-CH$$
$$CH_3-CH-CH$$

or by oxygen.

$$:O: \quad Ad_N \qquad :O:^{\ominus}$$
$$CH_3-CH_2-CH \quad \rightleftharpoons \quad \pm \; CH_3-CH_2-CH$$
$$:O:^{\ominus} \qquad\qquad :O:$$
$$CH_3-CH=CH \qquad\qquad CH_3-CH=CH$$

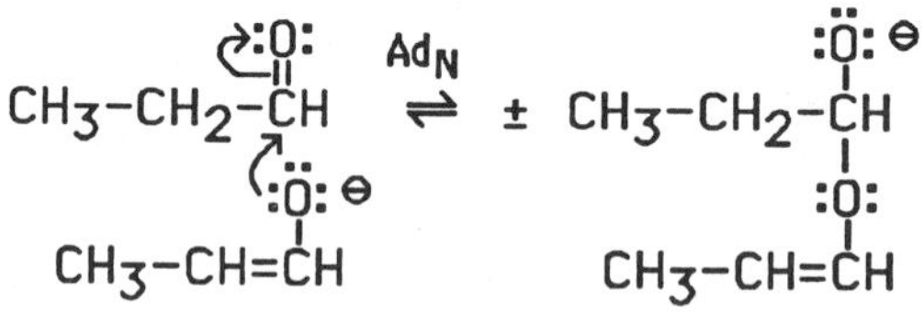

Evaluate and cross-check

Protonation of an anion either takes us to the enol or back to starting material. At this point we can again look forward to product and see not only that nucleophilic attack has occurred but also that carbon not oxygen has been the nucleophile (as expected, Section 8.4). The product of either nucleophilic attack is a secondary alkoxide, having a pK_{abH} of about 17, approximately the same as that of the aldehyde enolate; either route is energetically reasonable by the ΔpK_a rule.

Understand the system

We now have a new source, the alkoxide anion just formed.

Find possible routes

Using the new source, we find that there are just two paths that fit the medium, sources and sinks: protonation of an anion, p.t., and addition to a polarized multiple bond, Ad_N. A third path, path E_β, takes us back to an already explored point, so we shall omit that path and just consider those paths that get us to someplace new: protonation,

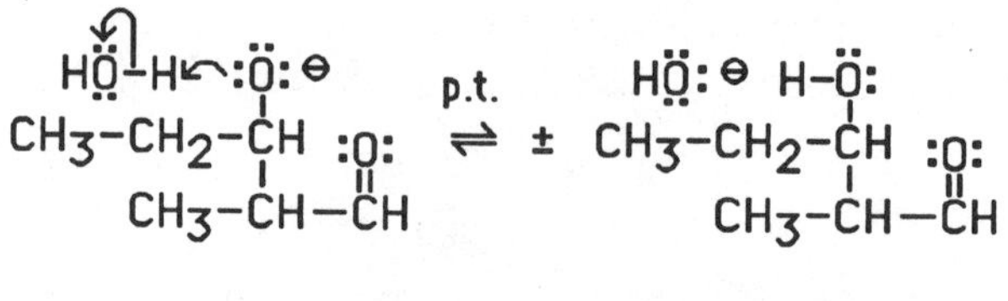

$$HO-H \quad :O:^{\ominus} \qquad HO:^{\ominus} \quad H-O:$$
$$CH_3-CH_2-CH \quad :O: \quad \rightleftharpoons \quad \pm \; CH_3-CH_2-CH \quad :O:$$
$$CH_3-CH-CH \qquad\qquad CH_3-CH-CH$$

or addition to the polarized multiple bond, path AdN.

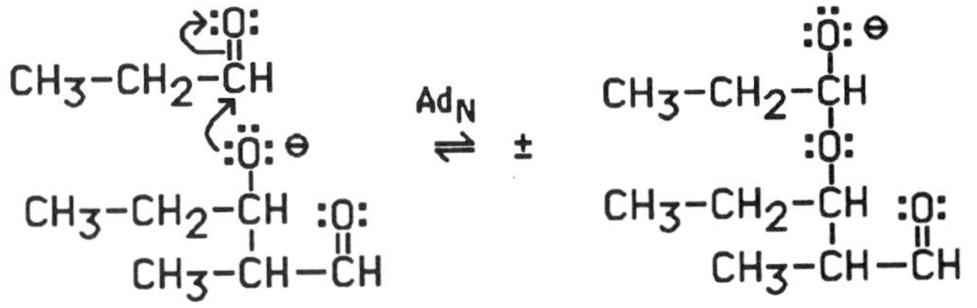

$$CH_3-CH_2-\overset{\overset{\displaystyle :O:}{\|}}{CH} \quad \overset{AdN}{\underset{\pm}{\rightleftharpoons}} \quad CH_3-CH_2-CH$$

(mechanism structures showing AdN addition)

$$CH_3-CH_2-CH \quad :O:$$
$$CH_3-CH-CH$$

Evaluate and cross-check

Protonation of an anion takes us to a neutral compound. We can again look forward to product and see that another nucleophilic attack is not needed and would be expected to be reversible since the pK_{abH} of the nucleophile and of the product are about the same. The proton-transfer step is predicted to have a favorable proton transfer K_{eq}.

Understand the system

An elimination of water gets us to the product. We now are close enough to product to restrict paths to elimination pathways. Resonance forms:

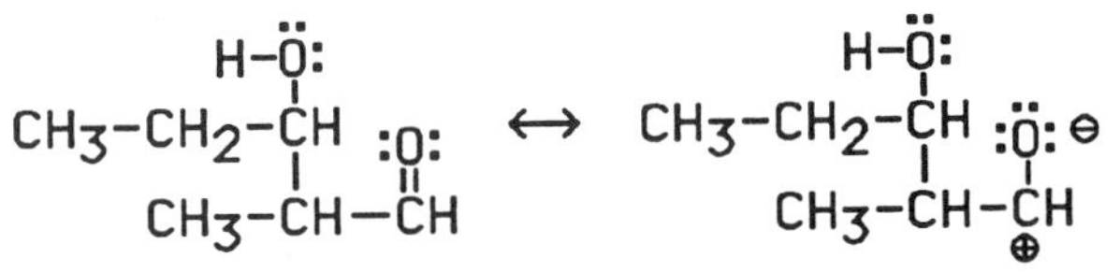

$$CH_3-CH_2-CH \quad :O: \quad \longleftrightarrow \quad CH_3-CH_2-CH \quad :O: \ominus$$
$$CH_3-CH-CH \qquad\qquad CH_3-CH-CH_\oplus$$

Find possible routes

There are just two elimination paths that fit the medium, sources and sinks: the E2 elimination,

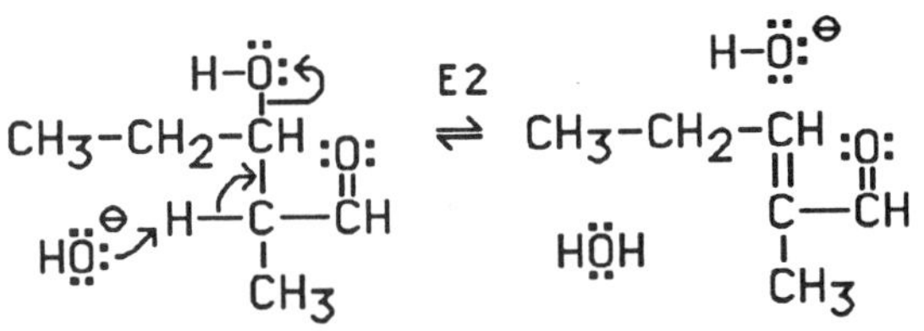

$$CH_3-CH_2-CH \quad :O: \quad \overset{E2}{\rightleftharpoons} \quad CH_3-CH_2-CH \quad :O:$$
$$HO:\!\!\!\ominus \quad H-C-CH \qquad\qquad HOH \qquad C-CH$$
$$CH_3 \qquad\qquad\qquad\qquad CH_3$$

or the E1cB (path p.t. followed by path Eβ).

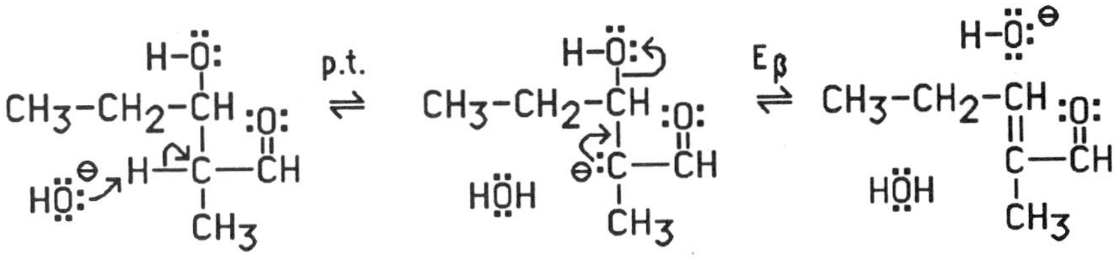

$$CH_3-CH_2-CH \quad :O: \quad \overset{p.t.}{\rightleftharpoons} \quad CH_3-CH_2-CH \quad :O: \quad \overset{E\beta}{\rightleftharpoons} \quad CH_3-CH_2-CH \quad :O:$$

Evaluate and cross-check

Both elimination processes are reasonable, but the E1cB path is more probable because the H is acidic by virtue of being next to the aldehyde, and the leaving group is poor, pK_{aHL} 15.7, so an adjacent charge is necessary to help it leave. The pK_a span on all steps has been within reason.

Overview

We are done, but let's take an overview of the surface (Figure 9.1). The reaction path is shown vertically in the center and the alternate routes are shown to the side. The path abbreviation for the forward reaction only is shown. To summarize the route: Proton transfer generates the nucleophile, which reacts by Ad_N2, and then the E1cB process gives us the product.

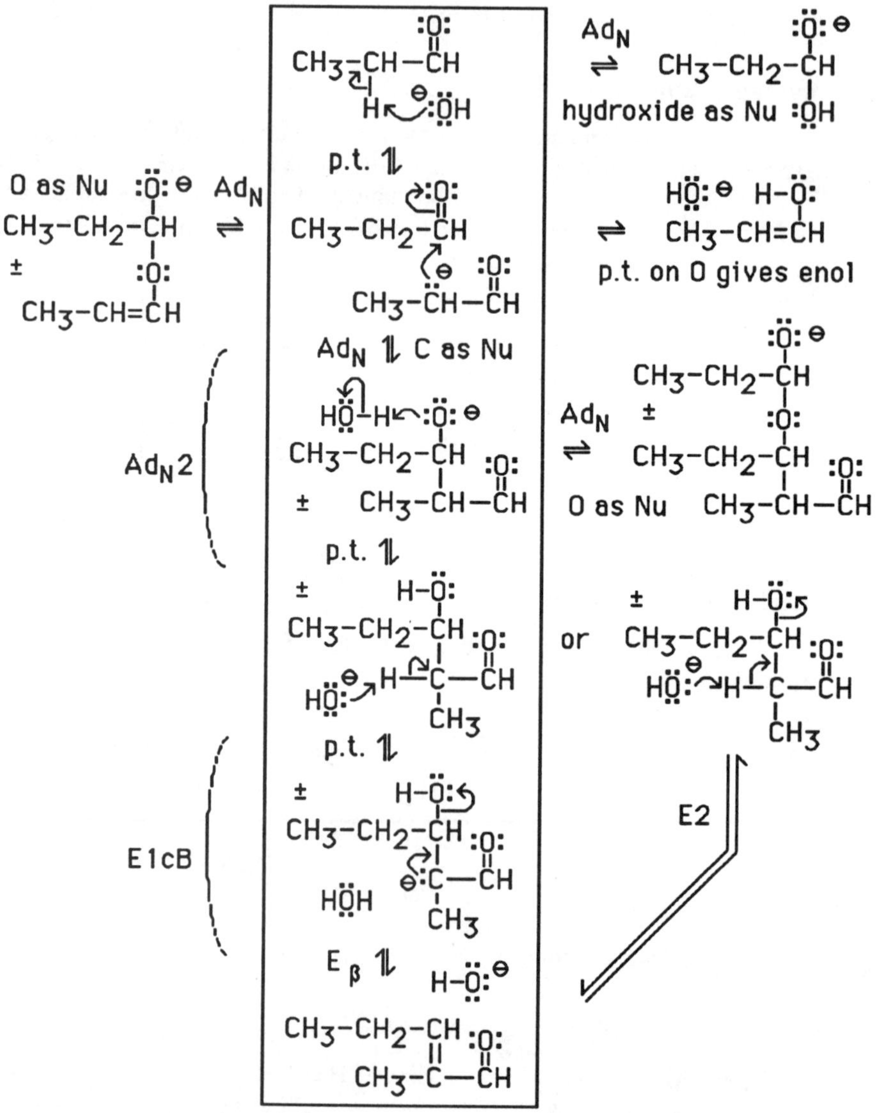

Figure 9.1 Aldol condensation overview. The most likely reaction path is shown boxed.

3. Ketal Bromination

Give a reasonable mechanism for the following reaction.

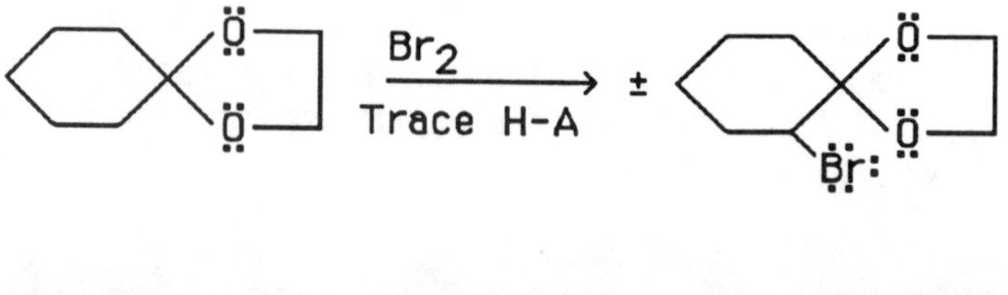

Understand the system

Balanced? No, HBr must be added to the right-hand side to balance the equation. Generic process? An apparent substitution of Br for an H of a nonreactive CH_2. Medium? Acidic. Sources? The ketal lone pairs. Sinks? Bromine is a Y−L. Acidic H's? None. Leaving groups? Bromide. Resonance forms? None

Every bond can be considered to be a combination of a nucleophile and an electrophile. The electrophile for the C−Br bond was bromine, the Y−L. A beginner might be tempted to use the CH bond as a nucleophile and get the problem over in one step, but CH bonds, being strong and not very polarized, make rotten electron sources. A better nucleophile must be generated because the CH_2 is not reactive.

Find possible routes

There is only one path, proton transfer to the lone pair.

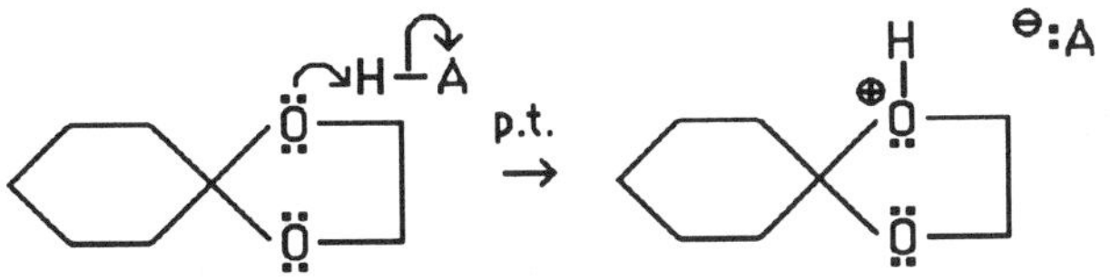

Evaluate and cross-check

The reaction is downhill in energy if the acid catalyst has a pK_a of less than -3.5, the pK_a of the protonated ether.

Understand the system

The protonated ether is now a good leaving group.

Find possible routes

Elimination would produce a pi bond, which could act as a nucleophile. Two elimination routes are possible, the E2,

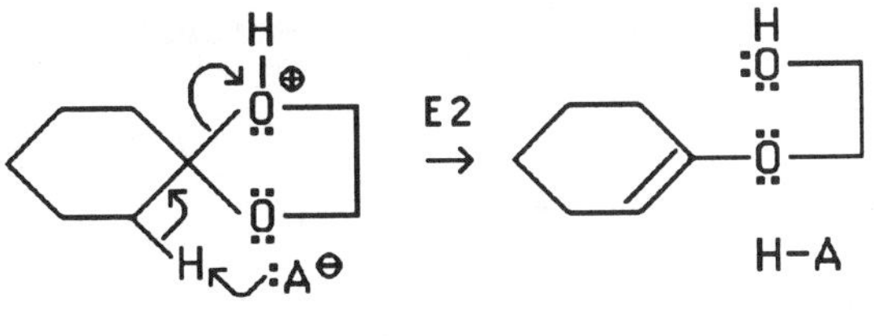

or the E1 (path D_N followed by D_E).

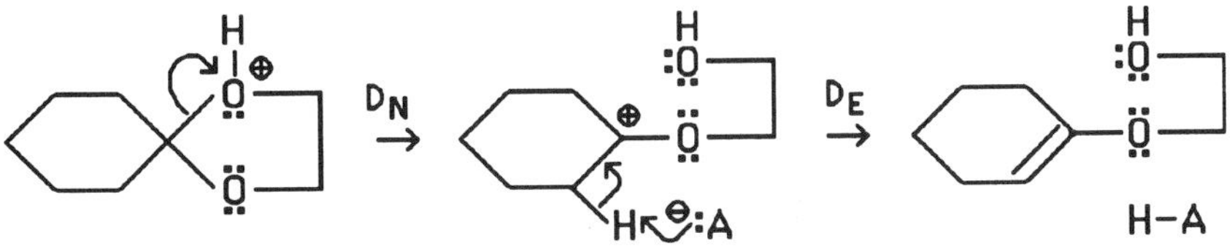

Evaluate and cross-check

Considering the lone pair stabilization of the carbocation formed and the fact that the leaving group is good, the E1 is a good route. Both are reasonable routes. A ΔH

calculation shows that we have gone uphill in energy. Bonds broken, C–H, C–O, C–C, minus bonds made, C=C, O–H, gives $(99 + 86 + 83) - (146 + 111) = 268 - 257 = +11$ kcal/mol ($+46$ kJ/mol).

Understand the system

The enol ether is a good allylic electron source, and bromine is a good electron sink (a Y–L).

Find possible routes

Two addition routes are possible, the Ad_E3,

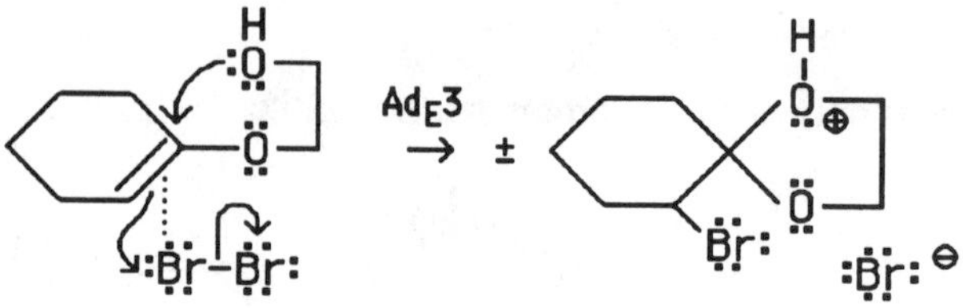

or the Ad_E2, which proceeds via a lone-pair-stabilized carbocation (path A_E followed by path A_N).

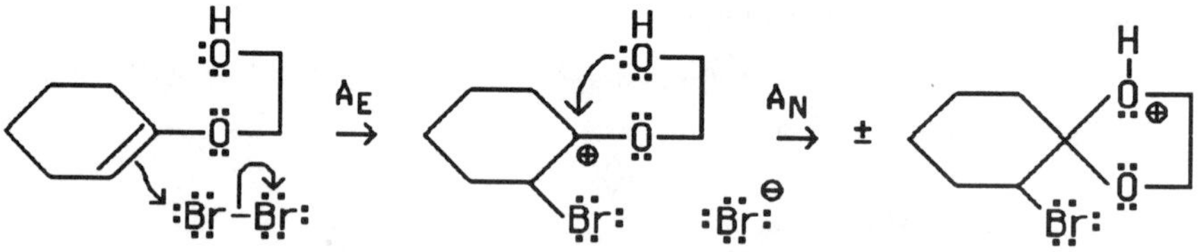

Evaluate and cross-check

The carbocation is relatively good; both routes are reasonable. Five-membered ring formation is expected to be fast. A third possibility is to add bromine by path NuL to form the bromonium ion which would then get attacked by the alcohol lone pair via an internal S_N2. However, the carbocation is stable enough not to require bridging by bromine, thus making this third route essentially the same as the Ad_E2.

Overview

A following proton transfer that regenerates the acid catalyst gives us the product. The solvent molecule that picks off the proton is often assumed and not drawn.

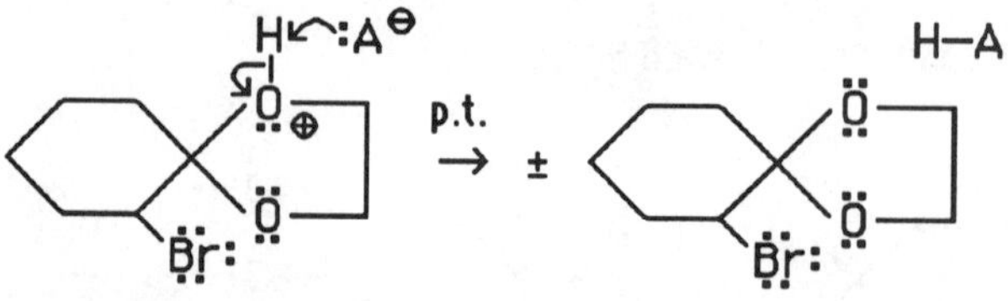

Now let's back off from the problem far enough to see the entire process as a whole (Figure 9.2). The mechanism for this transformation or "mechanistic sentence" can be made from our "mechanistic phrases" or path combinations: A proton transfer improves the leaving group, then an E1 (or E2) elimination followed by an Ad_E2 (or Ad_E3) addition, ending with a proton transfer. The reaction path is shown vertically in the center and the alternate routes are shown to either side.

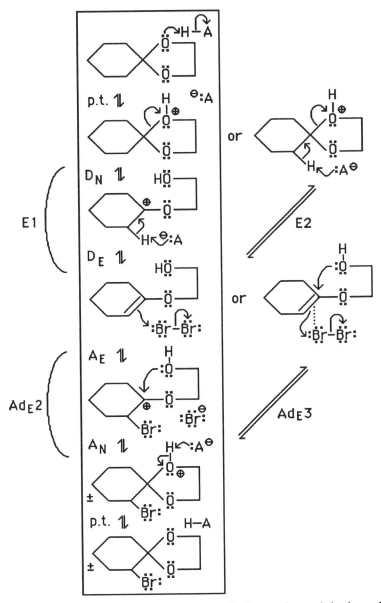

Figure 9.2 Ketal bromination overview. The most likely reaction path is shown boxed.

4. Organometallic on a Nitrile

Give a reasonable mechanism for the following reaction.

$$Ar-C\equiv N: \quad \xrightarrow[\text{2) }H_3\overset{..}{O}^{\oplus}]{\text{1) }CH_3MgBr} \quad Ar-\overset{\overset{\displaystyle :O:}{\|}}{C}-CH_3$$

Understand the system

Balanced? No, $Mg^{2\oplus}$, $Br^{\ominus}$, and some nitrogen species are missing from the right-hand side. Generic process? An addition has occurred, but the rest is unclear. Media?

Step 1 is highly basic, but step 2 is in acidic water, pK_a -1.7. Sources for step 1? The organometallic. Sinks? The nitrile is a polarized multiple bond. Acidic H's? None. Leaving groups? None. Resonance forms?

$$Ar-C\equiv N: \longleftrightarrow Ar-\overset{\oplus}{C}=\ddot{N}:^{\ominus} \qquad H_3C-MgBr \longleftrightarrow H_3C:^{\ominus} Mg^{2+} :\ddot{\underset{..}{Br}}:^{\ominus}$$

Find possible routes

There is only one path possible: the addition to a polarized multiple bond (path Ad_N).

$$\begin{array}{c} Ar-C\equiv\overset{\curvearrowright}{\ddot{N}:} \\ H_3C:^{\ominus}\!\!\nearrow Mg^{2+} :\ddot{\underset{..}{Br}}:^{\ominus} \end{array} \quad \overset{Ad_N}{\longrightarrow} \quad \begin{array}{c} Ar-C=\ddot{N}:^{\ominus} \\ | \\ H_3C \quad Mg^{2+} :\ddot{\underset{..}{Br}}:^{\ominus} \end{array}$$

Evaluate and cross-check

There is only one source and one sink and one path. The organometallic addition creates a basic anion whose pK_{abH} is not on the chart. By analogy to the corresponding carbanion, this sp^2 hybridized anion should be more stable than the sp^3 hybridized one but less stable than the delocalized anions. Therefore, this anion should be a weaker base than the anion of diethylamine, $Et_2N^{\ominus}$ pK_{abH} of 36, but should be a stronger base than the delocalized anion of aniline, $PhNH^{\ominus}$ pK_{abH} 27. The reaction of the organometallic, pK_{abH} 48, to produce this anion is definitely downhill in energy.

Understand the system

Now all the carbons of the product are accounted for. The medium is strongly basic, and the remaining sink is anionic and would repel most nucleophiles. The magnesium organometallic, although very reactive, is not a good enough source to attack an anionic sink. There appear to be no new paths at this time (always finish completely with a step before starting the next one). Now let's explore the second step. The medium is acidic water. The best source is the anionic addition product and the best sink is $H_3O^{\oplus}$.

Find possible routes

Proton transfer should be rapid and appears to be the only available path.

$$\begin{array}{c} Ar-C=\ddot{N}:^{\ominus}\!\!\curvearrowright H\overset{\curvearrowright}{\underset{\oplus}{O}}H_2 \\ | \\ H_3C \end{array} \quad \overset{p.t.}{\longrightarrow} \quad \begin{array}{c} Ar-C=\ddot{N}-H \quad :\ddot{O}H_2 \\ | \\ H_3C \end{array}$$

Evaluate and cross-check

The pK_{abH} of the nitrogen anion should be in the low 30's and the pK_a of acidic water is -1.7; therefore the proton transfer is irreversible.

Understand the system

The imine is a polarized multiple bond sink and can also serve as a lone pair source since the nitrogen lone pair is reasonably basic, pK_{abH} is approximately 5. We can get an overview of the general direction the reaction goes by looking at the final product, which has a double bond to oxygen rather than to nitrogen. We must add an oxygen and get rid of the nitrogen.

Find possible routes

The only nucleophile present is water. Two addition paths to the imine that would fit the reaction conditions are the hetero Ad_E3,

$$H_2\ddot{O}: \quad Ar-\underset{\underset{H_3C}{|}}{C}=\overset{}{N}-H \quad H-\overset{..}{O}H_2 \quad \overset{\text{Hetero}}{\underset{\rightarrow}{Ad_E3}} \quad H_2\overset{\oplus}{\ddot{O}} \quad Ar-\underset{\underset{H_3C}{|}}{\overset{|}{C}}-\overset{H}{N}-H \quad :\ddot{O}H_2$$

or the hetero Ad_E2 (path p.t., protonation, then path Ad_N, nucleophile attack).

$$Ar-\underset{\underset{H_3C}{|}}{C}=\overset{}{N}-H \quad H-\overset{..}{O}H_2 \quad \overset{p.t.}{\rightarrow} \quad H_2\ddot{O}: \quad Ar-\underset{\underset{H_3C}{|}}{C}=\overset{\oplus}{N}\diagdown^H \quad \overset{Ad_N}{\rightarrow} \quad \pm \quad H_2\overset{\oplus}{\ddot{O}} \quad Ar-\underset{\underset{H_3C}{|}}{\overset{|}{C}}-N\diagdown^H$$

Evaluate and cross-check

Both routes are known to occur in acidic media. (A third alternative, the Ad_N2, is ruled out by the ΔpK_a rule, for it would create too basic an anion, pK_{abH} 36, from a nucleophile of pK_{abH} -1.7.) The carbocation formed in the hetero Ad_E2 process is very stable, pK_a near 5. The proton transfer is irreversible; K_{eq} is very favorable, close to 10^{+7}; and such proton transfers are fast. Both additions form the same product, but the hetero Ad_E2 with its very favorable proton transfer and excellent sink, the protonated imine, appears slightly lower in energy. The overall process starts with $H_3O^\oplus$, pK_a -1.7, and ends with a protonated alcohol, pK_a about -3 to -4; a slight change in pK_a has occurred. We have broken a pi bond and made two sigma bonds. Most likely this addition was downhill in energy.

Understand the system

The protonated alcohol is acidic, and the nitrogen lone pair is reasonably basic, pK_{abH} of approximately 8 to 9.

Find possible routes

The one proton transfer path is mediated by solvent.

$$\overset{\oplus}{H\ddot{O}-H} \quad :\ddot{O}H_2 \quad Ar-\underset{\underset{H_3C}{|}}{\overset{|}{C}}-\overset{..}{N}-H \diagdown^H \quad \overset{p.t.}{\rightarrow} \quad Ar-\underset{\underset{H_3C}{|}}{\overset{\overset{H}{|}{:O:}}{C}}-\overset{..}{N}-H \diagdown^H \quad H-\overset{..}{O}H_2 \quad \overset{p.t.}{\rightarrow} \quad Ar-\underset{\underset{H_3C}{|}}{\overset{\overset{H}{|}{:O:}}{C}}-N\overset{\oplus}{\diagdown^H_{,H}} \quad :\ddot{O}H_2$$

Evaluate and cross-check

An intramolecular proton transfer process is possible, but since it involves a strained four-membered transition state, this alternate route is probably of higher energy. We can expect proton transfer to be extremely fast between groups that hydrogen bond to the water solvent. Very little needs to happen to turn a hydrogen bond into a "real" bond or vice versa. The proton transfer is irreversible with a K_{eq} close to 10^{+12}.

Understand the system

The protonated nitrogen is a fair leaving group, pK_{aHL} 9.2. We need to lose it and a proton to get to the final product.

Find possible routes

Three elimination paths are possible, the E2,

$$H_2\ddot{O}\!:\curvearrowright H\!-\!\ddot{O}\!: \quad\quad \xrightarrow{E2} \quad H_2\overset{\oplus}{\ddot{O}}\!-\!H \quad :\ddot{O}:$$

$$Ar\!-\!C\!-\!N\!-\!H \qquad\qquad\qquad Ar\!-\!C \quad :N\!-\!H$$

$$H_3C \qquad\qquad\qquad\qquad H_3C \qquad H$$

or the lone-pair-assisted E1 (path Eβ followed by path p.t.),

$$Ar\!-\!C\!-\!N\!-\!H \xrightarrow{E\beta} H_2\ddot{O}\!:\curvearrowright H\!-\!\overset{\oplus}{\ddot{O}}\!:\oplus \xrightarrow{p.t.} H_2\overset{\oplus}{\ddot{O}}\!-\!H \quad :\ddot{O}:$$

$$H_3C \qquad\qquad\qquad\qquad Ar\!-\!C \quad :NH_3 \qquad Ar\!-\!C \quad :NH_3$$

$$\qquad\qquad\qquad\qquad\qquad H_3C \qquad\qquad\qquad H_3C$$

or the E1cB (path p.t. followed by path Eβ).

$$H_2\ddot{O}\!:\curvearrowright H\!-\!\overset{\oplus}{\ddot{O}}\!: \xrightarrow{p.t.} H_2\overset{\oplus}{\ddot{O}}\!-\!H \quad :\ddot{O}:^{\ominus} \xrightarrow{E\beta} H_2\overset{\oplus}{\ddot{O}}\!-\!H \quad :\ddot{O}:$$

$$Ar\!-\!C\!-\!N\!-\!H \qquad\qquad Ar\!-\!C\!-\!NH_3 \qquad\qquad Ar\!-\!C \quad :N\!-\!H$$

$$H_3C \qquad H \qquad\qquad H_3C \qquad\qquad\qquad H_3C \qquad H$$

Evaluate and cross-check

To place this trail junction on a surface previously explored, look at point P on the surface in Figure 6.13; our desired goal is point R on the surface. The path combination through the lower right corner is the E1cB, whose first step is more difficult in strongly acidic media. The path combination through the upper left corner is the lone-pair-assisted E1, which is favored by strongly acidic media and good leaving groups. The diagonal route is the E2 path. Whereas the E2 produces $H_3O^\oplus$, pK_a -1.7, the lone-pair-assisted E1 produces a less stable and therefore more acidic cation, approximate pK_a -7. The E2 must have the O—H and C—N bonds coplanar for reaction. The lone-pair-assisted E1 needs to align just one of the two lone pairs but usually requires a better leaving group. The E1cB suffers from a poor proton transfer K_{eq} (estimated at about 10^{-12}), but the anion is a good enough source to kick out a fair to poor leaving group. In basic media the E1cB would be the best route, but as the medium is acidic the concentration of the required anionic intermediate is very small. All routes have problems; mechanistic studies have shown that in acidic media this elimination is the slow step.

In all cases a following proton transfer step is expected. This final proton transfer has a K_{eq} of $10^{+10.9}$ and is irreversible.

$$H_3N\!:\curvearrowright H\!-\!\overset{\oplus}{\ddot{O}}H_2 \xrightarrow{p.t.} H\!-\!\overset{}{N}\!-\!H \quad H_2\ddot{O}\!:$$

$$\qquad\qquad\qquad\qquad\qquad\qquad H$$

Overview

We are done, but let's take an overview of the surface (Figure 9.3). The reaction path is shown vertically in the center and the alternate routes are shown to either side. The path abbreviation for the forward reaction only is shown. The overall most probable route is then a nucleophilic addition to a polarized multiple bond, protonation upon workup, the hetero Ad$_E$2 addition, two proton transfers, an E2, and a final proton transfer.

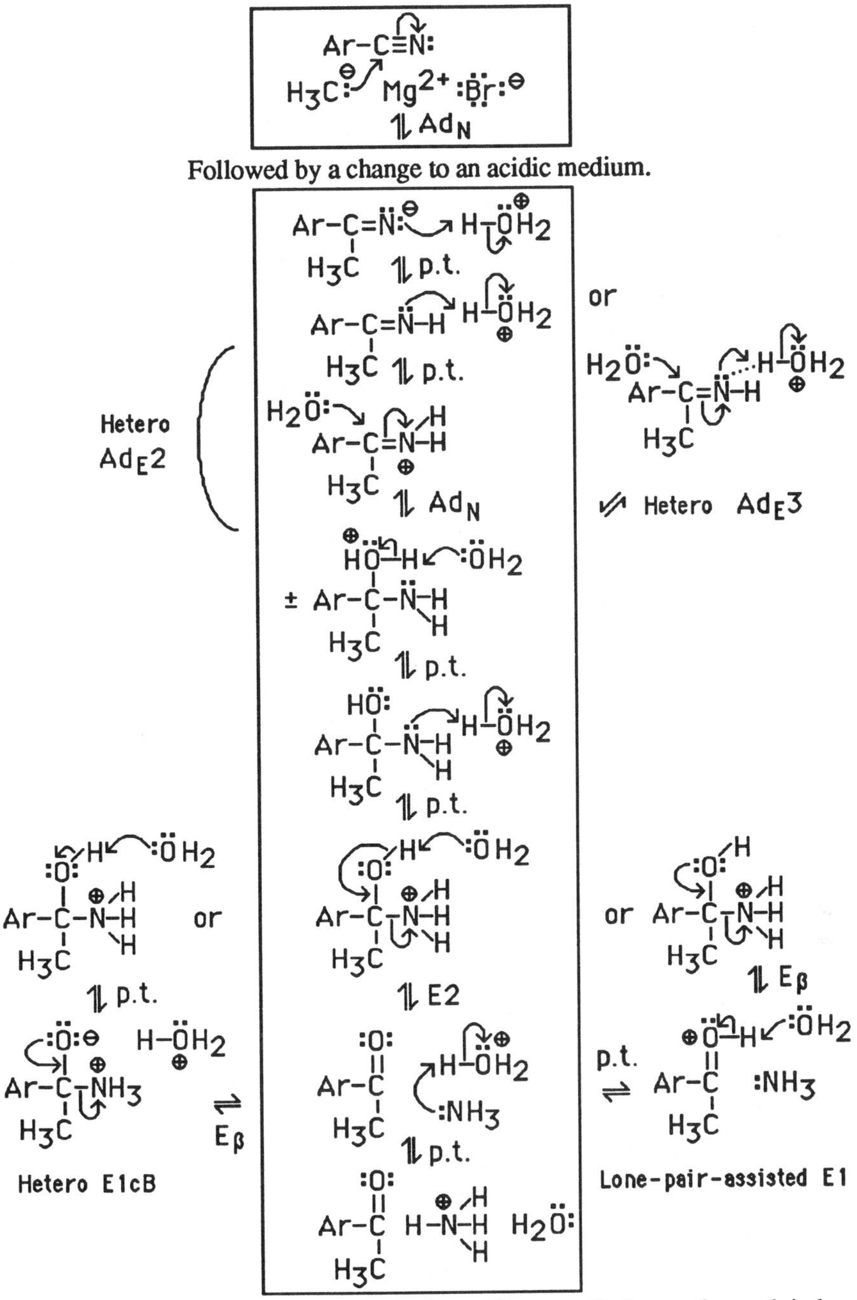

Figure 9.3 Organometallic on a nitrile overview. The most likely reaction path is boxed.

5. Glucose to Fructose

Give a reasonable mechanism for the following reaction.

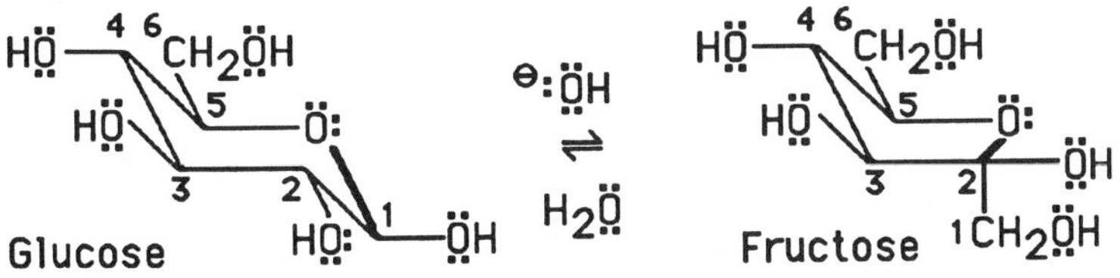

Understand the system

Because the reactant and the product are structural isomers of one another, all pieces are accounted for in this balanced reaction. No overall addition of a nucleophile has occurred (there is nothing new attached), and thus we can assume that hydroxide serves as a base. The carbons are numbered to aid in the reaction analysis. The C−O bond to carbon 1 has been broken and a new C−O bond made to carbon 2. However there is more; the line structure shorthand partially disguises that a C−H bond on carbon 2 has been broken and a new C−H bond made on carbon 1. The functional group at carbon 1 of glucose is a hemiacetal.

The novice may be tempted to get this mechanism over with quickly by swapping C−O and C−H bonds between carbons 1 and 2. Even though it can be drawn with arrows, such a 1,2 swap has never been observed in the real world. It is not a valid electron flow pathway. Now let's look where the known electron flow pathways lead us.

Find possible routes

First try proton transfer. All the hydroxyls are within a few pK_a units of the hydroxide base, so all are available for proton transfer. Only one hydroxyl, that on carbon one, bears anything that can serve as a leaving group. All the other hydroxyls can only deprotonate then reprotonate.

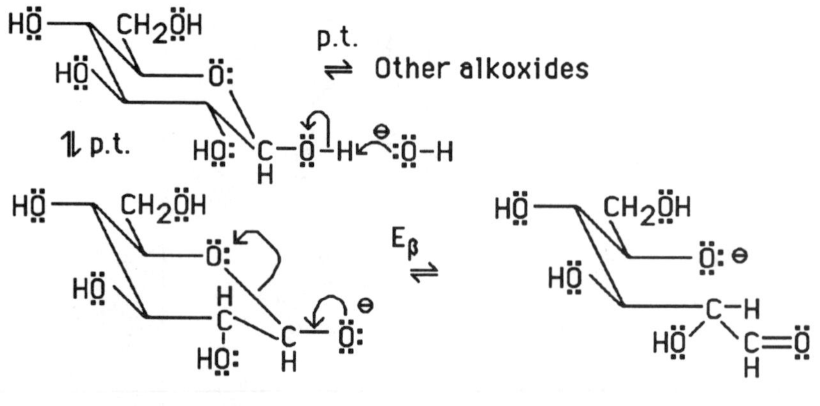

Evaluate and cross-check

We head in the only direction that we can go; deprotonate the carbon 1 hydroxyl and kick out the only leaving group. This E1cB path combination (path p.t. followed by path E_β) accomplishes one goal of the transformation; it breaks the C−O bond to carbon 1. The product alkoxide is in equilibrium with the alcohol form in this medium. The literature value for the pK_a of glucose is 12.3. The alkoxide at this pK_{abH} has kicked out a leaving group at about pK_{aHL} 14 (the pK_a of $HOCH_2CH_2OH$ is 14.2) which checks with the ΔpK_a rule.

Understand the system

The newly formed aldehyde acts as an ewg and makes the proton on carbon 2 acidic; loss of this proton accomplishes the second goal of the transformation.

Find possible routes

Although hydrate formation by attack of hydroxide on the aldehyde is fast and reversible, it is also a dead end; a ΔH calculation shows us that the hydrate is uphill from the aldehyde. The alkoxide that we just kicked out could attack the aldehyde to return to starting material. The alkoxide can also serve as a base and remove the proton next to the aldehyde.

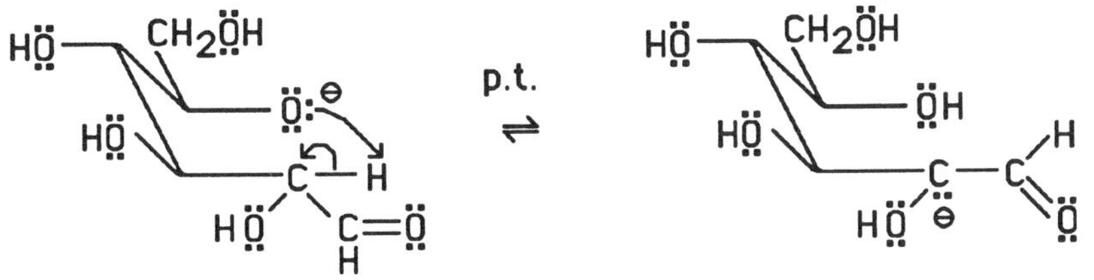

p.t.

$\rightleftharpoons$

Evaluate and cross-check

The proton transfer step is the only unexplored new route. It is intramolecular and therefore expected to be fast. The alkoxide, whose pK_{abH} is 14, has removed a proton adjacent to an aldehyde at approximately the same pK_a for a proton transfer K_{eq} of approximately 1.

Understand the system

The newly formed anionic intermediate is called an enediolate and has two resonance forms. It is internally hydrogen bonded. The enediolate is halfway along; two out of four goals have been achieved. Now another H is needed on carbon 1 and a C—O bond needs to be made to carbon 2. Because there is no electrophilic site at carbon 2, the C—O bond cannot yet be made.

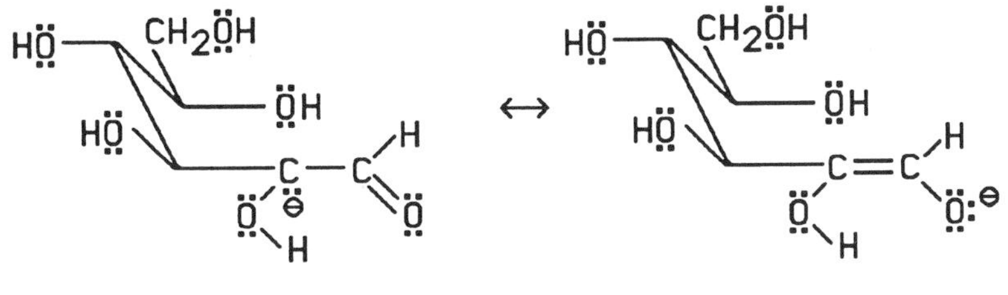

$\leftrightarrow$

Find possible routes

The internally hydrogen bonded enediolate intermediate undergoes rapid intramolecular proton transfer.

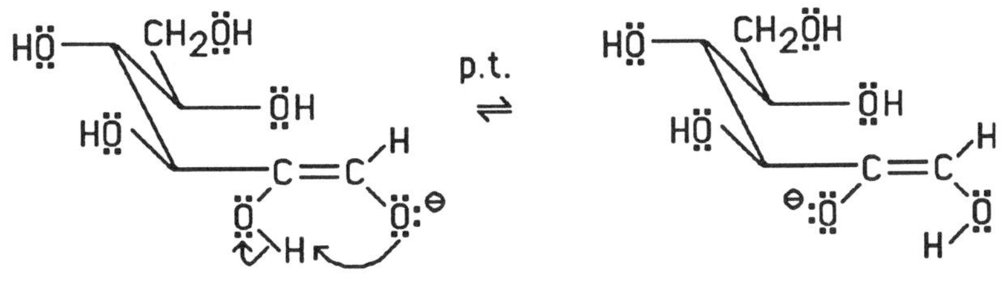

p.t.

$\rightleftharpoons$

Evaluate and cross-check

Since the transition state is a five-membered ring, this intramolecular process is expected to be exceedingly fast. This proton transfer step is between two groups with the about same pK_a for a proton-transfer K_{eq} close to 1.

Understand the system

The second resonance form should provide a hint toward product. Proton transfer to this anion not only puts the H where it is needed on carbon 1 but also generates a site for nucleophilic attack on carbon 2.

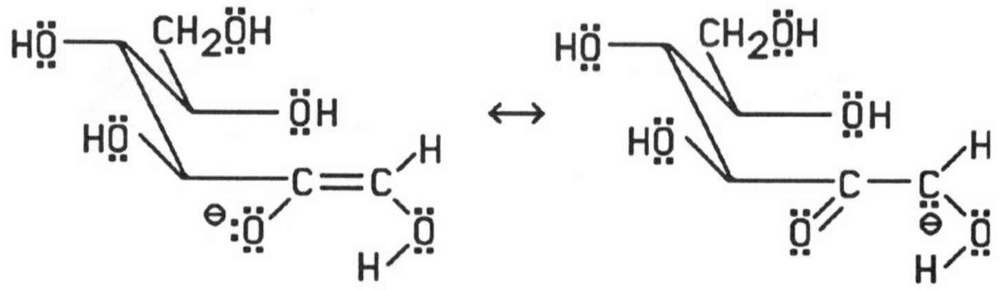

Find possible routes

An intramolecular proton transfer can occur.

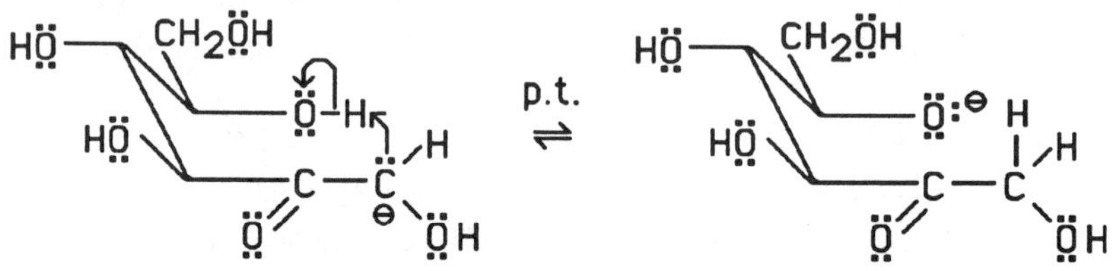

Evaluate and cross-check

Since the transition state is a seven-membered ring, this intramolecular process is expected to be reasonably fast. This proton-transfer step is between two groups with about the same pK_a for a proton-transfer K_{eq} of approximately 1. This proton-transfer step can also be mediated by the water solvent that is hydrogen-bonded to the anion.

Understand the system

The carbonyl is a new site for nucleophilic attack; the proton transfer has generated a good nucleophile.

Find possible routes

The C—O bond can now be made because we have a good nucleophile, a good sink, and a known path, Ad_N, for reaction.

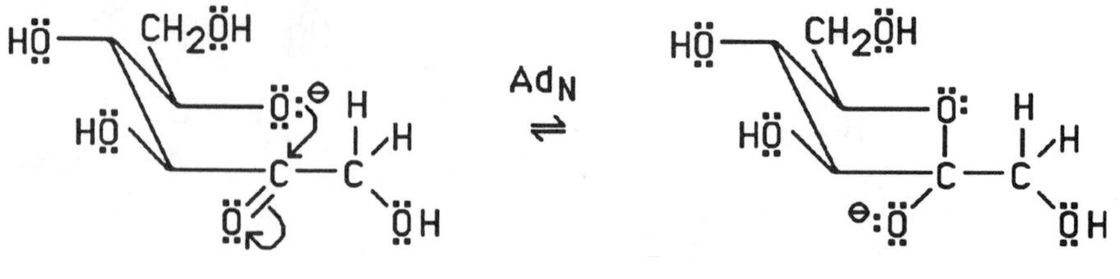

Overview

At this point we can see that protonation of the product yields fructose. These two paths together form the Ad_N2 path combination. Overall, the mechanism for this transformation was an E1cB elimination, three proton transfers, and an Ad_N2 addition (Figure 9.4).

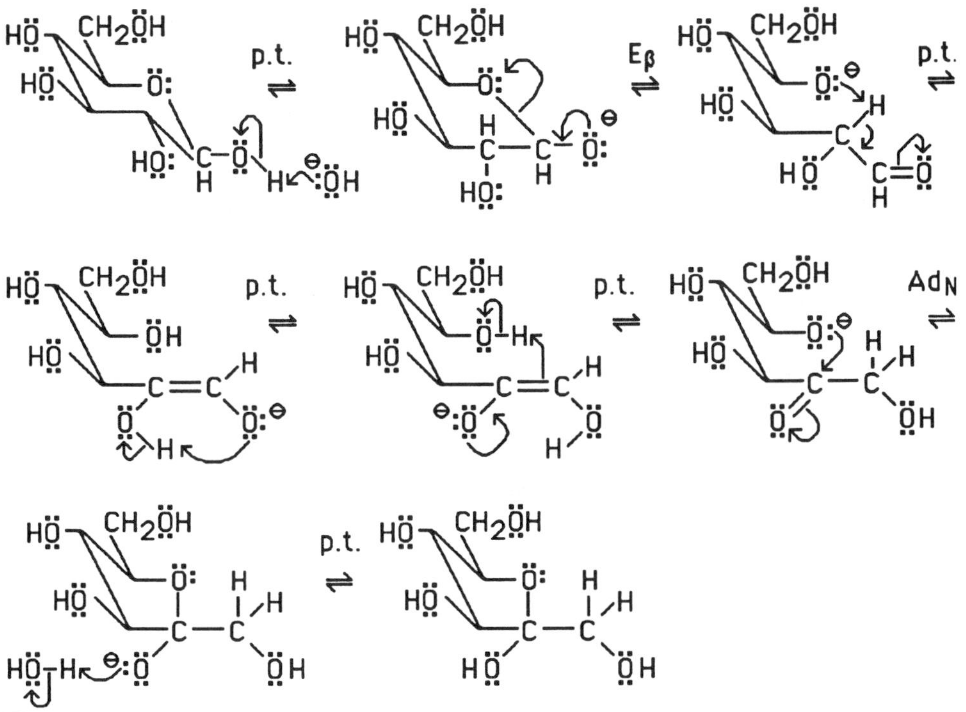

Figure 9.4 Glucose to fructose overview.

9.6 WORKED PRODUCT PREDICTION EXAMPLES

The first thing to consider in the prediction of organic reaction products is any reasonable proton transfer. Remember to consider both the original species and the product of any proton transfer when you go to classify species into the generic classes of sources and sinks. Look for leaving groups, polarized single bonds, and polarized multiple bonds. Pick the best source and sink and combine them via an appropriate pathway to yield a preliminary product. Cross-check the pathway for steric, electronic, solvent, and medium limitations.

The difficult part is to decide whether or not the preliminary product is really the final product or just an intermediate. The ΔpK_a rule can be helpful to determine whether the process has climbed in energy or not. If you have formed an unstable anion or cation, of course it is just an intermediate. If there remains an easily lost leaving group or another reactive electron sink, look for a good source to react with it. Basically look for any good process that can occur, if none are found, stop before you do anything rash.

Lastly, have faith in the predictive process and allow it to carry you to the answer. Do not try to shorten the process to get to the answer by combining or skipping steps. Don't panic and force an answer because you can't see the answer from the beginning. In the product prediction examples, if there are several routes to the same compound, for simplicity we will show only the more reasonable.

1. Predict the Product of the Reaction of a Carboxylate and 4-Bromophenacyl Bromide

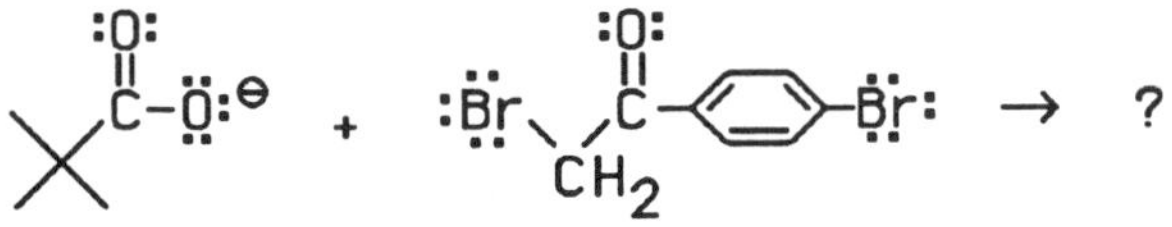

Understand the system

Medium? Mildly basic; $RCOO^\ominus$ has a pK_{abH} of 4.8. Sources? The carboxylate anion. Leaving groups? Bromide. Sinks? The ketone is a polarized multiple bond. The CH_2Br is a reactive sp^3 bound leaving group. The bromine on the ring is a leaving group bound to an aromatic ring bearing an ewg, but it is definitely the least reactive sink. Acidic H's? The CH_2 should be about as acidic as a 2-chloroketone, pK_a 16; thus there are no acidic H's within range of the only base, carboxylate. Resonance forms?

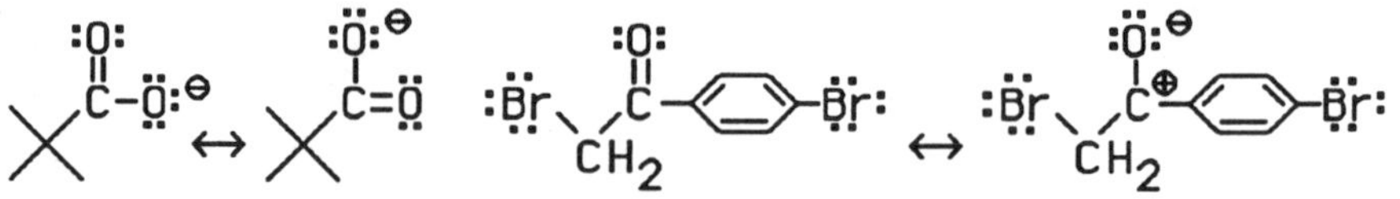

Find possible routes

There are two possible: addition to a polarized multiple bond (path Ad_N),

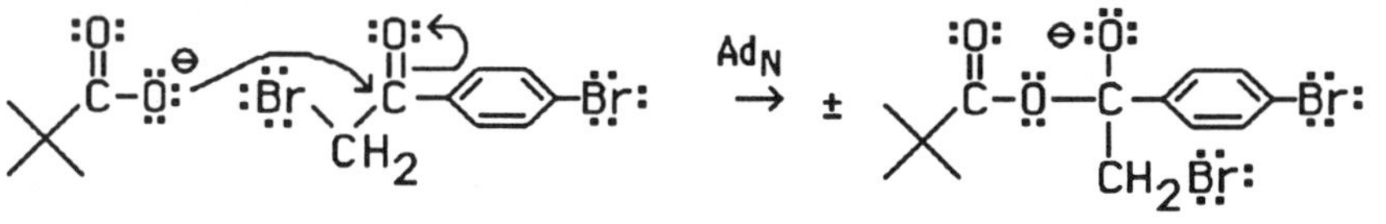

or the S_N2 substitution.

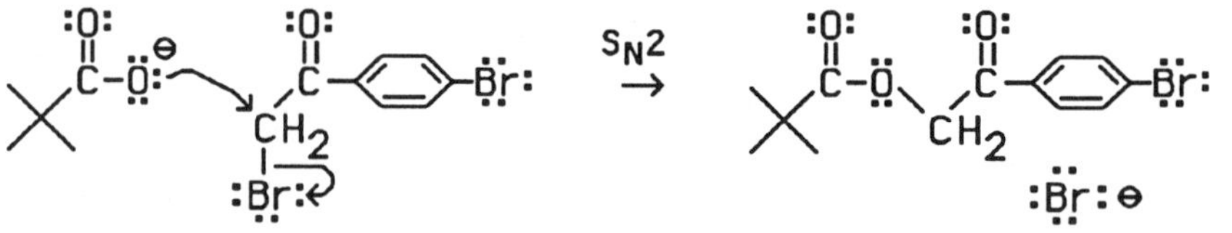

Evaluate and cross-check

The addition of the carboxylate to the ketone produces a more basic product anion of about pK_{abH} 10 to 13 from an incoming nucleophile of pK_{abH} 4.8. However, the S_N2 produces a product anion of pK_{abH} -9, definitely energetically favorable by the ΔpK_a rule. The S_N2 is predicted to be the lowest energy path. (A possible S_N1 is ruled out because the resultant carbocation would not be stable.)

All the paths from this point are definitely uphill; it is a good time to stop.

2. Predict the Product of the Reaction of an Ester and Ethoxide

Understand the system

Medium? Basic; ethoxide's pK_{abH} is 16. Sources? Ethoxide. Sinks? The ester is a carboxyl derivative. Acidic H's? The methyl next to the ester carbonyl, pK_a 24. Leaving groups? OEt from the ester (poor leaving group). Resonance forms?

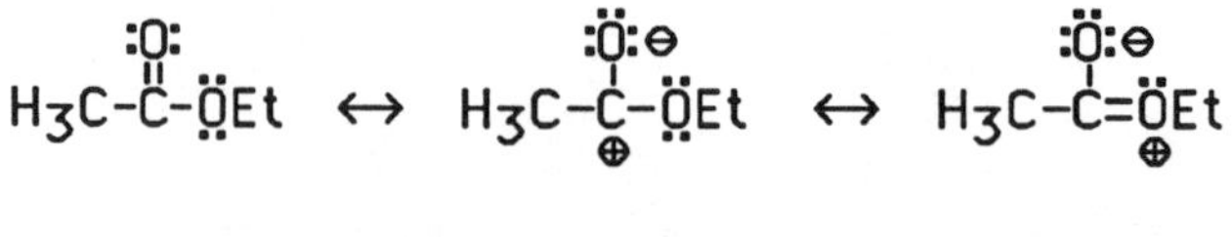

Find possible routes

Two routes are possible: Ad_N2 addition to a polarized multiple bond (Ad_N + p.t.),

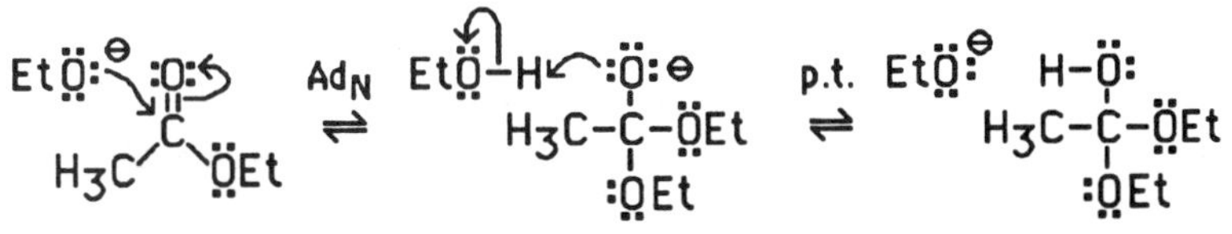

or deprotonation to form an anion.

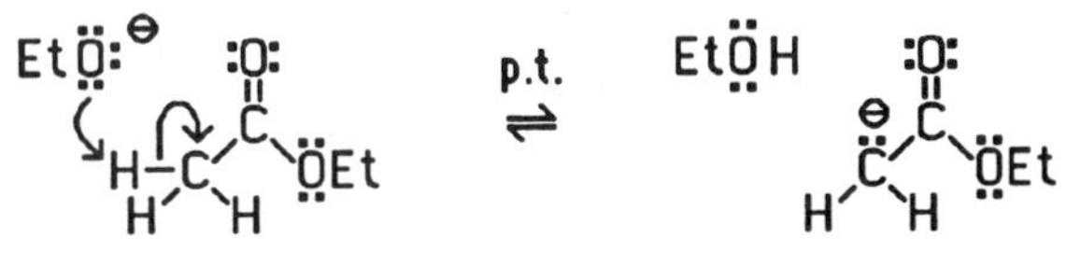

Evaluate and cross-check

Since the K_{eq} for proton transfer between the ester (pK_a = 24) and ethoxide (pK_{abH} = 16) is $10^{(16 - 24)} = 10^{-8}$ and therefore almost out of the useful range, it is reasonable to explore ethoxide as a nucleophile first.

Lone pair nucleophiles react with carboxyl derivatives by addition via path Ad_N, followed by beta elimination of the leaving group from the tetrahedral intermediate anion formed, path E_β. Because the leaving group is ethoxide, the same as the nucleophile, loss of either ethoxide returns us to the starting materials. This process is noticeable only if the nucleophile and leaving group are different (transesterification). The tetrahedral intermediate is too hindered to serve as a nucleophile, and if it is protonated it yields a species that is uphill on energy from the starting ester (by a ΔH calculation). There are no more basic medium paths except those that return to the starting ester.

After exploring the more probable process and ending up back at the start, we should now explore the energetically less favorable alternative, ethoxide as base even though the K_{eq} is so unfavorable.

Understand the system

The ester enolate can serve as a base or a nucleophile.

Find possible routes

Proton transfer easily occurs to the ester enolate, but that returns us to reactants. If the enolate is a nucleophile, a new product is formed by path Ad_N.

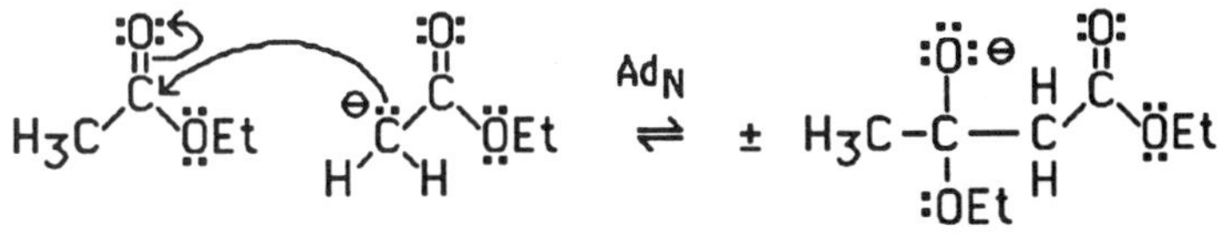

Evaluate and cross-check

Only one new path was found that did something new. The reaction forms a weaker base; the ester enolate pK_{abH} is 24; that of the tetrahedral intermediate is about 14.

Understand the system

The tetrahedral intermediate can lose the enolate and return; it is too hindered to serve as a good nucleophile.

Find possible routes

The tetrahedral intermediate can serve as a base (path p.t.),

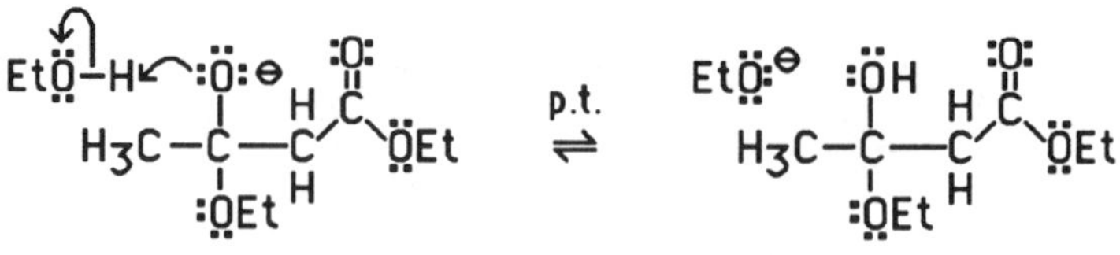

or it can lose a leaving group (path E_β).

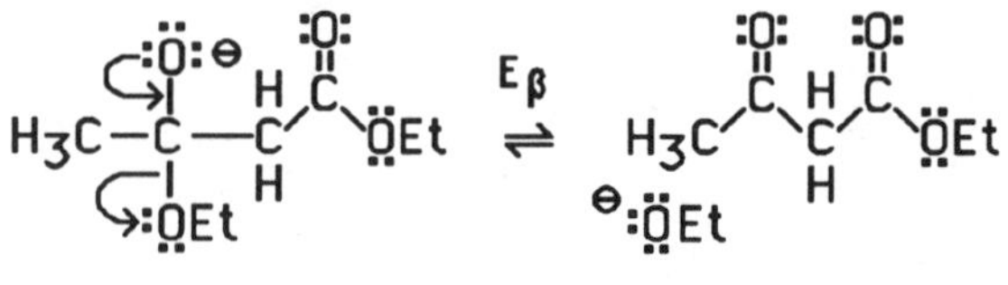

Evaluate and cross-check

Protonation gives the hemiketal, which by a ΔH calculation is energetically uphill. The hemiketal will simply return to reactants. The loss of the ester enolate would take us back to the previous structure. The leaving group trends tell us that ethoxide is the best leaving group and would be lost instead of the ester enolate. The loss of ethoxide has produced a slightly stronger base.

Understand the system

A ΔH calculation is possible at this point since we now have a neutral species; two esters have formed the ketoester and ethanol. We have broken a $C-H$ and a $C-O$ and in return formed a $C-C$ and an $O-H$ for $(99 + 86) - (83 + 111) = -9$ kcal/mol (-38 kJ/mol). This tells us that the reaction has probably gone downhill in energy, but we need to explore further to see if there is anything farther downhill.

Find possible routes

The ketoester can be attacked by ethoxide and return, or it can be deprotonated.

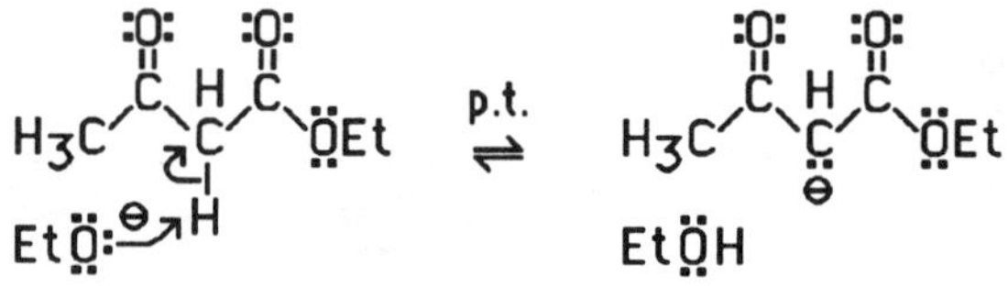

Evaluate and cross-check

Deprotonation has a very favorable K_{eq} of $10^{+5.3}$ because it produces a stable anion (pK_{abH} is 10.7) stabilized by two electron-withdrawing groups. Because it is so stabilized, this anion is a weak base and is expected to be energetically downhill from the original starting reagents. If this highly delocalized anion acted as a nucleophile and attacked another ester, it would just preferentially fall off again because it is a better leaving group than ethoxide. Therefore an exploration of the surface has produced a stable anion downhill from the reagents that appears not to be able to do anything more under the reaction conditions. This stable anion is the product of the reaction. Mechanistic studies have shown that the deprotonation step to form this highly delocalized anion is the driving force for the reaction.

Overview

The overall reaction has been a deprotonation to produce the nucleophile for an addition–elimination reaction that is followed by a deprotonation to yield a stabilized anion (Figure 9.5). The reaction path is shown vertically on the left, and the side routes are shown to the right. The path name for the forward reaction only is shown.

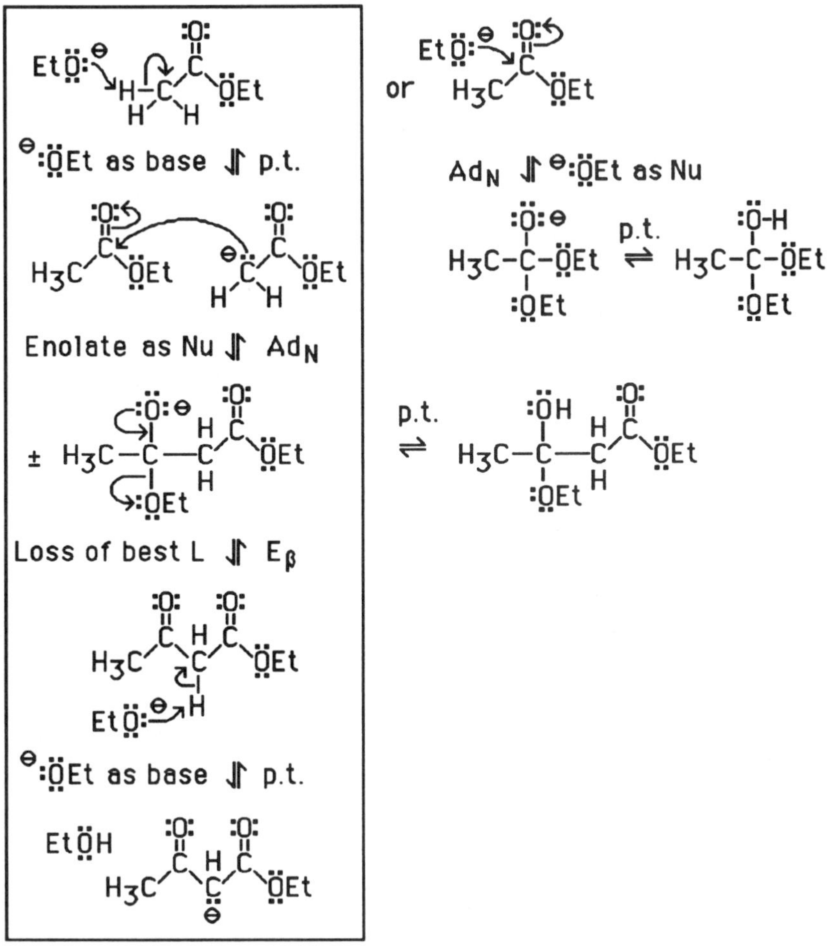

Figure 9.5 Ester and ethoxide overview. The most likely reaction path is boxed.

3. Predict the Product of the Reaction of a Ketal and Acidic Water

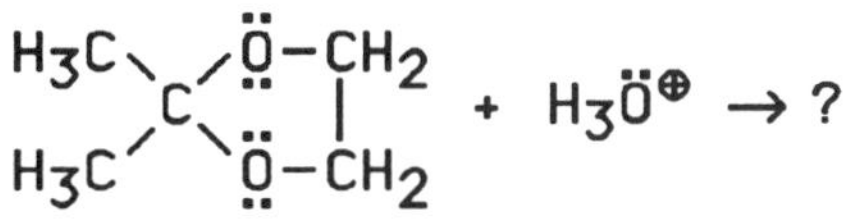

Understand the system

In this example, a ketal is mixed with an excess of acidic water. Medium? Acidic water, pK_a -1.7. Sources? The lone pairs on the oxygens. Sinks? The acidic H's of the acid. Leaving groups? None.

Find possible routes

The only path available at the start is proton transfer to the lone pair of the ketal.

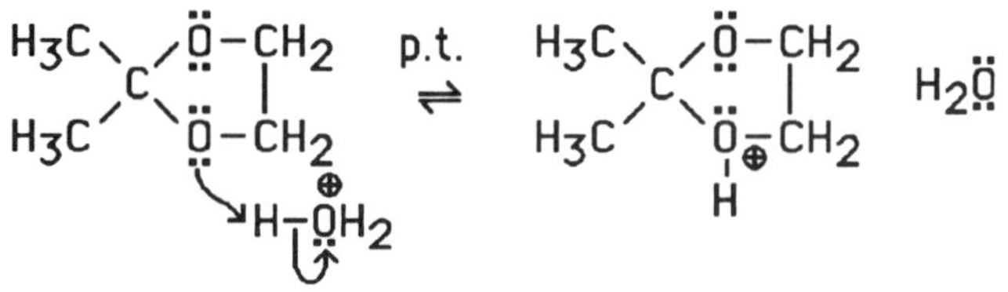

Evaluate and cross-check

There is only one source and one sink and one path. The proton transfer has a K_{eq} of $10^{(-3.5 - (-1.7))} = 10^{-1.8}$, slightly uphill, although still a good process.

Understand the system

Protonation of the ketal oxygen makes it a good leaving group with a $pK_{aHL} = -3.5$. Since the reaction is in an acidic medium, the leaving group is good, and the cation formed upon loss of the leaving group would also be relatively stable, the reacting species now enter the $S_N1/E1$ surface.

Find possible routes

The leaving group can ionize (path D_N).

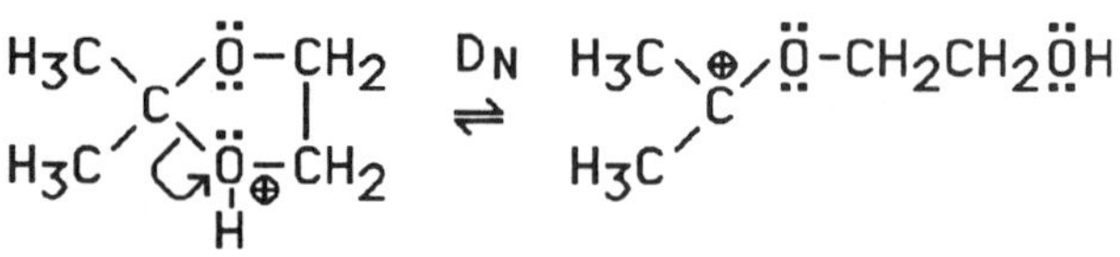

Either C—O bond to the protonated oxygen conceivably could break.

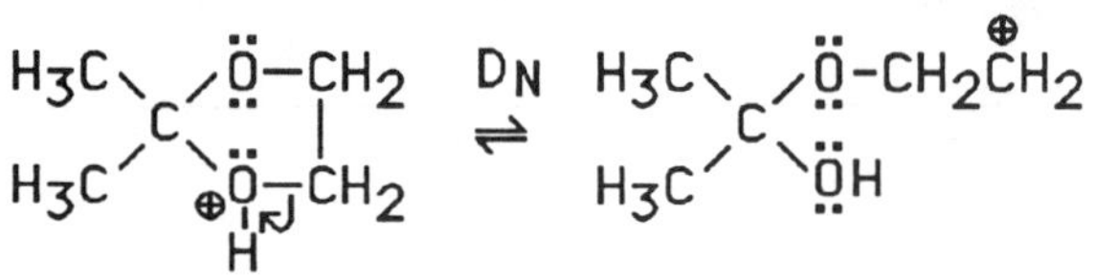

Evaluate and cross-check

The stability of both product cations must be checked: the lone-pair-stabilized tertiary cation is much more stable and therefore is favored over the unstabilized primary cation. Ionization of the leaving group (path D_N) creates a somewhat less stable cation (pK_a about -7) than the protonated ketal ($pK_a = -3.5$), so ionization of the leaving group is uphill in energy.

Understand the system

The lone-pair-stabilized tertiary cation is an excellent electron sink.

Find possible routes

This cation can lose an adjacent proton from the methyl (path D_E) to form the enol ether completing an E1 process, or the cation can get trapped by the most abundant nucleophile, water (path A_N), which completes an S_N1 process,

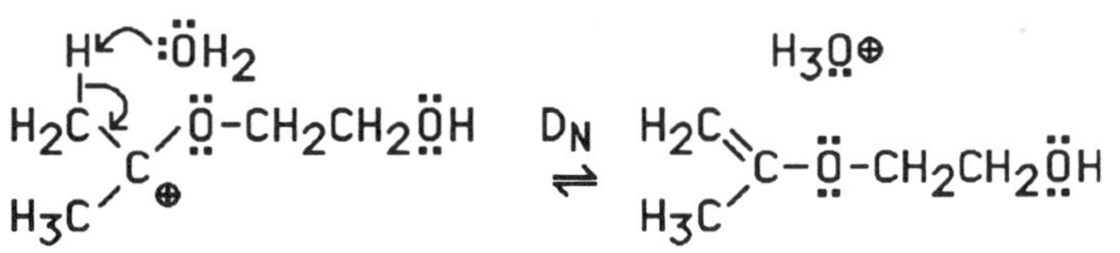

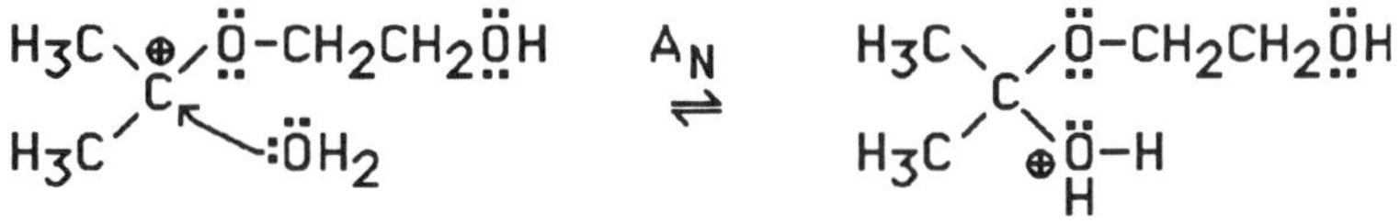

followed by proton loss to produce the hemiketal.

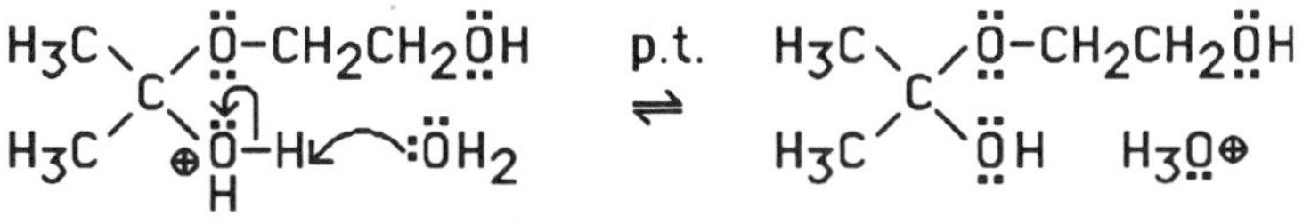

Evaluate and cross-check

Since both processes are reversible, the enol ether is uphill from the starting ketal [a ΔH calculation gives +11 kcal/mol (46 kJ/mol)], and there is an excess of nucleophile, S_N1 is favored. The proton transfer to water is favorable since each species has about the same pK_a.

Understand the system

Now we need to consider whether the hemiketal is downhill in energy from the ketal. A quick calculation of the ΔH of reaction yields zero, telling us that they are about the same in energy. It is best to keep exploring the energy surface to see if there is a lower-energy product.

Find possible routes

Protonation can occur on either oxygen's lone pairs.

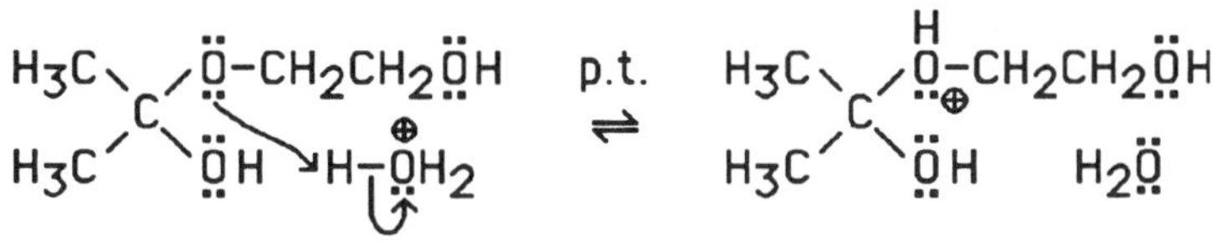

Evaluate and cross-check

Protonation on the alcohol lone pair is fine, but it returns us to a previous structure. Protonation on the slightly less basic ether lone pair converts it into a good leaving group. Similar to the first proton transfer step with the ketal, $K_{eq} = 10^{-1.8}$, the hemiketal is easily protonated in this acidic medium.

Understand the system

Protonation of the hemiketal oxygen makes it a good leaving group (pK_{aHL}= -3.5).

Find possible routes

Again, the reacting species enter the $S_N1/E1$ surface. The protonated hemiketal can lose the leaving group in a process identical with that of the ketal (path D_N) to give the lone-pair-stabilized tertiary cation.

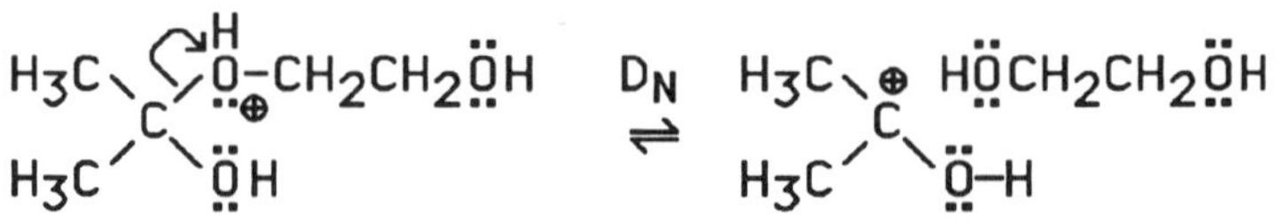

Again the lone-pair-stabilized cation can lose a proton from the methyl to form the enol (path D_E),

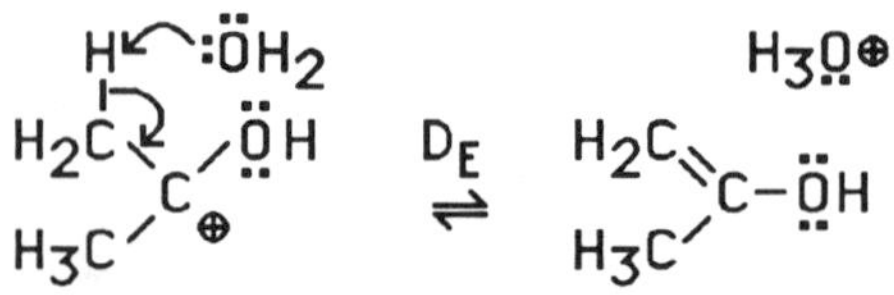

or can be trapped by water (path A_N) to yield, after deprotonation, the hydrate.

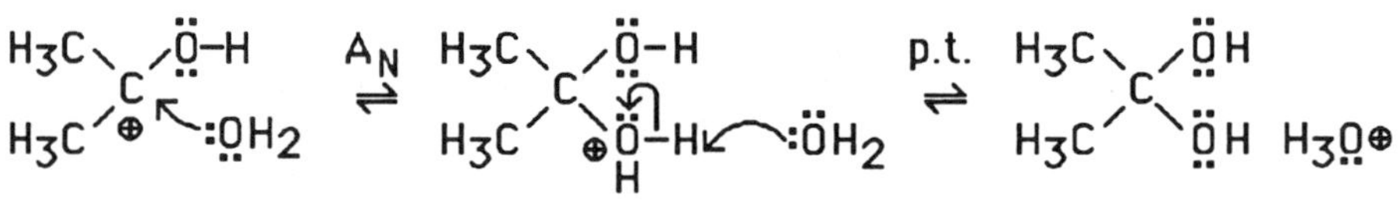

But additionally it can lose a proton from the oxygen to form the ketone (path D_E).

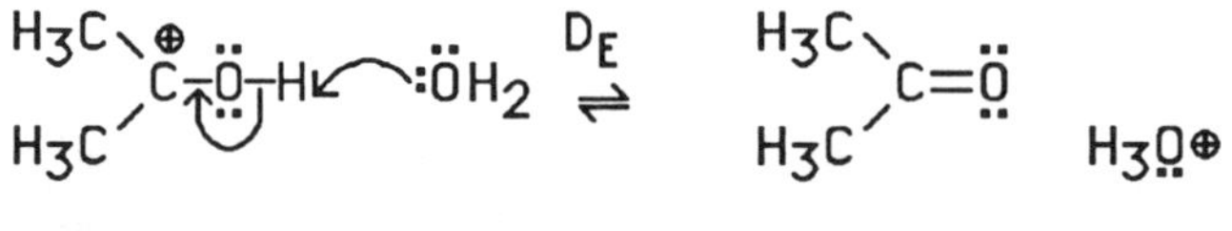

Evaluate and cross-check

Since an equilibrium process favors the most stable product, we need to again calculate ΔH to find out which of the three products are preferred. The hydrate is at about the same energy as the ketal, and the enol is calculated to be 11 kcal/mol (46 kJ/mol) higher, the ketone is 5 kcal/mol (21 kJ/mol) lower in energy than the ketal and is therefore the product.

Overview

Occasionally a reaction will be driven to a product uphill in energy by displacement of the equilibrium, commonly an excess of one reagent combined with the removal of one product. Ketals are formed from ketones in this manner: an excess of alcohol is used and the water formed is removed by distillation. The equilibria are the same as those discussed above, driven backward by mass balance effects. Watch reaction conditions for driven equilibria.

We have wandered over a rather complex energy surface (Figure 9.6), exploring every possible side reaction until there was nothing more that was reasonable to do. Looking at the stability of all the possibilities on the surface, we arrived at the correct, most favored product. The reaction path is shown vertically on the left and the side routes are shown to the right side. The path name for the forward reaction only is shown.

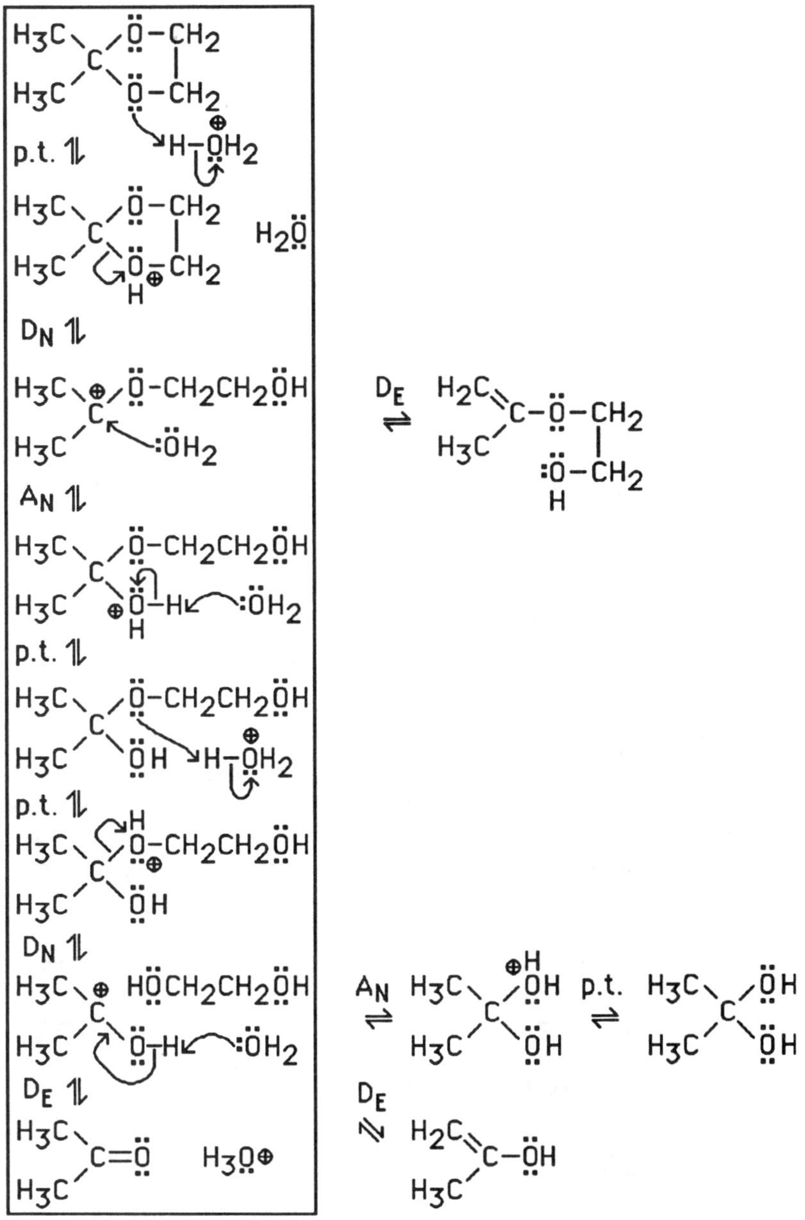

$$
\begin{array}{l}
H_3C \\
\quad\ \ C \\
H_3C \\
\end{array}
\begin{array}{l}
\ddot{O}-CH_2 \\
\ddot{O}-CH_2 \\
\end{array}
$$

H₃C, C, H₃C — Ö–CH₂ / Ö–CH₂ | ↷H–ÖH₂⊕

p.t. ⇅

H₃C, C, H₃C — Ö–CH₂ / ⤷Ö⊕CH₂ | H₂Ö̈
H

D_N ⇅

H₃C⊕ C / Ö–CH₂CH₂ÖH   D_E H₂C C–Ö–CH₂
H₃C ↖ :ÖH₂   ⇌   H₃C   :Ö–CH₂
                                    H

A_N ⇅

H₃C C / Ö–CH₂CH₂ÖH
H₃C ⊕Ö–H ↶ :ÖH₂
     H

p.t. ⇅

H₃C C / Ö–CH₂CH₂ÖH
H₃C ÖH ↷H–ÖH₂⊕

p.t. ⇅

H₃C C / ↶H Ö–CH₂CH₂ÖH
H₃C ÖH   ⊕

D_N ⇅

H₃C⊕ C HÖCH₂CH₂ÖH   A_N H₃C ⊕ÖH p.t. H₃C ÖH
H₃C ↷ Ö–H ↶:ÖH₂   ⇌   H₃C C ÖH ⇌   H₃C C ÖH

D_E ⇅   D_E ⇏ H₂C C–ÖH
H₃C   H₃C
    C=Ö H₃O⊕
H₃C

Figure 9.6 Ketal hydrolysis overview. The most likely reaction path is boxed.

4. Predict the Product of the Reaction of an Amide, Basic Water, and Heat

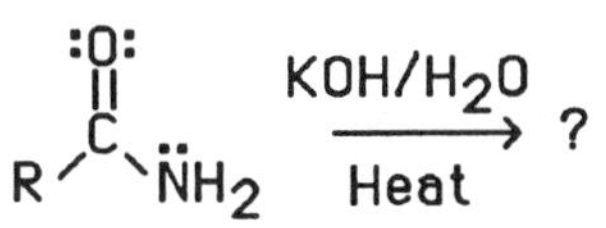

$$
\underset{R}{\overset{:O:}{\underset{}{\|}}}\!\!\!\overset{}{C}\!\!-\ddot{N}H_2 \quad \xrightarrow[\text{Heat}]{KOH/H_2O} \quad ?
$$

Understand the system

Medium? Basic water. Sources? Hydroxide anion is the best source. Leaving groups? None. Sinks? The amide is a carboxyl derivative. (Although the NH_2 places it in that class, it is a very poor leaving group.) Acidic H's? The NH_2 of the amide is pK_a 17. Bases? Hydroxide, $pK_{abH} = 15.7$. Resonance forms?

$$
\begin{array}{ccc}
\overset{\displaystyle :O:}{\underset{R}{\overset{\|}{C}}\!\!-\!\ddot{N}H_2} & \longleftrightarrow & \overset{\displaystyle :\ddot{O}:^{\ominus}}{\underset{R}{\overset{|}{C}}\!\!=\!\overset{\oplus}{N}H_2} & \longleftrightarrow & \overset{\displaystyle :\ddot{O}:^{\ominus}}{\underset{R}{\overset{\oplus}{C}}\!\!-\!\ddot{N}H_2}
\end{array}
$$

Find possible routes

There are two possible: We can use hydroxide as a nucleophile and attack the amide carbonyl (path Ad_N),

$$HO:^{\ominus} \;\text{attacking}\; \underset{R}{\overset{:O:}{\overset{\|}{C}}}\!\!-\!\ddot{N}H_2 \quad \overset{Ad_N}{\rightleftharpoons} \quad HO\!\!-\!\underset{R}{\overset{:\ddot{O}:^{\ominus}}{\overset{|}{C}}}\!\!-\!\ddot{N}H_2$$

or use hydroxide as a base and remove an amide NH to form the amidate anion (path p.t.).

$$\underset{R}{\overset{:O:}{\overset{\|}{C}}}\!\!-\!\underset{H}{\overset{\ominus}{N}}\!\!-\!H \cdots :\ddot{O}H^{\ominus} \quad \overset{p.t.}{\rightleftharpoons} \quad H\ddot{O}H \quad \underset{R}{\overset{:O:}{\overset{\|}{C}}}\!\!-\!\ddot{N}H^{\ominus}$$

Evaluate and cross-check

The K_{eq} for proton transfer is $10^{-1.3}$, which slightly favors reactants. If the amidate anion were to act as a nucleophile, the only site for attack would be another amide, and then it would just be kicked right back out again since it would be the best leaving group. If the amidate acts as a base, it returns to the reactants. A short exploration down the amidate route reveals it as a dead end. Hydroxide as a nucleophile looks more promising.

The ΔpK_a rule is hard to use because we do not have a good value for the pK_a of the product of the addition of hydroxide. Steric hindrance from R raises the pK_{abH} of an alkoxide anion (poorer intermolecular solvation), whereas the inductive/field effect from the nitrogen should lower the pK_{abH}. If these effects balance out, the product of nucleophilic attack would have a pK_{abH} of about 13 ($HOCH_2OH$ is pK_a 13.3).

Understand the system

There is no leaving group other than the original nucleophile, so we should examine the possibilities for proton transfer.

Find possible routes

Proton transfer can occur from water to the alkoxide anion or to the nitrogen lone pair or to both. These are four charge types of the tetrahedral intermediate.

$$
\begin{array}{ccc}
H\ddot{O}\!\!-\!\underset{R}{\overset{:\ddot{O}:^{\ominus}}{\overset{|}{C}}}\!\!-\!\ddot{N}H_2 & \overset{p.t.}{\rightleftharpoons} & H\ddot{O}\!\!-\!\underset{R}{\overset{:\ddot{O}-H}{\overset{|}{C}}}\!\!-\!\ddot{N}H_2 \\[2ex]
\updownarrow\; p.t. & & \updownarrow\; p.t. \\[2ex]
H\ddot{O}\!\!-\!\underset{R}{\overset{:\ddot{O}:^{\ominus}}{\overset{|}{C}}}\!\!-\!\overset{\oplus}{N}H_3 & \overset{p.t.}{\rightleftharpoons} & H\ddot{O}\!\!-\!\underset{R}{\overset{:\ddot{O}-H}{\overset{|}{C}}}\!\!-\!\overset{\oplus}{N}H_3
\end{array}
$$

Evaluate and cross-check

We are again hampered by the need to estimate pK_a's for these four charge types. The OH in the upper right structure was estimated at a pK_a of about 13. The $NH_3^\oplus$ in the lower right structure should have a pK_a of about 7 owing to the inductive/field effect of two hydroxyls (a simple protonated amine has a pK_a of 11). The OH in that structure should also be acidic, pK_a estimated at 10, because of the inductive/field effect of $NH_3^\oplus$. All these proton transfer reactions are reasonable.

If the medium were basic enough, we might expect the OH in the upper left structure to be deprotonated to give a dianion. It is difficult to gauge the reasonableness of the dianion since its pK_{abH} is difficult to estimate. Because of charge repulsion, it certainly has a pK_{abH} many units higher than 13, the estimated pK_{abH} of the upper left structure.

We seem to be at a proton transfer plateau. All four species are in equilibrium with each other. The top two species can do nothing except revert to starting materials, so in the absence of an obvious lower energy route, we must explore paths from both of the bottom two species.

Understand the system

The $NH_3^\oplus$ can serve as a fair leaving group (pK_{aHL} 9.2). **Note well: The unprotonated amine, pK_{aHL} 35, is not a leaving group at all.**

Find possible routes

Two elimination paths are possible: beta elimination from an anion (path E_β),

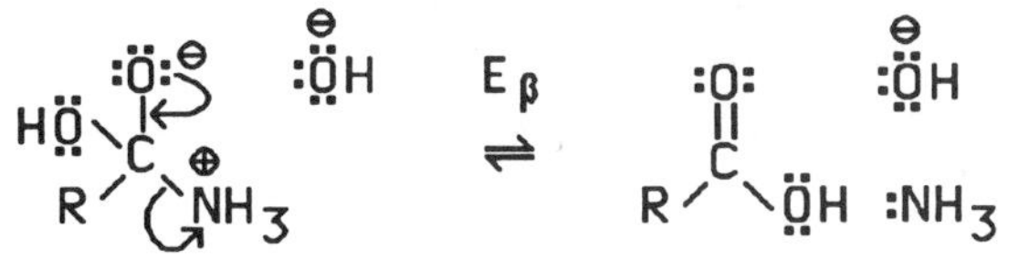

or E2 elimination.

Evaluate and cross-check

Both are viable routes, and thankfully they both go to the same product. The E2 forms a weaker base of pK_{abH} 9.2 from a starting anion of pK_{abH} 15.7. The beta elimination forms a slightly weaker base and might be the preferred route for elimination of a fair leaving group. We can spot an irreversible proton transfer step, $K_{eq} = 10^{+10.9}$, that creates an unreactive carboxylate anion. With this proton transfer we are done.

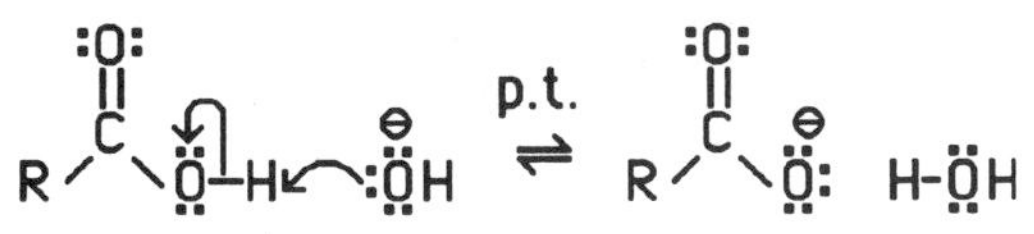

Overview

The overall reaction path is shown vertically, and the alternate paths are to the side (Figure 9.7).

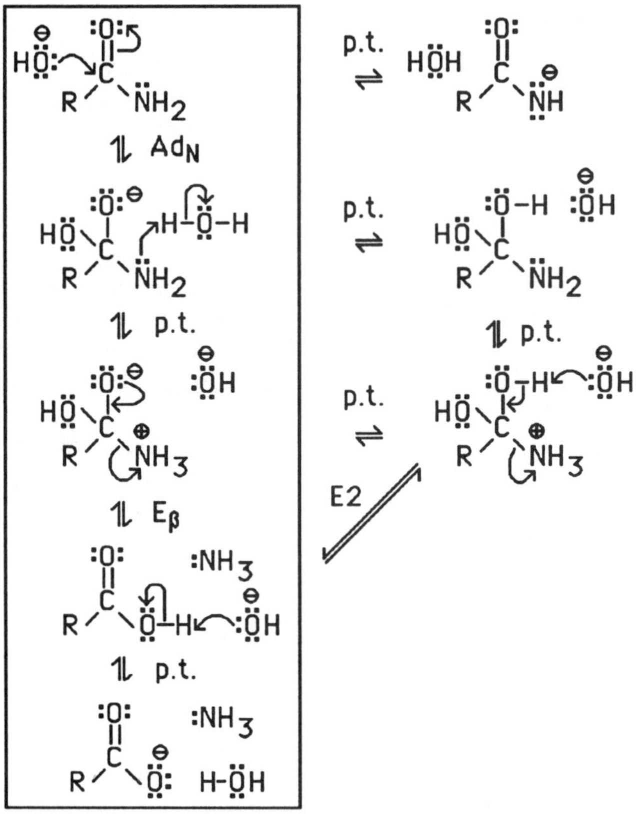

Figure 9.7 Amide hydrolysis overview. The most likely reaction path is boxed.

5. Predict the Product of the Reaction of a Carboxylic Acid and Thionyl Chloride

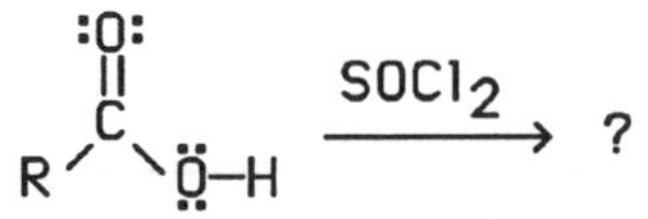

Understand the system

Medium? Acidic. Sources? The lone pairs on the carbonyl are the best source (much better than the lone pairs of the OH). Leaving groups? Chloride. Sinks? $SOCl_2$ is a Y—L, and the carboxylic acid is a carboxyl derivative. (Again, although the OH places it in that class, OH is a very poor leaving group.) The best sink is $SOCl_2$. Acidic H's? The carboxylic acid's OH. Bases? None. Resonance forms?

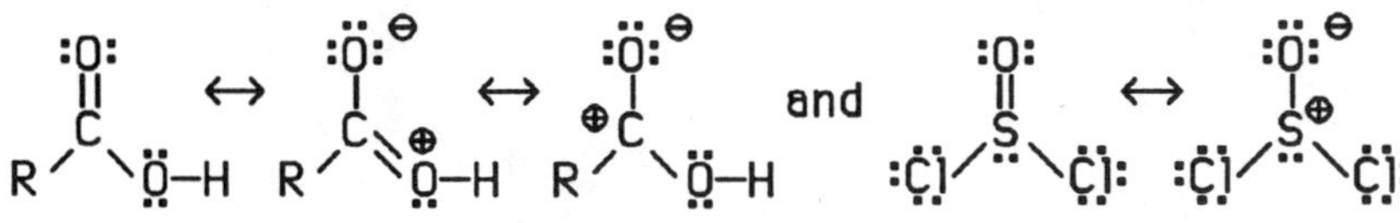

Find possible routes

We combine the best source and the best sink with the best path. The most probable is the S_N2 substitution. It is currently debated whether sulfur goes through a pentacoordinate intermediate (path pent.); however, both processes yield the same product.

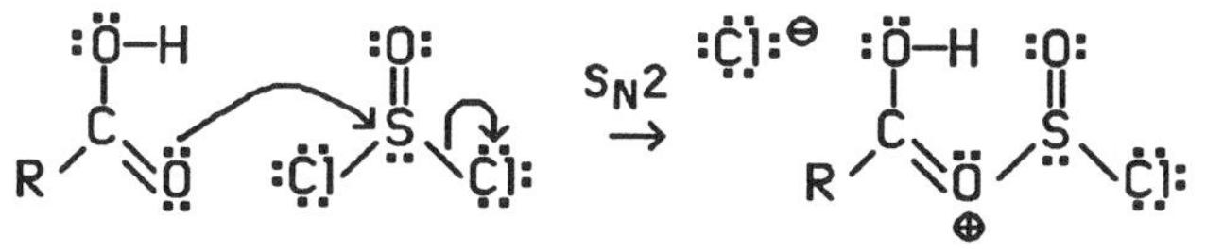

Evaluate and cross-check

If we had used the carboxylic acid's OH lone pair, the reaction would have produced an unstable cation, a positive oxygen between two electron-withdrawing groups, the carbonyl and the sulfoxide. The nucleophile's pK_{abH} is -6 and that of the leaving group is -7. The attack has broken a weak S—Cl bond. The reaction is most likely downhill in energy.

Understand the system

The carbocation formed is tertiary and substituted by two lone pair donors. Resonance forms:

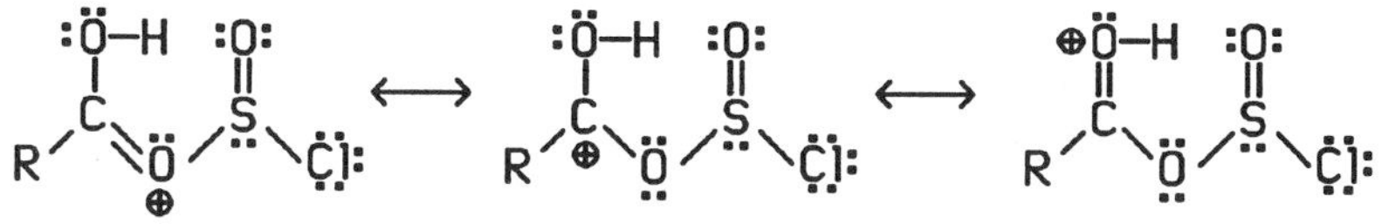

Find possible routes

The carbocation is an excellent acid and a good site for nucleophilic attack. The two alternatives are deprotonation (path p.t.),

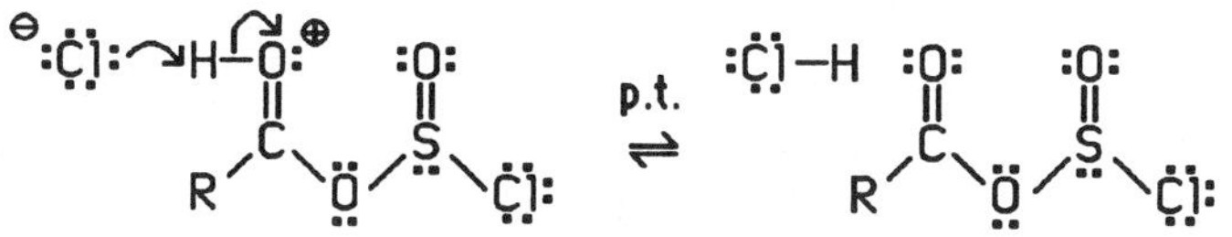

or attack (path Ad$_N$). The best nucleophile is now chloride.

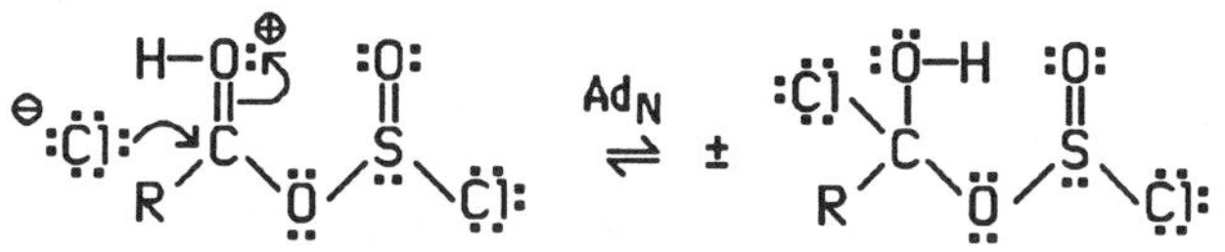

Evaluate and cross-check

Both processes should be favorable. The proton transfer occurs between two groups that have about the same pK_a to yield the acyl chlorosulfite. The nucleophilic attack to give the tetrahedral intermediate has neutralized charge also should be favorable. Both processes are consistent with a very acidic medium.

Understand the system

Let's explore the proton transfer route first. The reaction is generating HCl, pK_a -7, and the media is becoming strongly acidic The resonance forms for the acyl chlorosulfite are similar to those of the starting carboxylic acid.

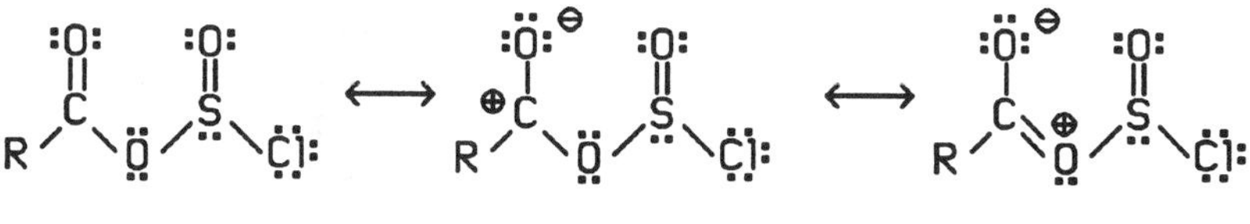

Find possible routes

The acyl chlorosulfite is a good site for nucleophilic attack. The best nucleophile is still chloride. The three alternatives are the Ad_N2 addition to a polarized multiple bond (path Ad_N then path p.t.),

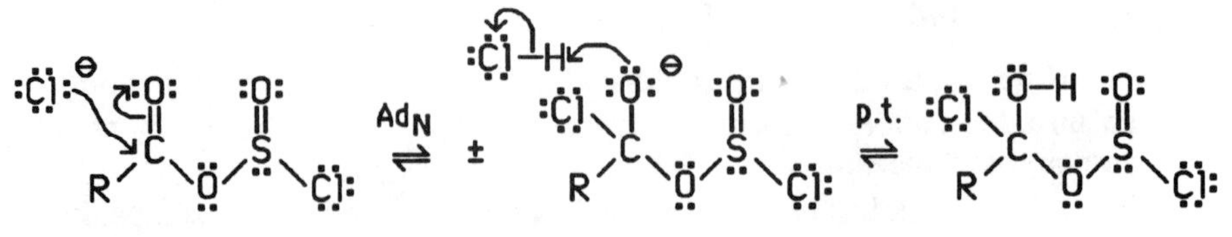

or the hetero Ad_E3 addition,

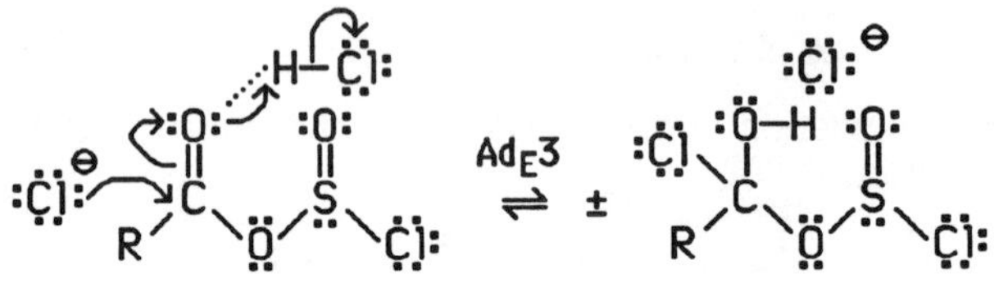

or the hetero Ad_E2 (path p.t. followed by path Ad_N).

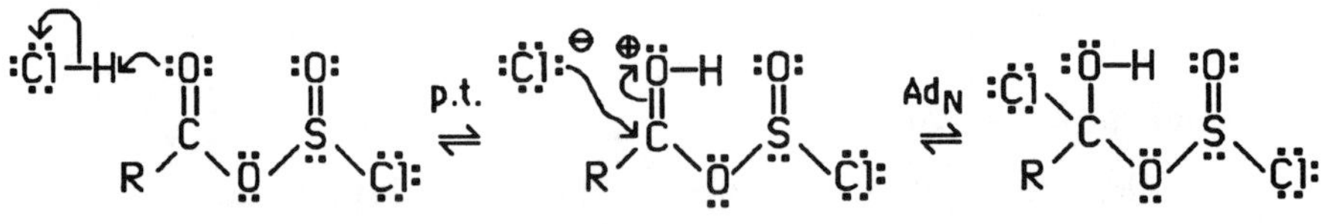

Evaluate and cross-check

The Ad_N2 is probably not a favored route because the nucleophilic attack of chloride, pK_{abH} of -7, to give the anionic tetrahedral intermediate, pK_{abH} about 10, goes against the ΔpK_a rule (a climb of 17 pK_a units) and forms a basic anion in a very acidic medium. The Ad_E3 and the hetero Ad_E2 are both reasonable routes by the ΔpK_a rule and are consistent with very acidic media.

Understand the system

We have come full circle back to the original tetrahedral intermediate. The tetrahedral intermediate has two attached leaving groups, the chloride and chlorosulfite. Loss of chloride reverts to reactants but loss of chlorosulfite yields new compounds.

Find possible routes

There are four possible elimination routes that produce the unstable chlorosulfite anion, which immediately loses the good leaving group chloride to produce sulfur dioxide gas. Chlorosulfite can be lost in a lone-pair-assisted E1 (path E_β followed by path p.t.),

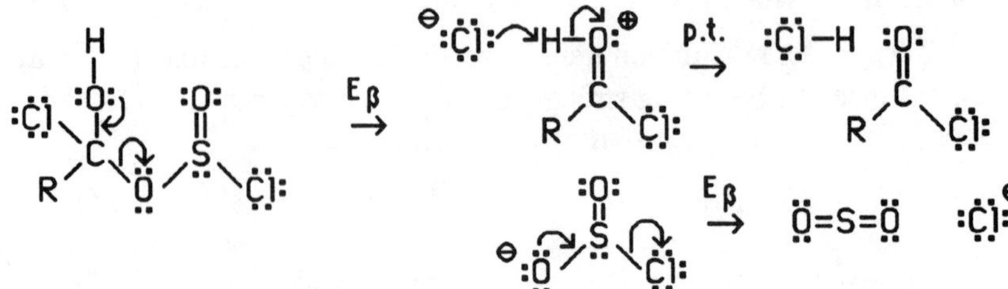

or an E2,

or by a *syn* internal elimination (path Ei),

or by an E1cB (path p.t. followed by path E$_\beta$).

Evaluate and cross-check

The E1cB is probably not a good route because the deprotonation by chloride, pK_{abH} -7, to give the anionic tetrahedral intermediate, pK_{abH} about 10, goes against the ΔpK_a rule (a climb of 17 pK_a units) and forms a basic anion in a strongly acidic medium.

The other three eliminations are compatible with a strongly acidic medium. Because chlorosulfite is so unstable, no pK_a is available for its conjugate acid, and it is difficult to rank as a leaving group. The closest we can get to chlorosulfite on the pK_a chart is benzenesulfinic acid, PhSO$_2$H, whose pK_a is 1.5; the substitution of an electronegative chlorine for phenyl will make the acid much more acidic, dropping the pK_a at least several units; thus chlorosulfite is a good leaving group. It is difficult to rank which elimination route is lowest in energy. The Ei elimination has the advantage of being intramolecular and therefore may be the fastest.

Overview

Now two gases have been evolved and should escape the reaction mixture driving the transformation to completion (Figure 9.8). The most probable route is in the center with alternative routes drawn to either side. The path name for the forward reaction only is shown. The low-probability routes have been omitted for clarity.

9.7 METHODS FOR TESTING MECHANISMS

So far we have seen how to generate mechanistic hypotheses. The next step in the scientific method is the testing of them to rule out ones incompatible with experimental observations. We use the tools presented in this section to **disprove** these hypotheses, not to try to prove one correct. A reaction mechanism always remains a hypothesis and therefore can never be proven true. Experimental design is important; the experiment should be set up such that it will clearly exclude one of the mechanistic alternatives. The experiment must be carefully carried out so as to get a clean result. We repeat the process until only one hypothesis remains, and within the limits of our abilities to test, fails to be disproven.

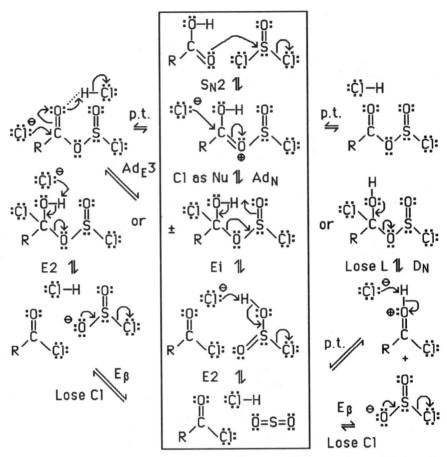

Figure 9.8 Acyl chloride formation overview. The most likely reaction path is boxed.

Initial Studies

The first thing that needs to be done in studying a reaction is to know what substances enter into the reaction and what are produced. The exact stoichiometry of the reaction must be known. Any stereochemistry associated with the starting material and product needs to be determined. Side products should be identified.

Catalysis, Inhibition

Often no reaction will occur until the requisite catalyst is added. Does the reaction require acid or base catalysis? Is a weak acid or base capable of catalyzing the reaction? Is a proton source, that is, trace water, required? Do Lewis acids catalyze the reaction? Can the reaction be inhibited by free radical inhibitors? Does the reaction require a free radical initiator or light? (See Chapter 10.)

Isolation or Detection of Intermediates

There are several spectroscopic methods that are capable of distinguishing reaction intermediates of a reaction in progress. Also, it is occasionally possible to isolate intermediates from especially mild reaction conditions. Other reagents can be added to the reaction to trap the supposed reactive intermediate specifically and preferentially. Reactive intermediates of unimolecular reactions can be characterized spectroscopically by running the reaction in a glass such as frozen argon.

Isotopic Labels

Labels are a convenient probe of a reaction mechanism, but care must be taken that whatever is chosen as a label does not alter the system under study. The label must be stable to the reaction conditions and not accidentally fall off into solution or scramble its position. Isotopic labels are a useful way of following an atom through a reaction. Common labels are ^2H, ^3H, ^{13}C, ^{14}C, ^{15}N, and ^{18}O. The radioactives, ^3H and ^{14}C, require extensive degradation schemes to determine the position of the label. With CMR spectroscopy, the location of a ^{13}C label is especially easy to carry out. Labels can be very useful in determining the origins of a particular group in a rearrangement, a cleavage, or a biosynthesis.

Doubly labeled compounds provide an interesting test for an intermolecular versus intramolecular transfer of a group. The group and another part of the same molecule are both labeled and the reaction is run with an equal amount of unlabeled material. The product of an intramolecular transfer would be either doubly labeled or unlabeled. The intermolecular process would produce a statistical mixture of products: About half the product would be singly labeled.

Example problem

How would you use a label to disprove that a methyl ester underwent basic hydrolysis in KOH/water by an S_N2 mechanism?

Answer: Label the oxygen in the KOH/water. An S_N2 on the methyl with labeled hydroxide will give the label in the methanol, whereas other mechanisms such as the addition-elimination path combination would not.

Stereochemistry and Chirality

Is stereochemistry lost or retained in the reaction? Is the process stereospecific? Does addition or elimination occur *anti* or *syn*? Is the chirality at the reacting center preserved, inverted, or lost?

Example problem

The addition of HBr to cyclohexene gives the anti addition product. Is this result more consistent with an Ad_E2 or an Ad_E3 mechanism.

Answer: The Ad_E3 is consistent with *anti* addition. The Ad_E2 would be expected to give a *syn/anti* mixture with *syn* addition preferred. The Ad_E2 intermediate ion pair collapses preferentially *syn*; the symmetrically solvated carbocation would have no preference.

Kinetics

Kinetics can yield information about the rate-determining step of a reaction. At a constant temperature, the concentration of a particular reactant is changed and the effect on the rate of reaction is noted. In this way the kinetic order with respect to each reactant is determined. These kinetic orders can then be compared to that expected for a particular mechanism. Often, however the kinetic order can easily fit several possibilities.

Example problem

The addition of HCl to 2-methyl-2-butene in nitromethane is first order in alkene and second order in HCl. Which of the possible addition mechanisms is consistent with the kinetic order?

Answer: The kinetic order is the sum of the orders for all species; the Ad_E3 would be third order. The Ad_E2 would be first order in HCl and first order in alkene, second order overall.

Solvent Effects

Solvent polarity can stabilize charged species, and can affect the rate of a reaction. If the reaction forms a charged intermediate or product, increased solvent polarity will speed up the reaction by lowering the energy of the charged species. However, if charge is neutralized or dispersed during the reaction, a more polar solvent will slow down the rate because reactant stabilization increases the reaction barrier.

Example problem

If the substitution of an alkyl chloride by an anionic nucleophile is greatly slowed as the reaction solvent becomes less polar, is this solvent effect more compatable with an S_N1 or S_N2 mechanism?

Answer: The S_N1 rate-determining step is the ionization of the leaving group to form a carbocation and a chloride ion. The solvent must stabilize these charged species, or the reaction will be slowed or stopped. The S_N2 reaction rate would be expected to increase slightly since the charge is dispersed in the transition state. The observed solvent effect is inconsistent with an S_N2 mechanism.

Substituent Effects

Electron-withdrawing or donating substituents can greatly alter the rate of a reaction and often cause a change in mechanism. A common way to study substituent effects is to connect to the site undergoing reaction, a benzene ring that has a meta or para substituent group. This substituent group is then varied, and the rate of the reaction is compared to the rate when the group is simply H. If electron-donating groups accelerate the rate of reaction, the conclusion is drawn that a positive charge is increased or a negative charge is decreased in the rate-determining step. If electron-withdrawing groups accelerate the rate, either a negative charge is increased or a positive charge is decreased in the rate-determining step.

Example problem

The rate of the substitution reaction of $Ar(CH_3)_2C-Cl$ with water is greatly favored by electron-donating groups on the aryl group. Is this substituent effect explained better by an S_N1 or S_N2 mechanism?

Answer: A positive charge is increased or a negative charge is decreased in the rate-determining step. The S_N1 rate-determining step is the formation of a carbocation that donors would stabilize. The neutral S_N2 transition state would not be expected to have a significant substituent effect.

Primary Deuterium Isotope Effects

The fact that a C—H bond is just slightly easier to break than a C—D bond can be used to probe reaction mechanisms that have C—H bond breaking in their rate-determining step. In this case the rate of the deuterium substituted reactant will be 2 to 7 times slower.

Example problem

The rate of the elimination reaction of deuterated 2-bromopropane with ethoxide is 6.7 times slower than the undeuterated. Is this observation consistent with an E1 mechanism?

Answer: No, since the E1 rate-determining step is leaving group loss and not C—H bond breakage, it would not show a primary deuterium isotope effect; an E2 reaction would.

Barrier Data

From the change in the rate constant as a function of temperature both the $\Delta H^\ddagger$ and $\Delta S^\ddagger$ and therefore $\Delta G^\ddagger$ can be determined. Since bond breaking in the transition state increases the $\Delta H^\ddagger$, and bond formation diminishes it, processes that have simultaneous bond breakage and formation are expected to have a low $\Delta H^\ddagger$. Transition states that require a specific spatial arrangement of reactants will tend to have a negative $\Delta S^\ddagger$ because the degree of disorder decreases in going to the transition state.

ADDITIONAL EXERCISES

(Refer to the Hints to Problems from Chapters 7, 8, and 9 if you need assistance).

9.1 Provide a mechanism for the following transformations.

(a)

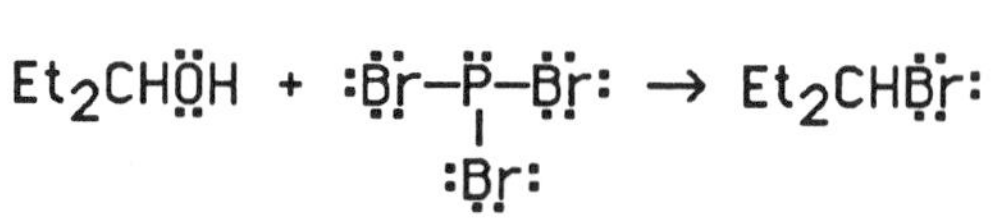

$$Et_2CH\ddot{O}H + :\ddot{B}r\!-\!\overset{\displaystyle |}{\underset{\displaystyle :\ddot{B}r:}{P}}\!-\!\ddot{B}r: \rightarrow Et_2CH\ddot{B}r:$$

(b)

$$H_2C=CH-CH_2-\overset{\displaystyle :\overset{..}{O}:}{\overset{\|}{C}}H \xrightarrow[H_2O]{KOH} H_3C-CH=CH-\overset{\displaystyle :\overset{..}{O}:}{\overset{\|}{C}}H$$

(c)

$$R-C{\equiv}C-R \xrightarrow{H_3\ddot{O}\oplus} R-\overset{\displaystyle :\overset{..}{O}:}{\overset{\|}{C}}-CH_2R$$

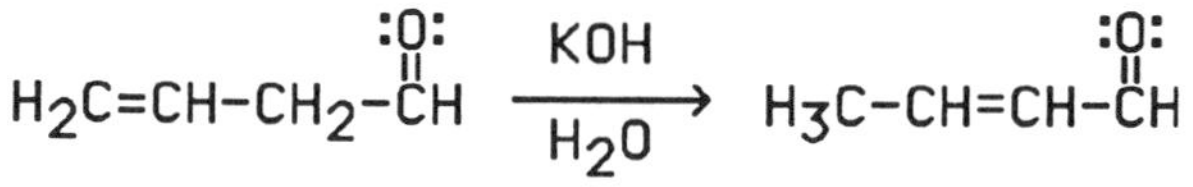

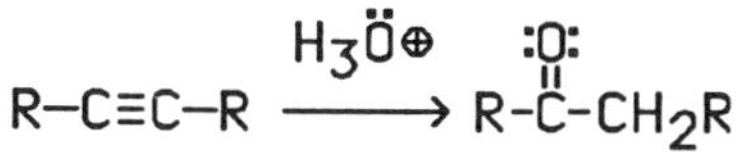

(d)

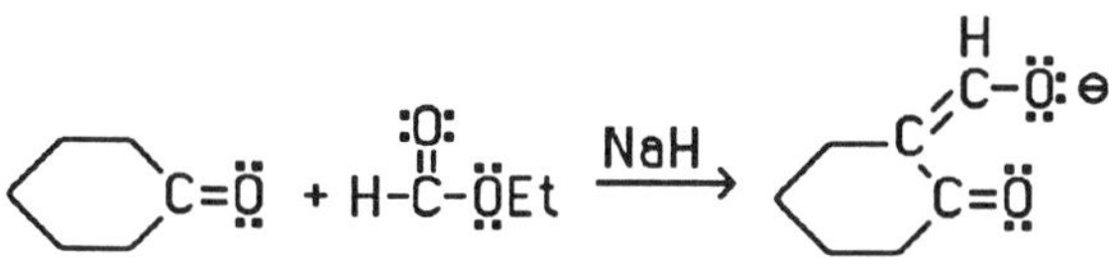

(e)

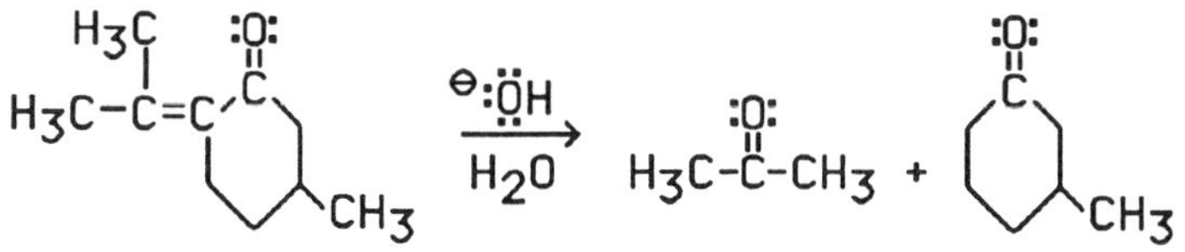

9.1 (continued)

(f)

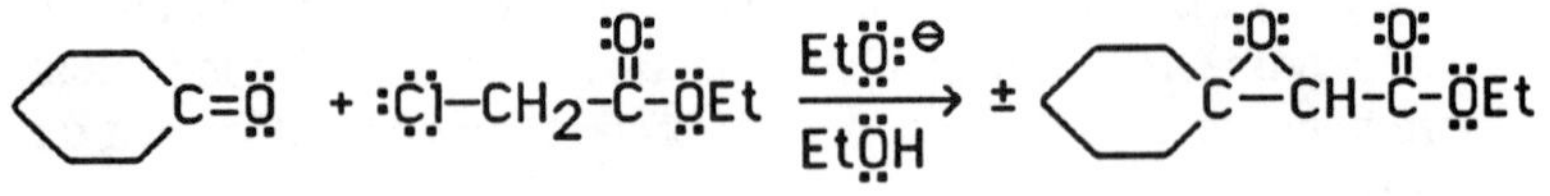

9.2 Provide a mechanism for the following transformations (more difficult).

(a)

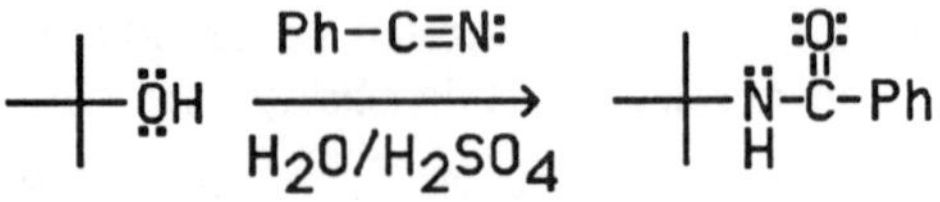

(b)

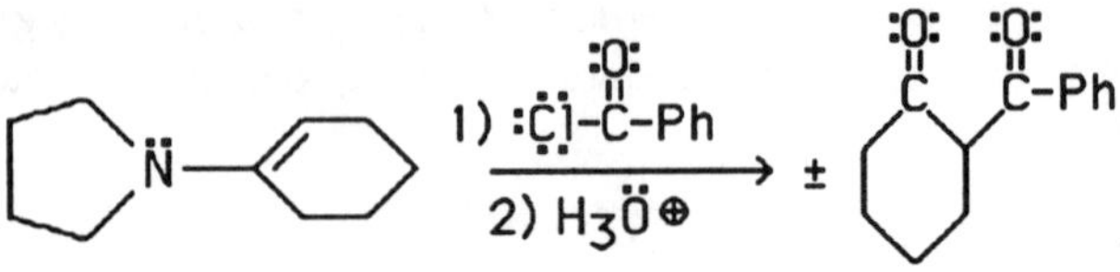

(c)

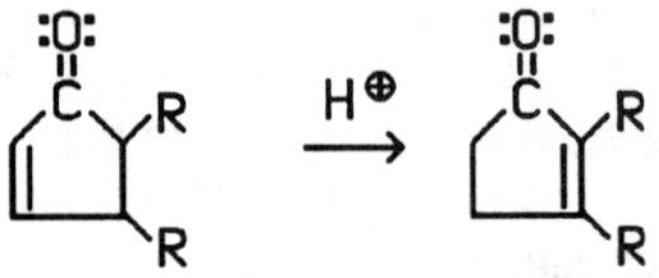

(d)

$$Ph-\overset{:O:}{\overset{\|}{C}}-CH_2CH_3 \xrightarrow[\text{3) } H_2\ddot{O}_2,\text{ warm}]{\begin{array}{l}1)R_2\ddot{N}:^{\ominus} Li^{\oplus}\\ 2)Ph-\ddot{S}e-\ddot{B}r:\end{array}} Ph-\overset{:O:}{\overset{\|}{C}}-CH=CH_2$$

(e)

$$R-\overset{:O:}{\overset{\|}{C}}-CH_2-\overset{:O:}{\overset{\|}{C}}-\ddot{O}Et \xrightarrow[\text{Heat}]{H_3O:^{\oplus}} R-\overset{:O:}{\overset{\|}{C}}-CH_3 + H\ddot{O}Et + \ddot{O}=C=\ddot{O}$$

(f)

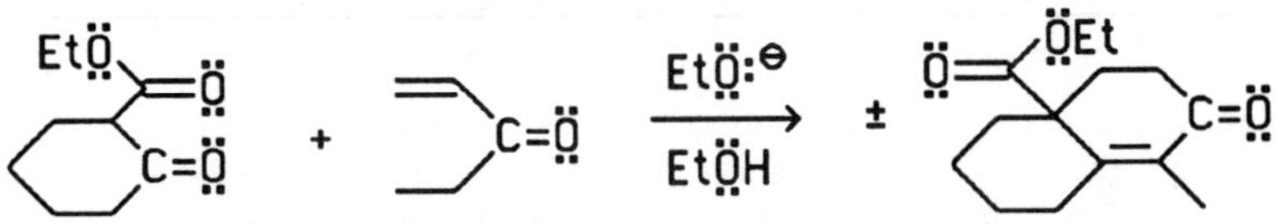

9.3 Give the product of the following reactions.

(a)

$$CH_3CH_2CH_2\ddot{B}r: + CH_3\ddot{S}:^{\ominus} \rightarrow$$

(b)

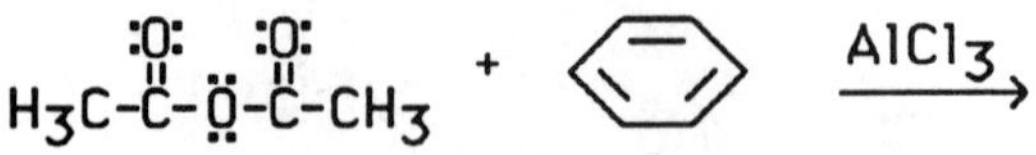

(c)

$$CH_3CH_2CH_2CH_2CH_2-\ddot{B}r: \xrightarrow{Ph_3P:}$$

(d)

9.3 (continued).

(e)

$$CH_3-\overset{\overset{\Large :O:}{\|}}{C}-\ddot{O}-H \;+\; RMgBr \longrightarrow$$

(f)

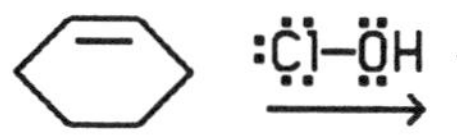

(g)

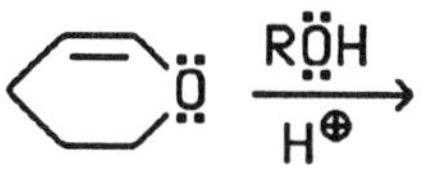

9.4 Give the product of the following reactions (more difficult).

(a)

$$Ph-\underset{CH_3}{\overset{H}{C}}-\ddot{N}(CH_3)_2 \xrightarrow[\text{2) Heat}]{\text{1) } H\ddot{O}-\ddot{O}H}$$

(b)

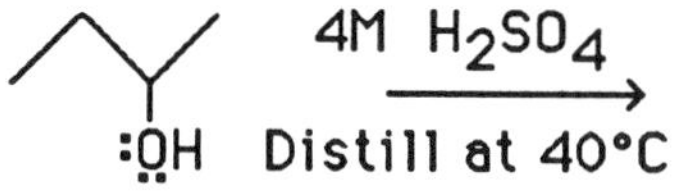

(c)

$$Ph-\ddot{N}=C=\ddot{O} \xrightarrow{Et\ddot{O}H}$$

(d)

$$H-\overset{\overset{:O:}{\|}}{C}-\ddot{N}(CH_3)_2 \xrightarrow[\text{2) } H_3\ddot{O}^{\oplus}]{\text{1) RMgBr}}$$

(e)

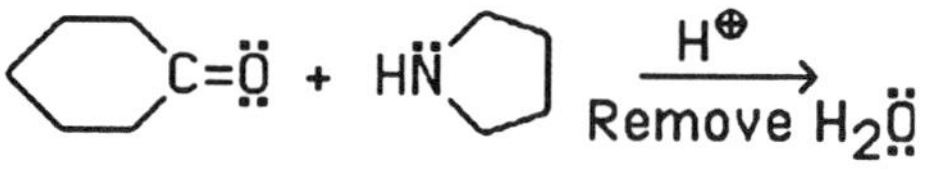

(f)

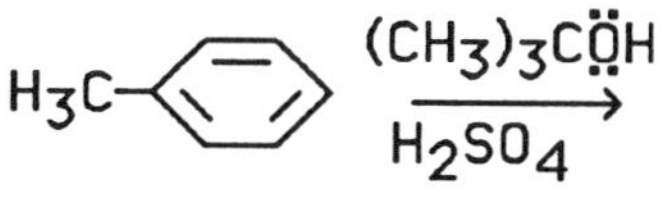

(g)

$$Ph-\overset{\overset{:O:}{\|}}{C}-\ddot{O}CH_2CH_3 \xrightarrow{LiAlH_4}$$

(h)

$$Ar_2HC-\overset{\overset{:O:}{\|}}{C}-\ddot{N}H_2 \;+\; H_3C-\overset{\overset{:O:}{\|}}{C}-\ddot{O}-\overset{\overset{:O:}{\|}}{C}-CH_3 \xrightarrow{\text{Heat}}$$

9.5 Provide a mechanism for the following transformations that is consistent with the experimental results provided.

(a) Similar systems run in ^{18}O water show no ^{18}O in the alcohol product.

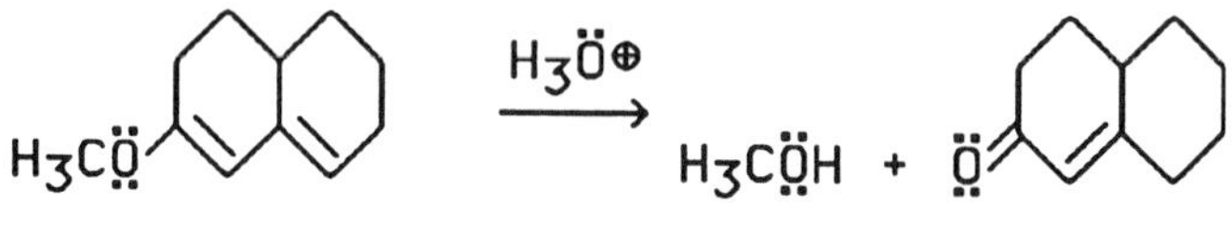

9.5 (continued).

(b) This reaction shows general acid and general base catalysis. The reaction is run in an acetate buffer, CH_3COOH and CH_3COONa. The rate decreases if the pH gets significantly more basic or more acidic.

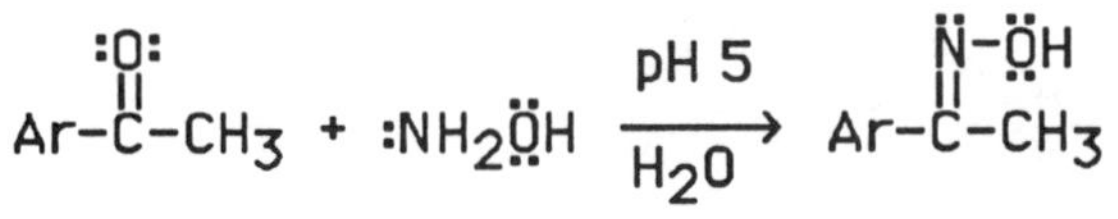

(c) Only this stereoisomer of product is formed.

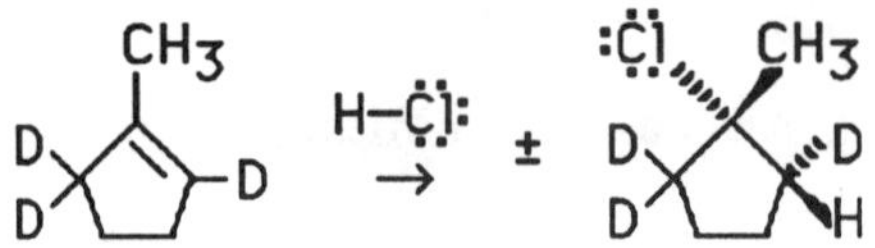

(d) Only this stereoisomer of product is formed at -72°C.

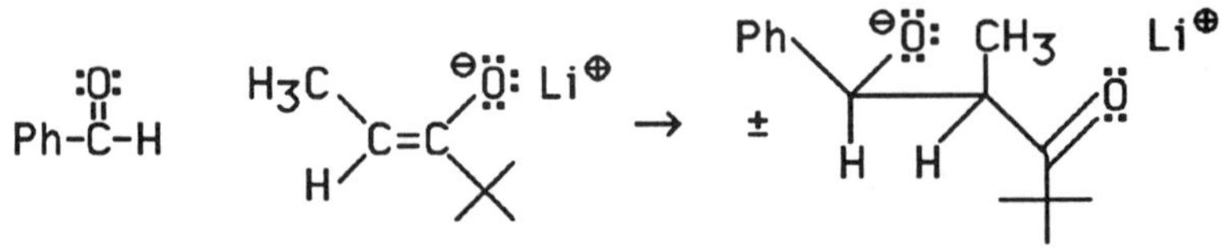

(e) Chiral optically active reactant gives a racemic product with overall retention of stereochemistry.

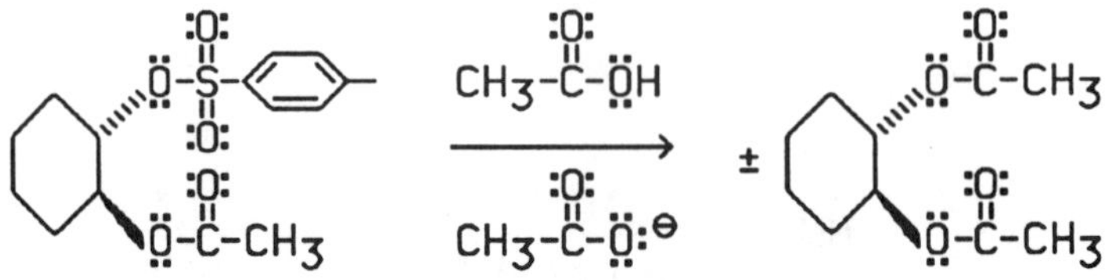

10

ONE-ELECTRON PROCESSES

10.1 RADICAL STRUCTURE AND STABILITY

One-electron processes belong to a conceptually different class of reactions and are best treated in this separate chapter. All one-electron processes require an initiation step, which is usually easily recognizable. These initiaton steps, which will be discussed in Section 10.3, should serve as a flag to alert you to shift into using one-electron pathways.

Because odd electron species, radicals, are usually unstable, their concentration in a reaction will be very low. Therefore reactions of radicals with other radicals will be less common than reactions of radicals with the even electron species that make up the bulk of the reaction mixture. **The most common radical reactions involve the radical reacting with an even electron species;** the reaction usually gives an even electron product and a new radical. These paths are discussed in Section 10.4.

Radical Structure

Simple alkyl radicals are believed to be nearly planar, shallow pyramids at the radical center, rapidly inverting with a barrier of <2 kcal/mol (8 kJ/mol). Thus all tetrahedral stereochemistry at the radical center is lost. In conjugated species, the radical is believed to be in a *p* orbital to maximize overlap with the rest of the pi system. Resonance forms allow us to predict the sites of radical reactivity in conjugated radicals. Examples are the methyl and allyl radicals.

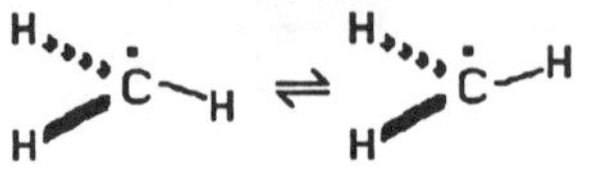

Methyl radical 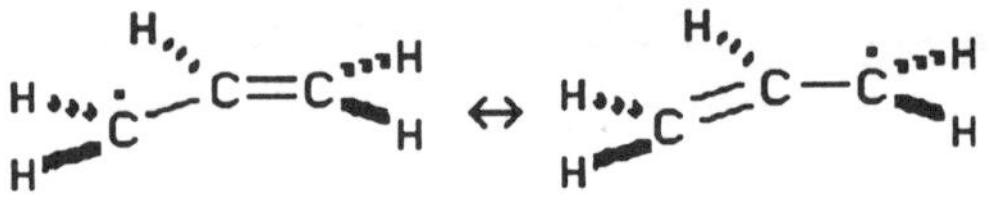   Allyl radical

Bond Dissociation Energies

The energy required to break homolytically the R—H bond into R· and H· radicals is related to the stability of the R· radical. Common C—H bond dissociation energies in kcal/mol (kJ/mol) are given in Table 10.1.

Table 10.1 C—H Bond Dissociation Energies

Radical Name	C—H Bond Broken	Dissociation Energy, kcal/mol (kJ/mol)
Break sp^3 C—H bond		
Methyl	CH_3—H	105 (439)
Ethyl	CH_3CH_2—H	98 (410)
Isopropyl	$(CH_3)_2CH$—H	95 (397)
tert-Butyl	$(CH_3)_3C$—H	93 (389)
Break sp^2 C—H bond		
Vinyl	$CH_2{=}CH$—H	110 (460)
Phenyl	C_6H_5—H	110 (460)
Acyl	CH_3CO—H	86 (360)
Delocalized radicals		
Acylmethyl	CH_3COCH_2—H	98 (410)
Cyanomethyl	$NCCH_2$—H	86 (360)
Benzyl	$C_6H_5CH_2$—H	88 (368)
Allyl	$CH_2{=}CHCH_2$—H	86 (360)
Misc. radicals		
Hydroxymethyl	$HOCH_2$—H	94 (393)
Trifluoromethyl	F_3C—H	107 (448)
Trichloromethyl	Cl_3C—H	96 (402)

Radical Stabilities

Although steric relief in going to the nearly planar radical contributes to the dissociation energy, it is generally believed that the stability of simple alkyl radicals follows the trend: tertiary > secondary > primary > methyl. Both vinyl and phenyl radicals are destabilized relative to methyl. Whereas lone pair donors stabilize radicals somewhat, groups that increase conjugation have the greatest stabilizing effect. Since a radical is a neutral species, solvent polarity generally has little effect.

To conclude, radicals can be stabilized by either a pi donor or pi acceptor because both delocalize the odd electron. Radicals that have both a pi donor and a pi acceptor are especially stable. The following resonance forms demonstrate the delocalization of a radical by a pi donor and by a pi acceptor.

$$R\ddot{\underset{..}{O}}-\dot{C}H_2 \longleftrightarrow R\overset{\oplus}{\underset{..}{O}}-\overset{\ominus}{C}H_2 \qquad :N\equiv C-\dot{C}H_2 \longleftrightarrow :\dot{N}=C=CH_2$$

Example problem

Which of the following, $(CH_3)_3C\cdot$, $CH_3CH_2\cdot$, $PhCH_2\cdot$, is the most stable radical; which is the least stable?

Answer: The delocalized radical, $PhCH_2\cdot$, is the most stable. The least stable is the primary radical, $CH_3CH_2\cdot$, because it is less substituted than the tertiary radical, $(CH_3)_3C\cdot$. Bond strengths are $PhCH_2$—H, 88 kcal/mol (368 kJ/mol); $(CH_3)_3C$—H, 93 kcal/mol (389 kJ/mol); CH_3CH_2—H, 98 kcal/mol (410 kJ/mol).

Interaction diagrams for radical species
(A supplementary, more advanced explanation)

The stabilization of a radical by conjugation with a carbon–carbon pi bond can be understood if we consider the interaction of a singly occupied p orbital with a carbon–carbon double bond to form the allylic system. Although the singly occupied orbital is not changed in energy, the two electrons in the π orbital are stabilized by the interaction (Figure 10.1).

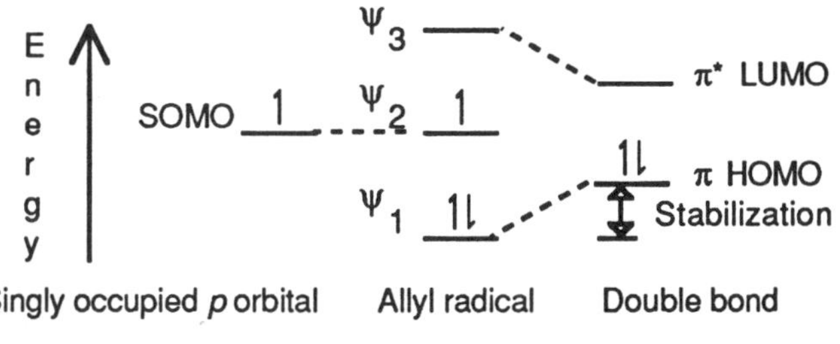

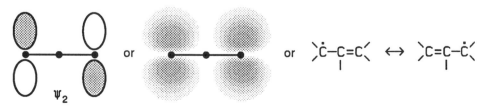

Figure 10.1 The interaction of a singly occupied p orbital with the π and π^* orbitals of a double bond to form the three molecular orbitals of an allylic radical.

The odd electron resides in Ψ_2 of the allyl system and is called the **Singly Occupied Molecular Orbital (SOMO)**. The SOMO describes the distribution of the odd electron over the molecule. The resonance forms are an attempt to describe this SOMO in the language of lines and dots (Figure 10.2).

Figure 10.2 Alternate descriptions of the allyl radical.

Orbital interaction diagrams can easily explain how an adjacent radical center is stabilized by a pi acceptor (Figure 10.3a). The pi acceptor has a low-lying empty orbital, LUMO, close in energy to the singly occupied molecular orbital, SOMO. The interaction of the SOMO with the empty LUMO forms a new pair of orbitals, bonding and antibonding. The new bonding orbital is lower in energy than either the original SOMO or LUMO. The single electron now resides in this new bonding orbital and is stabilized relative to a system that does not have this SOMO–LUMO interaction.

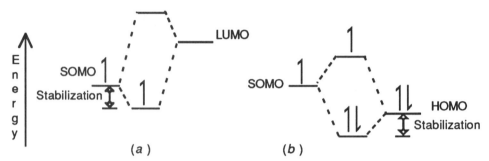

Figure 10.3 (a) The interaction of a singly occupied molecular orbital with a lowest unoccupied molecular orbital. (b) The interaction of a singly occupied molecular orbital with a highest occupied molecular orbital.

Similarly, orbital interaction diagrams can also the stabilization of a radical center by a pi donor (Figure 10.3b). The pi donor has an accessible full orbital, HOMO, close in energy to the orbital bearing the single electron, SOMO. The interaction of the SOMO with a full HOMO destabilizes one electron and stabilizes two for a net stabilization of one electron overall. We have created a pi bond with one electron in the antibonding orbital; the pi bond order is thus half (our resonance structures were unable to indicate this partial pi bond with lines and dots).

10.2   CHARGED RADICALS

Single Electron Transfer

$$A + e^{\ominus} \rightarrow A^{\ominus}\cdot \quad or \quad A - e^{\ominus} \rightarrow A^{\oplus}\cdot$$

Single electron transfer is the one-electron oxidation or reduction of a species; no covalent bond formation occurs. Charged radicals are formed by single electron transfer to or from another chemical species or an electrode. The chemical species must be picked, using standard electrode potentials, such that the electron transfer is energetically favorable. Single electron transfers using an electrode are more versatile since the electrical potential of the electrode can be accurately set over a wide range.

A radical anion is commonly produced chemically by a single electron transfer from an alkali metal such as lithium, sodium, or potassium. A radical anion has its extra electron in an antibonding orbital, which weakens the corresponding bonding orbital, and makes the radical anion more reactive than the neutral species.

$$\begin{array}{cc}
:\ddot{O}: & \ominus:\ddot{O}: \quad Na^{\oplus} \\
\| & | \\
R-\overset{}{C}-R + Na\cdot \rightarrow & R-\overset{}{\underset{\cdot}{C}}-R
\end{array}$$

A radical cation can be produced chemically by single electron transfer to commercially available stable amine radical cation salts. Again the radical ion is more reactive than the neutral.

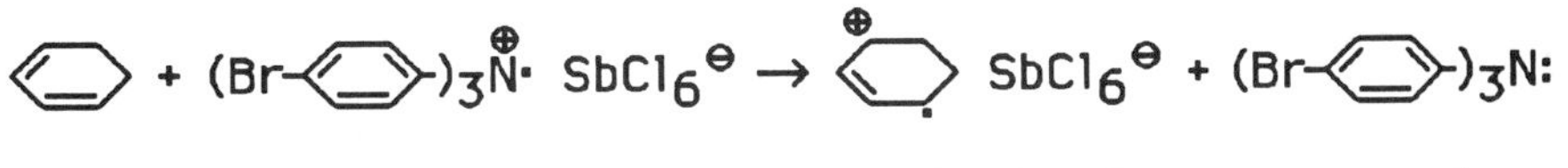

A molecular orbital explanation of single-electron transfer
(A supplementary, more advanced explanation)

Single-electron transfer reduction adds one electron into the LUMO of the molecule. The lower in energy the LUMO is, the easier it is to reduce; therefore conjugated systems are easier to reduce than unconjugated ones. Single-electron transfer oxidation removes one electron from the HOMO of the molecule. The higher in energy the HOMO, the easier it is to oxidize. For this reason conjugated systems are also easier to oxidize than unconjugated ones.

10.3 RADICAL PATH INITIATION

Homolytic Cleavage by Heat or Light

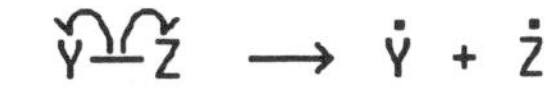

$$\widehat{Y-Z} \longrightarrow \dot{Y} + \dot{Z}$$

Since almost all stable species are even electron and spin paired (molecular oxygen is the obvious exception), radical processes must start by either single electron transfer or by homolytic cleavage. It is important to recognize these initiation steps in order to know when to use the one-electron paths discussed in this chapter rather than the two-electron paths described previously. Note that half-headed arrows are used to symbolize the movement of one electron.

Homolytic cleavage is the simple extension of a bond-stretching vibration (Figure 10.4). The process is **always** endothermic, for the barrier must be at least equal to the strength of the bond cleaved.

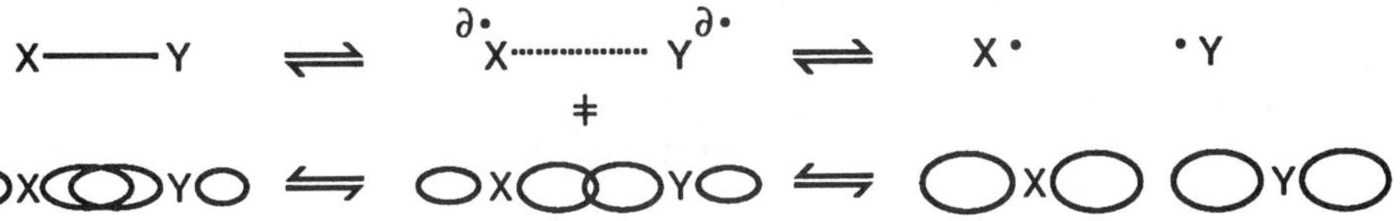

Figure 10.4 The homolytic cleavage of an X—Y bond with the transition state shown in the center.

The easiest bonds to break are weak sigma bonds between heteroatoms. The peroxide oxygen–oxygen single bond, RO—OR, and the halogen–halogen single bond, X—X, are easily broken by heat or light. Azo compounds, R—N=N—R, are also cleaved by heat or light into two R radicals and molecular nitrogen. A common radical initiator is azoisobutyronitrile (AIBN), $(CH_3)_2C(CN)$—N=N—$C(CN)(CH_3)_2$.

The following are some common radical initiators.

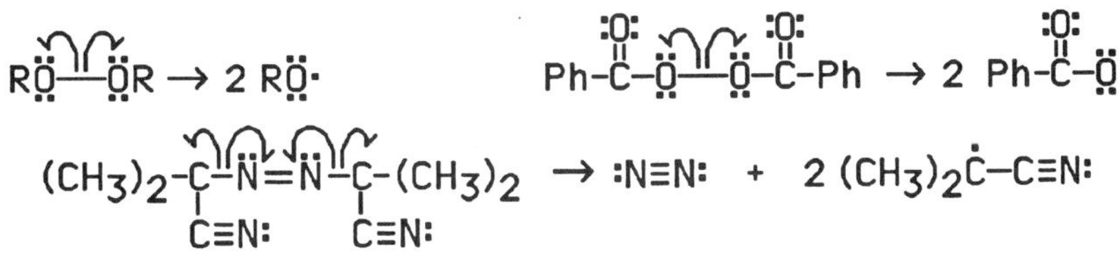

10.4   PATHS FOR RADICALS REACTING WITH NEUTRALS

Abstraction

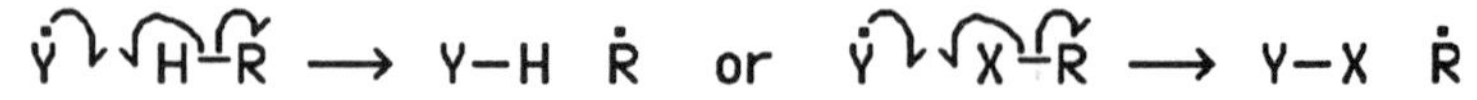

$$\overset{\curvearrowright}{Y}\overset{\curvearrowleft}{\frown}\overset{\curvearrowright}{H}\overset{\curvearrowleft}{\frown}R \longrightarrow Y-H \quad \dot{R} \quad \text{or} \quad \overset{\curvearrowright}{Y}\overset{\curvearrowleft}{\frown}\overset{\curvearrowright}{X}\overset{\curvearrowleft}{\frown}R \longrightarrow Y-X \quad \dot{R}$$

The initial radical abstracts an atom from a neutral molecule, giving a new radical (Figure 10.5). The atom abstracted is commonly a hydrogen or a halogen atom.

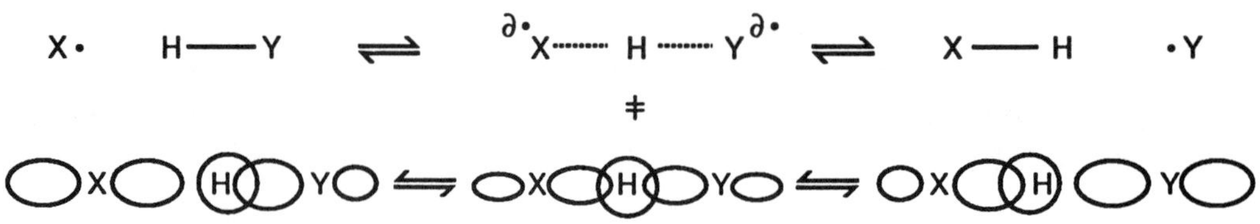

$$X\cdot \quad H{-\!\!\!-}Y \quad \rightleftharpoons \quad {}^{\partial\bullet}X{\cdots\cdots}H{\cdots\cdots}Y^{\partial\bullet} \quad \rightleftharpoons \quad X{-\!\!\!-}H \quad \bullet Y$$

Figure 10.5   The abstraction of a hydrogen atom with the transition state shown in the center.

Overlap: No overlap restrictions other than the incoming radical center must collide with the atom to be abstracted.

Selectivity: There is a preference to abstract the hydrogen atom that would produce the most stable radical (breaking the weakest bond is energetically favored). Table 10.1 can usually be used to predict which C—H bond abstraction is preferred. Exceptions occur with abstractions by very electronegative radicals such as a chlorine radical. With $CH_3CH_2CH_2$—ewg, a chlorine radical preferentially abstracts the CH_2 away from the ewg since that CH_2 bears less of a partial plus.

Energetics: Bond strength tables allow the calculation of the ΔH for the process. Highly exothermic abstractions tend to have poor selectivity. For example, C—H abstractions by a bromine radical (endothermic) are much more selective than C—H abstractions by a chlorine radical (exothermic).

Hydrogen abstraction can occur intramolecularly and is useful for functionalizing rather remote unactivated sites. With freely rotating alkyl systems, there is a great preference for a six-membered cyclic transition state. With rigid or conformationally restricted systems, molecular models are necessary to predict the site of intramolecular hydrogen abstraction.

Example problem

Calculate the ΔH for abstraction of an H from methane by a fluorine radical.

Answer: A methane C—H bond is broken, 105 kcal/mol (439 kJ/mol), and an H—F bond is formed, 135 kcal/mol (565 kJ/mol). The ΔH for the process is bonds broken - bonds made = 105 - 135 = -30 kcal/mol (126 kJ/mol). This abstraction is very exothermic.

Addition

$$\overset{\curvearrowright}{Y}\overset{\curvearrowleft}{\frown}CH_2{=\!\!=}\overset{\curvearrowleft}{C}H-R \longrightarrow Y-CH_2-\dot{C}H-R$$

The initial radical adds to a pi bond to form a new radical (Figure 10.6).

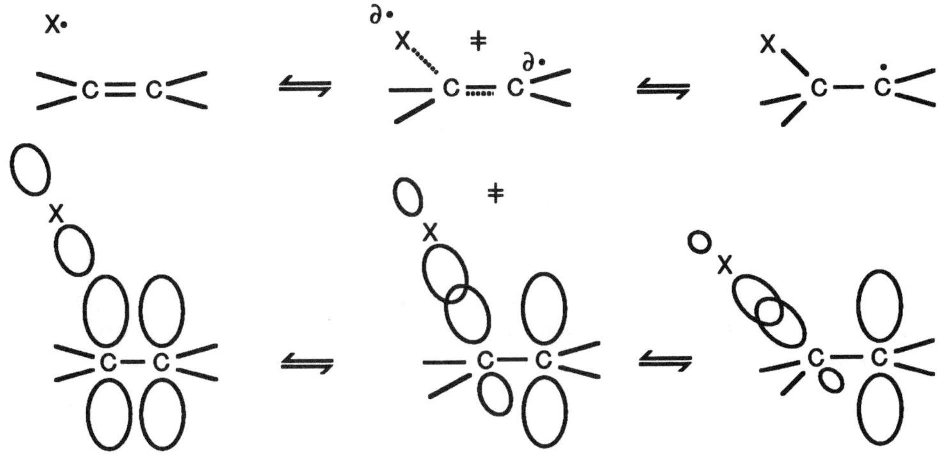

Figure 10.6 The addition of a radical to a pi bond with the transition state shown in the center.

Overlap: Calculations indicate that the preferred direction of attack of the radical on the end of the pi bond is at an angle of about 70° above the plane of the double bond (or approximately at 110° for the X$-$C$-$C angle in the figure).

Selectivity: Addition usually forms the more stable radical. Exceptions are relatively common since often the radical stability difference is less than a polar or steric effect.

Energetics: Since a pi bond is broken and usually a much stronger sigma bond is formed, this path is normally exothermic.

Cyclization is an intramolecular addition reaction. The preferred ring sizes for cyclization are 5 > 6 > 7 > 8; ring opening in these systems is rare. The closure to a four-membered ring has not been observed. Closure to a three-membered ring does occur but the ring opening is preferred by a factor of 12,000.

Example problem

Calculate the ΔH of an ethyl radical adding to ethene.

Answer: A C=C bond is broken, 146 kcal/mol (611 kJ/mol), and two C$-$C single bonds are formed, 83 kcal/mol (347 kJ/mol) for each C$-$C. The ΔH for the reaction is 146 - 166 = -20 kcal/mol (83 kJ/mol). The addition is exothermic.

10.5 OTHER RADICAL PATHS

Elimination or Fragmentation

$$\overset{\curvearrowleft}{Y} \underset{CH_2}{\frown} \overset{\curvearrowright}{CH}\text{-R} \rightarrow \dot{Y} \ CH_2\text{=}CH\text{-R} \quad \text{or} \quad \overset{\curvearrowleft}{R} \underset{CH_2}{\frown} \overset{\curvearrowright}{Z} \rightarrow \dot{R} \ CH_2\text{=}Z$$

This path is the reverse of the addition path.

Overlap: By microscopic reversibility, the bond cleaved must be coplanar with the orbital bearing the odd electron (Figure 10.6, but view the figure from right to left).

Energetics: This path is normally endothermic but can be favored by the formation of a strong pi bond and higher temperatures.

Decarboxylation is a very fast and common fragmentation reaction:

$$R\text{---}C\text{---}\ddot{O}: \quad \rightarrow \quad R\cdot \quad + \quad \ddot{O}=C=\ddot{O}$$

Decarbonylation

$$R\text{---}\dot{C} \quad \longrightarrow \quad \dot{R} \quad + \quad :C=\ddot{O} \quad \longleftrightarrow \quad \overset{\ominus}{C}\equiv\overset{\oplus}{O}$$

The loss of carbon monoxide is a reaction unique to radicals. If the radical formed upon decarbonylation is not stabilized, addition or abstraction can easily compete with this reaction.

Coupling

$$\dot{R} \quad \dot{R} \quad \longrightarrow \quad R\text{---}R$$

The coupling of two radicals to form a sigma bond is the reverse of the homolytic cleavage reaction and occurs without any activation barrier. Therefore, we might expect this reaction to be exceedingly fast. However, the rate of a reaction depends not only on the barrier height but also on the concentrations of the reacting species. Since the radical concentration in a usual reaction is low, radical–radical collisions are infrequent.

An obvious exception occurs when the reaction mixture is so inert that there are few species for radicals to react with except each other. Another exception is when both radicals are formed close together (for example, azo compound cleavage) and can react before they diffuse apart. Often collision with a solvent molecule will send one newly formed radical into the other, causing reaction; this is called the solvent cage effect. A third exception is that radicals react with molecular oxygen (which behaves as a diradical) to form R-O-O·, hydroperoxide radicals. Generally radical reactions are run in the absence of oxygen to prevent this reaction from occurring.

Radical Disproportionation

$$\dot{Y} \quad H\text{---}CH_2\text{---}\dot{C}H\text{---}R \quad \rightarrow \quad Y\text{---}H \quad CH_2=CH\text{---}R$$

Alternative to coupling, two radicals on collision can undergo disproportionation. Disproportionation occurs when one radical abstracts a hydrogen atom adjacent to the other radical center.

Rearrangement

$$Ph_2C\text{---}\dot{C}H_2 \quad \rightarrow \quad Ph_2C\text{---}CH_2 \quad \leftrightarrow \quad Ph_2C\text{---}CH_2 \quad \leftrightarrow \quad Ph_2\dot{C}\text{---}CH_2 \quad \rightarrow \quad Ph_2\dot{C}\text{---}CH_2$$

Rearrangements of radical species are rare. Migration occurs with unsaturated (aryl, vinyl, acyl) or halogen groups that can form a bridged intermediate. Rearrangement with unsaturated groups goes via intramolecular radical addition to form the three-membered ring followed by ring opening. The driving force for rearrangement is the formation of a more stable radical or the relief of steric strain.

10.6   RADICAL PATH COMBINATIONS

Chain reactions are characteristic of radicals. The radical chain is started with an *initiation* step. A chain *propagation* sequence occurs in which the radical species needed for the first propagation step is generated in the last step, and the sequence repeats many times. For a chain to be sustained, the overall process must be exothermic, and each individual step must have a low barrier. The chain undergoes *termination* whenever a coupling or disproportionation reaction removes the radical(s) needed to propagate the chain. If a propagation step is endothermic by more than 20 kcal/mol (84 kJ/mol) (the reaction barrier must be greater than this), this propagation step is slow enough that termination steps easily compete, stopping the chain. The average number of times a chain reaction repeats before termination, commonly in the thousands or more, is called the *chain length*.

The way to control a radical chain reaction is to control the initiation and termination steps. Radical chain reactions can be favored by adding radical initiators. Likewise chain reactions can be greatly diminished by adding compounds called *inhibitors* that react with radicals to increase chain termination. The sensitivity of the radical chain reaction to radical initiators and inhibitors provides a convenient way to test for this mechanism.

S_H2, Substitution Chain (Substitution, Homolytic, Bimolecular)

$$\dot{X} \curvearrowright H{-}R \longrightarrow X{-}H \quad \dot{R} \curvearrowright X{-}X \longrightarrow R{-}X \quad \dot{X}$$

The two propagation steps in the substitution chain are both radical abstractions. Radical chain halogenations illustrate different types of chain energetics. The $CH_3{-}X$ bond strengths are 108 (452), 83.5 (349), 70 (293), and 56 (234) for F, Cl, Br, and I, respectively in kcal/mol (kJ/mol). The total reaction is the sum of the propagation steps with species that appear on both sides of the transformation arrow cancelling out. Initiation or termination steps are not included. The initiation step for halogenations is usually heat or light induced homolytic cleavage of the X—X bond into two X· radicals.

Fluorination radical chain (explosive)

Propagation step	$CH_4 + F\cdot \rightarrow H{-}F + \cdot CH_3$	$\Delta H = -30 \ (-126)$
Propagation step	$\cdot CH_3 + F_2 \rightarrow F{-}CH_3 + F\cdot$	$\Delta H = -71 \ (-297)$
Total reaction	$CH_4 + F_2 \rightarrow F{-}CH_3 + H{-}F$	$\Delta H = -101 \ (-423)$

Chlorination radical chain (fast)

Propagation step	$CH_4 + Cl\cdot \rightarrow H{-}Cl + \cdot CH_3$	$\Delta H = +2.0 \ (+8)$
Propagation step	$\cdot CH_3 + Cl_2 \rightarrow Cl{-}CH_3 + Cl\cdot$	$\Delta H = -25.5 \ (-106)$
Total reaction	$CH_4 + Cl_2 \rightarrow Cl{-}CH_3 + H{-}Cl$	$\Delta H = -23.5 \ (-98)$

Bromination radical chain (slower)

Propagation step	$CH_4 + Br\cdot \rightarrow H{-}Br + \cdot CH_3$	$\Delta H = +18 \ (+75)$
Propagation step	$\cdot CH_3 + Br_2 \rightarrow Br{-}CH_3 + Br\cdot$	$\Delta H = -24 \ (-100)$
Total reaction	$CH_4 + Br_2 \rightarrow Br{-}CH_3 + H{-}Br$	$\Delta H = -6 \ (-25)$

Iodination radical chain (does not occur)

Propagation step	$CH_4 + I\cdot \rightarrow H{-}I + \cdot CH_3$	$\Delta H = +34 \ (+142)$
Propagation step	$\cdot CH_3 + I_2 \rightarrow I{-}CH_3 + I\cdot$	$\Delta H = -20 \ (-84)$
Total reaction	$CH_4 + I_2 \rightarrow I{-}CH_3 + H{-}I$	$\Delta H = +14 \ (+58)$

The first propagation step of the bromination radical chain is significantly endothermic. A small change in product radical stability is reflected in the barrier for reaction and consequently in the rate of reaction. Thus the first propagation step (and the chain reaction) that goes the fastest forms the most stable product radical. Bromination will select for allylic and benzylic > tertiary > secondary > primary > vinyl and phenyl. Allylic bromination is usually done with N–bromosuccinimide (NBS), which keeps the concentration of bromine low by reacting with the HBr formed in the first propagation step to produce the bromine needed for the second propagation step. This low bromine concentration suppresses the addition chain reaction, discussed next, by allowing time for the addition step to reverse before a bromine molecule is encountered.

The radical chain dehalogenation with Bu_3Sn-H is a very useful reaction. The ease of dehalogenation follows the $C-X$ bond strength: $I > Br > Cl > F$; the weakest $C-X$ bond is preferred. The initiator for this reaction is commonly AIBN.

Dechlorination with Bu_3SnH

Propagation step	$R-Cl + Bu_3Sn\cdot \rightarrow Bu_3Sn-Cl + R\cdot$	$\Delta H = -39$ (-163)
Propagation step	$R\cdot + Bu_3Sn-H \rightarrow R-H + Bu_3Sn\cdot$	$\Delta H = -19$ (-80)
Total reaction	$R-Cl + Bu_3Sn-H \rightarrow Bu_3Sn-Cl + R-H$	$\Delta H = -58$ (-243)

The oxidation radical chain (auto-oxidation) is very important in the spoiling of foodstuffs. The hydroperoxide products formed can easily undergo homolytic cleavage and thereby serve as an initiators for the chain. The mechanism of auto-oxidation is complex, for hydroperoxide radicals easily add to multiple bonds, and those products also can cleave. The following is a much simplified scheme.

Auto-oxidation

Propagation step	$O_2 + R\cdot \rightarrow R-O-O\cdot$
Propagation step	$R-O-O\cdot + R-H \rightarrow R-O-O-H + R\cdot$
Total reaction	$O_2 + R-H \rightarrow R-O-O-H$

Ad$_H$2, Addition Chain (Addition, Homolytic, Bimolecular)

The first propagation step in the addition chain is a radical addition. The second propagation step in the addition chain is a radical abstraction. Radical addition of HBr is a typical reaction. A bromine radical adds to the multiple bond to form the most stable of the possible radicals. The radical addition of HBr gives the opposite regiochemistry of addition as the polar addition of HBr.

Radical addition of HBr

Propagation	$CH_2=CH_2 + Br\cdot \rightarrow \cdot CH_2-CH_2-Br$	$\Delta H = -5$ (-21)
Propagation	$\cdot CH_2-CH_2-Br + H-Br \rightarrow H-CH_2-CH_2-Br + Br\cdot$	$\Delta H = -12$ (-50)
Total	$CH_2=CH_2 + H-Br \rightarrow H-CH_2-CH_2-Br$	$\Delta H = -17$ (-71)

The radical addition of HCl is not as useful since polymerization tends to compete with the addition chain. The radical addition of HI or HF does not occur because in each case a propagation step is too endothermic.

Example problem

Check the ΔH of the propagation steps for the radical chain addition of RSH to an alkene to see whether it is an energetically reasonable process.

Answer:

Step 1	$CH_2=CH_2 + RS\cdot \rightarrow \cdot CH_2-CH_2-SR$	$\Delta H = -2$ (-8)
Step 2	$\cdot CH_2-CH_2-SR + H-SR \rightarrow H-CH_2-CH_2-SR + RS\cdot$	$\Delta H = -16$ (-67)
Total	$CH_2=CH_2 + H-SR \rightarrow H-CH_2-CH_2-SR$	$\Delta H = -18$ (-75)

The radical chain addition of RSH to an alkene is an energetically reasonable process because there are no steps more than 20 kcal/mol (84 kJ/mol) endothermic, and the entire process is exothermic.

Polymerization Chain

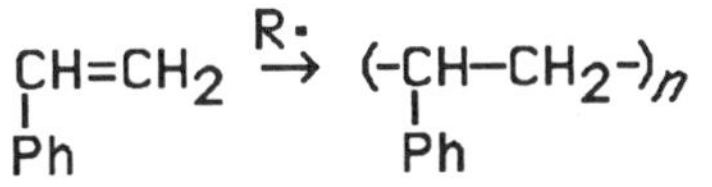

$$\underset{Ph}{\overset{|}{C}}H=CH_2 \xrightarrow{R\cdot} (\underset{Ph}{\overset{|}{-C}}H-CH_2-)_n$$

Polymerization is the repetition of an addition propagation step. The initiator radical adds to a multiple bond to produce another radical that adds again and again until termination occurs. Polymerization is favored by a high concentration of the multiple bond reactant. The preceding example was the polymerization of styrene to give polystyrene.

Polymerization of ethylene to polyethylene

Initiation step	$CH_2=CH_2 + R\cdot \rightarrow \cdot CH_2-CH_2-R$
Repeated propagation step	$CH_2=CH_2 + \cdot CH_2-CH_2-R \rightarrow \cdot (CH_2-CH_2)_2-R$
Total reaction	$n\ CH_2=CH_2 + R\cdot \rightarrow H-(CH_2-CH_2)_n-R$
(after termination step)	(and other terminations)

10.7 DISSOLVING METAL REDUCTIONS

Alkali metals added to protic solvents in the presence of another reactant provide the opportunity for both single electron transfer (s.e.t.) and two-electron processes to occur and are called dissolving metal reductions. One such mixed process adds hydrogen by way of single electron transfer, forming a radical anion, protonation of this radical anion to give the neutral radical, and a second single electron transfer to the radical, yielding an anion that then protonates. In this manner alkynes are reduced to *trans* alkenes. The second protonation gives the more stable *trans* alkene.

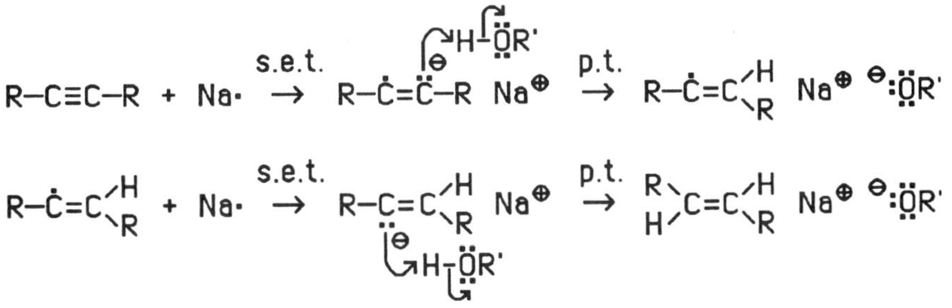

Aromatics and other conjugated systems can also be reduced.

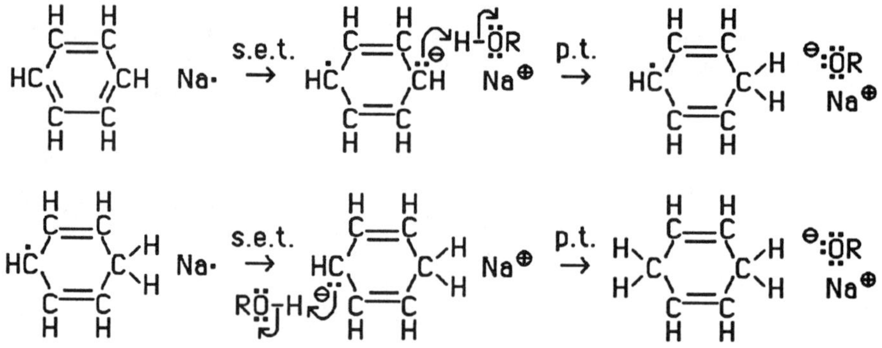

Another mixed process provides a way to remove halogens or other leaving groups. Single electron transfer forms a radical anion, which loses the leaving group anion to give the radical; a second single electron transfer to the radical yields the anion, which protonates.

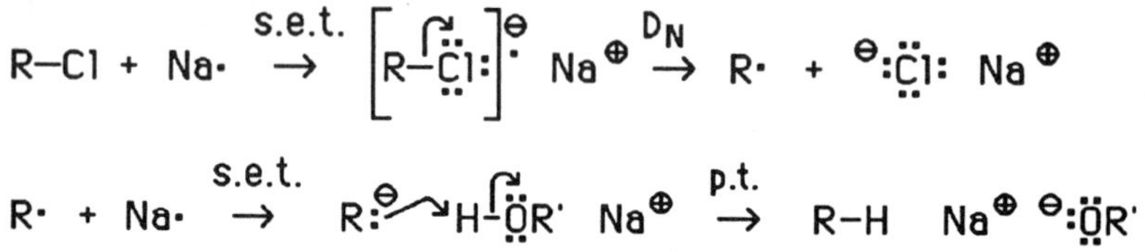

Carbon–carbon bond formation can be achieved by coupling radical anions.

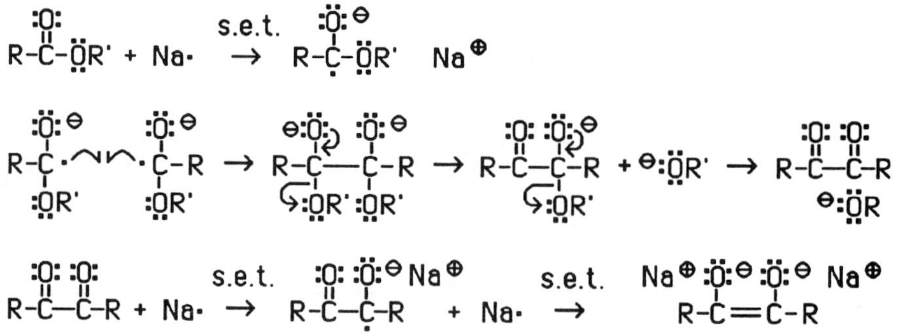

10.8   ELECTRON TRANSFER CATALYSIS

S$_{RN}$1 process (Substitution, Radical–Nucleophile, Unimolecular)

The S$_{RN}$1 process is a chain reaction that has both single electron transfers and two-electron processes. A single electron transfer forms a radical anion, which then loses the leaving group, forming the neutral radical. This radical is attacked by a nucleophile to form a new radical anion. This new radical anion serves as the electron source for the initial single electron transfer.

Initiation step	R—L + e$^{\ominus}$ → R—L$^{\ominus}$ then
Propagation step	R—L$^{\ominus}$ → R· + :L$^{\ominus}$
Propagation step	R· + :Nu$^{\ominus}$ → R—Nu$^{\ominus}$
Propagation step	R—Nu$^{\ominus}$ + R—L → R—L$^{\ominus}$ + R—Nu
Total reaction	R—L + :Nu$^{\ominus}$ → R—Nu + :L$^{\ominus}$

If R is alkyl, commonly the nucleophile and/or the R group will bear a nitro group so that the single electron transfer is favorable. If R is aryl, initiation will be by single electron transfer from an electrode or an alkali metal. Initiation by light, hv, is also possible. The $S_{RN}1$ process can be stopped with radical inhibitors. Systems that would not have reacted by an S_N1 or S_N2 process may react by an $S_{RN}1$ mechanism since the R−L bond is considerably weakened in the radical anion compared to the neutral.

$$\text{Ph-Br} \quad {}^{\ominus}\text{:}CH_2\overset{\overset{\displaystyle :O:}{\|}}{C}C(CH_3)_3 \quad \xrightarrow[-78°C,\ NH_3]{hv} \quad \text{Ph-}CH_2\overset{\overset{\displaystyle :O:}{\|}}{C}C(CH_3)_3 \qquad :\overset{..}{B}\overset{..}{r}:{}^{\ominus}$$

Other Electron Transfer Catalysis

The formation of organomagnesiums may occur by single electron transfer that triggers the loss of the halide followed by a second single electron transfer.

$$\text{Mg: } R{-}X \overset{s.e.t.}{\rightarrow} Mg\cdot^{\oplus} [R{-}X]^{\ominus\cdot} \overset{D_N}{\rightarrow} Mg\cdot^{\oplus} R\cdot \quad X{:}^{\ominus} \overset{s.e.t.}{\rightarrow} R{:}^{\ominus} Mg^{2+} \ X{:}^{\ominus}$$

The addition of some organometallics to carbonyl derivatives may occur by single electron transfer. Enones are relatively easier to reduce than simple ketones, making the single electron transfer more favorable. Organocopper reagents appear to use single electron transfer followed by radical coupling in reacting with enones. The intermediates do not escape the solvent cage.

$$Me_2Cu^{\ominus} \quad \underset{RCH=CH\overset{\overset{:O:}{\|}}{C}R}{} \rightarrow \left[Me_2Cu \quad \underset{R\dot{C}H-CH=\overset{\overset{\ominus:\overset{..}{O}:}{|}}{C}R}{} \right] \rightarrow \underset{\pm}{MeCu} \quad \underset{R\dot{C}H-CH=\overset{\overset{\ominus:\overset{..}{O}:}{|}}{C}R}{Me}$$

10.9 One-Electron Path Summary

Initiation Is Required

Single electron transfer (s.e.t.)
$$A + e^{\ominus} \rightarrow A\overset{\ominus}{\cdot} \quad \text{or} \quad A - e^{\ominus} \rightarrow A\overset{\oplus}{\cdot}$$

Homolytic cleavage by heat or light
$$\overset{\curvearrowright}{Y}\!\!\frown\!\!\overset{\curvearrowleft}{Z} \longrightarrow \dot{Y} + \dot{Z}$$

Radical Paths

Abstraction
$$\overset{\curvearrowright}{Y}\!\!\frown\!\!\overset{\curvearrowright}{H}\!\!\frown\!\!R \longrightarrow Y{-}H \ \dot{R} \quad \text{or} \quad \overset{\curvearrowright}{Y}\!\!\frown\!\!\overset{\curvearrowright}{X}\!\!\frown\!\!R \longrightarrow Y{-}X \ \dot{R}$$

Addition
$$\overset{\curvearrowright}{Y}\!\!\frown\!\!\overset{\curvearrowright}{CH_2}\!\!=\!\!\overset{\curvearrowright}{CH}{-}R \longrightarrow Y{-}CH_2{-}\dot{C}H{-}R$$

Elimination
$$\overset{\curvearrowright}{Y}\!\!\frown\!\!\overset{\curvearrowright}{CH_2}\!\!\frown\!\!\overset{\curvearrowright}{CH}{-}R \rightarrow \dot{Y} \ CH_2{=}CH{-}R \quad \text{or} \quad \overset{\curvearrowright}{R}\!\!\frown\!\!\overset{\curvearrowright}{CH_2}\!\!\frown\!\!\overset{\curvearrowleft}{Z} \rightarrow \dot{R} \ CH_2{=}Z$$

Decarbonylation

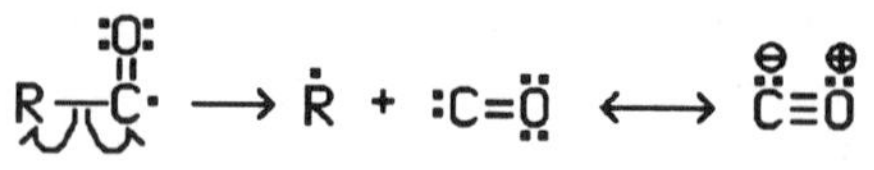

$$R-\overset{\overset{\displaystyle :\overset{..}{O}:}{\|}}{C}\cdot \longrightarrow \dot{R} + :C=\overset{..}{\overset{..}{O}} \longleftrightarrow \overset{\ominus}{\overset{..}{C}}\equiv\overset{\oplus}{O}$$

Coupling

$$\dot{R} \quad \dot{R} \longrightarrow R-R$$

Disproportionation

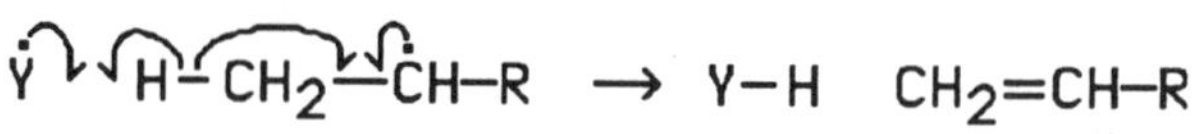

$$Y \quad H-CH_2-CH-R \longrightarrow Y-H \quad CH_2=CH-R$$

Rearrangement

Ph$_2$C–ĊH$_2$ Ph$_2$Ċ–CH$_2$ Ph$_2$Ċ–CH$_2$ Ph$_2$Ċ–CH$_2$ Ph$_2$Ċ–CH$_2$

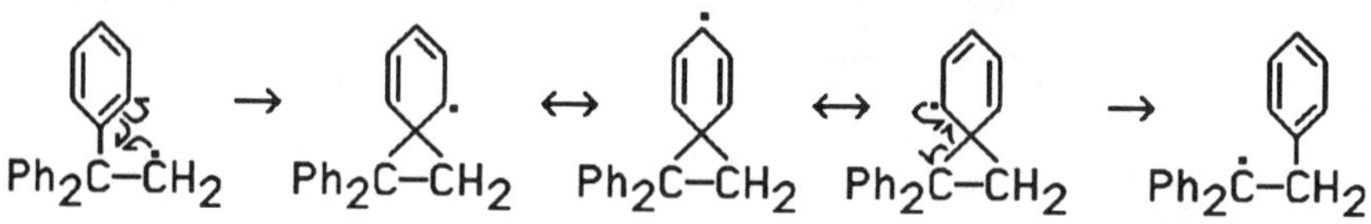

Radical Chain

Substitution, S_H2

$$\dot{X} \quad H-R \longrightarrow X-H \quad \dot{R} \quad X-X \longrightarrow R-X \quad \dot{X}$$

Addition, Ad_H2

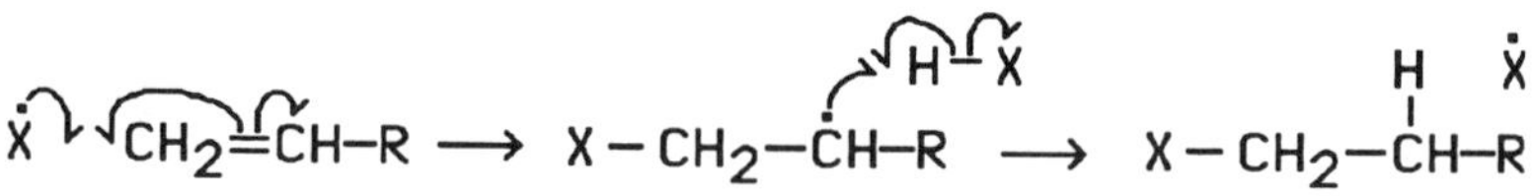

$$\dot{X} \quad CH_2=CH-R \longrightarrow X-CH_2-\dot{C}H-R \xrightarrow{H-X} X-CH_2-\overset{H}{\underset{|}{C}}H-R \quad \dot{X}$$

Polymerization

$$\underset{Ph}{\overset{|}{C}}H=CH_2 \xrightarrow{R\cdot} (-\underset{Ph}{\overset{|}{C}}H-CH_2-)_n$$

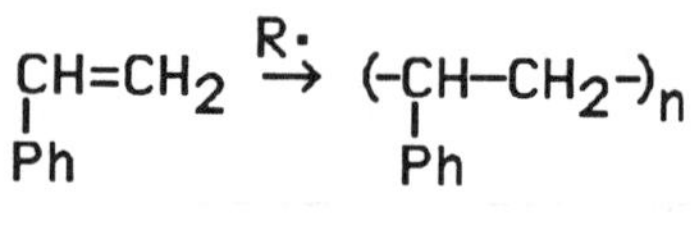

ADDITIONAL EXERCISES

10.1 Rank all species, beginning with the numeral 1, to designate the most stable radical.

·CH=CH$_2$ ·CH$_2$CH=CH$_2$ ·CH$_2$CH$_3$ ·CH(CH$_3$)$_2$ ·CH$_3$

10.2 Use radical stabilities to predict the regiochemistry of the radical addition chain reaction of HBr to propene (initiated by AIBN).

10.3 Calculate the ΔH for the abstraction of an H from ethane by a ·CCl$_3$ radical.

10.4 Calculate the ΔH of an iodine radical adding to a carbon–carbon double bond.

10.5 Check the ΔH of the propagation steps for the radical chain addition of HF to ethene to see whether it is an energetically reasonable process.

10.6 Use radical stabilities to predict the selectivity of the following radical chain substitution reaction (initiated by light).

$$Br_2 \ + \ C_6H_5CH_3 \ \rightarrow$$

10.7 Predict which you would expect to be more selective: radical chain chlorination or bromination of an alkane. Explain.

10.8 BHT is a radical chain terminator. Explain why H abstraction from oxygen yields a relatively unreactive radical.

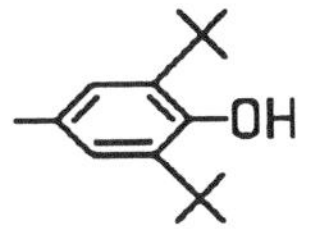

10.9 Predict the product of the following radical reaction (initiated by light).

$$Br_2 \ + \ (CH_3)_2CHCH_2CH_2CH_3 \ \rightarrow$$

10.10 Give a mechanism for this radical chain reaction (initiated by light).

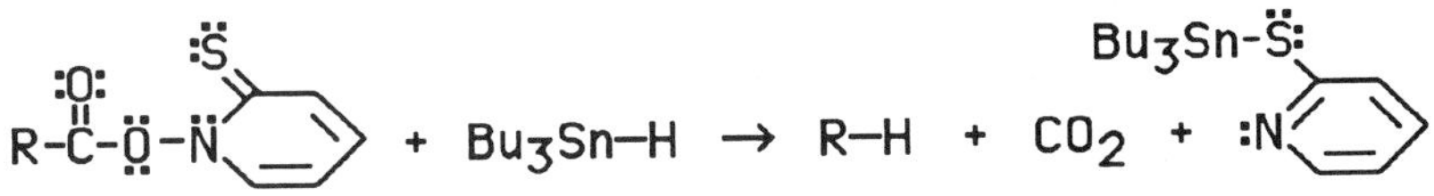

APPENDIX

(A COLLECTION OF IMPORTANT TOOLS)

BIBLIOGRAPHY

Undergraduate Texts

Pine, S. *Organic Chemistry*, 5th ed.; McGraw-Hill: New York, 1987.
Solomons, T. *Organic Chemistry*, 4th ed.; Wiley: New York, 1988.
Streitwieser, A.; Heathcock, C. *Introduction to Organic Chemistry*, 3rd ed.; Macmillan: New York, 1985.
Sykes, P. *A Guidebook to Mechanism in Organic Chemistry*, 6th ed.; Longman: New York, 1986.
Vollhardt, K. *Organic Chemistry*; Freeman: New York, 1987.

Graduate Texts

Carey, F.; Sundberg, R. *Advanced Organic Chemistry*, 3rd ed.; Plenum: New York, 1990.
Lowry, T.; Richardson, K. *Mechanism and Theory in Organic Chemistry*, 3rd ed.; Harper & Row: New York, 1987.
March, J. *Advanced Organic Chemistry*, 3rd ed.; Wiley: New York, 1985.

Molecular Orbital Theory

Fleming, I. *Frontier Orbitals and Organic Chemical Reactions*; Wiley: New York, 1976.
Jorgensen, W. and Salem, L. *The Organic Chemist's Book of Orbitals*; Academic Press: New York, 1973.

Physical Organic

Carpenter, B. *Determination of Organic Reaction Mechanisms*; Wiley: New York, 1984.
Isaacs, N. *Physical Organic Chemistry*; Longman: London, 1987.
Klumpp, G. *Reactivity in Organic Chemistry*; Wiley: New York, 1982.

Biochemical Mechanisms

Fersht, A. *Enzyme Structure and Mechanism*; Freeman: New York, 1985.
Walsh, C. *Enzymatic Reaction Mechanisms*; Freeman: New York, 1979.

Inorganic

Cotton, F.; Wilkinson, G. *Advanced Inorganic Chemistry*, 5th ed.; Wiley: New York, 1988.
Huheey, J. *Inorganic Chemistry: Principles of Structure and Reactivity*, 3rd ed.; Harper & Row: New York, 1983.

Bond Strengths

Benson, S. W. *J. Chem. Educ.* **1965**, *42*, 502.
Cottrell, T. *The Strengths of Chemical Bonds*, 2nd ed.; Butterworths: London, 1958.

pK_a Values

Dean, J., ed., *Lange's Handbook of Chemistry*, 13th ed.; McGraw-Hill: New York, 1985.
March, J. *Advanced Organic Chemistry*, 3rd ed.; Wiley: New York, 1985 (primary source).

Prediction of pK_a

Perrin, D.; Dempsey, B.; Serjeant,E. *pK_a Prediction for Organic Acids and Bases*; Chapman and Hall: London, 1981.

Hard–Soft Acid–Base Theory

Pearson, R. G. *J. Chem. Educ.* **1968**, *45*, 581.

Nucleophilicity

Pearson, R. G. *J. Am. Chem. Soc.* **1968**, *90*, 319.

The ΔpK_a Rule and Computer Prediction of Organic Reactions

Salatin, T. D.; Jorgensen, W. L., *J. Org. Chem.* **1980**, *45*, 2043.

ABBREVIATIONS USED IN THIS TEXT

Ac	Acetyl $CH_3C=O$
Ar	Any aryl (aromatic) group
b	Brønsted base, proton acceptor
$\partial+$	A partial positive charge
$\partial-$	A partial negative charge
don	Electron-donating group
E	Electrophile, Lewis acid
Et	Ethyl CH_3CH_2
ewg	Electron-withdrawing group
G	Unspecified group
HA	Brønsted acid, proton donor
HOMO	Highest occupied molecular orbital
i-Pr	Isopropyl $(CH_3)_2CH$
L	Leaving group
LUMO	Lowest unoccupied molecular orbital
M	Metal atom
Me	Methyl CH_3
MO	Molecular orbital
n-Bu	Normal-butyl group $CH_3CH_2CH_2CH_2$
Nu	Nucleophile
Ph	Phenyl group, C_6H_5, a monosubstituted benzene
R	Any alkyl chain
t-Bu	Tertiary butyl $(CH_3)_3C$
Ts	Toluenesulfonyl, $CH_3C_6H_4SO_2$
X	Chlorine, bromine, or iodine
Y, Z	Heteroatoms, commonly oxygen, nitrogen, or sulfur
±	Racemic mixture
→→	Multistep process
‡	Transition state
......	Partially broken bond (or weak complexation)

For the path acronyms, please see the path summary in this Appendix.

Functional Group Glossary

NAME	FUNCTIONAL GROUP	EXAMPLE
Acetal		
Acid anhydride		
Acyl halide		
Acyloin		
Alcohol		
Aldehyde		
Alkane		
Alkene		
Alkoxide		
Alkyl halide		
Alkyne		
Allene (a cumulene)		
Amide		
Amidate		
Amine		
Amine oxide		
Aromatic ring	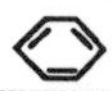	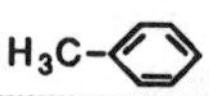
Aryl halide	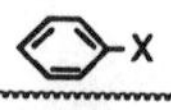	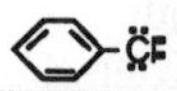
Borane		
Borate ester		
Carbamate		
Carbene		
Carbodiimide		
Carbonate		

Carboxylate	$\overset{:\overset{..}{O}:}{\underset{}{-C}}-\overset{..}{\underset{..}{O}}:\ominus$	$H_3C-\overset{:\overset{..}{O}:}{\underset{}{C}}-\overset{..}{\underset{..}{O}}:\ominus$
Carboxylic acid	$\overset{:\overset{..}{O}:}{\underset{}{-C}}-\overset{..}{\underset{..}{O}}-H$	$CH_3-\overset{:\overset{..}{O}:}{\underset{}{C}}-\overset{..}{\underset{..}{O}}-H$
Diazonium	$-\overset{\oplus}{N}\equiv N:$	$C_6H_5\overset{\oplus}{N}\equiv N:$
Diene	$\rangle C=C-C=C\langle$	$H_2C=CH-HC=CH_2$
Disulfide	$C-\overset{..}{\underset{..}{S}}-\overset{..}{\underset{..}{S}}-C$	$CH_3-\overset{..}{\underset{..}{S}}-\overset{..}{\underset{..}{S}}-CH_3$
Enamine	$\rangle C=C\overset{\overset{..}{N}\langle}{\langle}$	$H_2C=C\overset{N(CH_3)_2}{\underset{CH_3}{\diagdown}}$
Enediol	$\overset{H\overset{..}{O}:\ \overset{..}{O}H}{\underset{}{-C=C-}}$	$\overset{H\overset{..}{O}:\ \overset{..}{O}H}{H_3C-C=C-CH_3}$
Enol	$\rangle C=C\overset{\overset{..}{O}H}{\diagdown}$	$H_2C=C\overset{\overset{..}{O}H}{\underset{CH_3}{\diagdown}}$
Enol ether	$\rangle C=C-\overset{..}{\underset{..}{O}}-C$	$H_2C=C\overset{\overset{..}{O}-CH_3}{\underset{CH_3}{\diagdown}}$
Enolate	$\rangle C=C-\overset{..}{\underset{..}{O}}:\ominus$	$\overset{\ominus:\overset{..}{O}:}{H_2C=C-CH_3}$
Enone	$\rangle C=C-\overset{:\overset{..}{O}:}{\underset{}{C}}-C$	$H_2C=CH-\overset{:\overset{..}{O}:}{\underset{}{C}}-CH_3$
Epoxide	$\rangle C\overset{\overset{..}{O}}{\underline{\diagdown\diagup}}C\langle$	$\overset{H_3C}{H}\diagup C\overset{\overset{..}{O}}{\underline{\diagdown\diagup}}C\diagdown\overset{CH_3}{H}$
Ester	$\overset{:\overset{..}{O}:}{\underset{}{-C}}-\overset{..}{\underset{..}{O}}-C$	$CH_3-\overset{:\overset{..}{O}:}{\underset{}{C}}-\overset{..}{\underset{..}{O}}-CH_3$
Ether	$C-\overset{..}{\underset{..}{O}}-C$	$CH_3-\overset{..}{\underset{..}{O}}-CH_3$
Halohydrin	$\overset{H\overset{..}{O}:\ :\overset{..}{X}:}{\rangle C-C\langle}$	$\overset{H\overset{..}{O}:\ :\overset{..}{C}l:}{H_3CHC-CHCH_3}$
Hemiacetal	$\overset{}{H}\diagup C\overset{\overset{..}{O}H}{\underset{\overset{..}{O}C}{\diagdown}}$	$\overset{H_3C}{H}\diagup C\overset{\overset{..}{O}H}{\underset{\overset{..}{O}CH_3}{\diagdown}}$
Hemiketal	$\overset{C}{C}\diagup C\overset{\overset{..}{O}H}{\underset{\overset{..}{O}C}{\diagdown}}$	$\overset{H_3C}{H_3C}\diagup C\overset{\overset{..}{O}H}{\underset{\overset{..}{O}CH_3}{\diagdown}}$
Hydrate	$\rangle C\overset{\overset{..}{O}H}{\underset{\overset{..}{O}H}{\diagdown}}$	$\overset{H_3C}{H}\diagup C\overset{\overset{..}{O}H}{\underset{\overset{..}{O}H}{\diagdown}}$
Imine	$\rangle C=\overset{..}{N}-$	$(CH_3)_2C=\overset{..}{N}-CH_3$
Isocyanate	$-\overset{..}{N}=C=\overset{..}{\underset{..}{O}}$	$C_6H_5-\overset{..}{N}=C=\overset{..}{\underset{..}{O}}$
Ketal	$\overset{C}{C}\diagup C\overset{\overset{..}{O}C}{\underset{\overset{..}{O}C}{\diagdown}}$	$\overset{H_3C}{H_3C}\diagup C\overset{\overset{..}{O}CH_3}{\underset{\overset{..}{O}CH_3}{\diagdown}}$
Ketene	$\rangle C=C=\overset{..}{\underset{..}{O}}$	$H_2C=C=\overset{..}{\underset{..}{O}}$
Ketone	$\overset{:\overset{..}{O}:}{\underset{}{C}-C-C}$	$\overset{:\overset{..}{O}:}{CH_3-C-CH_3}$
Nitrile	$C-C\equiv N:$	$CH_3-C\equiv N:$

Nitro compound	C–N⁺(=O)–O⁻	H₃C–N⁺(=O)–O⁻
Organometallic	C–M	CH₃–Li
Orthoester	C–C(OC)(OC)(OC)	H₃C–C(OCH₃)(OCH₃)(OCH₃)
Oxime	C=N–O–H	(CH₃)₂C=N–O–H
Peroxide	C–O–O–C	H₃C–O–O–CH₃
Phenol	Ar–O–H	C₆H₅–O–H
Phosphate ester	(O=)(CO)(CO)P–OC	(O=)(H₃CO)(H₃CO)P–OCH₃
Selenoxide	C–Se(=O)–C	C₆H₅–Se(=O)–C₆H₅
Sulfate ester	CO–S(=O)(=O)–OC	H₃CO–S(=O)(=O)–OCH₃
Sulfide	C–S–C	H₃C–S–CH₃
Sulfinic acid	C–S(=O)–OH	C₆H₅–S(=O)–OH
Sulfonate ester	C–S(=O)(=O)–OC	H₃C–S(=O)(=O)–OCH₃
Sulfone	C–S(=O)(=O)–C	H₃C–S(=O)(=O)–CH₃
Sulfonic acid	C–S(=O)(=O)–OH	H₃C–S(=O)(=O)–OH
Sulfonyl halide	C–S(=O)(=O)–X	H₃C–S(=O)(=O)–Cl
Sulfoxide	C–S(=O)–C	H₃C–S(=O)–CH₃
Thioester	C(=O)–S–C	CH₃–C(=O)–S–CH₃
Thiol	C–S–H	H₃C–S–H
Urea	N–C(=O)–N	H₂N–C(=O)–NH₂
Vinyl halide	C=C–X	H₂C=CH–Cl
Xanthate	O–C(=S)–S	H₃C–O–C(=S)–SCH₃
Ylide	⁻Nu–L⁺	H₂C⁻–S⁺(CH₃)₂

COMPOSITE pK_a CHART

The Acidic H is in Boldface[1]

Oxygen acids

pK_a	Acid	Base
	FSO_3H	$FSO_3^{\ominus}$
-12	$RNO_2H^{\oplus}$	RNO_2
-10	$HClO_4$	$ClO_4^{\ominus}$
	H_2SO_4	$HSO_4^{\ominus}$
-9	R-C(OH$^{\oplus}$)=CF	R-C(O)=CF
-8	R-C(OH$^{\oplus}$)-H	R-C(O)-H
-7	R-C(OH$^{\oplus}$)-R	R-C(O)-R
-6.5	$ArSO_3H$	$ArSO_3^{\ominus}$
-6	CH_3SO_3H	$CH_3SO_3^{\ominus}$
-6.5	R-C(OH$^{\oplus}$)-OR	R-C(O)-OR
-6.4	$Ar\overset{\oplus}{O}H_2$	$Ar\ddot{O}H$
-6	R-C(OH$^{\oplus}$)-OH	R-C(O)-OH
-6	Ar-$\overset{H}{\underset{\oplus}{O}}$-R	$Ar\ddot{O}R$
-3.5	R-$\overset{H}{\underset{\oplus}{O}}$-R	$R\ddot{O}R$
-2.4	$CH_3CH_2\overset{\oplus}{O}H_2$	$CH_3CH_2\ddot{O}H$
-1.7	$H_3\overset{\oplus}{O}$	$H_2\ddot{O}$
-1.5	Ar-C(OH$^{\oplus}$)-$\ddot{N}H_2$	Ar-C(O)-$\ddot{N}H_2$
-1.5	$(CH_3)_2S=\overset{\oplus}{O}H$	$(CH_3)_2S=\ddot{O}$
-1.4	HNO_3	$NO_3^{\ominus}$
-0.5	R-C(OH$^{\oplus}$)-$\ddot{N}H_2$	R-C(O)-$\ddot{N}H_2$
0.5	CF_3-C(O)-$\ddot{O}H$	CF_3-C(O)-$\ddot{O}^{\ominus}$
0.7	pyridine-N$^{\oplus}$-$\ddot{O}H$	pyridine-N$^{\oplus}$-$\ddot{O}^{\ominus}$
1.5	Ph-S(O)-$\ddot{O}H$	Ph-S(O)-$\ddot{O}^{\ominus}$
1.7	O_2N-CH_2-C(O)-$\ddot{O}H$	O_2N-CH_2-C(O)-$\ddot{O}^{\ominus}$
1.8	$(CH_3)_3\overset{\oplus}{N}CH_2$-C(O)-$\ddot{O}H$	$(CH_3)_3\overset{\oplus}{N}CH_2$-C(O)-$\ddot{O}^{\ominus}$
2.0	$HSO_4^{\ominus}$	SO_4^{2-}

pK_a	Acid	Base
2.2	H_3PO_4	$H_2PO_4^{\ominus}$
2.4	$H_3\overset{\oplus}{N}CH_2$-C(O)-$\ddot{O}H$	$H_3\overset{\oplus}{N}CH_2$-C(O)-$\ddot{O}^{\ominus}$
2.5	$N\equiv C$-CH_2-C(O)-$\ddot{O}H$	$N\equiv C$-CH_2-C(O)-$\ddot{O}^{\ominus}$
2.5	CH_3-C(O)--C(O)-$\ddot{O}H$	CH_3-C(O)--C(O)-$\ddot{O}^{\ominus}$
2.9	Cl-CH_2-C(O)-$\ddot{O}H$	Cl-CH_2-C(O)-$\ddot{O}^{\ominus}$
3.1	CF_3CH_2-C(O)-$\ddot{O}H$	CF_3CH_2-C(O)-$\ddot{O}^{\ominus}$
3.3	$H\ddot{O}$-$N=\ddot{O}$	$^{\ominus}\ddot{O}$-$N=\ddot{O}$
3.4	O_2N-C$_6$H$_4$-C(O)-$\ddot{O}H$	O_2N-C$_6$H$_4$-C(O)-$\ddot{O}^{\ominus}$
3.6	CH_3-C(O)-CH_2-C(O)-$\ddot{O}H$	CH_3-C(O)-CH_2-C(O)-$\ddot{O}^{\ominus}$
3.6	CH_3OCH_2-C(O)-$\ddot{O}H$	CH_3OCH_2-C(O)-$\ddot{O}^{\ominus}$
4.2	C$_6$H$_5$-C(O)-$\ddot{O}H$	C$_6$H$_5$-C(O)-$\ddot{O}^{\ominus}$
4.5	CH_3O-C$_6$H$_4$-C(O)-$\ddot{O}H$	CH_3O-C$_6$H$_4$-C(O)-$\ddot{O}^{\ominus}$
4.6	$(CH_3)_3\overset{\oplus}{N}$-$\ddot{O}H$	$(CH_3)_3\overset{\oplus}{N}$-$\ddot{O}^{\ominus}$
4.8	CH_3-C(O)-$\ddot{O}H$	CH_3-C(O)-$\ddot{O}^{\ominus}$
6.4	H_2CO_3	$HCO_3^{\ominus}$
7.2	$H_2PO_4^{\ominus}$	HPO_4^{2-}
7.2	O_2N-C$_6$H$_4$-$\ddot{O}H$	O_2N-C$_6$H$_4$-$\ddot{O}^{\ominus}$
10.0	C$_6$H$_5$-$\ddot{O}H$	C$_6$H$_5$-$\ddot{O}^{\ominus}$
10.2	CH_3O-C$_6$H$_4$-$\ddot{O}H$	CH_3O-C$_6$H$_4$-$\ddot{O}^{\ominus}$
10.3	$HCO_3^{\ominus}$	CO_3^{2-}
11.6	$H\ddot{O}$-$\ddot{O}H$	$^{\ominus}\ddot{O}$-$\ddot{O}H$
12.2	$(CH_3)_2C=N$-$\ddot{O}H$	$(CH_3)_2C=N$-$\ddot{O}^{\ominus}$
12.4	HPO_4^{2-}	PO_4^{3-}
12.4	$CF_3CH_2\ddot{O}H$	$CF_3CH_2\ddot{O}^{\ominus}$
13.3	$H\ddot{O}CH_2\ddot{O}H$	$HOCH_2\ddot{O}^{\ominus}$
14.2	$H\ddot{O}CH_2CH_2\ddot{O}H$	$HOCH_2CH_2\ddot{O}^{\ominus}$
15.5	$CH_3\ddot{O}H$	$CH_3\ddot{O}^{\ominus}$
15.7	$H_2\ddot{O}$	$H\ddot{O}^{\ominus}$
16	$CH_3CH_2\ddot{O}H$	$CH_3CH_2\ddot{O}^{\ominus}$
18	$(CH_3)_2CH\ddot{O}H$	$(CH_3)_2CH\ddot{O}^{\ominus}$
19	$(CH_3)_3C\ddot{O}H$	$(CH_3)_3C\ddot{O}^{\ominus}$

---

[1]Modified with permission from March, J. *Advanced Organic Chemistry*, 3rd ed; pp.220–222. Copyright ©1985 by Wiley–Interscience, New York.

Nitrogen acids

pKa	Acid	Base
-10	RC≡NH (+)	RC≡N:
-5	Ar$_3$NH (+)	Ar$_3$N
1	Ar$_2$NH$_2$ (+)	Ar$_2$NH
1.0	O$_2$N-C$_6$H$_4$-NH$_3$ (+)	O$_2$N-C$_6$H$_4$-NH$_2$
4.6	C$_6$H$_5$-NH$_3$ (+)	C$_6$H$_5$-NH$_2$
5.2	pyridinium N-H (+)	pyridine N:
5.4	CH$_3$O-C$_6$H$_4$-NH$_3$ (+)	CH$_3$O-C$_6$H$_4$-NH$_2$
5.8	HO-NH$_3$ (+)	HO-NH$_2$
7.0	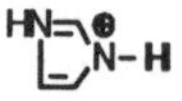 imidazolium N-H (+)	 imidazole N:
7.9	H$_2$N-NH$_3$ (+)	H$_2$N-NH$_2$
8.5	PhSO$_2$NH$_2$	PhSO$_2$NH (−)
9.2	NH$_4$ (+)	NH$_3$
9.3	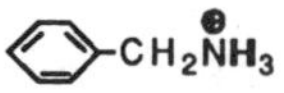 C$_6$H$_5$-CH$_2$NH$_3$ (+)	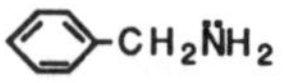 C$_6$H$_5$-CH$_2$NH$_2$
5.8	HOCH$_2$NH$_3$ (+)	HOCH$_2$NH$_2$
9.6	(CH$_2$-C)$_2$NH	(CH$_2$-C)$_2$N (−)
9.8	(−)O-C-CH$_2$NH$_3$ (+)	(−)O-C-CH$_2$NH$_2$
10.6	EtNH$_3$ (+)	EtNH$_2$
10.7	Et$_3$NH (+)	Et$_3$N
11	Et$_2$NH$_2$ (+)	Et$_2$NH
13.6	(H$_2$N)$_2$C=NH$_2$ (+)	(H$_2$N)$_2$C=NH
17	R-C-NH$_2$	R-C-NH (−)
25.8	((CH$_3$)$_3$Si)$_2$NH	((CH$_3$)$_3$Si)$_2$N (−)
27	PhNH$_2$	PhNH (−)
35	NH$_3$	NH$_2$ (−)
36	Et$_2$NH	Et$_2$N (−)

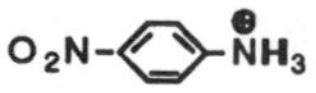

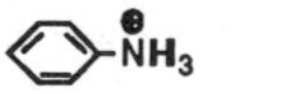

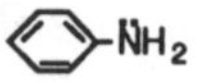

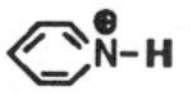

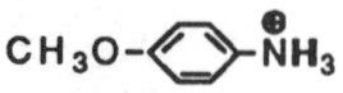

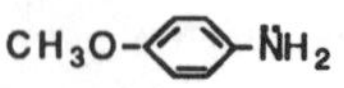

Carbon acids

pKa	Acid	Base
-5	HC(CN)$_3$	C(CN)$_3$ (−)
3.6	H$_2$C(NO$_2$)$_2$	HC(NO$_2$)$_2$ (−)
5	H$_2$C(-C-H)$_2$	HC(-C-H)$_2$ (−)
9	H$_2$C(-C-CH$_3$)$_2$	HC(-C-CH$_3$)$_2$ (−)
9	N≡C-CH$_2$-C-CH$_3$	N≡C-CH-C-CH$_3$ (−)
9.2	HC≡N	(−)C≡N
10	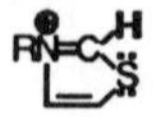 RN thiazolium C-H	 RN thiazolylidene (−)
10.2	H$_3$C-NO$_2$	H$_2$C-NO$_2$ (−)
10.7	CH$_3$-C-CH$_2$-C-OEt	CH$_3$-C-CH-C-OEt (−)
11.2	H$_2$C(C≡N)$_2$	HC(C≡N)$_2$ (−)
12.5	H$_2$C(SO$_2$CH$_3$)$_2$	HC(SO$_2$CH$_3$)$_2$ (−)
13	H$_2$C(-C-OEt)$_2$	HC(-C-OEt)$_2$ (−)
13.5	H$_3$C-C-O-C-CH$_3$	H$_2$C-C-O-C-CH$_3$ (−)
13.6	HCCl$_3$	(−)CCl$_3$
~14	H$_3$C-C-SR	H$_2$C-C-SR (−)
15.9	H$_3$C-C-CH$_2$Ph	H$_3$C-C-CHPh (−)
16	cyclopentadiene CH$_2$	CH (−)
16	H$_3$C-C-CH$_2$Cl	H$_3$C-C-CHCl (−)
~16	H$_3$C-C-Cl	H$_2$C-C-Cl (−)
16.7	H$_3$C-C-H	H$_2$C-C-H (−)
19.2	H$_3$C-C-CH$_3$	H$_2$C-C-CH$_3$ (−)
23	H$_3$C-SO$_2$CH$_3$	H$_2$C-SO$_2$CH$_3$ (−)
24	H$_3$C-C-OR	H$_2$C-C-OR (−)
25	H$_3$C-C≡N	H$_2$C-C≡N (−)
25	HC≡CH	(−)C≡CH
~28	H$_3$C-C-NR$_2$	H$_2$C-C-NR$_2$ (−)
31.1	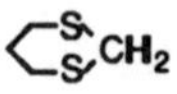 dithiane CH$_2$	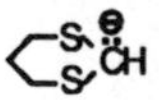 dithiane CH (−)
31.5	Ph$_3$CH	Ph$_3$C: (−)

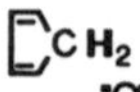

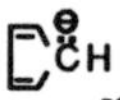

Carbon acids (continued)

pK_a	Acid	Base
33.5	Ph_2CH_2	$Ph_2\overset{\ominus}{C}H$
35	$H_3C-\overset{:O:}{\underset{\|}{S}}-CH_3$	$H_2\overset{\ominus}{C}-\overset{:O:}{\underset{\|}{S}}-CH_3$
35	$H_3C-\overset{\oplus}{P}Ph_3$	$H_2\overset{\ominus}{C}-\overset{\oplus}{P}Ph_3$
41	H_3C-Ph	$H_2\overset{\ominus}{C}-Ph$
43	$H_3C-HC=CH_2$	$H_2\overset{\ominus}{C}-HC=CH_2$
43	HPh	$\ominus{:}Ph$
44	$H_2C=CH_2$	$H\overset{\ominus}{C}=CH_2$
48	CH_4	$\ominus{:}CH_3$
50	H_3C-CH_3	$H_2\overset{\ominus}{C}-CH_3$
51	$H_2C(CH_3)_2$	$H\overset{\ominus}{C}(CH_3)_2$
>52	$HC(CH_3)_3$	$\ominus{:}C(CH_3)_3$

Miscellaneous acids

pK_a	Acid	Base
−10	HI	$I^{\ominus}$
−9	HBr	$Br^{\ominus}$
−7	HCl	$Cl^{\ominus}$
−7	$R\overset{\oplus}{S}H_2$	RSH
−5.3	$R_2\overset{\oplus}{S}H$	R_2S
2.7	$Ph_3\overset{\oplus}{P}H$	$Ph_3P{:}$
3.2	HF	$F^{\ominus}$
3.3	$CH_3-\overset{:O:}{\underset{\|}{C}}-SH$	$CH_3-\overset{:O:}{\underset{\|}{C}}-\overset{..}{S}^{\ominus}$
3.9	H_2Se	$HSe^{\ominus}$
6.5	$Ph\overset{..}{S}H$	$Ph\overset{..}{S}^{\ominus}$
7.0	$H_2\overset{..}{S}$	$H\overset{..}{S}^{\ominus}$
8.7	$Et_3\overset{\oplus}{P}H$	$Et_3P{:}$
10.6	$Et\overset{..}{S}H$	$Et\overset{..}{S}^{\ominus}$
35	H_2	$H^{\ominus}$

Acid Strength

Strong acids have low pK_a's

Base Strength

Strong bases have high pK_{abH}'s

Proton Transfer

Proton transfer reactions usually form the weaker base.

$$\log K_{eq} = pK_{abH} - pK_{aHA}$$

$$K_{eq} = 10^{(pK_{abH} - pK_{aHA})}$$

Simply take the pK_{abH} of the base and subtract from it the pK_a of the acid to get the exponent of the K_{eq}. If the K_{eq} is equal to or greater than approximately 10^{-8} (negative by 8 pK_a units), the proton transfer is within the useful range. If the proton transfer K_{eq} is greater than 10^{+7}, it can for all practical purposes be considered irreversible. Proton transfer is the first thing that happens in many common reactions.

The ΔpK_a Rule

The leaving group or anion produced should be no more than about 8 pK_a units more basic than the incoming nucleophile or base. Reactions tend to form the weaker base. No reaction has huge jumps upward in the pK_{abH} of its intermediates. The energy drops if the pK_{abH} drops significantly.

Hard and Soft Acids and Bases

The HSAB principle: Hard bases favor binding with hard acids; soft bases favor binding with soft acids.

When a pair of molecules collide, two attractive forces lead to reaction: the hard–hard attraction (opposite charges attracting each other), and the soft–soft attraction (the interaction of filled orbitals with empty orbitals).

BOND STRENGTH TABLE[2]

Average Single Bond Energies in kcal/mol (kJ/mol)

	I	Br	Cl	S	P	Si	F	O	N	C	H
H	71 (297)	87 (364)	103 (431)	83 (347)	77 (322)	76 (318)	135 (565)	111 (464)	93 (389)	99 (414)	104 (435)
C	51 (213)	68 (285)	81 (339)	65 (272)	63 (264)	72 (301)	116 (485)	86 (360)	73 (305)	83 (347)	
N			46 (192)				65 (272)	53 (222)	39 (163)		
O	48 (201)	48 (201)	52 (218)		141 (590)	108 (452)	45 (188)	47 (197)			
F	58 (243)	60 (251)	61 (255)	68 (285)	117 (490)	135 (565)	37 (155)				
Si	56 (234)	74 (310)	91 (381)			53 (222)					
P	44 (184)	63 (264)	78 (326)		48 (201)						
S		52 (218)	61 (255)	60 (251)							
Cl	50 (209)	52 (218)	58 (243)								
Br	42 (176)	46 (192)									
I	36 (151)										

Average Multiple Bond Energies in kcal/mol (kJ/mol)

C=C	146 (611)	C≡C	200 (837)
C=N	147 (615)	C≡N	213 (891)
C=O	177 (741)		
C=S	128 (536)		
N=N	100 (418)	N≡N	226 (946)
N=O	145 (607)		
O=O	119 (498)		

Note: this table contains average values that should be considered very approximate (± several kcal/mol); only the values that correspond to bond strengths of simple diatomic molecules (like HCl) have little error.

[2]Modified with permission of Macmillan Publishing Company from Streitwieser, A.; Heathcock, C. *Introduction to Organic Chemistry*, 3rd ed., pp. 1153. Copyright ©1985 by Macmillan Publishing Co., New York. Additional values from Cottrell, T. *The Strength of Chemical Bonds*, 2nd ed.; Butterworths: London, 1958, and from Benson, S. W. *J. Chem. Educ.* **1965**, 42, 502.

THE TWELVE GENERIC ELECTRON SOURCES

Generic Class	Symbol	Examples
Nonbonding Electrons (4.2)		
Lone pairs on heteroatoms	Z:	$I^{\ominus}$, $HO^{\ominus}$, H_2O, H_3N, $t\text{-BuO}^{\ominus}$
(Nu / base dual behavior)		$CH_3COO^{\ominus}$
Metals	M:	Na, Li, Mg, Zn metals
Electron-rich sigma bonds (4.3)		
Organometallics	R–M	$RMgBr$, RLi, $R_2Cu^{\ominus}$ $Li^{\oplus}$
Metal hydrides	$MH_4^{\ominus}$ (Nu)	$LiAlH_4$, $NaBH_4$,
	MH (Bases)	NaH, KH
Electron-rich pi bonds (4.4)		
Allylic sources	C=C–Z:	Enols: C=C–OH
		Enolates: $C=C-O^{\ominus}$
		Enamines: $C=C-NR_2$
Allylic alkyne sources	C≡C–Z:	$EtC≡C\text{-}NEt_2$
Simple pi bonds (4.5)		
Alkenes	C=C	$H_2C=CH_2$
Dienes	C=C–C=C	$H_2C=CH\text{-}HC=CH_2$
Alkynes	C≡C	HC≡CH
Allenes	C=C=C	$H_2C=C=CH_2$
Aromatic rings (4.6)		
Aromatics	ArH	

THE EIGHTEEN GENERIC ELECTRON SINKS

Generic class	Symbol	Examples
Electron-deficient species (5.2)		
Carbocations	$\geqslant C^{\oplus}$	$^{\oplus}C(CH_3)_3$
Lewis acids	$\geqslant A$	BF_3, BH_3, $AlCl_3$
Metal ions	$M^{2\oplus}$	$Hg(OAc)_2$
Weak single bonds (5.3)		
Acids	H–L (or H–A)	HCl, $ArSO_3H$, CH_3COOH
Weak single bonds	Y–L	RS–SR, HO–OH, Br–Br,
between heteroatoms		HO–Cl
Leaving groups bound to	$\geqslant C-L$	CH_3CH_2–Br, $(CH_3)_3C$–Br,
sp^3 carbon		$(CH_3)_2CH$–Cl, CH_3–I
		Ketals $R_2C(OCH_3)_2$ (2 poor Ls)
Polarized multiple bonds without leaving groups (5.4)		
Heteroatom-carbon	C=Y	Aldehydes: RHC=O
multiple bonds		Ketones: $R_2C=O$
		Imines: $R_2C=NR$
	C≡Y	Nitriles: RC≡N
Conjugate acceptors	C=C–ewg	Enones: C=C–C=O,
		Conjugated esters: C=C–COOR
	C≡C–ewg	RC≡C–COOR
Heterocumulenes	C=C=Y	Ketenes: $R_2C=C=O$
	Z=C=Y	Ph-N=C=O, RN=C=NR, O=C=O
	C=C=C–ewg	$(CH_3)_2C=C=CHCOOCH_3$
	ewg–C=C=C–ewg	ArCOCH=C=CHCOAr

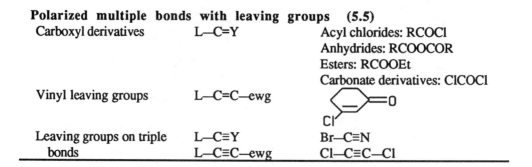

Polarized multiple bonds with leaving groups   (5.5)

Carboxyl derivatives	L—C=Y	Acyl chlorides: RCOCl
		Anhydrides: RCOOCOR
		Esters: RCOOEt
		Carbonate derivatives: ClCOCl
Vinyl leaving groups	L—C=C—ewg	
Leaving groups on triple bonds	L—C≡Y	Br—C≡N
	L—C≡C—ewg	Cl—C≡C—Cl

FLOW CHARTS FOR THE CLASSIFICATION OF ELECTRON SOURCES AND SINKS

For some individuals, a graphical presentation of a process in the form of a flow chart greatly helps them understand and visualize the overall process (Figure A.1).

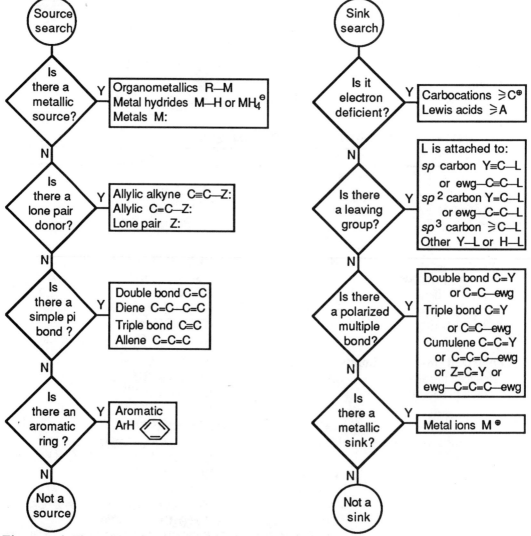

Figure A.1 Flow charts for the classification of electron sources and sinks.

PATHWAY SUMMARY

Major Paths

p.t., proton transfer to and from an anion or lone pair

Deprotonations

$$\overset{\ominus}{b}: \curvearrowright H \quad \underset{Y-}{\overset{b-H}{\searrow}} \quad \rightarrow \quad \overset{\ominus}{\ddot{Y}-} \quad \text{or} \quad \overset{\ominus}{b}: \curvearrowright H \quad \underset{-\overset{|}{\underset{|}{C}}-ewg}{\overset{b-H}{\searrow}} \rightarrow \quad \overset{\ominus\,..}{-\overset{|}{\underset{|}{C}}-ewg}$$

$$\overset{\ominus}{A}: \curvearrowright H \quad \underset{\overset{\oplus}{Y}-}{\searrow} \quad \rightarrow \quad \ddot{Y}- \quad \text{or} \quad \overset{\ominus}{A}: \curvearrowright H \quad \underset{\overset{\oplus}{Z}=C\langle}{\searrow} \rightarrow \quad \ddot{Z}=C\langle$$

Protonations

$$b-H \quad \overset{\ominus}{b}: H \quad \rightarrow \quad Y- \quad \text{or} \quad b-H \quad \overset{\ominus}{b}: H \quad \rightarrow \quad -\overset{|}{\underset{|}{C}}-ewg$$

$$A-H \quad \overset{\ominus}{A}: H \quad \rightarrow \quad \overset{\oplus}{Y}- \quad \text{or} \quad A-H \quad \overset{\ominus}{A}: H \quad \ddot{Z}=C\langle \rightarrow \quad \overset{\oplus}{Z}=C\langle \longleftrightarrow \overset{\oplus}{Z}-C\langle$$

D$_N$, ionization of a leaving group

$$\rangle C-L \rightarrow \overset{\oplus}{C}\langle \quad :L^{\ominus} \quad \text{or} \quad \overset{Y}{\underset{||}{C}}-L \rightarrow \overset{Y}{\underset{||}{\overset{\oplus}{C}}} \quad :L^{\ominus}$$

$$\rangle C-\overset{\oplus}{L} \rightarrow \overset{\oplus}{C}\langle \quad :L \quad \text{or} \quad \overset{Y}{\underset{||}{C}}-\overset{\oplus}{L} \rightarrow \overset{Y}{\underset{||}{\overset{\oplus}{C}}} \quad :L$$

A$_N$, trapping of an electron deficient species by a nucleophile

$$\overset{\ominus}{Nu}: \curvearrowright \overset{\oplus}{C}\langle \rightarrow Nu-C\langle \quad \text{or} \quad \overset{\ominus}{Nu}: \curvearrow \overset{Y}{\underset{||}{\overset{\oplus}{C}}} \rightarrow Nu-\overset{Y}{\underset{\backslash}{C}}$$

A$_E$, electrophile addition to a multiple bond

$$E^{\oplus} \quad -\overset{}{C}=C- \rightarrow \overset{E}{\underset{}{\rangle C}}-\overset{\oplus}{C}-$$

D$_E$, electrophile loss from a cation to form a pi bond

$$E \quad \rangle C-\overset{\oplus}{C}\langle \rightarrow E^{\oplus} \quad \rangle C=C\langle$$

1,2R, rearrangement of a carbocation

$$R \quad \rangle C-\overset{\oplus}{C}\langle \rightarrow \overset{\oplus}{\rangle C}-C\overset{R}{\langle}$$

S$_N$2,   the S$_N$2 substitution

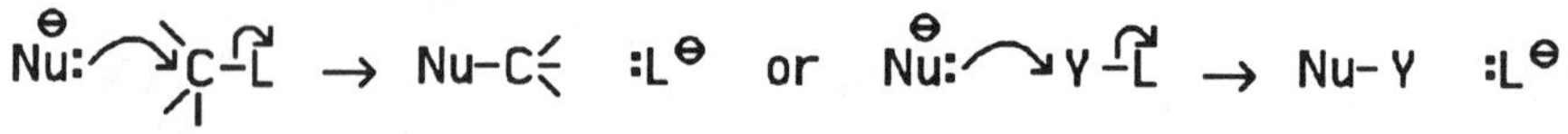

$$Nu:\!\!\overset{\ominus}{} \curvearrowright C\!-\!L \rightarrow Nu\!-\!C \quad :L^{\ominus} \quad \text{or} \quad Nu:\!\!\overset{\ominus}{} \curvearrowright Y\!-\!L \rightarrow Nu\!-\!Y \quad :L^{\ominus}$$

E2,   the E2 elimination

$$\overset{\ominus}{b}: \quad \underset{L}{\overset{H}{-C-C-}} \rightarrow \underset{L:^{\ominus}}{\overset{b\,\,H}{-C=C-}}$$

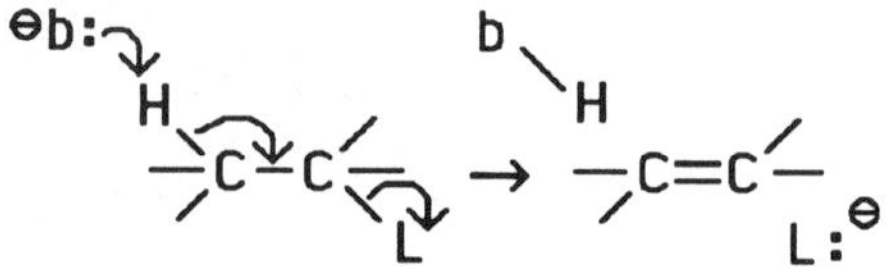

Ad$_E$3,   the Ad$_E$3 addition

$$\underset{\text{pi-complex}}{\overset{A\cdots H}{-C=C-}} \rightarrow \underset{\overset{\ominus}{A}:}{\overset{A\,H}{-C=C-}} \rightarrow \underset{A}{\overset{H}{-C-C-}} \quad \overset{\ominus}{A}:$$

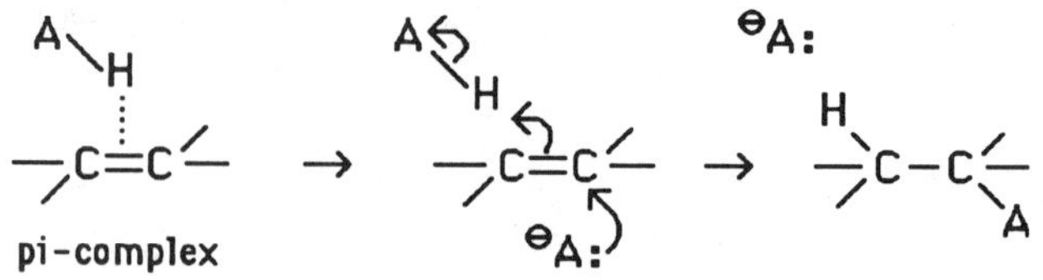

Ad$_N$,   nucleophilic addition to a polarized multiple bond

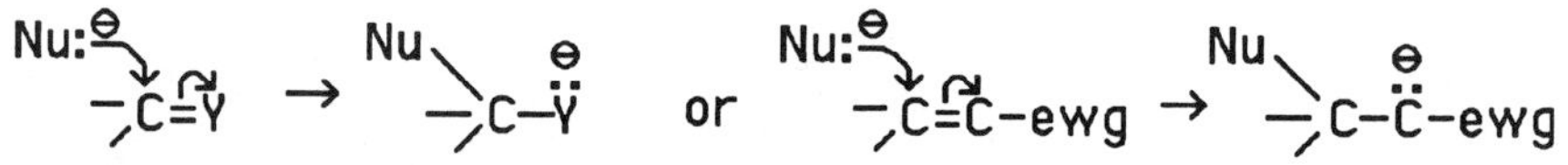

$$Nu:^{\ominus} \overset{}{-C=Y} \rightarrow \underset{}{\overset{Nu}{-C-\ddot{Y}}}{}^{\ominus} \quad \text{or} \quad Nu:^{\ominus} -C=C\text{-ewg} \rightarrow \underset{}{\overset{Nu}{-C-\ddot{C}}}{}^{\ominus}\text{-ewg}$$

E$_\beta$,   beta elimination from an anion or lone pair

$$L \curvearrowright \overset{}{-C-\ddot{Y}}{}^{\ominus} \rightarrow L:^{\ominus} \quad -C=Y \quad \text{or} \quad L \curvearrowright \overset{}{-C-\ddot{C}}{}^{\ominus}\text{-ewg} \rightarrow L:^{\ominus} \quad -C=C\text{-ewg}$$

6e,   concerted six-electron pericyclic

Thermal rearrangements:

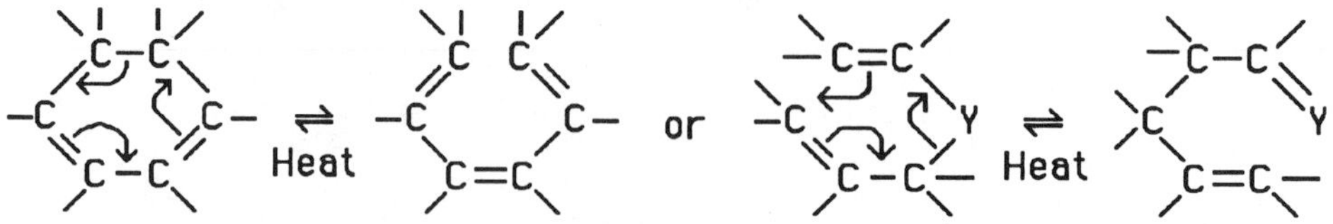

Thermal cycloadditions or cycloreversions:

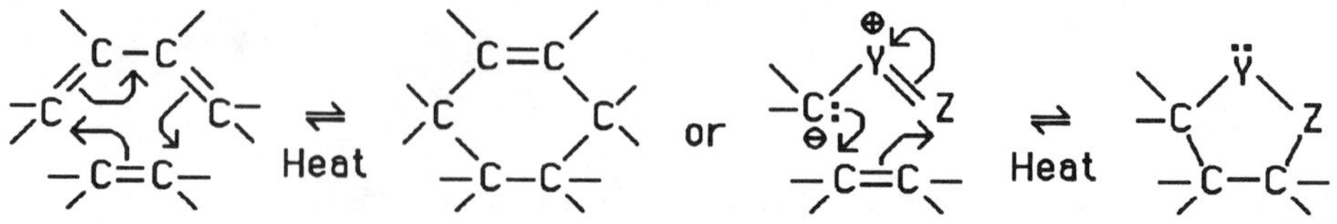

Metal-chelate-catalyzed additions

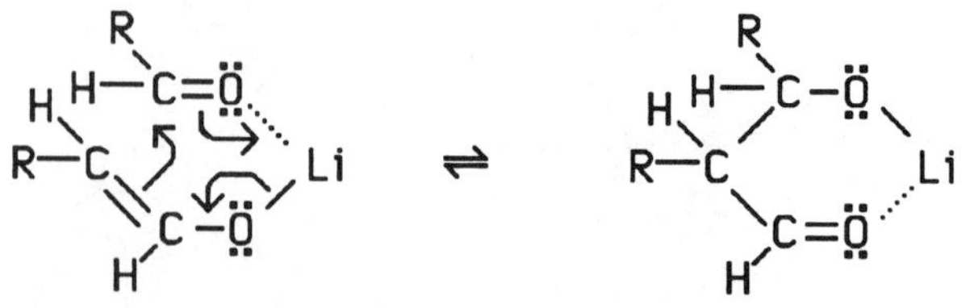

Minor Paths

Ei, internal *syn* elimination

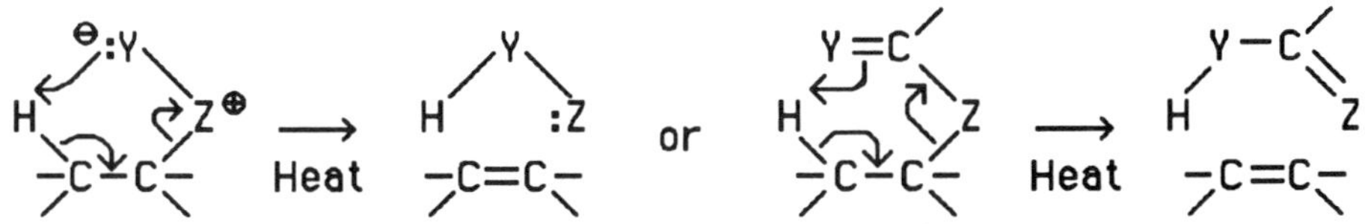

AdgB, general base-catalyzed addition to a polarized multiple bond

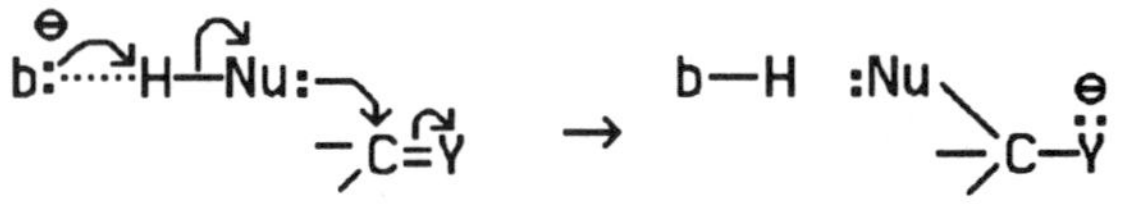

EgA, general acid-catalyzed beta elimination

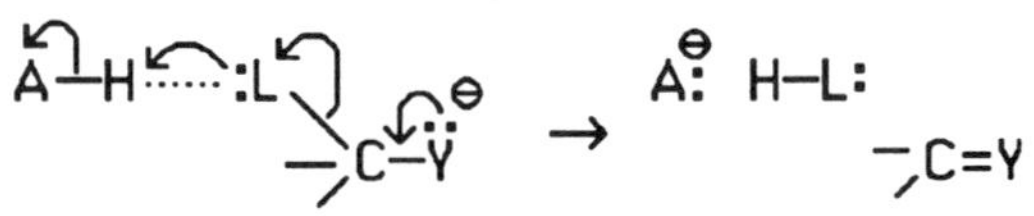

NuL, Nu—L additions

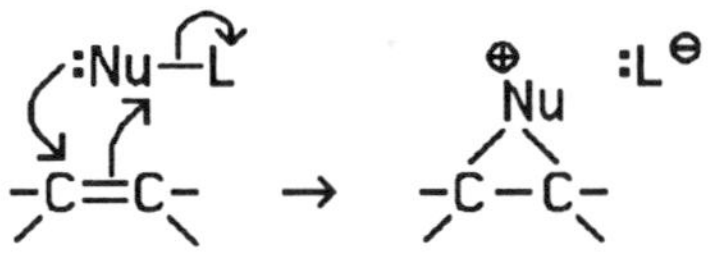

1,2RL, rearrangement with loss of leaving group

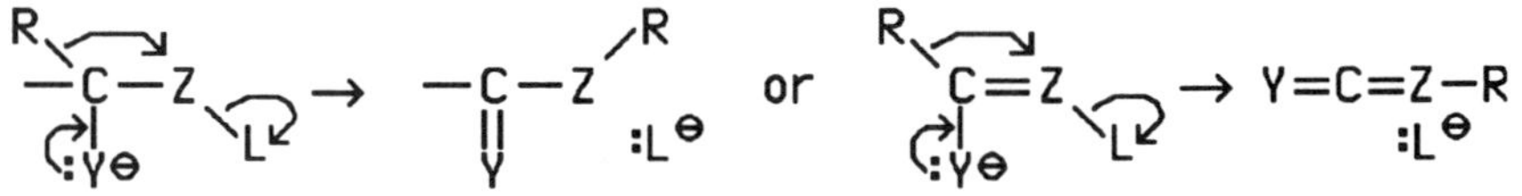

4e, four-center, four-electron (three cases only)

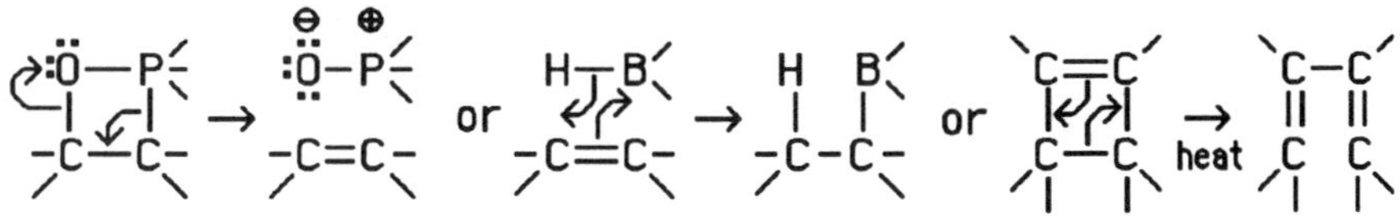

pent., substitution via a pentacovalent intermediate
(not for carbon or other first-row elements)

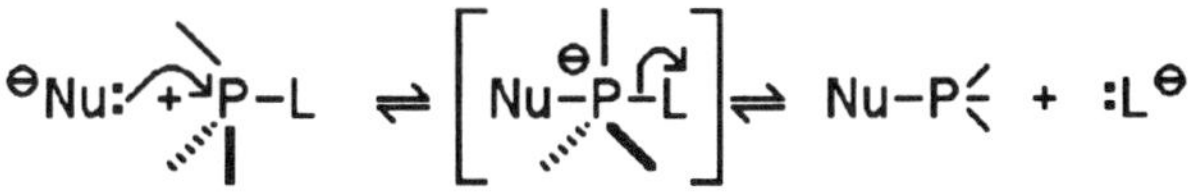

$$^\ominus Nu\!:\!\overset{\curvearrowright}{\diagup}\!P\!-\!L \;\rightleftharpoons\; \left[Nu\!-\!\overset{|}{\underset{|}{P}}\!\overset{\ominus}{\curvearrowleft}\!L \right] \rightleftharpoons\; Nu\!-\!P\!\!\diagdown \;+\; :L^\ominus$$

H$^\ominus$ t., hydride transfer to a cationic center

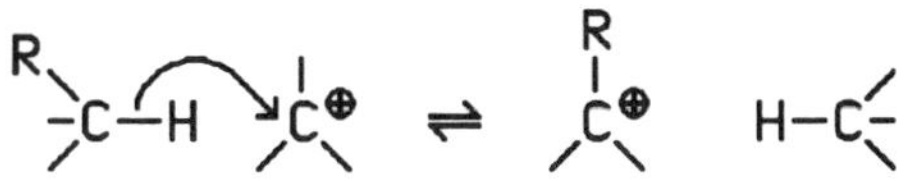

Path Combinations

S_N1 (substitution, nucleophilic, unimolecular) (path D_N followed by path A_N)

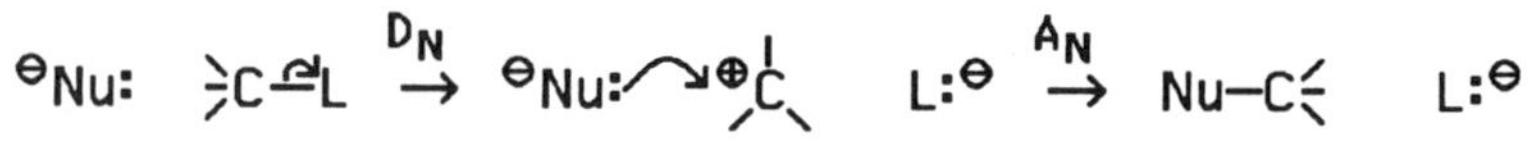

E1 (elimination, unimolecular) (path D_N followed by path D_E)

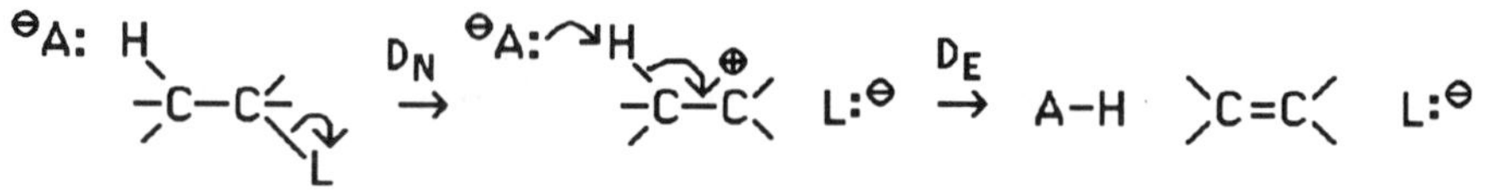

Ad_E2 (addition, electrophilic, bimolecular) (path A_E followed by path A_N)

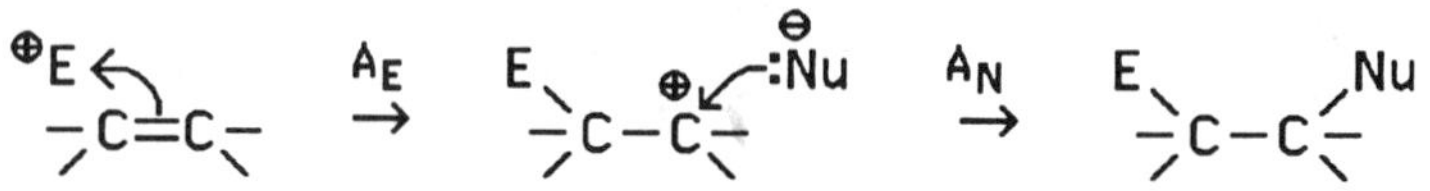

Hetero Ad_E2 (path p.t. followed by path Ad_N)

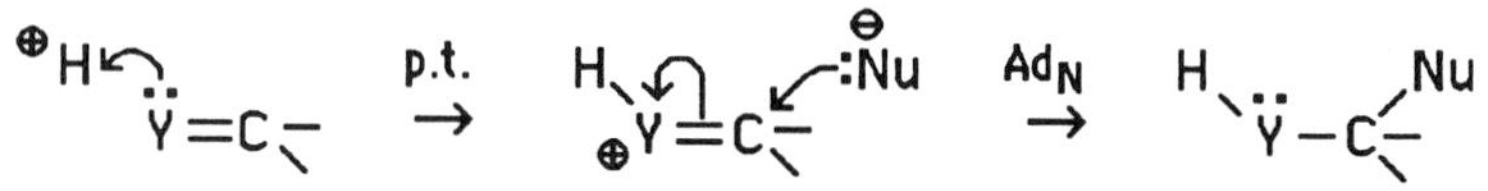

Lone-pair-assisted E1 (path E_β followed by path p.t.)

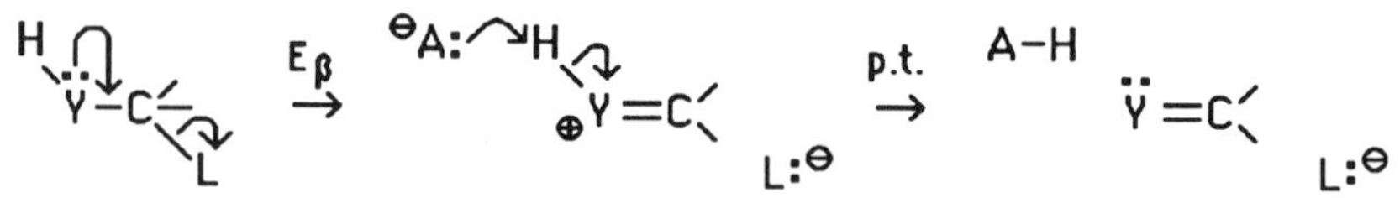

Electrophilic aromatic substitution (path A_E followed by path D_E)

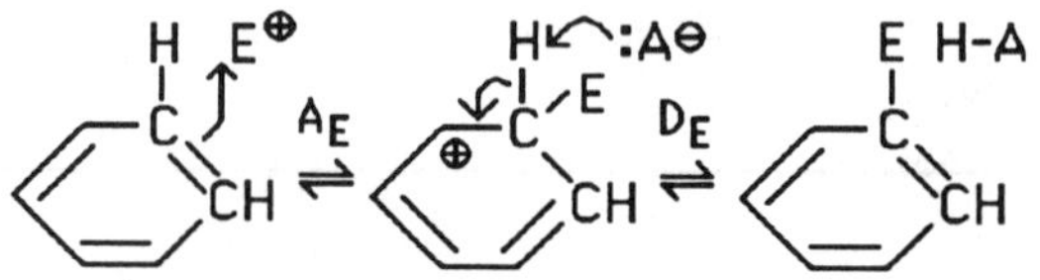

E1cB (elimination, unimolecular, conjugate base) (path p.t. followed by path E_β)

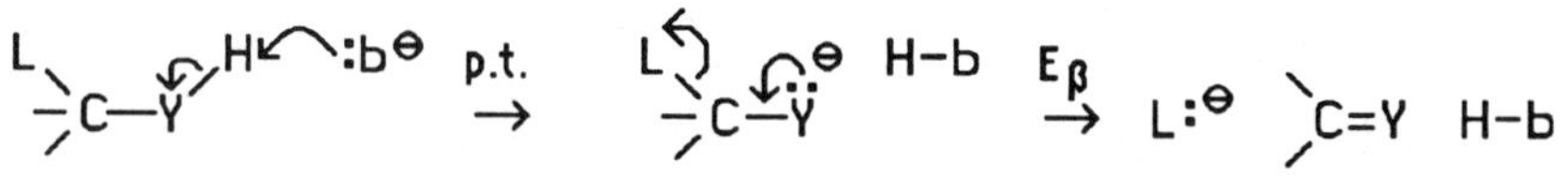

or

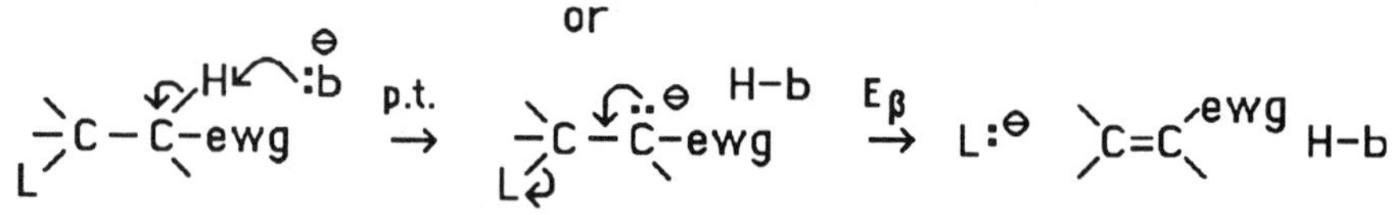

Addition–elimination (path Ad_N followed by path E_β)

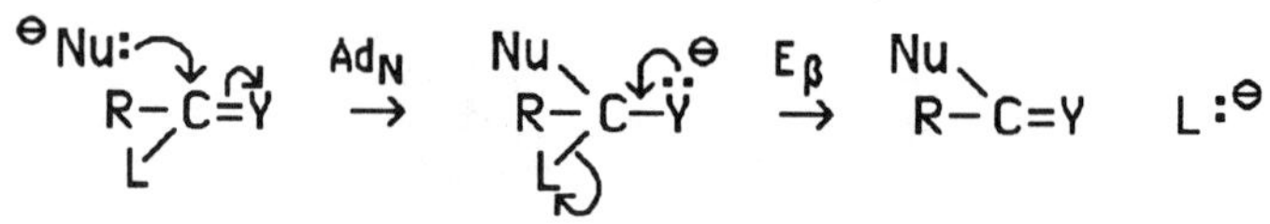

Ad$_N$2 (addition, nucleophilic, bimolecular) (path Ad$_N$ followed by path p.t.)

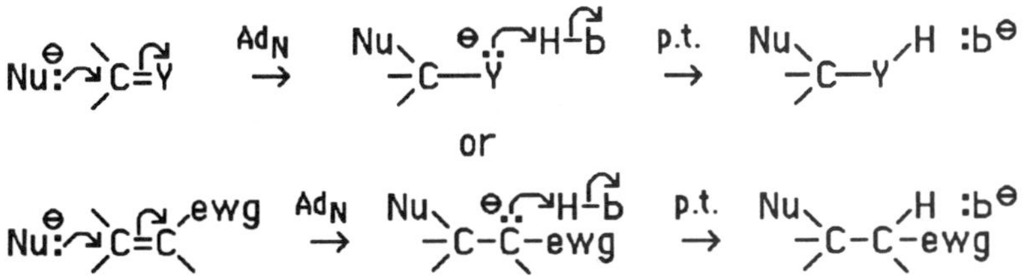

or

Tautomerization (H—C—C=Z $\rightleftharpoons$ C=C—Z—H)

Acid catalysis (path p.t. followed by path D$_E$):

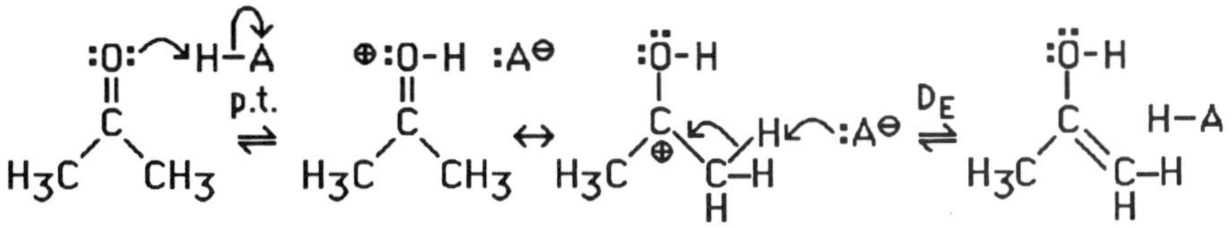

Base catalysis (path p.t. twice):

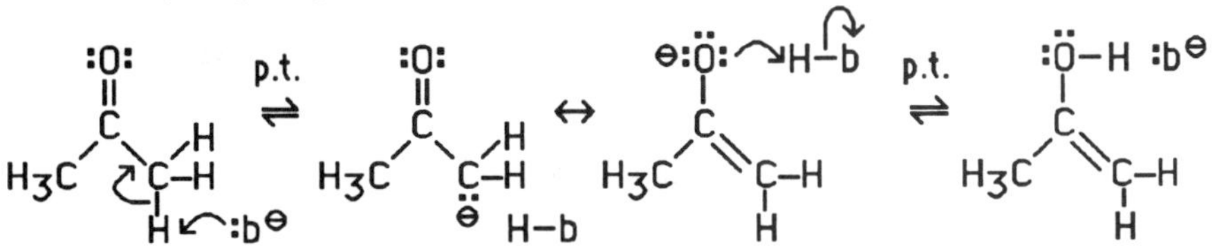

SUBSTITUTION VS. ELIMINATION DECISION MATRICES

(Section 8.5)

Examples			Unhindered primary	
Weak base	Strong base		Weak base	Strong base
I$^\ominus$ RS$^\ominus$	HO$^\ominus$ EtO$^\ominus$	Good nucleophile	Subst.	Subst.
H$_2$O EtOH	t-BuO$^\ominus$ R$_2$N$^\ominus$	Poor nucleophile	Subst.	Mixture

Secondary			Tertiary	
Weak base	Strong base		Weak base	Strong base
Subst.	Mixture	Good nucleophile	Mixture	Elim.
Subst.	Elim.	Poor nucleophile	Subst.	Elim.

ADDITION/ELIMINATION ENERGY SURFACES

(The reverse processes are given in parentheses)

Basic Polarized Multiple-Bond Addition/Elimination Paths

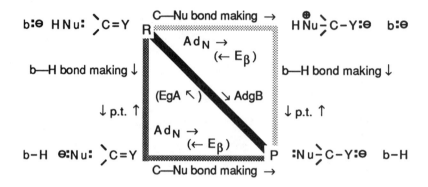

Acidic Polarized Multiple-Bond Addition/Elimination Paths

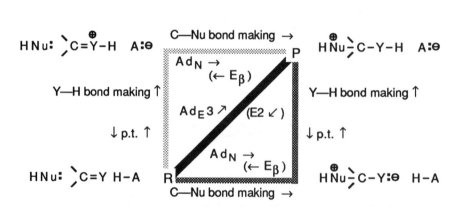

Carbon–Carbon pi Bond Addition/Elimination Mechanisms

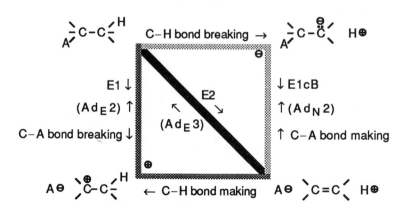

TRENDS GUIDE

Carbocation Stability Trend

Lone pair donors are best, pi bond resonance next, substitution last. The effects are additive; three of a lesser type of stabilization are at least as good as one of the better type.

Carbocation name	Structure	Comments on Stabilization
Stable carbocations		
Guanidinium ion	$(H_2\overset{..}{N})_3\overset{\oplus}{C}$	Three N lone pairs stabilize, pK_a = 13.6
Tropylium cation	 ⬡⊕	Stabilized by aromaticity
Protonated amide	$H_3C-\overset{\overset{..}{O}H}{\underset{}{\overset{\oplus}{C}}}\overset{..}{N}H_2$	N and O lone pairs and one R, pK_a = 0.5
Moderately stable		**(Easily formed)**
Protonated carboxylic acid	$H_3C-\overset{\overset{..}{O}H}{\overset{\oplus}{C}}\overset{..}{O}H$	Two O lone pairs and one R, pK_a = −6
Triphenylmethyl cation	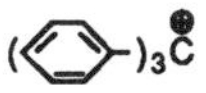 $(\bigcirc\!\!-)_3\overset{\oplus}{C}$	Resonance with three phenyl groups
Protonated ketone	$H_3C-\overset{\overset{..}{O}H}{\overset{\oplus}{C}}CH_3$	An O lone pair donor and two R's, pK_a = −7
Diphenylmethyl cation	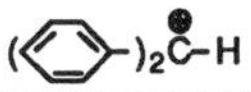 $(\bigcirc\!\!-)_2\overset{\oplus}{C}-H$	Resonance with two phenyl groups
Average stability		**(Common, of similar stability)**
Tertiary cation	$(H_3C)_3\overset{\oplus}{C}$	Three alkyl groups stabilize
Benzyl cation	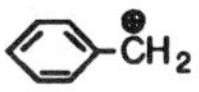 $\bigcirc\!\!-\overset{\oplus}{C}H_2$	Resonance with phenyl group
Primary allyl cation	$H_2C=CH-\overset{\oplus}{C}H_2$	Resonance with a pi bond only
Acylium ion	$H_3C-\overset{\oplus}{C}=\overset{..}{O}$	A vinyl cation stabilized by an O lone pair
Moderately unstable		**(Infrequent)**
Secondary cation	$(H_3C)_2\overset{\oplus}{C}-H$	Stabilized by two alkyl groups
Secondary vinyl cation	$H_2C=\overset{\oplus}{C}-CH_3$	*sp* hybrid, Less stable than secondary cation
Unstable		**(Very rare)**
Primary cation	$H_3C-\overset{\oplus}{C}H_2$	Only one alkyl group stabilizes
Primary vinyl cation	$H_2C=\overset{\oplus}{C}H$	Less stable than primary cation due to sp hybrid
Phenyl cation	 $\bigcirc\!\!-\overset{\oplus}{C}$	A bent vinyl cation
Methyl cation	$\overset{\oplus}{C}H_3$	No stabilization

Donor Trend

Excellent donors

−$\overset{\ominus}{\overset{..}{C}}H_2$ Anionic

−$\overset{\ominus}{\overset{..}{N}}H$ Anionic, more electronegative than C

−$\overset{..}{\overset{..}{O}}{:}^{\ominus}$ Anionic, more electronegative than N

Good donors

−$\overset{..}{N}(CH_3)_2$ Neutral

−$\overset{..}{N}H_2$ Neutral

−$\overset{..}{\overset{..}{O}}H$ Neutral, more electronegative than N

−$\overset{..}{\overset{..}{O}}CH_3$ Neutral, more electronegative than N

−$\overset{..}{N}H-\overset{\overset{\textstyle :\!O\!:}{\|}}{C}-CH_3$ Amide, carbonyl decreases lone pair availability

−$\overset{..}{\overset{..}{S}}CH_3$ Sulfur, poor 2p—3p pi bond

---

Poor donors

−Ar   Delocalization of charge into the aromatic ring

−R   Hyperconjugation only, no lone pairs to donate

−H   No substituent at all

Very poor donors   (pi donors whose overall effect is electron withdrawing)

−$\overset{..}{\overset{..}{C}}{:}$ Cl, electronegative, poor 2p—3p pi bond

−$\overset{..}{\overset{..}{O}}-\overset{\overset{\textstyle :\!O\!:}{\|}}{C}-CH_3$ Electronegative, carbonyl decreases lone pair availability

---

Carbanion Stability Trend

Stable carbanions have low pK_{abH}'s.

Electron-Withdrawing Group Trends

CH_3−ewg has low pK_a for good ewg.

Leaving Group Trends

Good L has pK_{aHL} of 0 or less.
Fair L has pK_{aHL} of 1 to 12.
Poor L has pK_{aHL} of 13 to 30.

Alkene Stability Trends

The more substituted, the more stable.

Allylic Electron Source Reactivity Trends

The better the donor, the more reactive.

Aromatic Ring Reactivity Trends

The better the donor, the more reactive the ring is to electrophilic attack.
Donors direct electrophiles ortho–para.
The better the electron-withdrawing group, the less reactive the ring is to electrophilic attack. Electron-withdrawing groups direct electrophiles meta.

Lone Pair Nucleophilicity Trends

Softness, basicity, and solvation are important.

Relative nucleophilicity toward CH_3I (in CH_3OH):

CH_3OH	1
$F^{\ominus}$	500
$CH_3COO^{\ominus}$	20,000
$Cl^{\ominus}$	23,000
Et_2S	220,000
NH_3	320,000
$PhO^{\ominus}$	560,000
$Br^{\ominus}$	620,000
$CH_3O^{\ominus}$	1,900,000
Et_3N	4,600,000
$CN^{\ominus}$	5,000,000
$I^{\ominus}$	26,000,000
Et_3P	520,000,000
$PhS^{\ominus}$	8,300,000,000
$PhSe^{\ominus}$	50,000,000,000

Organometallic Reactivity Trends

The greater the electronegativity difference, the more reactive.

	Most reactive
1.62	R-Na
1.57	R-Li
1.24	R-MgX
0.9	R_2Zn
0.86	R_2Cd
0.65	$R_2Cu^{\ominus}Li^{\oplus}$
0.55	R_2Hg
0.22	R_4Pb
	Least reactive

Carboxyl Derivative Trends

Donor groups deactivate carboxyl derivatives toward nucleophilic attack.

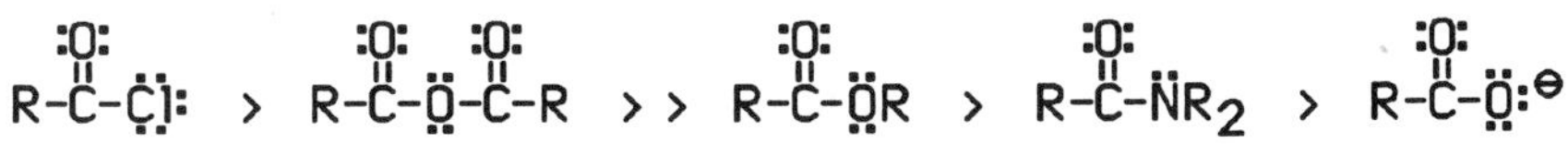

$$R-\overset{\overset{\displaystyle :O:}{\|}}{C}-\ddot{\underset{..}{C}}l: \; > \; R-\overset{\overset{\displaystyle :O:}{\|}}{C}-\ddot{O}-\overset{\overset{\displaystyle :O:}{\|}}{C}-R \; >> \; R-\overset{\overset{\displaystyle :O:}{\|}}{C}-\ddot{O}R \; > \; R-\overset{\overset{\displaystyle :O:}{\|}}{C}-\ddot{N}R_2 \; > \; R-\overset{\overset{\displaystyle :O:}{\|}}{C}-\ddot{O}:^{\ominus}$$

Conjugate Acceptor Reactivity Trends

The better the electron withdrawing group, the more reactive.

THERMODYNAMICS AND KINETICS

Free Energy, ΔG, and K_{eq}

$\Delta G° = -RT \ln K_{eq}$

At room temperature, every 1.37 kcal/mol (5.73 kJ/mol) changes the equilibrium constant by a factor of ten.

$\Delta G°$ kcal/mol	K_{eq}	Reactant	Product	$\Delta G°$ (kJ/mol)
+5.46	0.0001	99.99	0.01	+22.84
+4.09	0.001	99.9	0.1	+17.11
+2.73	0.01	99	1	+11.42
+1.37	0.1	91	9	+5.73
+1.0	0.18	85	15	+4.18
+0.5	0.43	70	30	+2.09
0	1	50	50	0
-0.5	2.33	30	70	-2.09
-1.0	5.41	15	85	-4.18
-1.37	10	9	91	-5.73
-2.73	100	1	99	-11.42
-4.09	1,000	0.1	99.9	-17.11
-5.46	10,000	0.01	99.99	-22.84
-9.56	10^7	Essentially complete		-40.00

The Relationship of Free Energy to Enthalpy and Entropy

$\Delta G = \Delta H - T\Delta S$

Enthalpy, Heat of Reaction, ΔH

$\Delta H = \Delta H_{(bonds\ broken)} - \Delta H_{(bonds\ made)}$

Kinetics, $\Delta G^{\ddagger}$

The $\Delta G^{\ddagger}$ of a reaction that proceeds at a reasonable rate at room temperature is 20 kcal/mol (84 kJ/mol). The diffusion controlled limit of 10^{10} liters/mol-sec corresponds to a reaction upon every collision. At room temperature, dropping the $\Delta G^{\ddagger}$ by 1.37 kcal/mol (5.73 kJ/mol) increases the rate tenfold. The more reactive a species is, the less selective it is. The more stable a compound is, the less reactive it is.

GENERATION OF ALTERNATE PATHS, REACTION CUBES

One of the more difficult tasks in working mechanism or product prediction problems is deciding whether the set of alternate reaction paths is complete or not. Second only to that is keeping track of those paths and putting them in some logical order. Quite often reactants have four charge types, but rarely do more than three occur in the same medium; one is usually an acidic medium charge type, one for a basic medium, and two can often occur in either medium. Products likewise can have four charge types, again with the same media preferences. Frequently, these charge types are interrelated by proton transfer (Figure A.2).

A data structure for describing the interrelation of eight variables — these charge types of both reactants or products — is a simple cube (Figure A.3). Each charge type can then be related to any of the other seven by either an edge, face diagonal, or body diagonal. The horizontal axis is the reaction coordinate for bond formation between the nucleophile and the

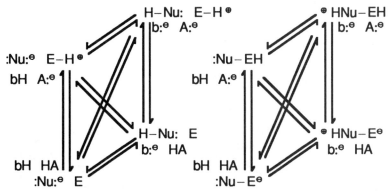

Figure A.2 Possible interconversions of charge types of reactants (*left*) and products (*right*).

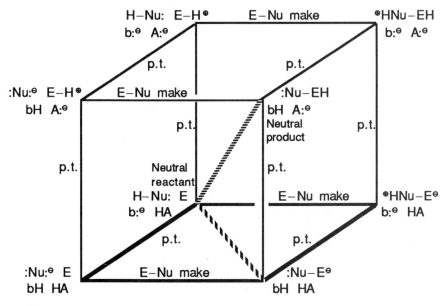

Base of cube is the simple energy surface

Face diagonal process makes E–Nu bond and alters charge type by one proton transfer

Body diagonal process makes E–Nu bond and alters charge type by two proton transfers

Figure A.3 A generic reaction cube illustrating the interaction of various charge types of reactants and products. p.t. = proton transfer.

electrophile. The base of the cube (and likewise the top, front, and back) form the planar projection onto a horizontal plane of the simple energy surface relating two charge types of the reactants with two charge types of the products.

Figure A.4 shows the usefulness of the reaction cube as a data structure. Additions to carbonyls often occur between different charge types, and frequently three-dimensional energy surfaces are used to clarify the various equilibria. We have seen two faces of this cube before as individual energy surfaces. The bottom face of the cube is Figure 6.12, polarized multiple bond addition/elimination mechanisms in basic media. The back face of the cube is Figure 6.13, polarized multiple bond addition/elimination mechanisms in acidic media.

One or more corners of this cube can be of higher energy because of a wide pK_a span. When we add the dimension of energy to the reaction cube, in essence, it becomes a four-dimensional energy surface. Each corner can be ranked according to its probability of occurrence under the particular reaction conditions, and the most probable lowest energy route would then be expected to avoid those high-energy corners. Another way to view a reaction cube is that each face is the planar projection of a three-dimensional energy surface. Conditions that raise the energy of a particular corner in the three-dimensional energy surface likewise affect that corner of the reaction cube.

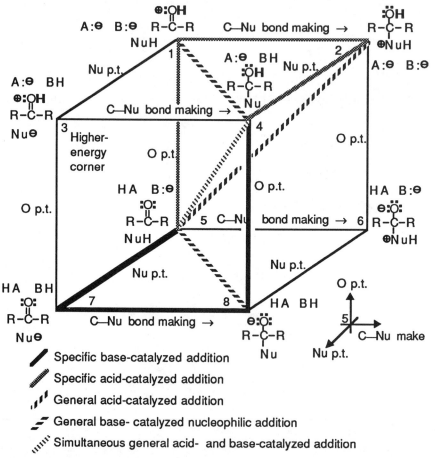

Figure A.4 The reaction cube for the addition of a nucleophile to a carbonyl in protic media (for hemiketal/hemiacetal, NuH = ROH) where O p.t. is proton transfer to oxygen, and Nu p.t. is proton transfer to the nucleophile. The axes for point 5 are shown in the lower right. (Reprinted with permission from P. H. Scudder, *J. Org. Chem.*, **1990**, *55*, 4238-4240. Copyright ©1990 by the American Chemical Society.)

There is a need for interrelating the various mechanisms in a way that all alternatives are easily visible so that none are left unexplored. With 12 edges, 12 face diagonals, and 4 body diagonals, there are 28 possible interactions that require checking. Specific acid or base processes lie along the edges of the cube. The general acid-catalyzed , Ad_E3, or general base-catalyzed processes constitute the face diagonals. The body diagonal is the simultaneous general acid, general base catalysis (push–pull catalysis).

The charge types on a reaction cube can be so arranged such that basic media routes lie primarily or exclusively in one hemisphere and acidic media routes thereby lie in the other. Reactions proceeding through an intermediate can be represented by cubes sharing a common face, edge, or vertex corresponding to the intermediate charge type(s).

Hints

TO PROBLEMS FROM CHAPTERS 7, 8, AND 9

CHAPTER 7

Problem	Source class (first step only)	Sink class (first step only)
7.6a	Z as b	$\geqslant$C—L
7.6b	Z as Nu	$\geqslant$C—L
7.6c	C=C	HA
7.7a	Z as Nu	$\geqslant$C—L
7.7b	Nu—L	C=Y
7.7c	C=C—Z	C=C—ewg
7.8a	Z as b	Y=C—L
7.8b	Z as b	HA
7.8c	Z as b	HA
7.9a	R—M	Y=C=Y
7.9b	Z as Nu	$\geqslant$C—L
7.9c	C=C	$\geqslant$A
7.10a	C=C	HA
7.10b	Z as Nu	Y=C—L
7.10c	ArH	$\geqslant$C$^{\oplus}$

CHAPTER 8

Problem	Site	Base	Nucleophile	Minor variable
8.1a	Primary	Weak	Good	
8.1b	Primary	Strong	Poor	
8.1c	Primary	Strong	Poor	
8.1d	Hindered primary	Weak	Good	
8.1e	Secondary	Weak	Poor	
8.1f	Secondary	Strong	Good	Heated
8.1g	Primary	Strong	Poor	
8.1h	Secondary	Weak	Good	
8.1i	Secondary	Weak	Good	
8.1j	Hindered secondary	Weak	Good	Heated
8.1k	Tertiary	Weak	Poor	
8.1l	Tertiary	Strong	Good	
8.1m	Tertiary	Weak	Good	

Problem	Source class (for first step)	Sink class (for first step)	Decision type
8.2a	R—M	Y=C—L	Multiple addition
8.2b	Z—C=C	C=C—ewg	Ambident electrophile
8.2c	Z—C=C	Y=C—L	Ambident nucleophile
8.2d	Z	$\geqslant$C—L	Intermolecular versus intramolecular
8.2e	Z	C=Y	Thermodynamic versus kinetic enolate

Problem	Source class (first step)	Sink class (first step)	Decision type
8.4a	R—M	Y=C—L	Multiple addition
8.4b	R—M	Y=C—L	Multiple addition
8.4c	Z	$\geqslant$C—L	Substitution/elimination
8.4d	R—M	C=C—ewg	Ambident electrophile
8.4e	R—M	C=C—ewg	Ambident electrophile
8.4f	Z—C=C	Y—L	Ambident nucleophile
8.4g	Z	$\geqslant$C—L	Substitution/elimination

CHAPTER 9

Problem	Source class (first step)	Sink class (first step)	First step pathway
9.1a	Z	Y—L	pent. or S_N2
9.1b	Z	Y=C	p.t.
9.1c	C≡C	HA	Ad_E3
9.1d	Z	Y=C	p.t.
9.1e	Z as Nu	C=C—ewg	Ad_N
9.1f	Z	Y=C—L	p.t.
9.2a	Z	HA	p.t.
9.2b	Z—C=C	Y=C—L	Ad_N
9.2c	Z	HA	p.t.
9.2d	Z	C=Y	p.t.
9.2e	Z	HA	p.t.
9.2f	Z	Y=C—L	p.t.
9.3a	Z	$\geqslant$C—L	S_N2
9.3b	Z	$\geqslant$A	A_N
9.3c	Z	$\geqslant$C—L	S_N2
9.3d	Z	Y=C—L	Ad_N
9.3e	R—M	HA	p.t.
9.3f	Nu—L	C=C	NuL
9.3g	Z—C=C	H—A	p.t.
9.4a	Z	Y—L	S_N2
9.4b	Z	HA	p.t.
9.4c	Z	C=C=Y	Ad_N
9.4d	R—M	Y=C—L	Ad_N
9.4e	Z	C=Y and HA	Ad_E3
9.4f	Z	HA	p.t.
9.4g	$MH_4^{\ominus}$	Y=C—L	Ad_N
9.4h	Z	Y=C—L	Ad_N
9.5a	Z—C=C—C=C	HA	A_E
9.5b	Z	C=Y	Ad_E3
9.5c	C=C	HA	Ad_E3
9.5d	C=C—Z	C=Y	6e
9.5e	Z	$\geqslant$C—L	S_N2

Exercise Answers

CHAPTER 1: ADDITIONAL EXERCISES

1.1 Draw Lewis structures with resonance for the following neutral compounds:
O—O—O H_2C—N—N O—C—O H—O—N—O H—O—NO_2.

(Lewis structures with resonance forms shown)

(The acids, H—A, have extra resonance forms that corresponds to $H^{\oplus} A^{\ominus}$)

1.2 Draw Lewis structures with resonance for the following charged species:
$H_2CNH_2^{\oplus}$ $OCN^{\ominus}$ $HOCO_2^{\ominus}$ $H_2COH^{\oplus}$ $H_2CCHCH_2^{\oplus}$.

(Lewis structures with resonance forms shown)

Note that $H_2CCHCH_2^{\oplus}$ is electron deficient and cannot satisfy all shells.

1.3 What is the polarization of the indicated bond?

Br—Br	HO—Cl	H_2B—H	H_2C=O	I—Cl
None	∂- ∂+	∂+ ∂-	∂+ ∂-	∂+ ∂-

1.4 Circle the electrophiles and underline the nucleophiles in the following group.

(BF₃)   (H⊕)   Ne   NH₃   ⊖C≡N

1.5 Draw the Lewis structure(s) that would be the product of the arrows.

(a)

(Lewis structures with reaction arrows shown)

(b)

(c)

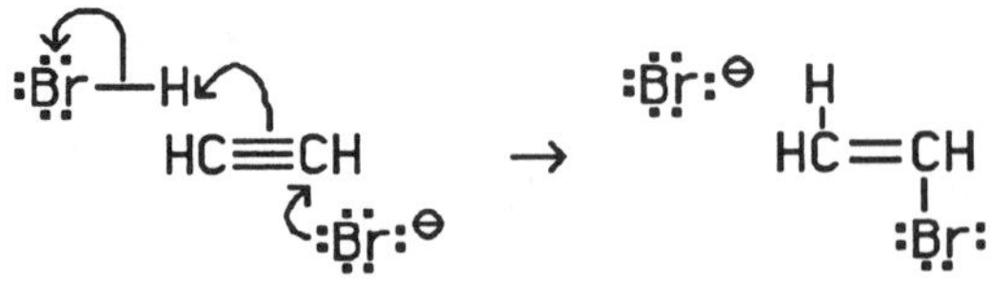

(d)

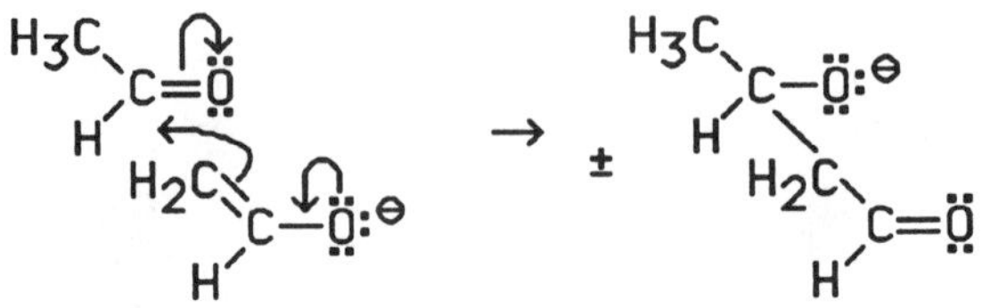

(A new chiral center is formed; attack on the C=O could occur equally from the top or the bottom. The ± indicates the product is racemic, a 1:1 mix of enantiomers.)

(e)

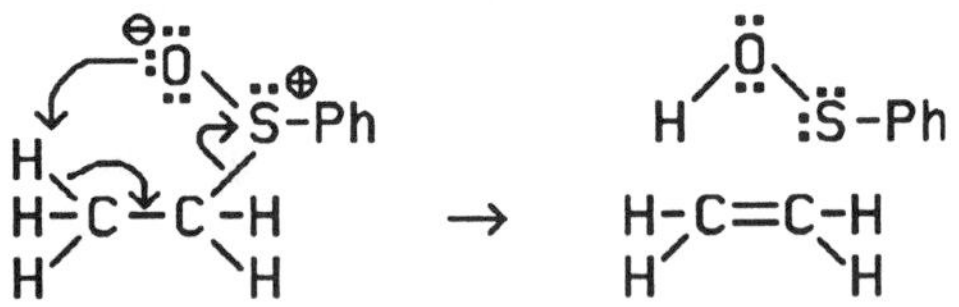

1.6 Give the curved arrows necessary for the following reactions. **Every arrow must start from either a lone pair or bond.**

(a)

(b)

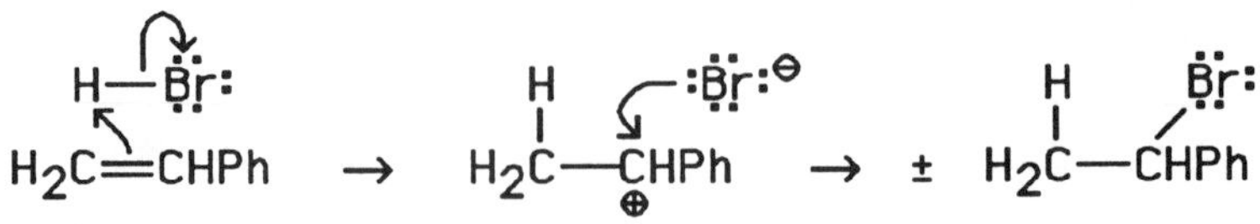

(c)

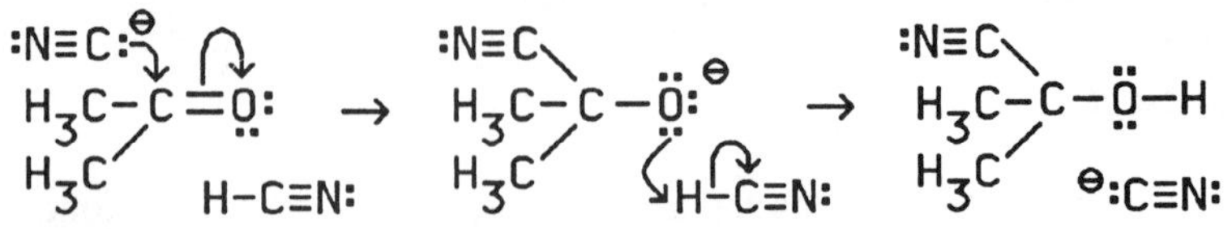

(d)

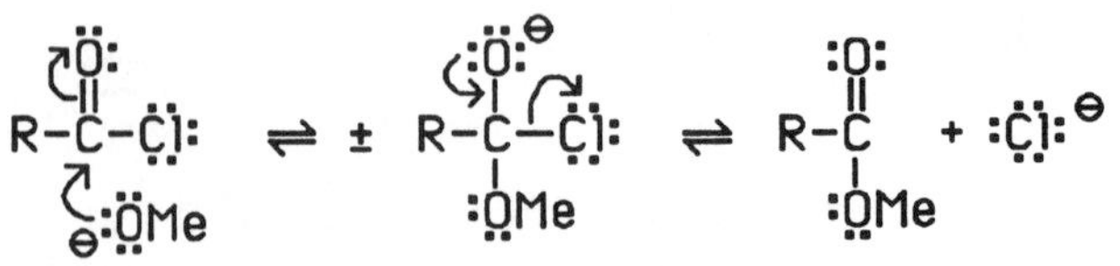

(e)

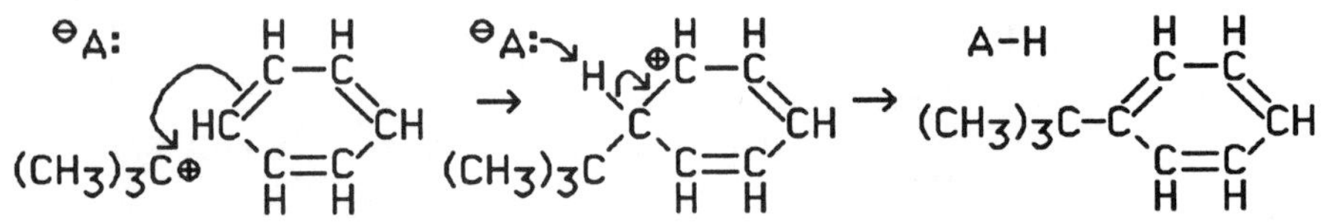

1.7 Draw full Lewis structures for the following line structure drawings. Common errors in this problem are H's missing or five bonded carbons.

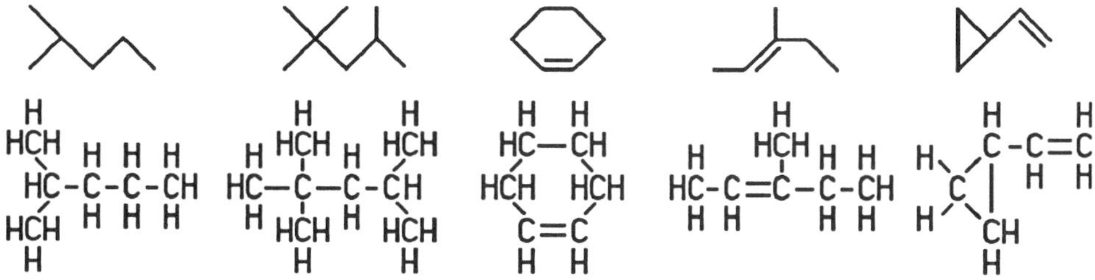

1.8 The carbon atom that bears a significant partial positive charge is circled on the following structures. The last one has two sites by resonance.

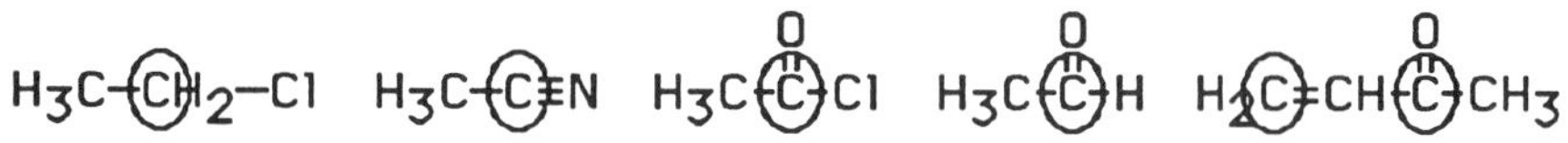

1.9 The carbon atom that bears a negative or significant partial negative charge is shown on the following resonance structures.

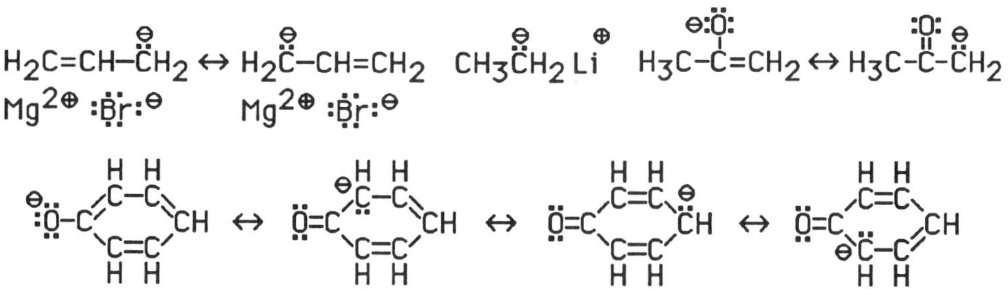

1.10 Give the hybridization of the carbons in the following structures.

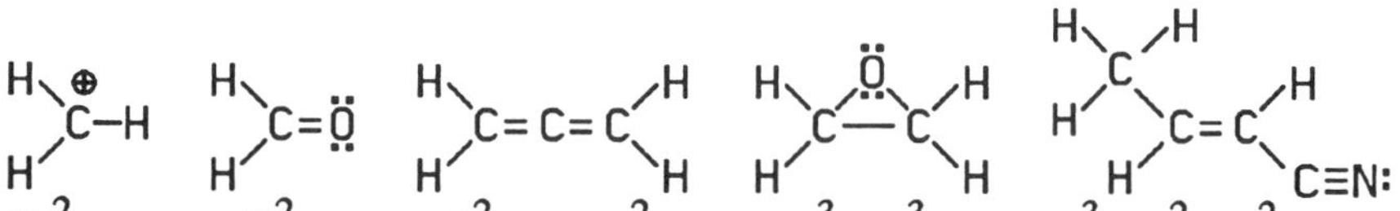

*sp*² *sp*² *sp*² *sp* *sp*² *sp*³ *sp*³ *sp*³ *sp*² *sp*² *sp*

1.11 Which of the following structures are chiral?

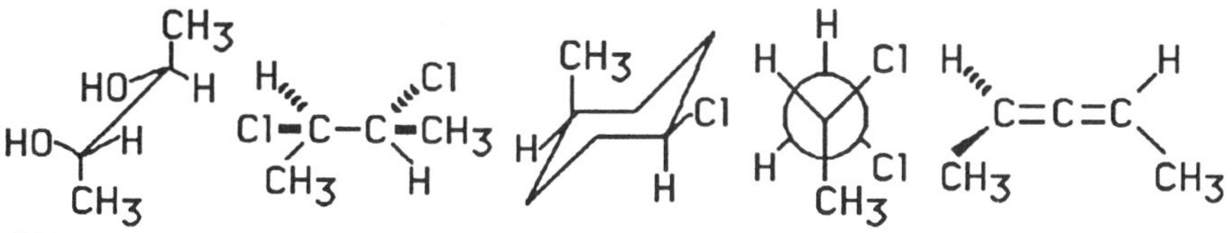

Chiral Not chiral Not chiral Chiral Chiral

Each structure that is not chiral has an internal mirror plane.

1.12 The pi overlap is shown for the following systems: an amide, carbon dioxide, an ester, and a vinylogous amide. Notice the two perpendicular allylic systems in CO_2; a lone pair in a *p* orbital from each oxygen is conjugated to the adjacent C=O pi bond.

1.13 The aromatic compounds in the following list are circled.

1.14 The conjugated systems in the following list are circled.

$$H_3C-\underset{H}{C}=\underset{H}{C}-C=CH_2 \qquad H_2C=C=CH_2 \qquad H_2C=\underset{H}{C}-\overset{H}{\underset{H}{C}}-C=CH_2$$

$$H_2C=\underset{H}{C}-C\equiv N\colon \qquad H_2C=\underset{H}{C}-\ddot{C}\colon \qquad H_2C=\underset{H}{C}-C=\ddot{O} \qquad H_3C-C\equiv C-CH_3$$

1.15 The number of electrons in the pi system of the following compounds is

$$H_2C=\underset{H}{\overset{\oplus}{C}}-CH_2 \quad \text{(one pi bond, 2 electrons)}$$

$$H_3C-\underset{H}{C}=\underset{H}{C}-C=CH_2 \quad \text{and} \quad H_2C=\underset{H}{C}-\overset{:O:}{\underset{\parallel}{C}}-H \quad \text{(two pi bonds, 4 electrons)}$$

$$H_3C-\overset{:O:}{\underset{\parallel}{C}}-\ddot{O}H \quad \text{and} \quad H_2C=\underset{H}{C}-\ddot{O}H \quad \text{(one pi bond and a lone pair, 4 electrons)}$$

$$H_2C=\underset{H}{C}-\underset{H}{C}=\overset{\ominus}{\underset{}{\ddot{C}H_2}} \quad \text{(two pi bonds and a lone pair, 6 electrons)}$$

CHAPTER: 2 ADDITIONAL EXERCISES

2.1 According to the following energy diagram:

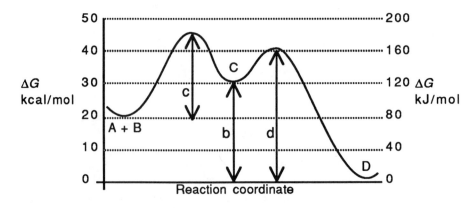

(a) The rate-determining step of A → D is conversion of A and B to C.

(b) The the $\Delta G°$ of C → D is about -30 kcal/mol (-120 kJ/mol), shown as b in the energy diagram.

(c) The $\Delta G^{\ddagger}$ of A → C is about 25 kcal/mol (100 kJ/mol), shown as c in the energy diagram.

(d) The $\Delta G^{\ddagger}$ of D → C is about 40 kcal/mol (160 kJ/mol), shown as d in the energy diagram.

(e) C is not stable enough to be isolated at room temperature because the energy well it is in has only a 10 kcal/mol (40 kJ/mol) barrier, which is easily traversed at 25°C.

2.2   According to the following energy diagram:

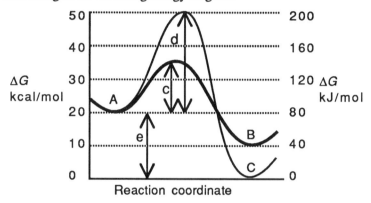

(a) The kinetic product from A is B because it is the product that would be formed by traversing the lowest energy barrier.

(b) The thermodynamic product from A is C because it is the most stable of the two products.

(c) The $\Delta G^{\ddagger}$ for A → B is about 15 kcal/mol (60 kJ/mol) and is shown as c in the energy diagram.

(d) The $\Delta G^{\ddagger}$ for A → C is about 30 kcal/mol (120 kJ/mol) and is shown as d in the energy diagram.

(e) The $\Delta G°$ for A → C is about -20 kcal/mol (-80 kJ/mol) and is shown as e in the energy diagram.

(f) At 0°C. the most likely product from A is B because the barrier to C would be too high for significant formation of C.

(g) The approximate **equilibrium** ratio of C/B at room temperature is determined by their ΔG difference of approximately -10 kcal/mol (-40 kJ/mol). To get the exponent of K_{eq} divide -10 kcal/mol by -1.37 (or -40 kJ/mol by -5.73). The K_{eq} is 10^{+7} = [C]/[B].

2.3   Thermodynamic stability, the $\Delta G°$ reason: A is the most stable of all possibilities; there is nowhere to go but up. Kinetic stability, the $\Delta G^{\ddagger}$ reason: B is **not** the most stable of all possibilities, but the barriers are too high to traverse with the energy present.

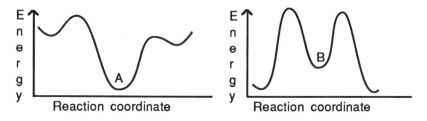

2.4 Explain why possibility B and not possibility A occurs even though they are technically the same process. Models do help.

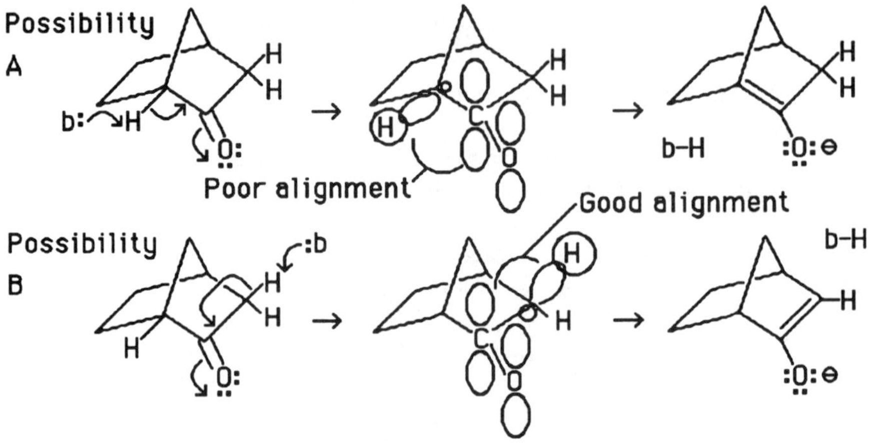

For A, the orbital alignment is very bad. The C–H bond and the pi bond are practically perpendicular. Since we are trying to form a very twisted pi bond, this process is expected to be very difficult. For B, the orbital alignment is good. The C–H bond and the carbon *p* orbital of the pi bond are almost in the same plane The new pi bond should form easily.

2.5 The law of microscopic reversibility states that the forward and reverse processes follow the same lowest energy path.

2.6 Use a Δ*H* calculation to determine which of the two possible products is the most stable.

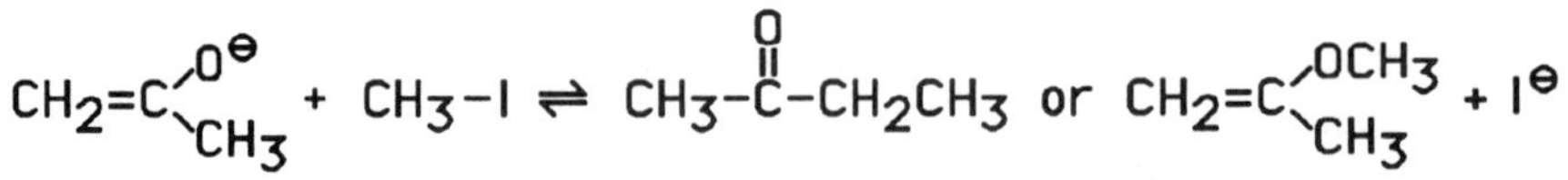

We must find out which of the two possible products has the stronger bonds. We need to look at only the bonds that differ. The ketone has a C=O and two C–C for a total of 177 + 83 + 83 = 343 kcal/mol (1435 kJ/mol). The enol ether has a C=C and two C–O for a total of 146 + 86 + 86 = 318 kcal/mol (1331 kJ/mol). The ketone has the stronger bonds and is therefore the thermodynamically preferred product.

2.7 The relationship between selectivity and reactivity is that more reactive species are less selective.

2.8 In each molecule the boldface carbon atom is softer because it bears a lesser partial positive charge. In the first, iodine is less electronegative than oxygen; in the second, the CH_2 is farther away from the electronegative oxygen

$I-\textbf{C}H_2CH_2CH_2CH_2-OSO_2Ph \qquad H_2\textbf{C}=CH-CHO$

2.9 A methyl on a cyclohexane ring prefers to be equatorial (*left structure*) because the methyl group is more hindered in the axial conformational isomer (*right structure*). The axial methyl tends to bump into the other axial hydrogens.

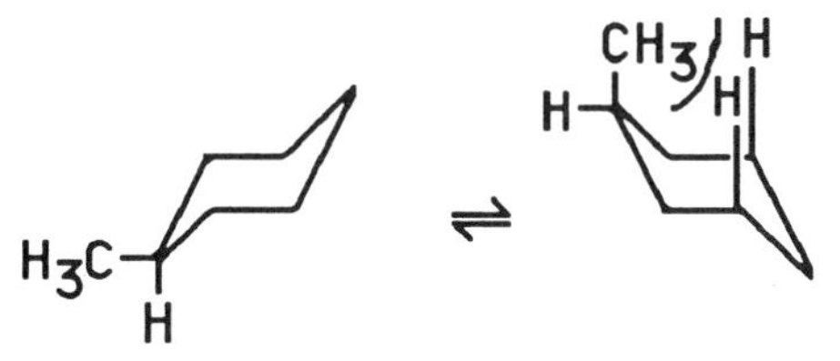

2.10 Using the bond strength table in the Appendix, calculate the ΔH of reaction for each of the following processes. ΔH = bonds broken - bonds made.

(a)

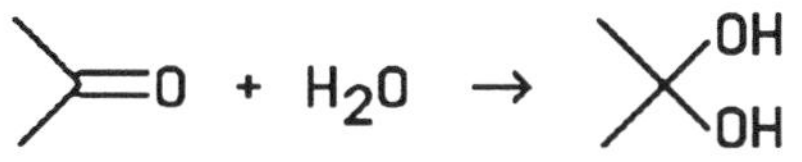

Broken (C=O) - made (two C−O) = (177) - (86 + 86) = +5 kcal/mol (21 kJ/mol).

(b)

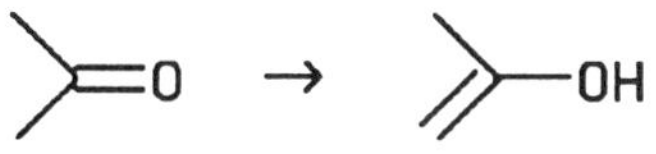

Broken (C=O, C−C, and C−H) - made (C−O , O−H, and C=C) = (177 + 83 +99) - (86 + 111 + 146) = 359 - 343 = +16 kcal/mol (67 kJ/mol).

2.11 At room temperature, the free energy difference for the 85:15 ratio between A $\rightleftharpoons$ B of +1 kcal/mol (+4 kJ/mol) can be read off Table 2.1 or calculated by $\Delta G° = -RT\ln K_{eq}$ (convert °C into °K).

2.12 For a reversible reaction with a $\Delta G°$ of -1 kcal/mol (-4.2 kJ/mol) run at -50°C, K_{eq} is calculated by $K_{eq.} = 10^{-\Delta G°/2.303RT}$ to be 9.5 This corresponds to an D:C ratio of 90.5:9.5 respectively. At 25°C the K_{eq} is 5.4, which corresponds to an D:C ratio of 85:15. At 100°C the K_{eq} is 3.8, which corresponds to an D:C ratio of 79:21.

2.13 If the concentration of one reactant of a bimolecular reaction is at 1.0 M and the other at 0.1 M, the reaction is pseudo-first order with $k' = (0.014)(1.0) = 0.014$. From Table 2.2 or from $\ln(c\%/c) = k't$, it takes 4.1 min at 25°C to be 97% complete if $\Delta G^{\ddagger}$ is 20 kcal/mol (84 kJ/mol).

2.14 If the relative rates of two related unimolecular reactions are 10,000 to 1 under identical conditions at 25°C (convert °C into K by adding 273), the difference in barrier heights is $k_2/k_1 = e^{-\Delta\Delta G^{\ddagger}/RT}$ or $-RT\ln(k_2/k_1) = \Delta\Delta G^{\ddagger} = 5.5$ kcal/mol (23 kJ/mol). Another way to get the answer quickly is that the difference is 10^{+4}; just multiply four times 1.37 kcal/mol per factor of magnitude (or 5.73 kJ/mol per factor of magnitude) to get the answer.

2.15 For the following allylic species, attack by an electrophile at carbon would occur perpendicular to the plane of the allylic system to overlap with the p orbital on carbon. Attack by an electrophile at oxygen would occur in the plane of the allylic system on the lone pairs not involved with the allyl resonance; the stabilization afforded by the allylic system is maintained throughout the reaction coordinate.

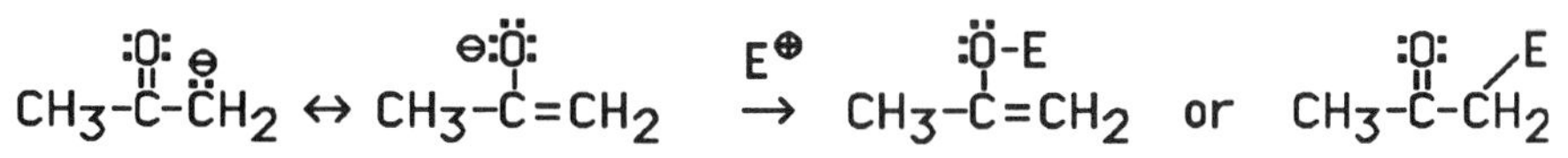

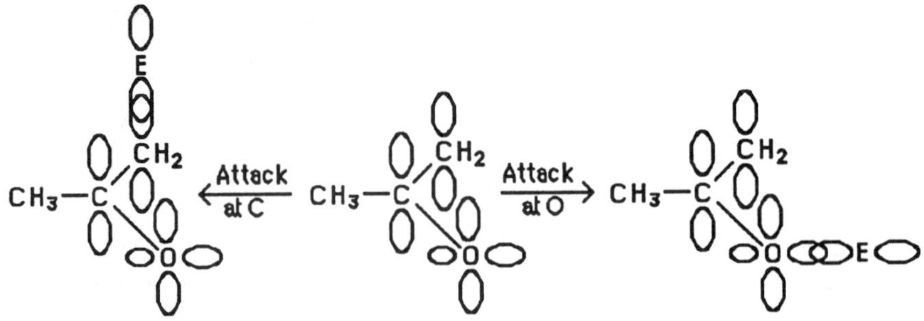

CHAPTER 3: ADDITIONAL EXERCISES

3.1 The numerical value of K_{eq} for the following reactions can be calculated by $\log K_{eq} = pK_{abH} - pK_{aHA}$.

PhONa + PhSH $\rightleftharpoons$ PhSNa + PhOH
$\log K_{eq} = 10 - 6.5 = +3.5$, thus $K_{eq} = 10^{+3.5}$

$\log K_{eq} = 35 - 19.2 = +15.8$, thus $K_{eq} = 10^{+15.8}$

$(CH_3)_3COK + HC\equiv CH \rightleftharpoons (CH_3)_3COH + HC\equiv CK$
$\log K_{eq} = 19 - 25 = -6$, thus $K_{eq} = 10^{-6}$

3.2 Trends: all species are ranked beginning with the numeral 1.
Rank the most stable carbanion by lowest pK_{abH}

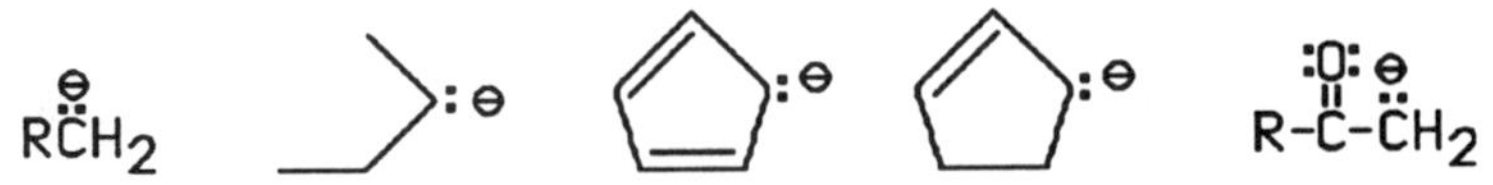

	$R\overset{\ominus}{C}H_2$				$R\text{-}\overset{O}{\overset{\|}{C}}\text{-}\overset{\ominus}{C}H_2$
pK_{abH}:	50	51	16	43	19.2
Rank:	4	5	1	3	2

The most stable carbocations are lone pair stabilized. Pi bond resonance stabilizes to a lesser degree. Alkyl group substitution stabilizes the least.

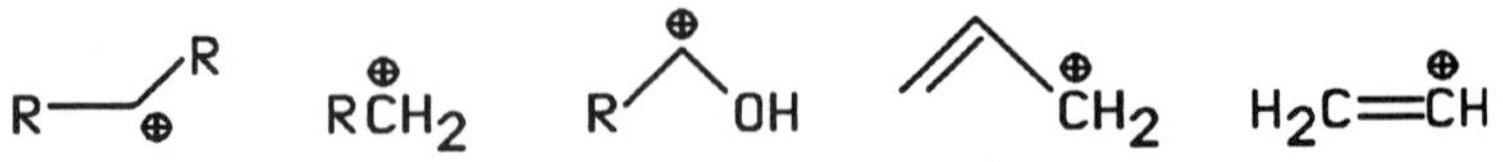

Rank:	3	4	1	2	5

The best electron-withdrawing group will make CH3—ewg acidic.

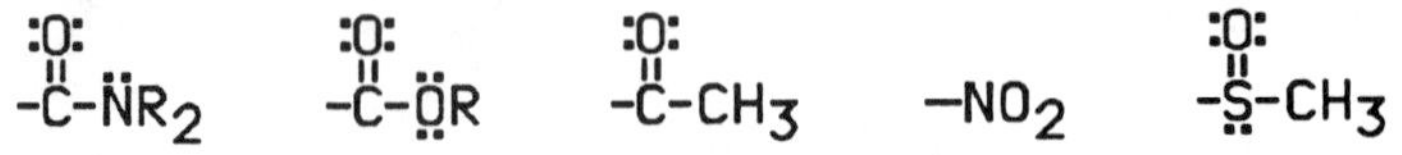

	$-\overset{O}{\overset{\|}{C}}\text{-}NR_2$	$-\overset{O}{\overset{\|}{C}}\text{-}\ddot{O}R$	$-\overset{O}{\overset{\|}{C}}\text{-}CH_3$	$-NO_2$	$-\overset{O}{\overset{\|}{S}}\text{-}CH_3$
CH3-ewg pK_a	30	25	19.2	10.2	35
Rank:	4	3	2	1	5

The best base has the highest pK_{abH}

	$^{\ominus}CN$	$PhNH_2$	$^{\ominus}NH_2$	$HO^{\ominus}$	$(CH_3)_3CO^{\ominus}$
pK_{abH}:	9.2	4.6	35	15.7	19
Rank:	5	4	1	3	2

The best electron donors are anions. Less electronegative atoms are good.

	$-O^{\ominus}$	$-OR$	$-R$	$-NR_2$	$-Cl$
Rank:	1	3	4	2	5

The best leaving group attached to R has the lowest pK_{aHL}.

	$R-OH_2^{\oplus}$	$R-Br$	$R-Cl$	$R-OCH_2CH_3$	$R-NR_2$
pK_{aHL}:	-1.7	-9	-7	16	36
Rank:	3	1	2	4	5

The best acid will have the lowest pK_a.

	$H_3O^{\oplus}$	HF	MeOH	HCN	NH_3
pK_a:	-1.7	3.2	15.5	9.2	35
Rank:	1	2	4	3	5

The most polar solvent can hydrogen bond and has little "grease."

	$CH_3CH_2CH_2CH_2CH_3$	EtOEt	H_2O	CH_3OH	$CH_3CH_2CH_2OH$
Rank:	5	4	1	2	3

3.3 Using the ΔpK_a rule, predict whether each of the following equilibria would favor reactants or products. The equilibria would favor the formation of the weaker base.

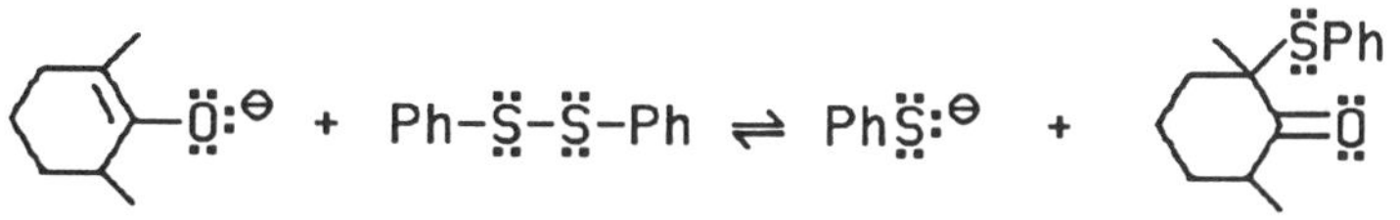

$pK_{abH}= 19.2$ 6.5   favors products.

$^{\ominus}OCH_3 + H_3CCOCl \rightleftharpoons Cl^{\ominus} + H_3CCOOCH_3$
$pK_{abH} = 15.5$ -7   favors products.

$HS^{\ominus} + H_3C\text{-}I \rightleftharpoons H_3C\text{-}SH + I^{\ominus}$
$pK_{abH} = 7.0$ -10   favors products.

3.4 The most acidic H is circled in each of the following compounds.

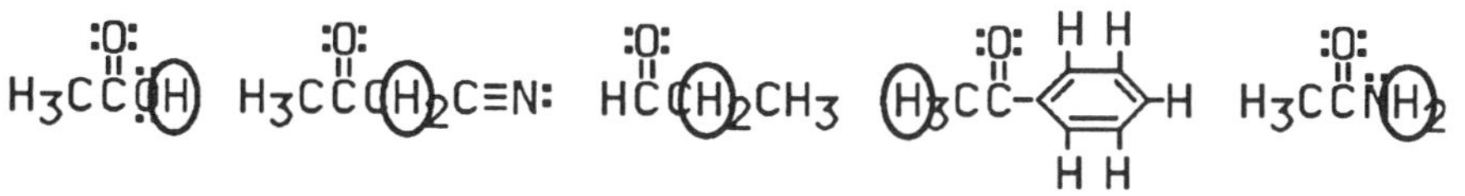

3.5 Use the ΔpK_a rule to determine which of the two following alternatives is the lower energy process. The reaction would favor the formation of the weaker base.

$CH_3COOEt + HO^{\ominus}$ reacting to form $H_2O + {}^{\ominus}CH_2COOEt$
        $pK_{abH} = 15.7$ going up 8.3 units to   24   is less favored than

$CH_3COOEt + HO^{\ominus}$ reacting to form $EtOH + CH_3COO^{\ominus}$
        $pK_{abH} = 15.7$ going down 10.9 units to   4.8.

3.6 The leaving group trend tells us that bromide pK_{aHL} -9 is a better leaving group than chloride pK_{aHL} -7. Displacing the better leaving group bromide is energetically preferred over displacing the poorer leaving group chloride. Of the two choices,

        $ClCH_2CH_2CH_2Br + CN^{\ominus}$ forms $ClCH_2CH_2CH_2CN$.

3.7 Use the carbocation stability trend to decide which of the following two alternatives is the lower energy process.

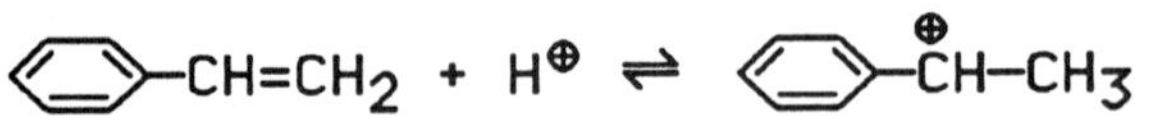

is much better than

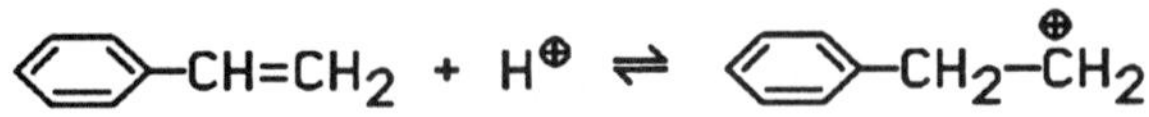

because the secondary benzylic carbocation is more stable than the primary.

3.8 Use your knowledge of anion stability to decide which of the following two alternatives is the lower energy process. The two choices are the extended enolate (vinylogous enolate) and the enolate. The extended enolate (*top*) spreads the charge over two carbons and an oxygen with three resonance forms. The enolate (*bottom*) spreads the charge over one carbon and an oxygen with two resonance forms. Therefore the extended enolate is favored.

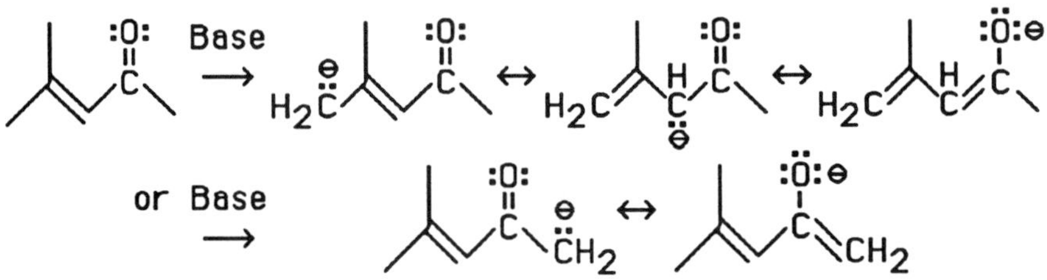

3.9 Use your knowledge of proton transfer equilibria to decide which of the following two alternatives is the lower energy process. Proton transfer equilibria favor the formation of the weaker base. The carboxylate anion at pK_{abH} 2.5 is definitely the weaker base compared to the enolate at a pK_{abH} of about 19.

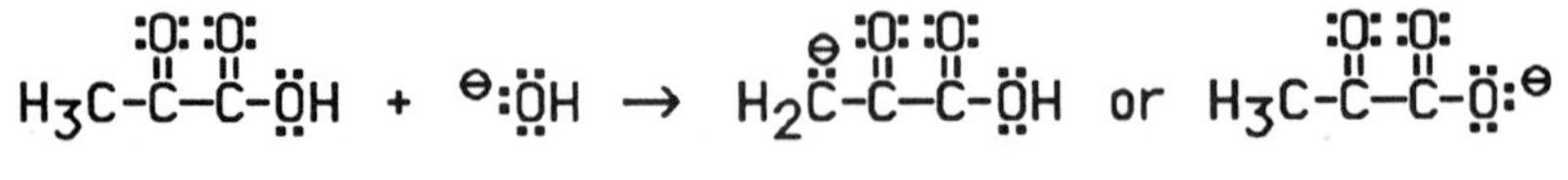

3.10 The first of the following two alternatives is the lower energy process. The reaction favors the formation of the weaker base. The first reaction forms an anion of pK_{abH} 10.7 and is favored over the second reaction, which forms an anion of pK_{abH} 19.2.

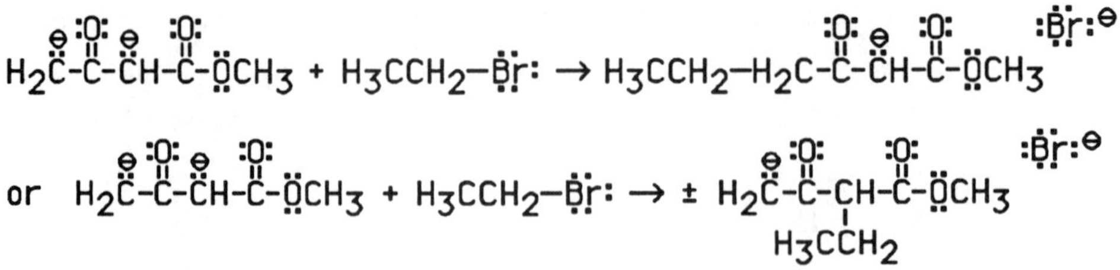

3.11 The second of the following two alternatives is the lower energy process. The first alternative is the formation of a more basic anion of pK_{abH} of about 10 to 13 from a nucleophile with a pK_{abH} of 4.8. The second alternative forms an anion of pK_{abH} -9 and is favored by the ΔpK_a rule.

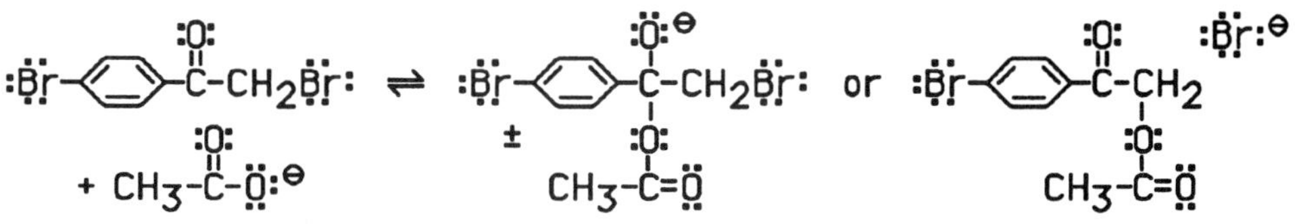

3.12 Ascorbic acid, vitamin C, has one very acidic H ($pK_{a1} = 4.1$), and all the others are not very acidic ($pK_{a2} = 11.8$). The circled acidic H is as acidic as a carboxylic acid H ($pK_a = 4.8$) because it is a vinylogous carboxylic acid H.

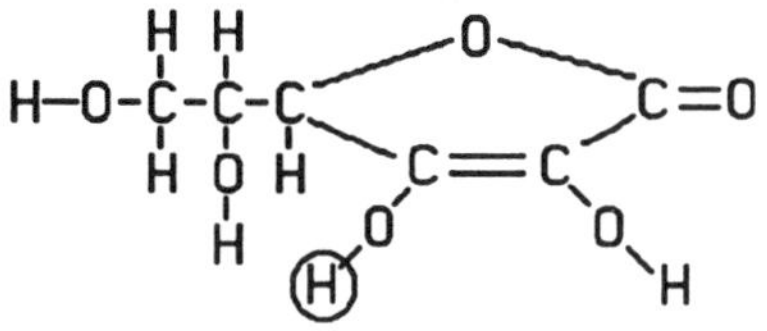

3.13 The most acidic is the protonated nitrile; least acidic is the protonated amine.

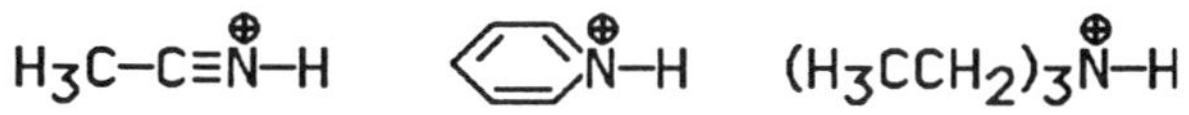

$pK_a = -10$ $pK_a = 5.2$ $pK_a = 10.7$

The reason for this acidity trend is that lone pairs in orbitals having more %s character are more stable and therefore less basic. The conjugate acid of a weak base is a strong acid.

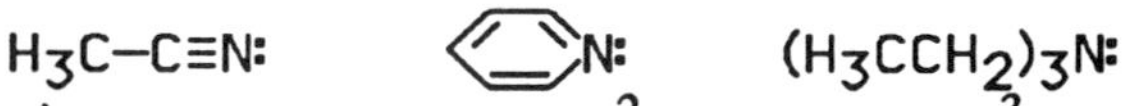

Lone pair hybridization:	sp	sp^2	sp^3
%s character:	50	33	25

3.14 Account for the different rates in the ionization of the leaving group X.

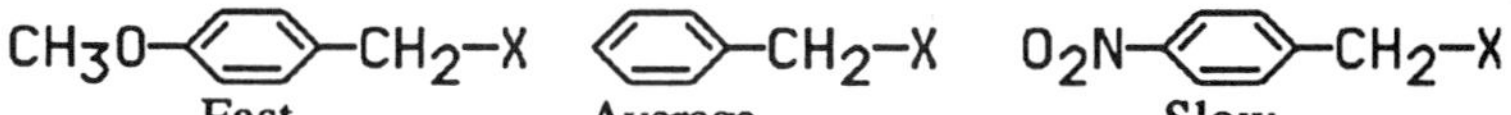

Fast   Average   Slow

The ionization of the leaving group as $X^\ominus$ leaves behind carbocations that differ in stability. The most stable carbocation will be formed the easiest. The methoxy lone pair donor stabilizes the carbocation, whereas the nitro electron-withdrawing group destabilizes the carbocation.

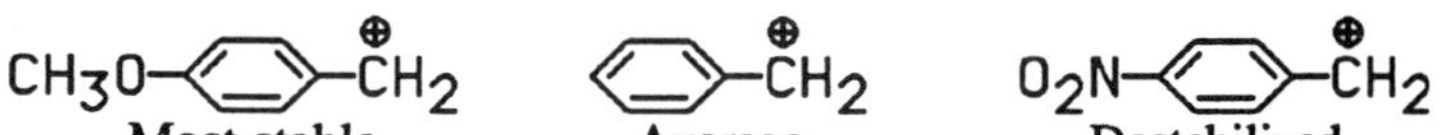

Most stable   Average   Destabilized

3.15 Explain why the loss of the leaving group for

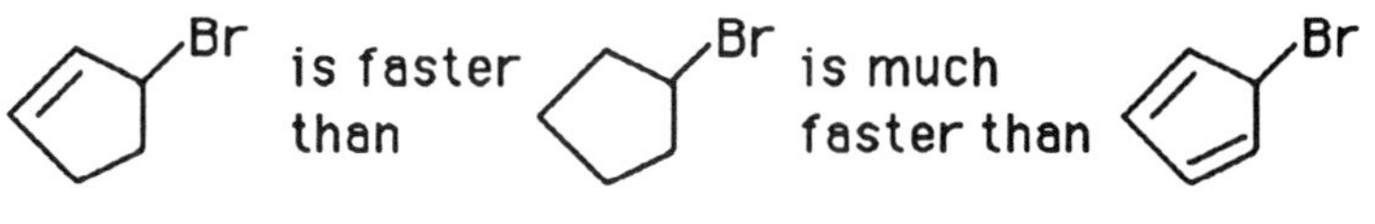

is faster than   is much faster than

Again we look to carbocation stabilities to explain the trend.

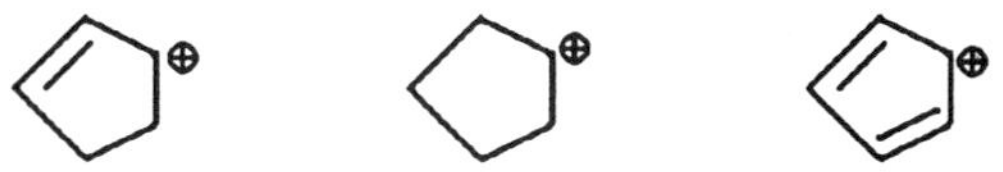

The carbocation on the left is more stable than the one in the center because it is resonance stabilized. The carbocation on the right is very destabilized because it is antiaromatic.

CHAPTER 4: ADDITIONAL EXERCISES

4.1 Give the generic class of each of the following electron sources.

LiAlH₄	MeLi	NH₃	MeOCH=CH₂	Et₂NLi
MH₄	RM	Z (weak b)	Z—C=C	Z (good b)
Mg	MeCOO⊖	EtMgBr	Me₂NCH=CH₂	H₂O
M	Z	RM	Z—C=C	Z (weak b)
NaH	CH₃C≡CH	Me₂C=CH₂	t-BuO⊖	PhCH₃
MH	C≡C	C=C	Z (good b)	ArH

4.2 Give the generic class of each of the following electron sources.

PhNH₂	⊖C≡N	Cl⊖	⊖CH₂COOCH₃	CH₃OH
Z (or b)	Z (good Nu)	Z	Z—C=C	Z (weak b)
NaBH₄	⊖O—CH=CH₂	HO⊖	Zn	NaOEt
MH₄	Z—C=C	Z (Nu or b)	M	Z (Nu or b)
⊖SC≡N	H₂C=C=CH₂	PhSH	H₂C=CHCH=CH₂	F⊖
Z	C=C=C allene	Z (good Nu)	C=C—C=C diene	Z (weak b)

4.3 Draw Lewis structures for the species in problem 4.1 and indicate on the structures which electron pair(s) would be the start of the electron flow in a reaction. If there is more than one lone pair on an atom, any will do. For PhCH₃ any ring pi bond will do.

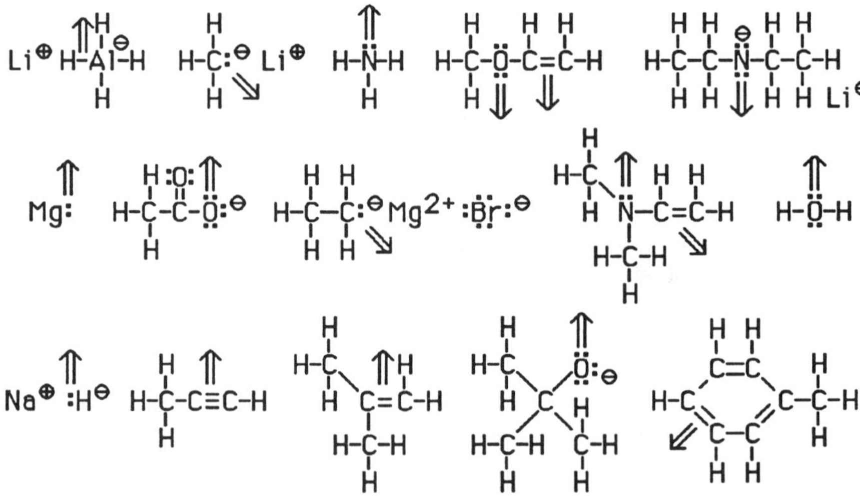

4.4 Draw Lewis structures for the species in problem 4.2 and indicate on the structures which electron pair(s) would be the start of the electron flow in a reaction. If there is more than one lone pair on an atom, any will do.

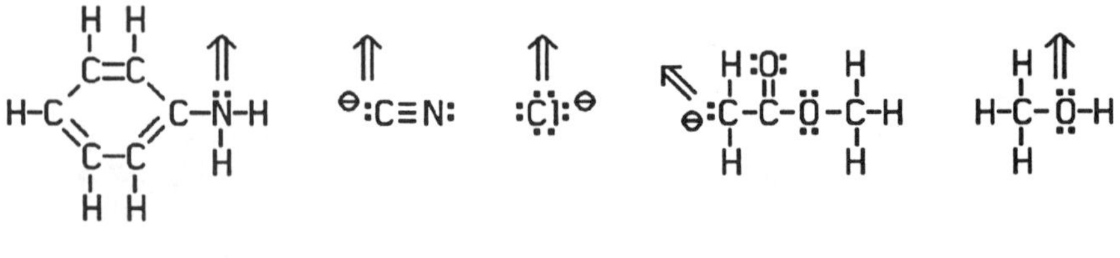

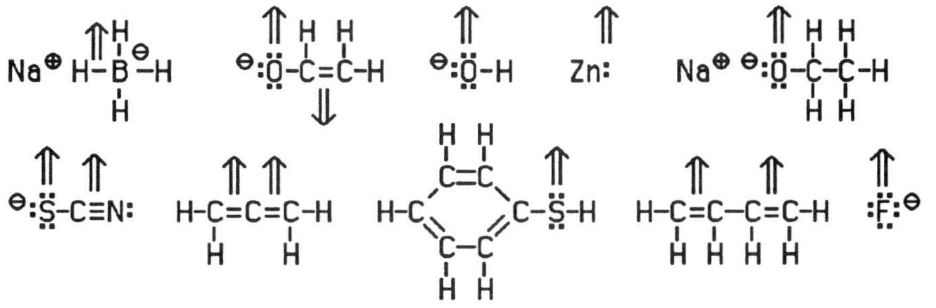

$Na^{\oplus}$ H–$\overset{H}{\underset{H}{B}}$–H $^{\ominus}$:Ö–C=C–H $^{\ominus}$:Ö–H   Zn: $Na^{\oplus}$ $^{\ominus}$:Ö–$\overset{H}{\underset{H}{C}}$–$\overset{H}{\underset{H}{C}}$–H

$^{\ominus}$:S̈–C≡N: H–C=C=C–H   H–C$\underset{C=C}{\overset{C=C}{\big\langle}}$C–S̈–H   H–C=C–C=C–H   :F̈:$^{\ominus}$

4.5 The one atom on each molecule that is best for electrophilic attack is circled; it gives the most stable cation upon attack.

⟨benzene⟩–CH=CH–CH=(C)H₂   (H₃C)₂N̈CCH₃   H₂C=CHCCH₃   (H₃C)₂C=(C)H₂

4.6 All species are ranked beginning with the numeral 1.

The best nucleophile to react with CH_3I in CH_3OH is ranked by softness first, then the most basic within identical attacking atoms.

	$PhSe^{\ominus}$	$PhO^{\ominus}$	$PhS^{\ominus}$	$CH_3COO^{\ominus}$	CH_3OH
Rank:	1	3	2	4	5

The most reactive organometallic is the most ionic. Mercury is the most electronegative (more than cadmium) so dimethylmercury is least reactive

	MeK	MeMgBr	Me_2Hg	Me_2Cd	MeLi
Rank:	1	3	5	4	2

The best allylic electron source bears the best donor.

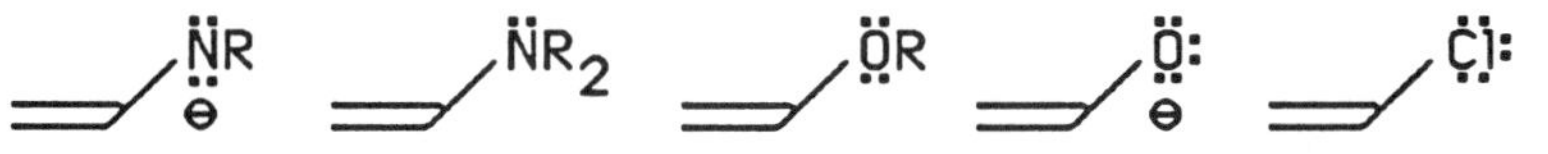

	N̈R$^{\ominus}$	N̈R₂	ÖR	Ö:$^{\ominus}$	C̈l:
Rank:	1	3	4	2	5

The most electron-rich aromatic ring is the most reactive toward electrophilic attack. Therefore donors activate, and electron-withdrawing groups deactivate.

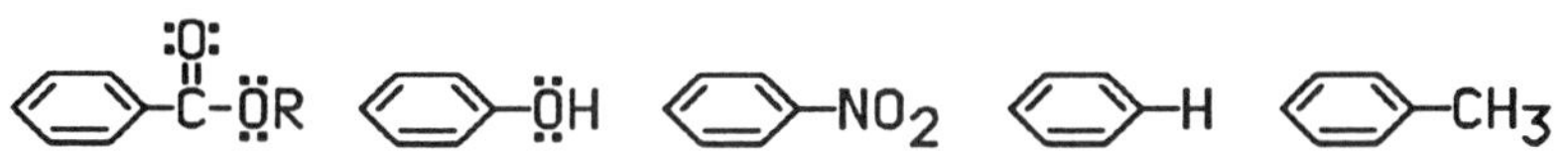

	⟨Ph⟩–C(=Ö)–ÖR	⟨Ph⟩–ÖH	⟨Ph⟩–NO_2	⟨Ph⟩–H	⟨Ph⟩–CH_3
Rank:	4	1	5	3	2

4.7 Markovnikov's rule: The formation of the more stable carbocation will determine the site of electrophilic attack.

4.8 The electron-donor groups on PhOH and PhCl direct electrophilic attack ortho and para. The electron-withdrawing groups on $PhCF_3$, $PhNO_2$, and PhCOOH direct electrophilic attack meta.

4.9 In a protic solvent, the softer $CH_3S^{\ominus}$ is a better nucleophile than $CH_3O^{\ominus}$ for reaction with a soft electrophile like an alkyl halide. Tight solvation of methoxide decreases its ability to act as a nucleophile. Therefore $CH_3CH_2CH_2Br + CH_3S^{\ominus}$ displacing bromide to form $CH_3CH_2CH_2SCH_3$ is the faster process in methanol.

4.10 In a protic solvent, the more basic, less electronegative NH_3 is a better nucleophile than H_2O for reaction with a harder electrophile like an acyl halide. Therefore $CH_3COCl + NH_3$ displacing chloride to form CH_3CONH_2 is the faster process.

CHAPTER 5: ADDITIONAL EXERCISES

5.1 Give the generic class of each of the following sinks.

PhNCO	BF$_3$	HOBr	i-PrBr	Me$_2$CO
Y=C=Z	$\geqslant$A	Y–L	$\geqslant$C–L	C=Y

Ac$_2$O	CH$_3$Br	AlCl$_3$	CH$_2$=CHCOCH$_3$	CO$_2$
Y=C–L	$\geqslant$C–L	$\geqslant$A	C=C–ewg	Y=C=Z

CH$_3$C≡N	(CH$_3$)$_2$N=CH$_2^{\oplus}$	AgNO$_3$	CH$_3$COCl	(CH$_3$)$_3$C$^{\oplus}$
C≡Y	C=Y or $\geqslant$C$^{\oplus}$	M$^{\oplus}$	Y=C–L	$\geqslant$C$^{\oplus}$

5.2 Give the generic class of each of the following sinks.

H$_2$SO$_4$	PhSO$_3$CH$_3$	CH$_3$SO$_2$Cl	SOCl$_2$	CH$_2$=CHCN
HA	$\geqslant$C–L	Y–L	Y–L	C=C–ewg

COCl$_2$	HOAc	H$_2$O$_2$	CH$_3$COOEt	CH$_3$OSO$_3$CH$_3$
Y=C–L	HA	Y–L	Y=C–L	$\geqslant$C–L

PhCHO	EtOCOCl	Cl$_2$	Cl-C≡N	H$_2$C=C=O
C=Y	Y=C–L	Y–L	L–C≡Y	C=C=Z

5.3 Draw Lewis structures for all the species in problem 5.1 and designate which atom(s) on each would be attacked by the electron source in a reaction.

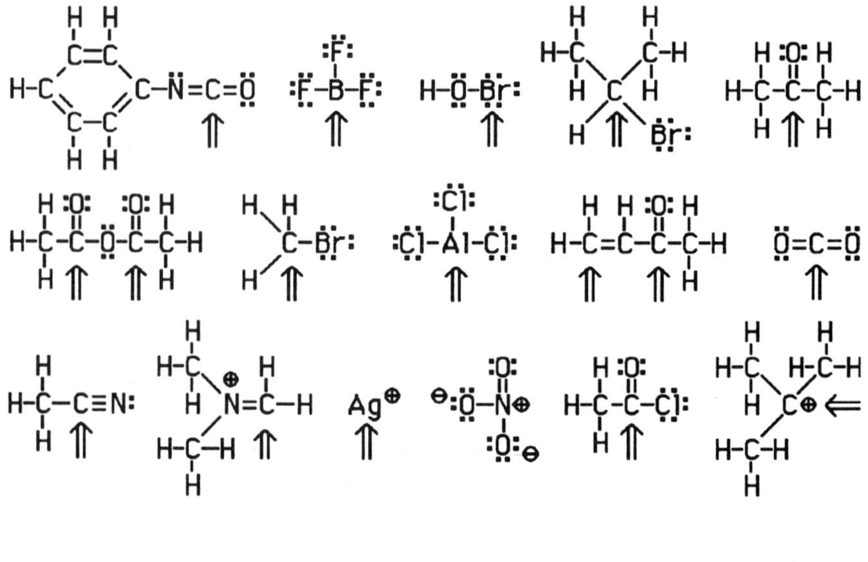

5.4 Draw Lewis structures for all the species in problem 5.2 and designate which atom(s) on each would be attacked by the electron source in a reaction.

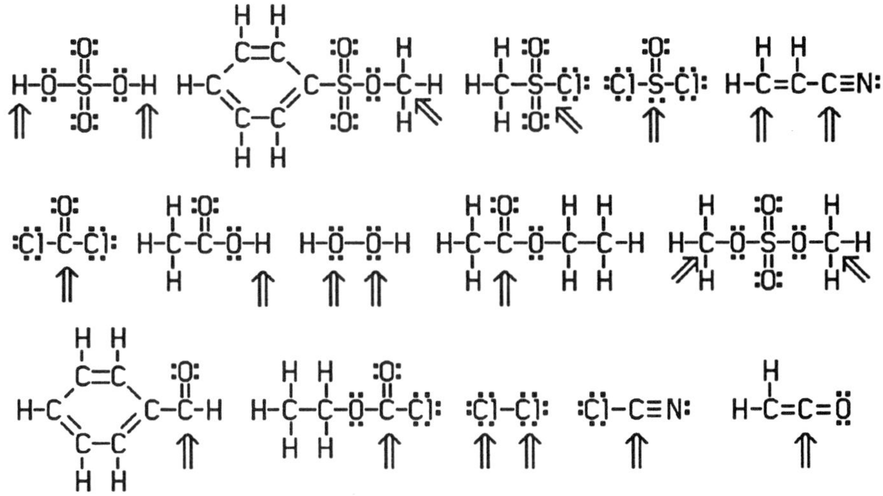

5.5 Circle the best atom on each molecule for nucleophilic attack (L = leaving group).

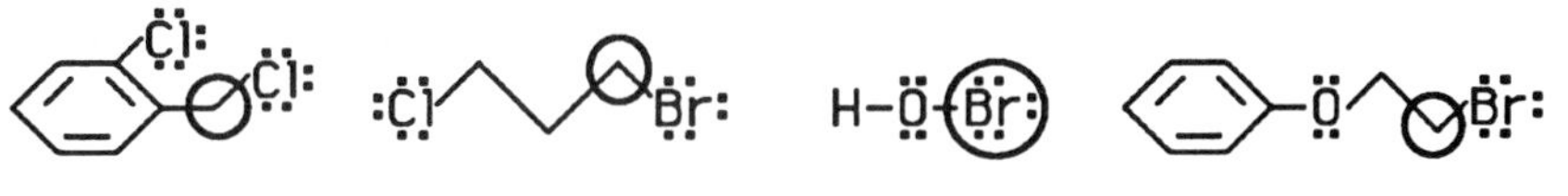

| Benzylic L better than aromatic L | Bromide better L than chloride | Electronegativity of O makes Br $\partial+$ end | Bromide better L than phenolate |

5.6 All species are ranked beginning with the numeral 1.

The most reactive toward nucleophilic attack bears the best leaving group.

	Me—OCH$_3$	Me—Br	Me—NMe$_2$	Me—I	Me—Cl
Rank:	4	2	5	1	3

The most reactive conjugate acceptor bears the best electron-withdrawing group.

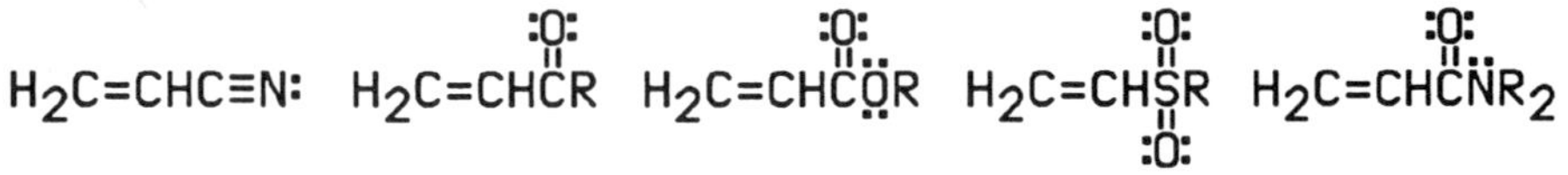

Rank:	4	1	3	2	5

The most reactive toward nucleophilic attack bears the poorest donor. The ketone cannot be ranked properly using the leaving group trend since it has no leaving group.

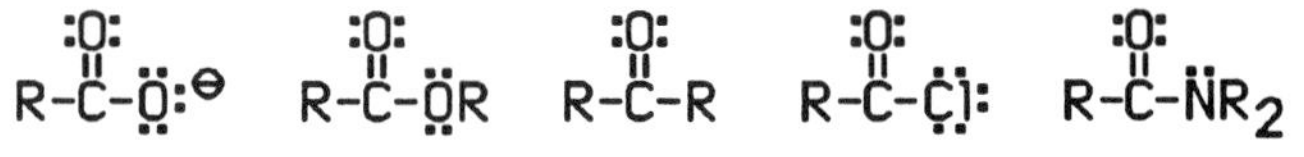

Rank:	5	3	2	1	4

5.7 Ketals, R$_2$C(OR)$_2$, fail to react (act as electron sinks) in basic media because RO$^\ominus$ (pK_{aHL} 15.5 to 19) is not good enough to serve as a leaving group. They react rapidly in acidic media because the oxygen can protonate, giving ROH (pK_{aHL} about -2.4), which is a good leaving group.

5.8   Sodium borohydride reduces ketones quickly but reacts very slowly to reduce esters. Since an aldehyde is more reactive than a ketone, it will react very quickly with sodium borohydride.

5.9   Circle the best atom on each molecule for nucleophilic attack (see Section 5.6).

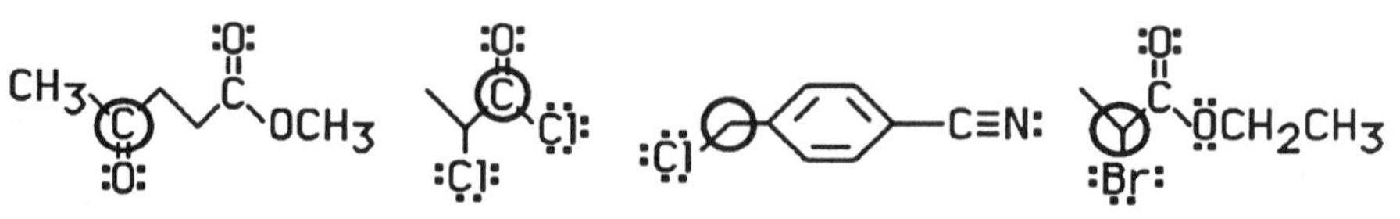

5.10   Thioesters are much more reactive than regular esters because sulfur has poor 2p-3p overlap with the carbonyl. There is little delocalization of the leaving group lone pairs into the carbonyl (similar to the acyl chloride). The SR group is both a poorer donor and a better leaving group (RS$^\ominus$ at p$K_{a\text{HL}}$ 10.6) than the OR group (RO$^\ominus$ at p$K_{a\text{HL}}$ 16).

CHAPTER: 6 ADDITIONAL EXERCISES

6.1   Go back to the worked examples in Chapter 1, Section 1.3 and place the name of the path over each transformation arrow.

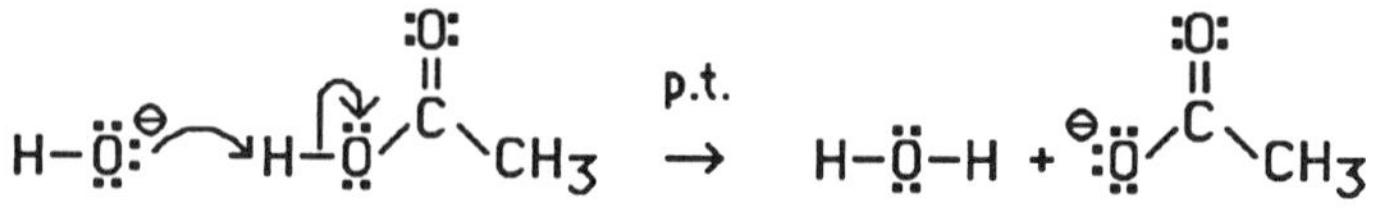

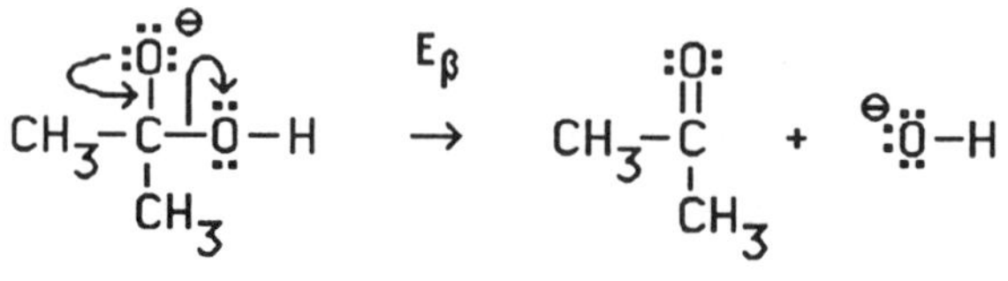

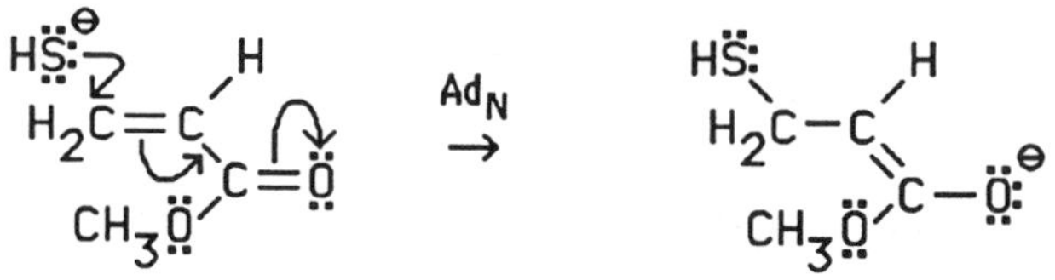

(an extra arrow is used to go to the major resonance form with the charge on O).

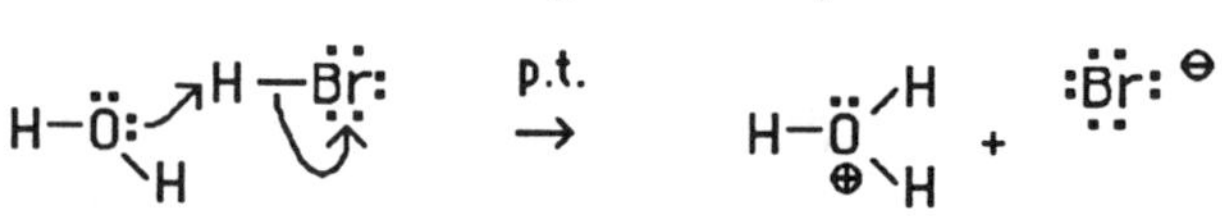

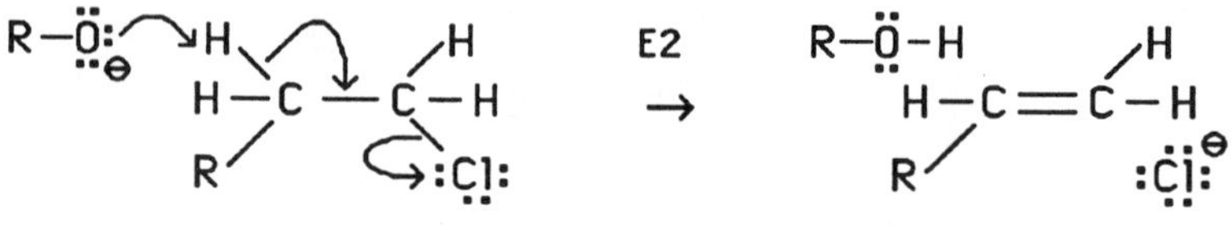

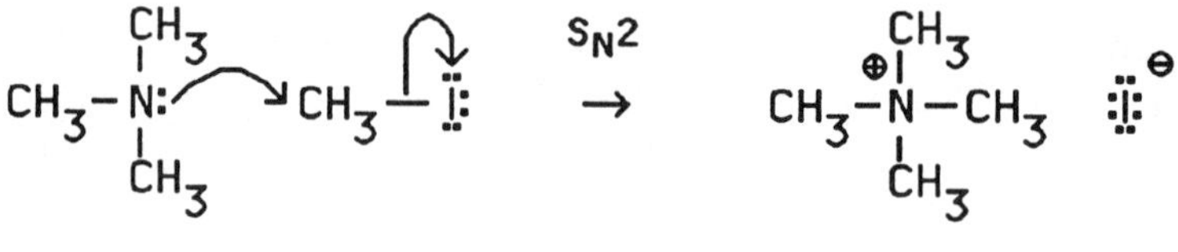

6.2 Go back to Chapter 1, exercises 5 and 6 and place the name of the path over each transformation arrow.

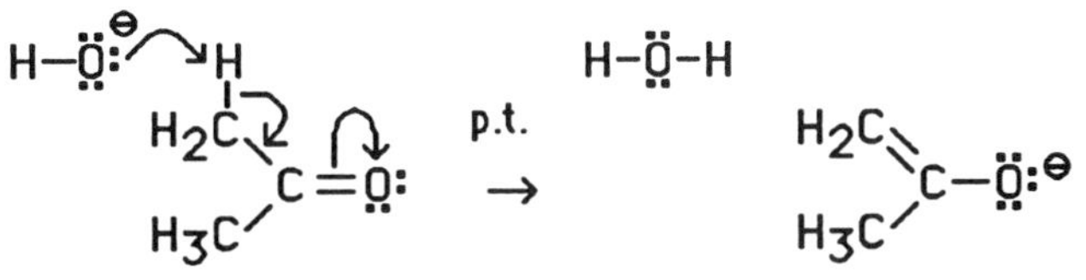

(An extra arrow is used to go to the major resonance form with the charge on O.)

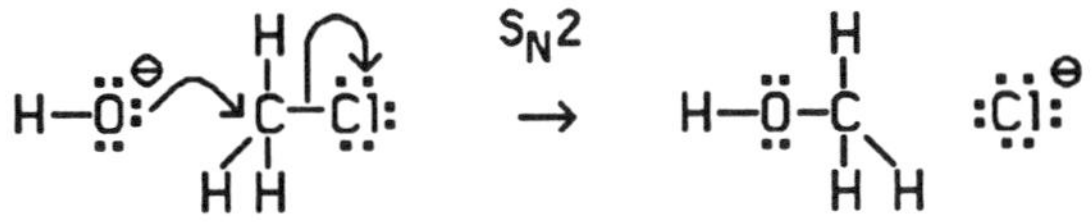

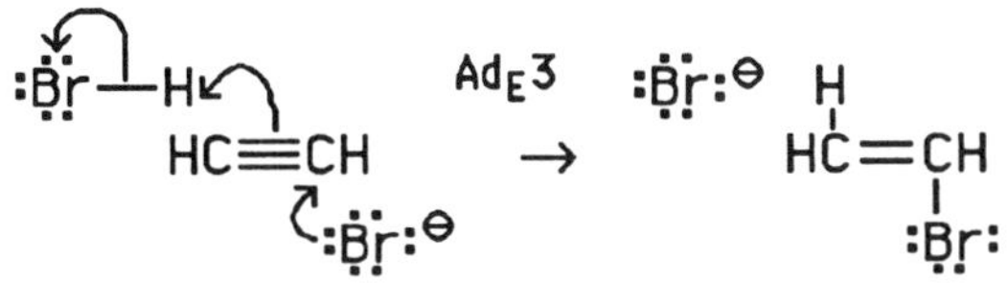

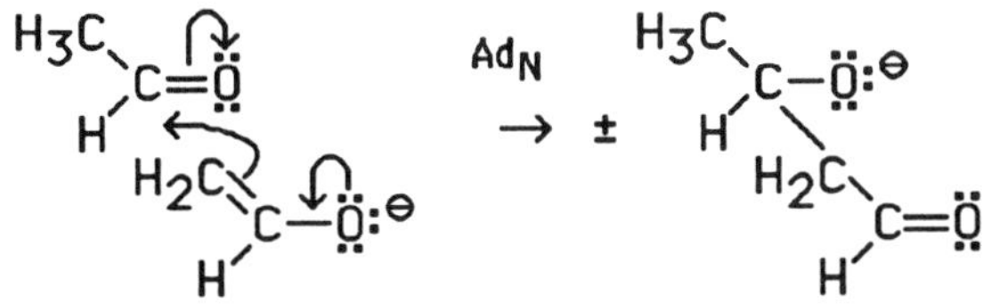

(An extra arrow is needed because the allylic source attacks on carbon, and the resonance form with the charge on oxygen was shown.)

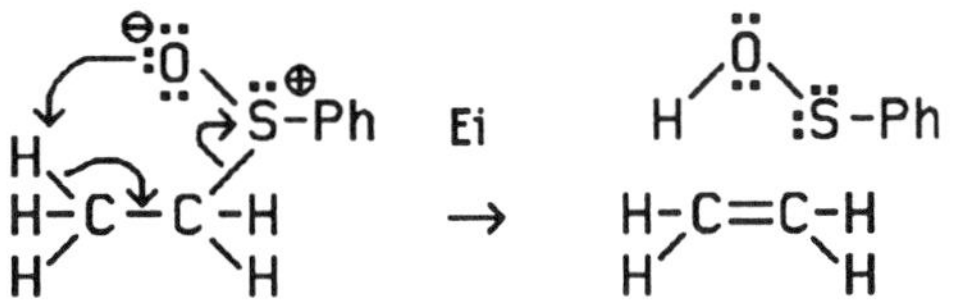

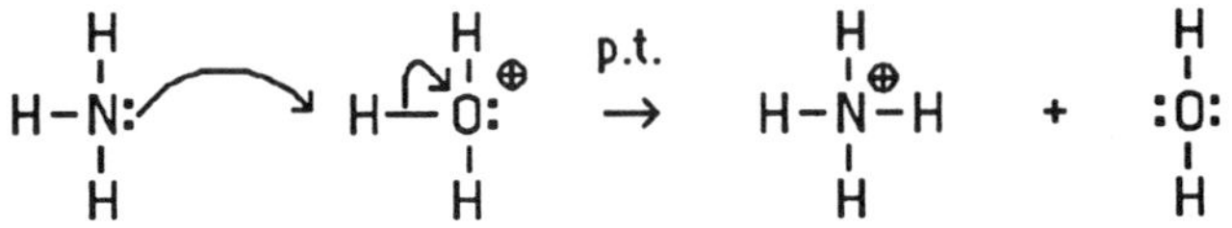

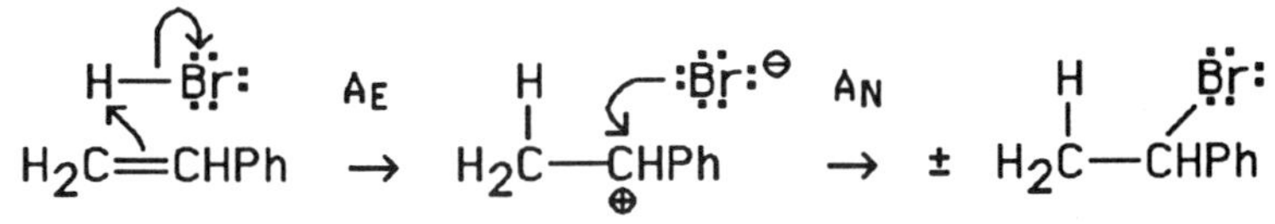

(This is the Ad$_E$2 path combination.)

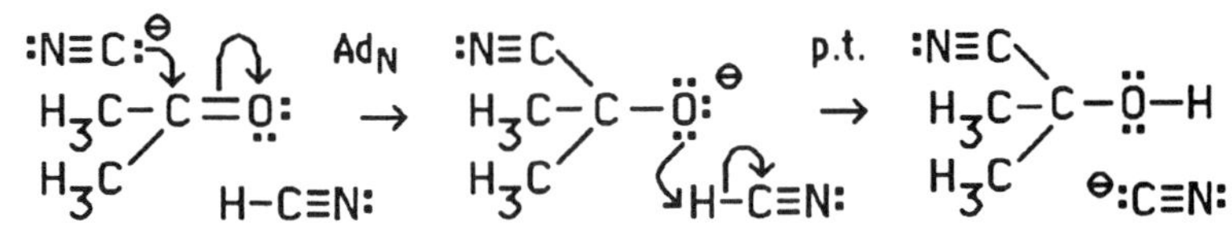

(This is the Ad$_N$2 path combination.)

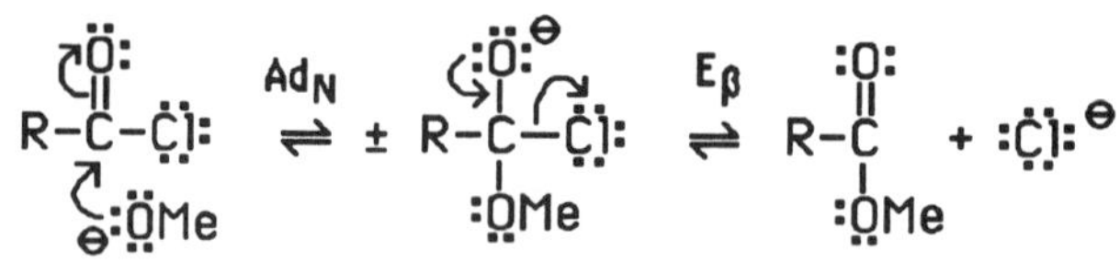

(This is the addition–elimination path combination.)

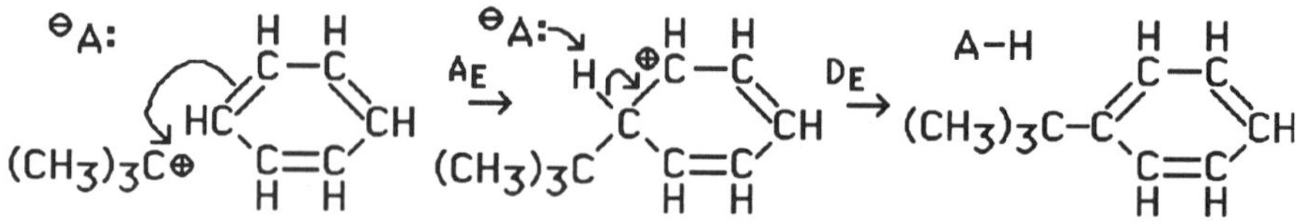

(This is the electrophilic aromatic substitution path combination.)

6.3 and **6.4** For a general example of the process indicated refer to Chapter 6 or the Appendix pathway summary.

6.5 The dozen most common pathways by acronym are p.t., D_N, A_N, A_E, D_E, 1,2R, S_N2, E2, Ad_E3, Ad_N, E_β, and 6e.

6.6 The ten common path combinations are the S_N1, Ad_N2, E1, Ad_E2, E1cB, addition–elimination, aromatic electrophilic substitution, hetero Ad_E2, tautomerization, and lone-pair-assisted E1.

6.7 Four acidic media additions of an electrophile to a multiple bond are the Ad_E2, hetero Ad_E2, A_E, and Ad_E3.

6.8 The three basic media nucleophilic additions to multiple bonds are Ad_N, AdgB, and Ad_N2.

6.9 Two nucleophilic substitutions of a group bound to a sp^3 hybridized carbon are the S_N1 and the S_N2.

6.10 The two substitutions of a group bound to a sp^2 hybrid carbon are addition–elimination and electrophilic aromatic substitution.

6.11 The five eliminations of an H and a leaving group from a neutral molecule are E1, E2, E1cB, lone-pair-assisted E1, and Ei.

6.12 The overlap limitation of the E2 process is that the C–H bond and the C–L bond must be close to coplanar. The overlap limitation for the rearrangement of a carbocation is that the migrating group must be able to overlap both the orbital it is migrating from and the orbital it is migrating to.

6.13 A general acid catalyst is most effective if its pK_a is close to the pH of the medium. If its pK_a is much higher than the pH, it is too weak an acid to provide assistance in the reaction. If its pK_a is too low, it is a stronger acid, ionizes easily, and there is little unionized form present to be an acid catalyst.

6.14 The electron sink is a leaving group bound to an sp^3 carbon, and the electron source is a heteroatom lone pair. The reaction is a substitution. With the sink as a guide, the possible paths are the S_N1 and the S_N2. The path restriction for the S_N1 requires a stable cation; a primary cation is not stable. This reaction is the S_N2.

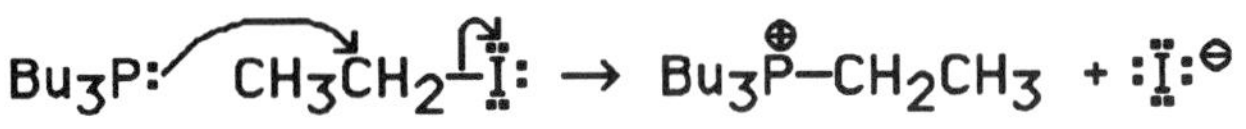

$$Bu_3P: \quad CH_3CH_2-I: \rightarrow Bu_3\overset{\oplus}{P}-CH_2CH_3 + :I:^\ominus$$

6.15 The product of an S_N2 is

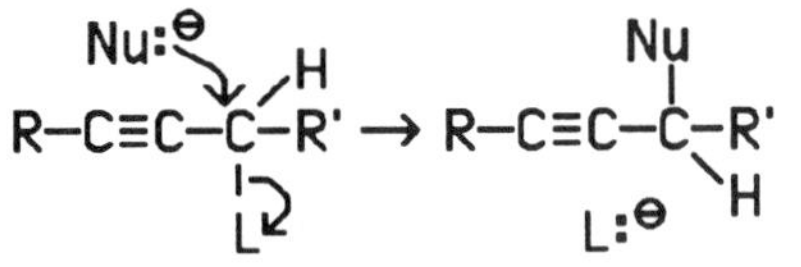

The product of an S_N2' is

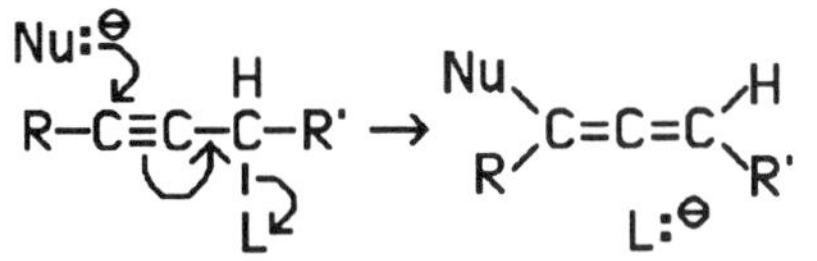

CHAPTER 7: ADDITIONAL EXERCISES

7.1   All species are ranked beginning with the numeral 1.

The most reactive for S_N1 would give the most stable carbocation upon loss of the leaving group. The tertiary carbocation is more stable than the unsubstituted allylic.

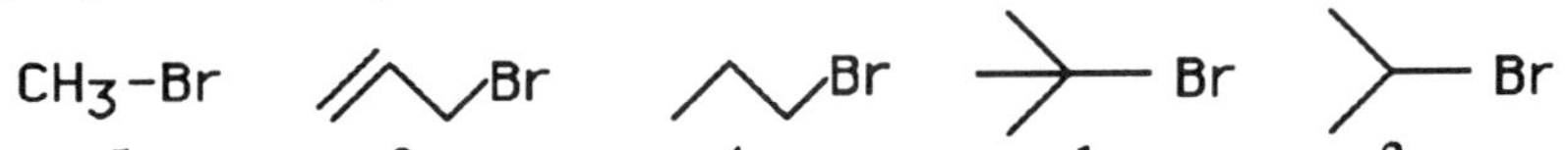

The most reactive for S_N2 are allylic and benzylic. The next most reactive is methyl followed by sites of increasing steric hindrance.

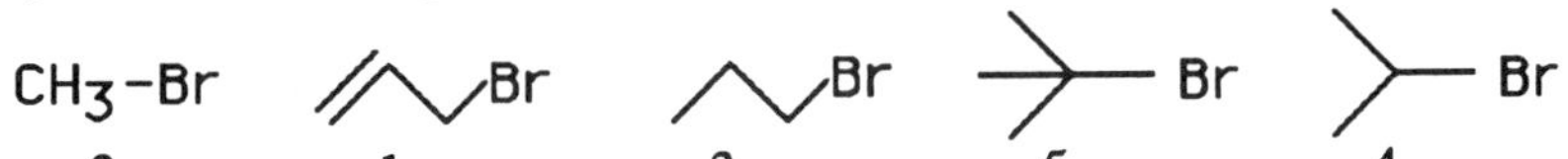

7.2   The E1 process favors the formation of the more substituted alkene. The E2 regiochemistry is controlled by minimization of steric interactions in the transition state. The E1cb regiochemistry is controlled by the loss of the most acidic proton.

7.3   Using the energy surfaces in Section 7.2f, predict how increasing the solvent polarity should affect the competition between S_N1 and S_N2. The corner of the surface that has the most charges and is therefore lowered the most by increased solvation is the carbocation corner. Increasing the solvent polarity should tilt the surface toward more S_N1.

7.4 Draw the substitution product and decide whether the reaction is S_N1 or S_N2.

(a) S_N1; good cation, weak Nu:

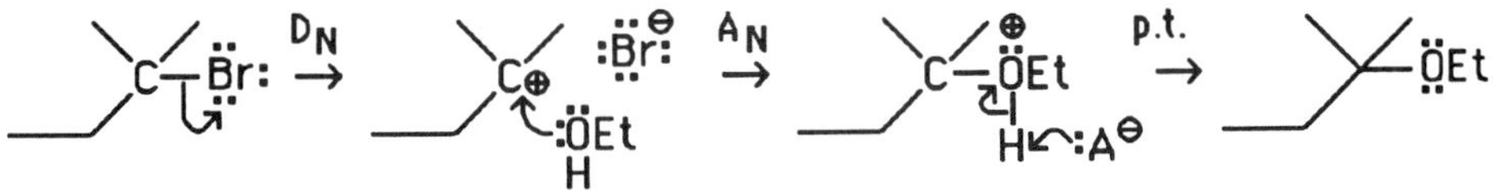

(b) S_N2; poor cation, good Nu:

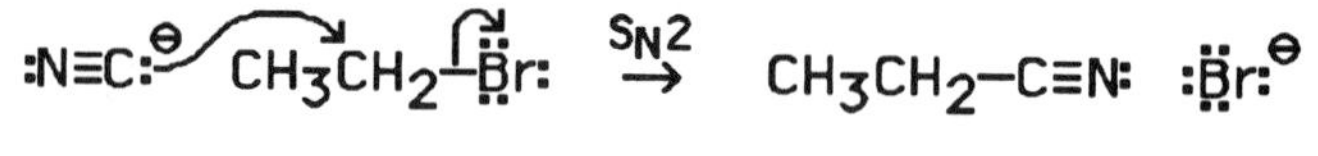

$$:N\equiv C:^{\ominus} \curvearrowright CH_3CH_2-\ddot{B}r: \xrightarrow{S_N2} CH_3CH_2-C\equiv N: \quad :\ddot{B}r:^{\ominus}$$

7.5 Draw the elimination product and decide whether the reaction is E1, E2, or E1cB.
(a) E2; poor cation, strong base:

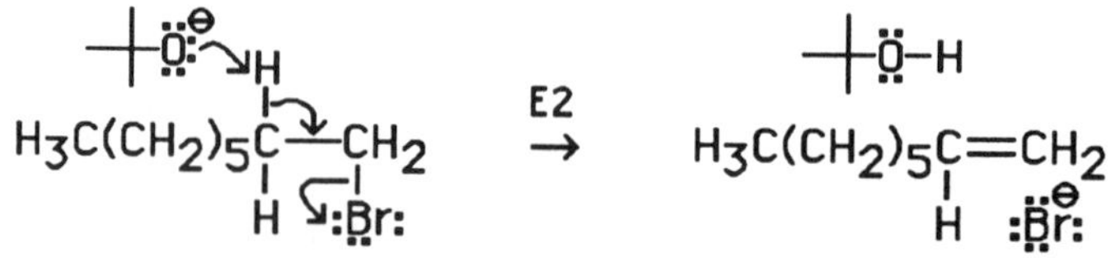

(b) E1; good cation, weak base, and after protonation of the alcohol by the acid, a good L:

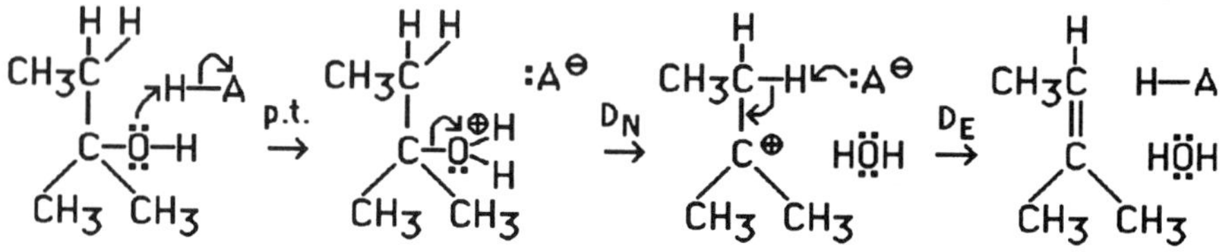

(c) E1cB; ewg makes H acidic, the L is poor:

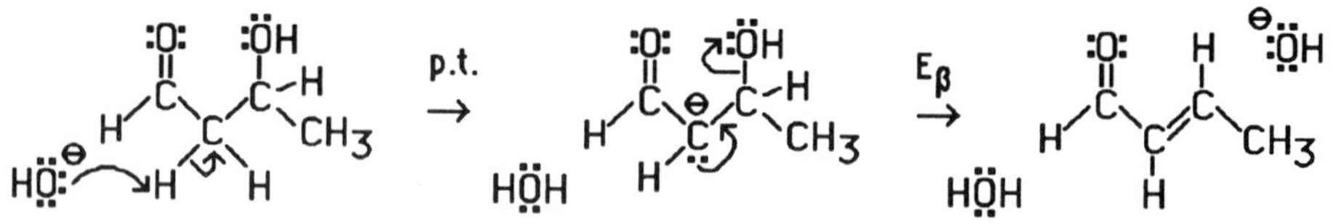

7.6 Classify each reactant into its generic class of source and sink, then use the appropriate electron flow pathway to rationalize the product of these **one-step** reactions. These mechanisms are simple in that there is only one source and one sink present in the reaction mixture. Find the source and the sink, then combine them with the appropriate pathways discussed in this chapter.

Problem   7.6a
Source class   Z as b   *Sink class* $\geqslant$C–L
Reaction type Elimination   *Pathway*   E2
Comments   Hindered site and strong base.

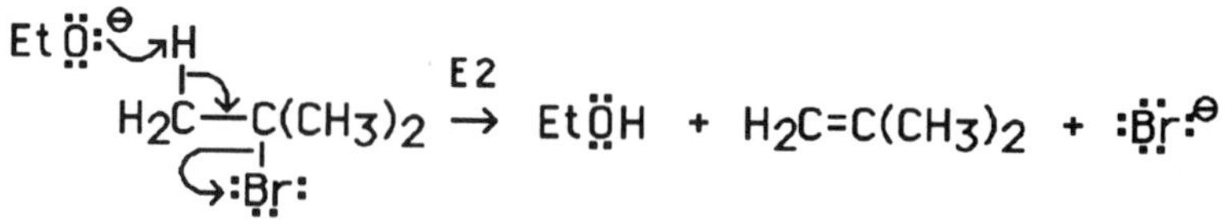

Problem   7.6b
Source class   Z as Nu   *Sink class* $\geqslant$C–L
Reaction type Substitution   *Pathway* S_N2
Comments   Primary site and unhindered nucleophile.

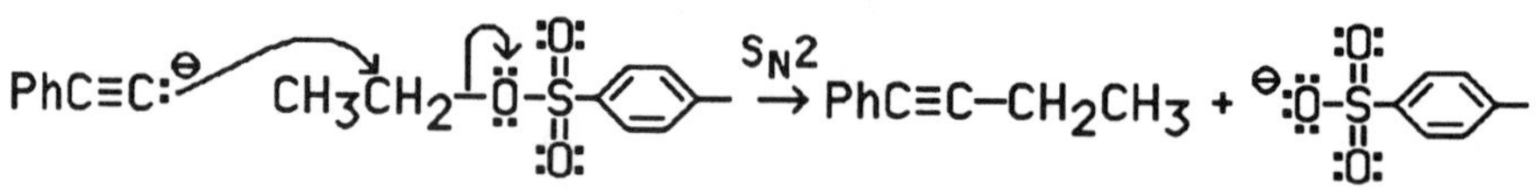

Problem   7.6c
Source class   C=C   *Sink class*   HA
Reaction type Addition   *Pathway*   Ad$_E$3
Comments   *Anti* addition is characteristic of the Ad$_E$3.

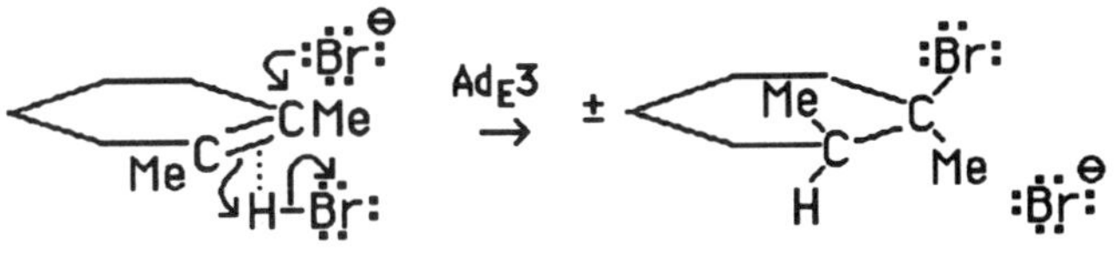

7.7 Classify each reactant into its generic class of source and sink, then use the appropriate electron flow pathway to rationalize the product of these **two-step** reactions. There is only one source and one sink present in the reaction mixture. Find the source and the sink, then combine them with the appropriate pathways. Draw the product of the first step, then examine the product of the overall reaction to determine which bonds must be made or broken in the second step.

Problem   7.7a
Source class   Z as Nu   *Sink class* $\geqslant$C–L (first step)
Reaction type Substitution   *Pathway*   D$_N$ then A$_N$, the S$_N$1
Comments   Diphenylmethyl cation is reasonably stable.

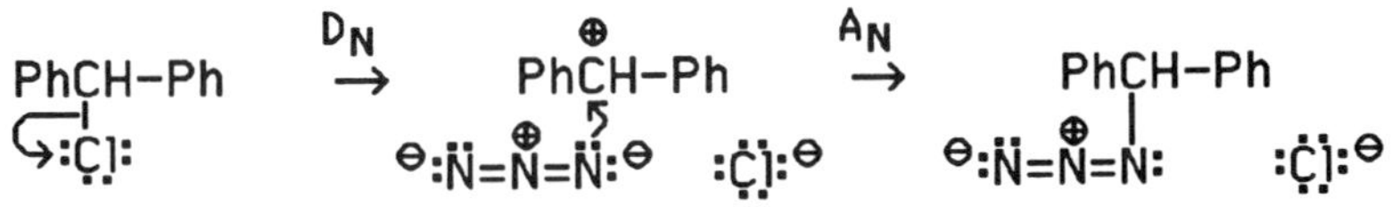

Problem   7.7b
Source class   Nu–L   *Sink class*   C=Y (first step)
Reaction type Addition   *Pathway*   Ad$_N$2 addition, then S$_N$2
Comments   Intermediate alkoxy anion is stable.

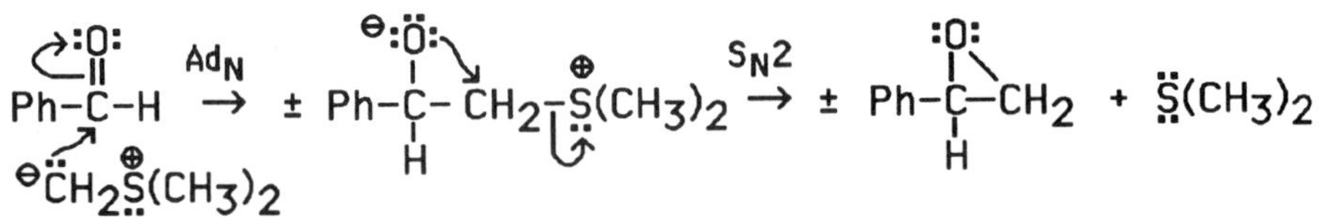

Problem   7.7c
Source class   C=C–Z   *Sink class*   C=C–ewg (first step)
Reaction type Addition   *Pathway*   Ad$_N$, then p.t.; the Ad$_N$2 addition
Comments   Proton transfer goes via solvent.

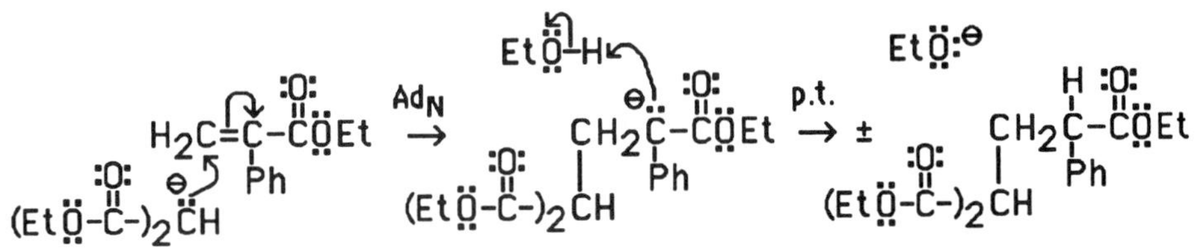

7.8 Classify each reactant into its generic class of source and sink, then use the appropriate electron flow pathway to rationalize the product of these **two-step** reactions.

The first step is a favorable proton transfer that generates the electron source for the reaction; there is only one electron sink.

Problem   **7.8a** (first step is proton transfer)
Source class   Vinylogous Z–C=C   *Sink class* $\geqslant$C–L
Reaction type   Substitution   *Pathway*   p.t., deprotonation, then path S_N2
Comments   Proton transfer $K_{eq} = 10^{+3}$.

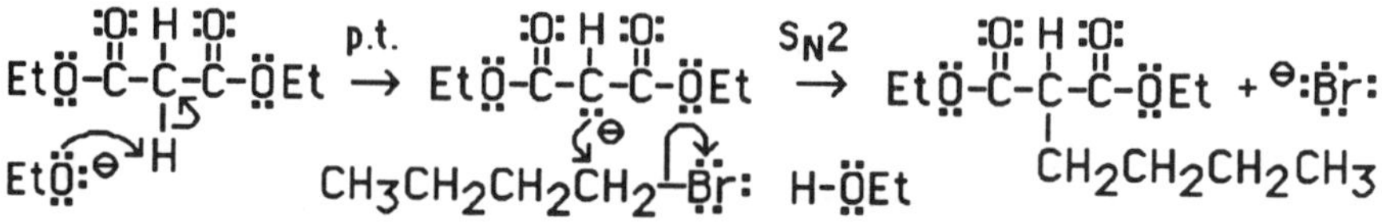

Problem   **7.8b** (first step is proton transfer)
Source class   Z as b   *Sink class* $\geqslant$C–L
Reaction type   Elimination   *Pathway*   p.t., then path E_β; E1cb elimination
Comments   Approximate proton transfer $K_{eq} = 10^{+2}$ to 10^{+3}.

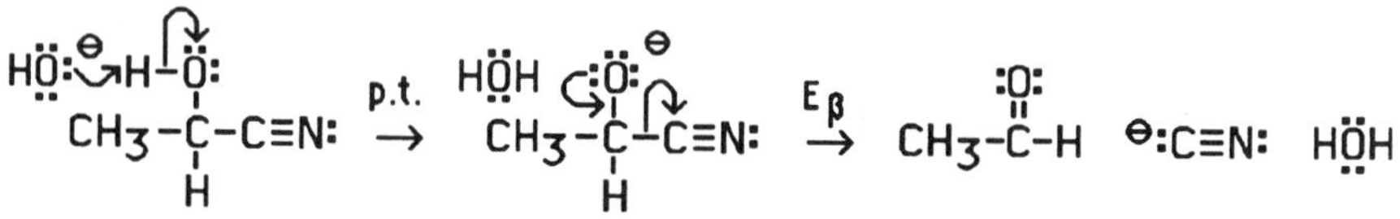

Problem   **7.8c** (first step is proton transfer)
Source class   Z as b   *Sink class* $\geqslant$C–L
Reaction type   Substitution   *Pathway*   p.t., then S_N2 substitution
Comments   Proton transfer creates a good leaving group, $K_{eq} = 10^{+6.6}$.

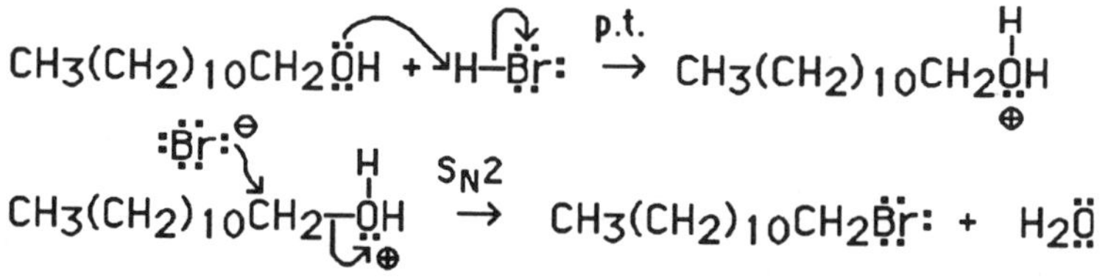

7.9   Classify each reactant into its generic source and sink, then use the appropriate electron flow pathway to predict the product of these **one-step** reactions.

Problem   **7.9a**
Source class   R–M   *Sink class*   Y=C=Y
Reaction type   Addition   *Pathway* Ad_N, addition to a polarized multiple bond
Comments pK_{abH} drops from 43 to 4.8, very favorable.

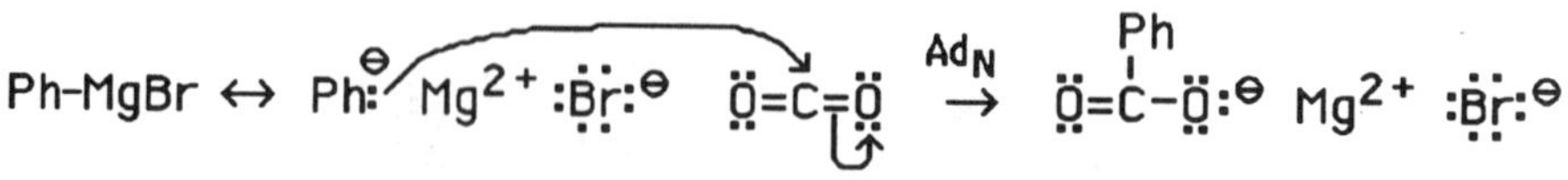

Problem   **7.9b**
Source class   Z as Nu   *Sink class* $\geqslant$C–L
Reaction type   Substitution   *Pathway* S_N2 substitution
Comments pK_{abH} drops from +2.7 to -10, very favorable.

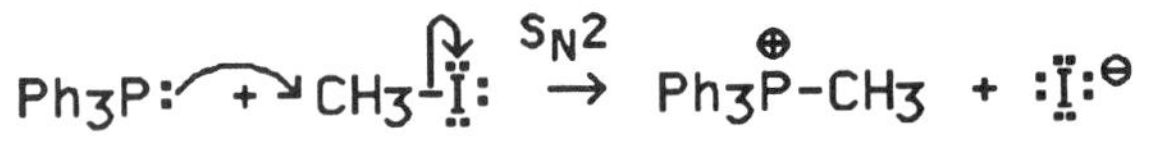

$$Ph_3P: \quad + \quad CH_3-I: \quad \xrightarrow{S_N2} \quad Ph_3\overset{\oplus}{P}-CH_3 \quad + \quad :\overset{\ominus}{I}:$$

Problem   7.9c
Source class   C=C   *Sink class*   ⩾A
Reaction type Addition   *Pathway*   4e, four-center, four-electron
Comments   Boron pi-complex forms, then H is donated to the largest ∂+.

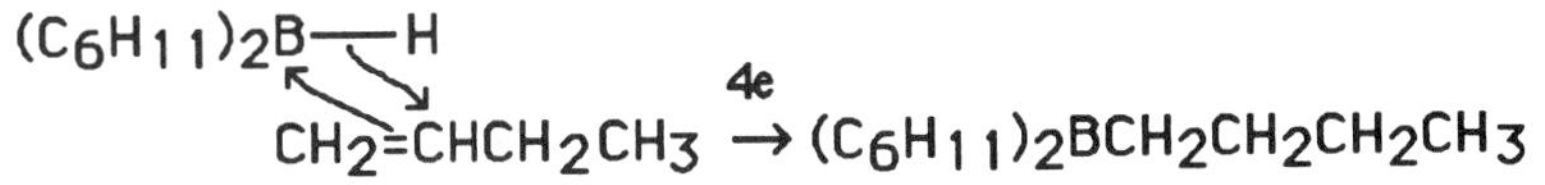

$$(C_6H_{11})_2B-H$$
$$CH_2=CHCH_2CH_3 \xrightarrow{4e} (C_6H_{11})_2BCH_2CH_2CH_2CH_3$$

7.10 Classify each reactant into its generic class of source and sink, then use the appropriate electron flow pathway to predict the product of these **two-step** reactions. Use a common path combination.

Problem   7.10a
Source class   C=C (first step only)   *Sink class*   HA (first step only)
Reaction type Addition   *Pathway* A_E, then A_N; the Ad_E2 addition
Comments   Secondary benzylic cation is moderately stable.

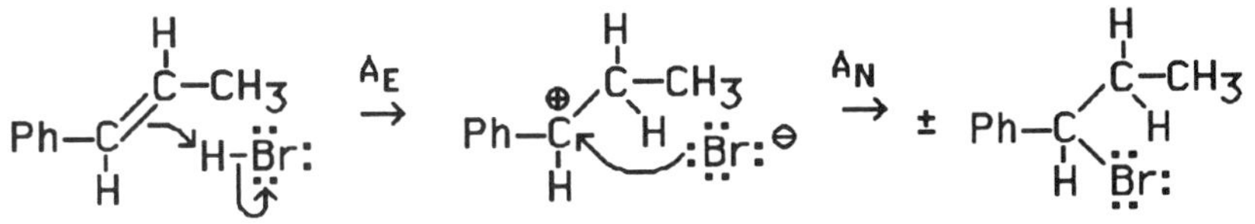

$$Ph-C(H)=C(CH_3)(H) \quad H-\overset{..}{Br}: \quad \xrightarrow{A_E} \quad Ph-\overset{\oplus}{C}(H)(H)...C(H)(CH_3) \quad :\overset{..}{Br}:^{\ominus} \quad \xrightarrow{A_N} \quad \pm \quad Ph-C(H)(H)...C(H)(CH_3)\overset{..}{Br}:$$

Problem   7.10b
Source class   Z as Nu (first step)   *Sink class*   Y=C–L (first step)
Reaction type Substitution   *Pathway* Ad_N, then E_β; addition-elimination
Comments   Reaction forms the weaker base.

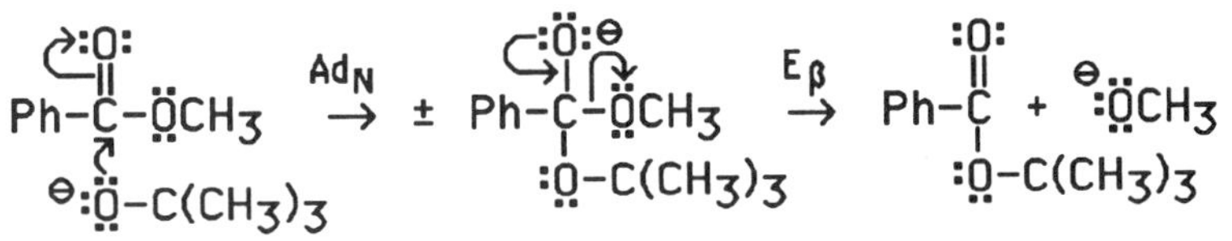

$$Ph-\overset{\overset{\displaystyle :O:}{\|}}{C}-\overset{..}{O}CH_3 \quad \overset{\ominus}{:}\overset{..}{O}-C(CH_3)_3 \quad \xrightarrow{Ad_N} \quad \pm \quad Ph-\overset{\overset{\displaystyle :O:^{\ominus}}{|}}{C}-\overset{..}{O}CH_3 \quad :O-C(CH_3)_3 \quad \xrightarrow{E_\beta} \quad Ph-\overset{\overset{\displaystyle :O:}{\|}}{C} \quad + \quad \overset{\ominus}{:}\overset{..}{O}CH_3 \quad :O-C(CH_3)_3$$

Problem   7.10c
Source class   ArH (first step)   *Sink class*   ⩾$C^{\oplus}$ (first step)
Reaction type Substitution   *Pathway* $A_E + D_E$, electrophilic aromatic substitution
Comments   Second step restores aromatic stabilization. The proton is removed by the solvent.

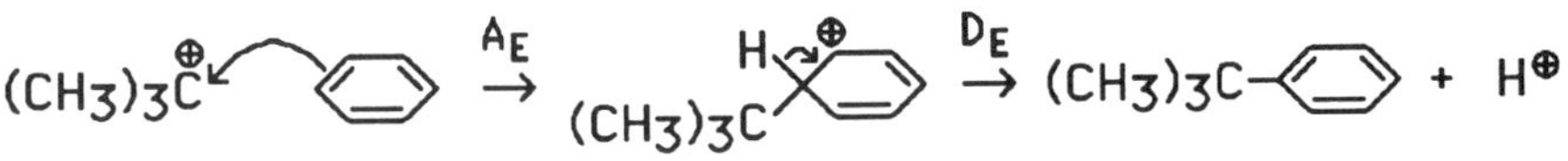

$$(CH_3)_3\overset{\oplus}{C} \quad \bigcirc \quad \xrightarrow{A_E} \quad \overset{H}{\underset{(CH_3)_3C}{\overset{\oplus}{\bigcirc}}} \quad \xrightarrow{D_E} \quad (CH_3)_3C-\bigcirc \quad + \quad H^{\oplus}$$

CHAPTER 8: ADDITIONAL EXERCISES

8.1   For each of the following reactions decide whether substitution or elimination predominates. First identify the site hindrance, the quality of the nucleophile, and the strength of the base; then use the substitution/elimination matrices to determine the preferred product.

	Substrate	Reagent	Base	Nu	Main product
(a)	$CH_3CH_2CH_2CH_2OH$, primary	HBr	Weak	Good	Substitution
(b)	$CH_3CH_2CH_2CH_2OTs$, primary	$t\text{-BuO}^{\ominus}$	Strong	Poor	Substitution
(c)	$CH_3CH_2CH_2CH_2Br$, primary	$t\text{-BuO}^{\ominus}$	Strong	Poor	Elimination
(d)	$(CH_3)_2CHCH_2Br$, hindered primary	$I^{\ominus}$	Weak	Good	Substitution
(e)	$(CH_3)_2CHBr$, secondary	EtOH	Weak	Poor	Substitution
(f)	$(CH_3)_2CHBr$, secondary, 55°C	$EtO^{\ominus}$	Strong	Good	Elimination
(g)	$(CH_3)_2CHCH_2CH_2Cl$, primary	$[(CH_3)_2CH]_2N^{\ominus}$	Strong	Poor	Elimination
(h)	$(CH_3)_2CHBr$, secondary	$CH_3COO^{\ominus}$	Weak	Good	Substitution
(i)	$(CH_3)_2CHBr$, secondary	$EtS^{\ominus}$	Weak	Good	Substitution
(j)	$(CH_3)_2CHCHBrCH_3$, hindered secondary, 55°C	$CH_3COO^{\ominus}$	Weak	Good	Elimination
(k)	$CH_3CH_2CBr(CH_3)_2$, tertiary	CH_3OH	Weak	poor	Substitution
(l)	$(CH_3CH_2)_3CCl$, tertiary	$CH_3O^{\ominus}$	Strong	Good	Elimination
(m)	$(CH_3)_3CBr$, tertiary	$CH_3COO^{\ominus}$	Weak	Good	Elimination

8.2 For each of the following reactions a decision discussed in this chapter is required. Choose which product would be preferred. All of the reactions occur in basic media.

Problem   8.2a
Source class   R—M   *Sink class*   Y=C—L
Decision type   Multiple addition

$$EtÖ-\overset{:O:}{\underset{}{\overset{\|}{C}}}-ÖEt \quad \xrightarrow{EtMgBr} \quad Et-\overset{:O:}{\underset{}{\overset{\|}{C}}}-ÖEt \quad \xrightarrow{EtMgBr} \quad Et-\overset{:O:}{\underset{}{\overset{\|}{C}}}-Et \quad \xrightarrow{EtMgBr} \quad Et-\overset{\ominus:Ö:}{\underset{Et}{\overset{|}{C}}}-Et$$

Rationale: The intermediates (the products of single and double addition–elimination) are more reactive than the starting material. The organometallic will react faster with them than with the starting material.

Problem   8.2b
Source class   Z—C=C   *Sink class*   C=C—ewg
Decision type   Ambident electrophile

Rationale: The enolate has a delocalized charge (soft) and is a stable anion (pK_{abH} 13). Thus the addition is reversible, giving the thermodynamic product.

Problem   8.2c
Source class   Z—C=C   *Sink class*   Y=C—L
Decision type   Ambident nucleophile

Rationale: The ester electrophile is reasonably soft and is a relatively stable anion (pK_{abH} 24), and the addition is reversible. The thermodynamic product is expected.

Problem   8.2d
Source class   Z   *Sink class* ⩾C-L
Decision type Intermolecular vs. intramolecular

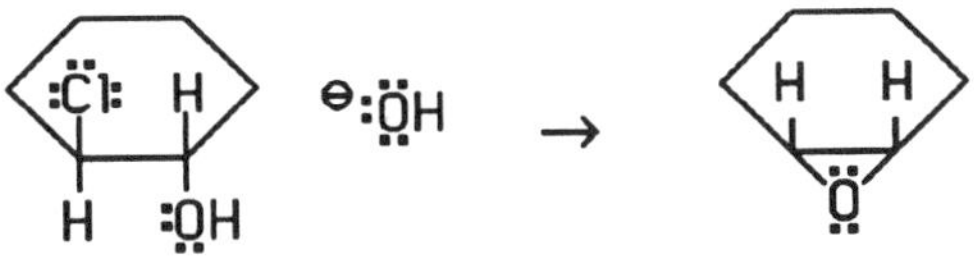

Rationale: Intramolecular attack in this case is much faster than intermolecular, for the nucleophile and electrophile collide more frequently.

Problem   8.2e
Source class   Z   *Sink class* C=Y
Decision type Thermodynamic vs. kinetic enolate

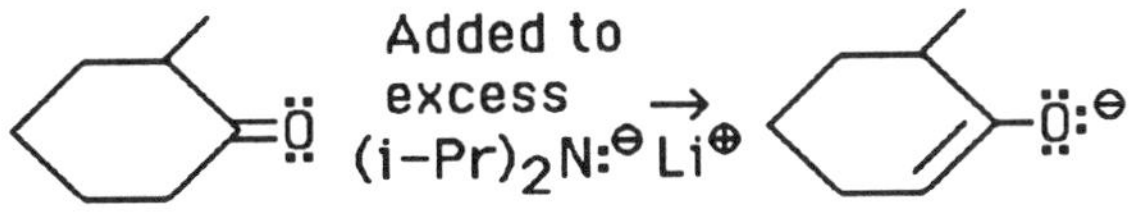

Rationale: These are kinetic conditions; only the most accessible H will be removed.

8.3 Provide a mechanism for the formation of the preferred products in problem 8.2. The source and sink for each part was discussed at the beginning of the answer for problem 8.2.

(a) Addition–elimination occurs twice; then a final nucleophilic addition occurs.

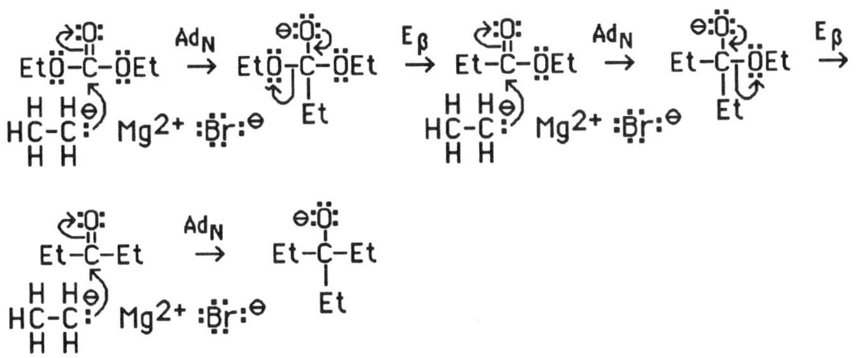

(b) Nucleophilic addition to the polarized multiple bond occurs by path Ad$_N$.

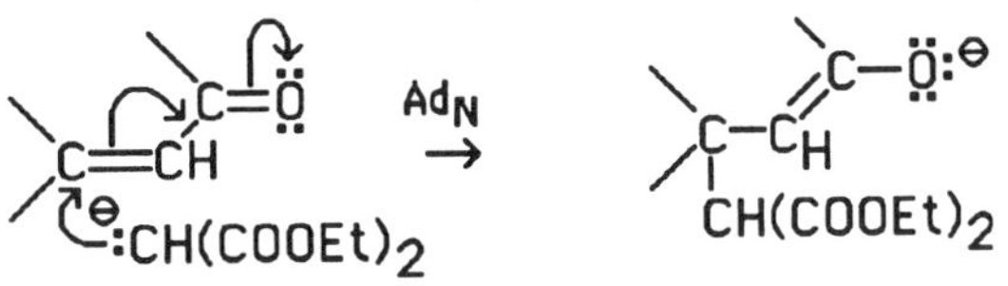

(c) Path Ad$_N$, then E$_\beta$; addition–elimination occurs.

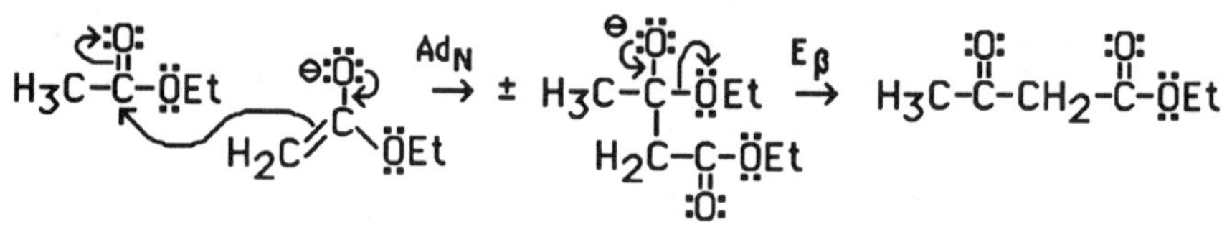

d) Deprotonation, path p.t., is followed by an intramolecular S_N2.

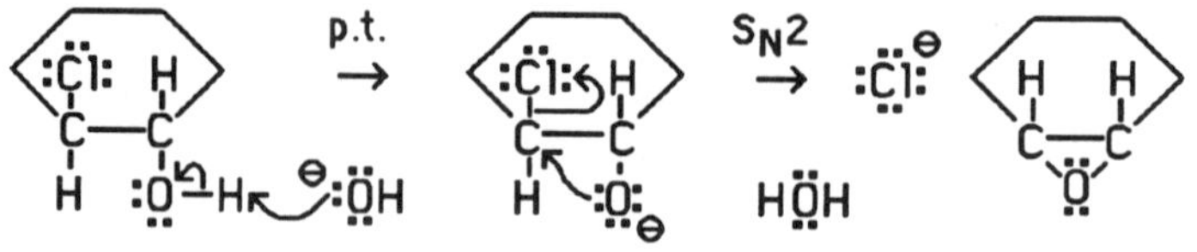

(e) Deprotonation, path p.t.

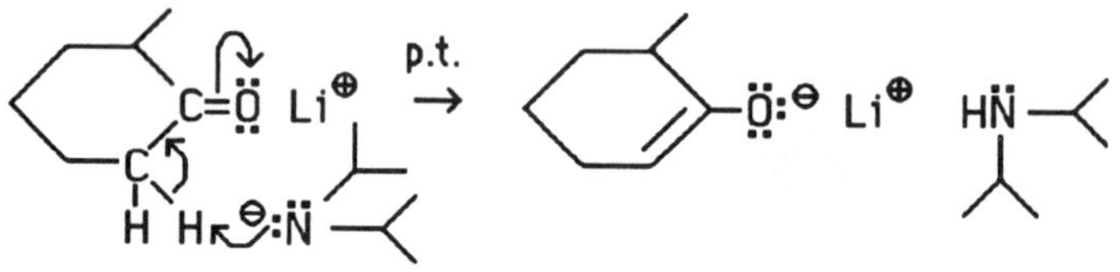

8.4 Each of the following reactions has only one electron source and sink. Give the product of the following reactions; a decision discussed in this chapter is required. All of the following reactions occur in basic media.

Problem   8.4a
Source class   R−M (for first step)   *Sink class*   Y=C−L (for first step)
Decision type Multiple addition
Pathways AdN + $E\beta$, addition–elimination. No second attack occurs.

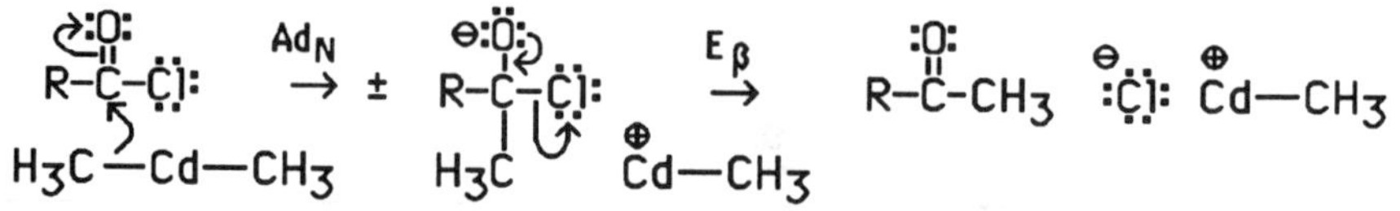

Rationale: Organocadmium reagents are more covalent and less reactive and will not add to ketones at a reasonable rate. Acyl chlorides are more reactive than ketones and will react with organocadmium reagents.

Problem   8.4b
Source class   R−M (for first step)   *Sink class*   Y=C−L (for first step)
Decision type Multiple addition
Pathways AdN + $E\beta$, addition–elimination followed by a second addition, path AdN

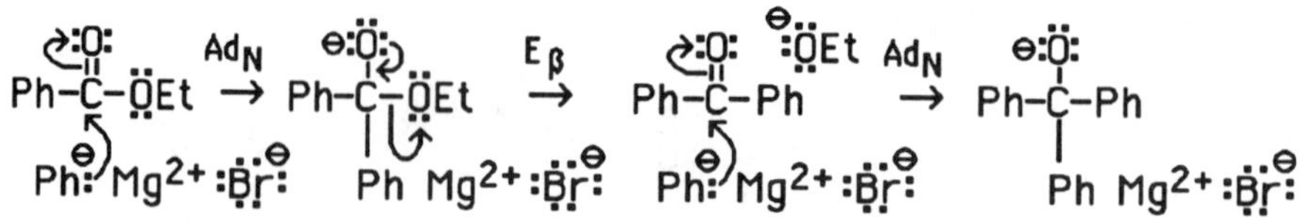

Rationale: Organomagnesium reagents easily add to ketones. Esters are less reactive than ketones, and therefore the organometallic reacts faster with the ketone product than with the ester.

Problem   **8.4c**
Source class   Z   *Sink class* $\geqslant$C–L
Decision type   Substitution/elimination
Pathway   The S$_N$2 substitution

$$HO^{\ominus} + CH_3CH_2CH_2CH_2\text{–Cl} \xrightarrow{S_N2} CH_3CH_2CH_2CH_2\text{–OH} + \overset{\ominus}{:}\text{Cl}:$$

Rationale: The primary site is unhindered, and the nucleophile is good. Substitution is expected to predominate over elimination.

Problem   **8.4d**
Source class   R–M   *Sink class*   C=C–ewg
Decision type   Ambident electrophile
Pathways   Path Ad$_N$, nucleophilic addition to a polarized multiple bond

Rationale: The addition is not reversible and the nucleophile is hard; 1,2 addition is preferred.

Problem   **8.4e**
Source class   R–M   *Sink class*   C=C–ewg
Decision type   Ambident electrophile
Pathways   Path Ad$_N$, nucleophilic addition to a polarized multiple bond

Rationale: The combination of an organomagnesium and a copper salt forms an organocopper by transmetallation. The addition is not reversible and the nucleophile is soft. Conjugate (1,4) addition is preferred.

Problem   **8.4f**
Source class   Z–C=C   *Sink class*   Y–L
Decision type   Ambident nucleophile
Pathway   Substitution via a pentacovalent intermediate

Rationale: The Si–Cl bond is polarized (electronegativity difference is 3.16 - 1.90 = 1.26), and therefore the electrophile is relatively hard. By the ΔpK_a rule we expect this reaction to be very much downhill in energy (pK_{abH} 19.2 going to pK_{abH} -7) and thus probably to be irreversible.

Problem   8.4g
Source class   Z   *Sink class* ≥C—L
Decision type Substitution/elimination
Pathway   E2 elimination.

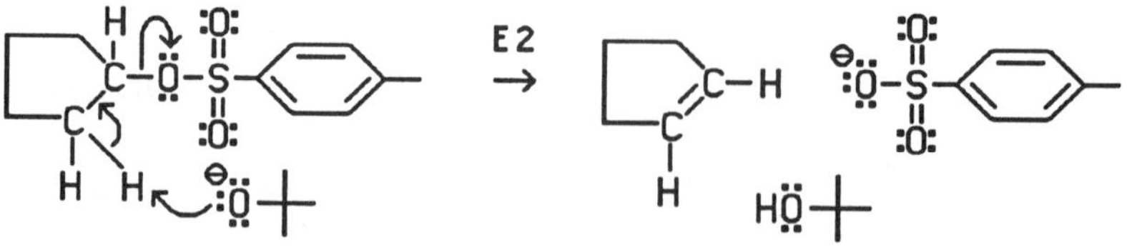

Rationale: The anion is a poor nucleophile and a good base. The secondary site is slightly hindered. Elimination is expected to predominate over substitution.

CHAPTER 9 ADDITIONAL EXERCISES

Problem 9.1a

$$Et_2CHOH + PBr_3 \rightarrow Et_2CHBr$$

Understand the system. A substitution has occurred. Balanced? No, $HOPBr_2$, an acid, is needed on the right to balance. Medium: Acidic, since the reaction produces an acid. Sources: The alcohol lone pair. Leaving groups: Bromide. Sinks: The phosphorus–bromine bond is a Y—L.
Find possible routes. There are two that fit the conditions, S_N2 substitution,

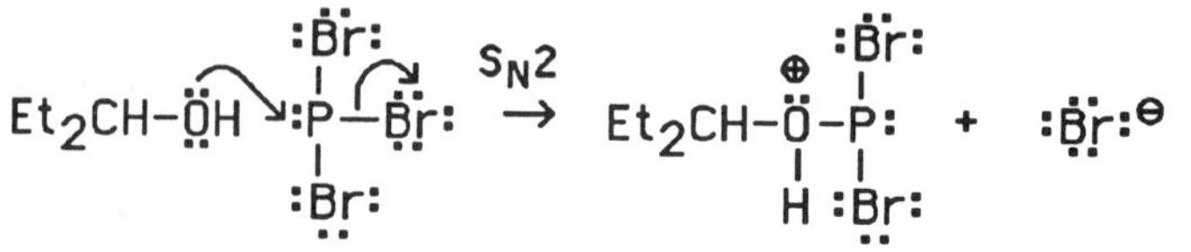

and substitution via a pentacovalent intermediate.

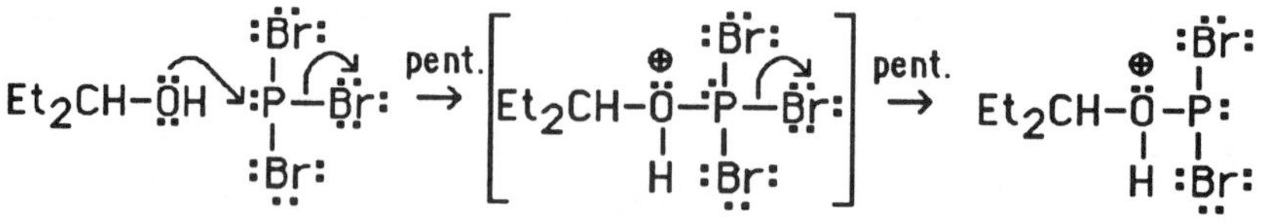

Evaluate and cross-check. Either route is possible for phosphorus. Since bromide is such a good leaving group, the pentacovalent intermediate, if it occurs, will be short lived. The product is expected to be a relatively strong acid ($R_2OH^{\oplus}$ has a pK_a of -3.5) and is reasonable for an acidic medium.
Understand the system. We have a new sink, the alcohol hydroxyl has been turned into a good leaving group. In addition, the H on the positive oxygen is now acidic.
Find possible routes. Using the new sink, we find that there could possibly be four paths that fit the medium, sources, and sinks: deprotonation to form a lone pair,

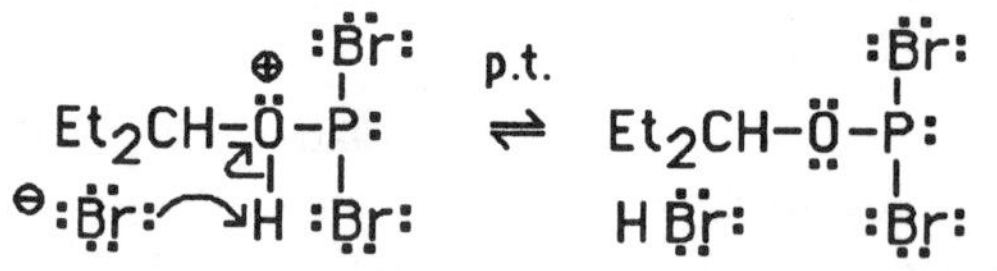

or E2 elimination,

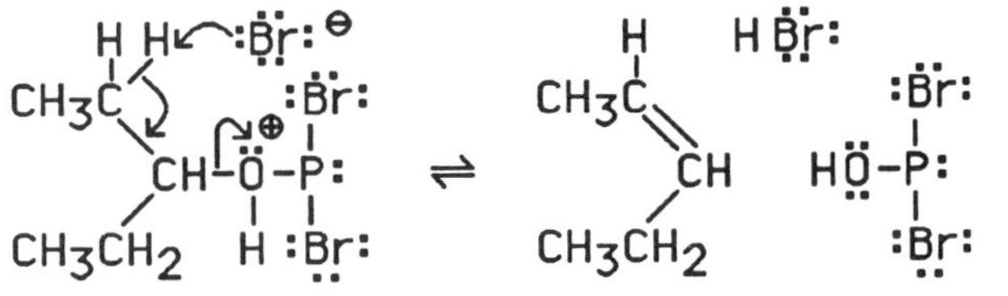

H H⤴ :B̈r: ⊖   H   H B̈r:
CH₃C⤸   CH₃C   :B̈r:
        :B̈r: ‖ |
    CH⌐Ö–P: ⇌   CH   HÖ–P:
        | | / |
CH₃CH₂ H :B̈r: CH₃CH₂   :B̈r:

or the S$_N$1 substitution (path D$_N$, then A$_N$),

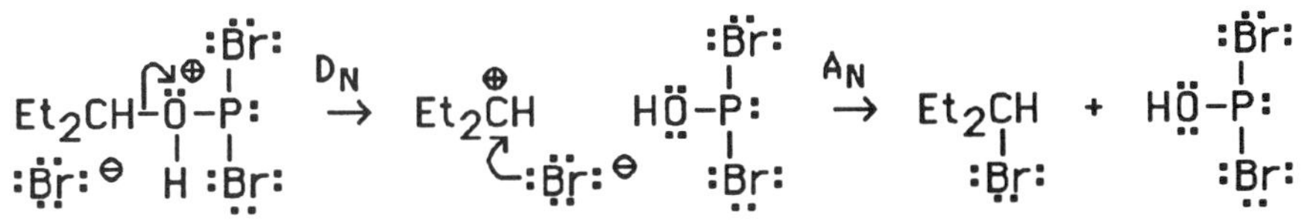

:B̈r: :B̈r: :B̈r:
              ⊕ | D$_N$ ⊕ | A$_N$ |
Et₂CH⌐Ö–P: →   Et₂CH   HÖ–P: →   Et₂CH + HÖ–P:
:B̈r: ⊖ H :B̈r: ⌐:B̈r: ⊖ :B̈r: :B̈r: :B̈r:

or S$_N$2 substitution.

:B̈r: :B̈r:
              ⊕ | S$_N$2 |
Et₂CH⌐Ö–P: →   Et₂CH + HÖ–P:
⊖:B̈r:⌐ H :B̈r: :B̈r: :B̈r:

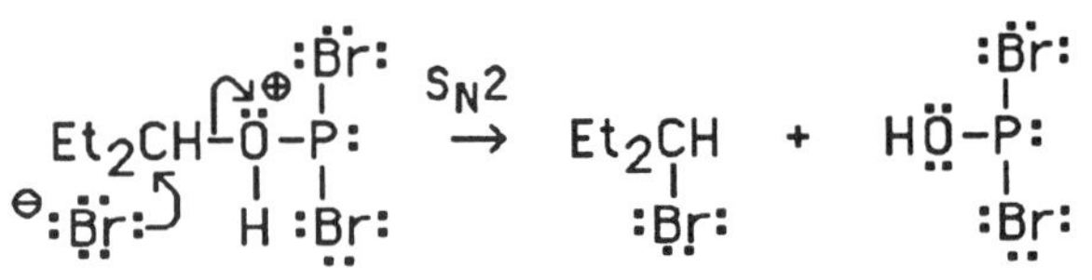

Evaluate and cross-check. The deprotonation step makes the leaving group poorer, and since we wish eventually to do a substitution, this route is less promising. However, the leaving group is still good enough to be displaced by a bromide nucleophile in an S$_N$2 to yield the product. The E2 path is suspect since bromide is a poor base; from Section 8.5 we expect substitution. The S$_N$1 produces a secondary carbocation, which we know from the carbocation trend is not especially stable; this S$_N$1 process is less likely. The S$_N$2 path seems the best. The pK_a span of all the intermediates is reasonable and appropriate for an acidic medium.

Overview: Our reasonable route is just two S$_N$2 reactions, but it is not the only way to get to product. The first step could easily go through a pentacovalent intermediate, and our last S$_N$2 step has two side routes that are also quite acceptable. Even if an alternate route's rate-determining barrier were a couple of kcal/mol (10 kJ/mol) higher in energy, many of the more energetic molecules could easily traverse that alternate route.

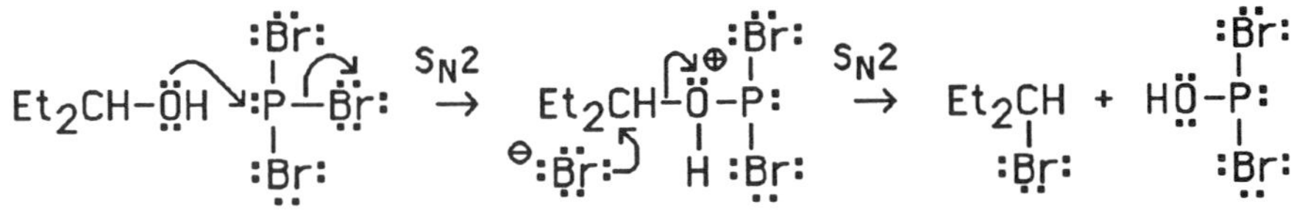

:B̈r: :B̈r: :B̈r:
              | S$_N$2   ⊕ | S$_N$2 |
Et₂CH–ÖH :P–B̈r: →   Et₂CH⌐Ö–P: →   Et₂CH + HÖ–P:
              | ⊖:B̈r:⌐ H :B̈r: |
:B̈r: :B̈r: :B̈r:

Problem 9.1b

          :Ö: :Ö:
          ‖   KOH   ‖
H₂C=CH–CH₂–CH ——→ H₃C–CH=CH–CH
                   H₂O

Understand the system. Balanced? Yes. The product is an isomer of the starting material, no addition or elimination has occurred. Medium: Definitely basic; the predominate anion is hydroxide, pK_{abH} 15.7, whose pK_a would give a useful proton transfer K_{eq} up to about pK_a 24. Sources: Alkene, carbonyl lone pair, water lone pair, and hydroxide anion. Best source: Hydroxide anion, a lone pair source that can behave as a nucleophile or as a base. Sinks: Polarized multiple bond, the aldehyde carbonyl. Acidic H's: Water and the CH_2 next to the aldehyde, which is within the pK_a range of the hydroxide base. Resonance forms:

$$\overset{\displaystyle :\overset{..}{O}:}{H_2C=CH-CH_2-\overset{|}{C}H} \;\longleftrightarrow\; \overset{\displaystyle :\overset{..}{\overset{..}{O}}:{}^{\ominus}}{H_2C=CH-CH_2-\overset{|}{\underset{\oplus}{C}}H}$$

Find possible routes. There are just two that fit the medium, sources, and sinks: deprotonation to form a delocalized anion, path p.t.,

$$H_2C=CH-\overset{|}{C}H-\overset{:\overset{..}{O}:}{\overset{|}{C}H} \quad\overset{p.t.}{\rightleftharpoons}\quad H_2C=CH-\overset{\ominus}{C}H-\overset{:\overset{..}{O}:}{C}H$$

$$\underset{H\;\curvearrowright\;\overset{\ominus}{:}\overset{..}{O}H}{} \qquad\qquad\qquad \underset{H\overset{..}{O}H}{}$$

or nucleophilic addition to a polarized multiple bond, path AdN.

$$H_2C=CH-CH_2-\overset{\curvearrowleft:\overset{..}{O}:}{\overset{||}{C}H} \quad\overset{AdN}{\rightleftharpoons}\quad H_2C=CH-CH_2-\overset{:\overset{..}{\overset{..}{O}}:{}^{\ominus}}{\underset{:\overset{..}{O}H}{\overset{|}{C}H}}$$

$$\underset{\overset{\ominus}{:}\overset{..}{O}H}{}$$

Evaluate and cross-check. We need to estimate pK_a's to use the ΔpK_a rule or calculate a K_{eq}. For the first p.t., the pK_a of the CH can be estimated by starting with the pK_a of a simple aldehyde, 16.7, and guessing how much it is lowered owing to the added delocalization of the anion by the double bond. From the pK_a chart, we see that adding a phenyl group to the structurally similar ketone lowers the acidity from 19.2 to 15.9, 3.3 pK_a units. We know that the delocalization of a charge by a simple double bond and by a phenyl group are similar, the delocalization by phenyl being slightly greater. We can then conclude that our aldehyde has a pK_a close to 14 or 15, down a couple of pK_a units from a simple aldehyde. The proton transfer K_{eq} should be favorable.

For the AdN, the pK_a of the OH of the hydrate can be estimated by starting with the pK_a of $HOCH_2OH$, 13.3, and knowing from the pK_a chart that the replacement of an R for an H raises the pK_a of an alcohol by about one unit. Since the pK_{abH} for both products can only be crudely estimated, and both values come out to be around 14 to 15, the products are too close in energy to distinguish. A use of "product hindsight" shows that no additional groups are attached, thus making the nucleophilic attack path less probable. If this were a product prediction problem and not a mechanism problem we would have to explore both routes.

Understand the system. We now have a new source, the resonance stabilized anion just formed. Resonance forms:

$$\overset{\ominus}{H_2\overset{..}{C}}-CH=CH-\overset{:\overset{..}{O}:}{C}H \;\longleftrightarrow\; H_2C=CH-\overset{\ominus}{C}H-\overset{:\overset{..}{O}:}{C}H \;\longleftrightarrow\; H_2C=CH-CH=\overset{:\overset{..}{\overset{..}{O}}:{}^{\ominus}}{C}H$$

Using the new source, we find that there are just two paths that fit the medium, sources, and sinks: protonation of an anion and addition to a polarized multiple bond, AdN. At this point we can again look forward to product and see that no nucleophilic attack has occurred; therefore we should consider the proton transfer path first. There are three sites at which protonation could occur.

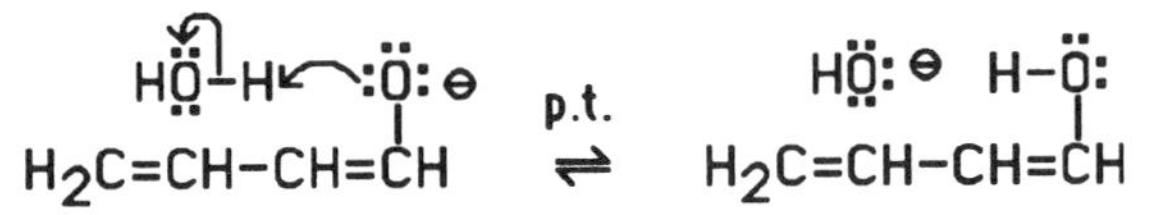

or

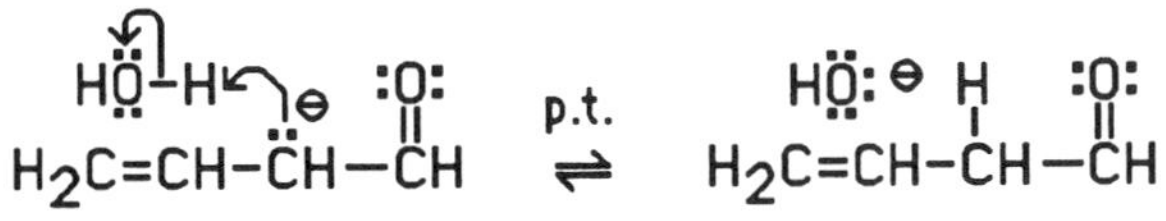

or

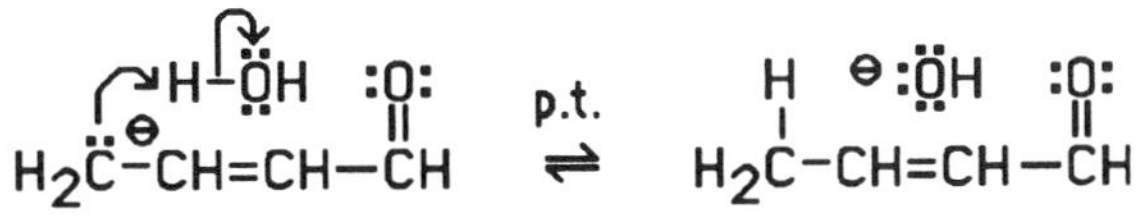

Evaluate and cross-check. Protonation on oxygen yields the enol, which from a ΔH calculation is shown to be higher in energy than the starting aldehyde. Protonation on the center carbon yields the reactants. Finally protonation on the terminal carbon yields the desired product. Since the starting aldehyde is not conjugated and the product aldehyde is, the product is about 3.5 kcal/mol (15 kJ/mol) more stable (Section 1.7).

Overview: Our final route is just two proton transfers.

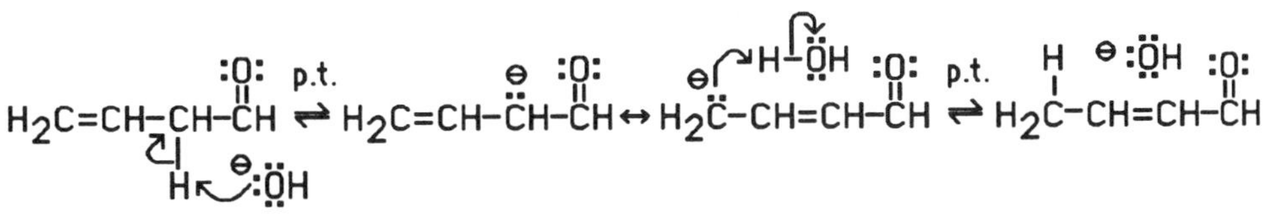

Problem 9.1c

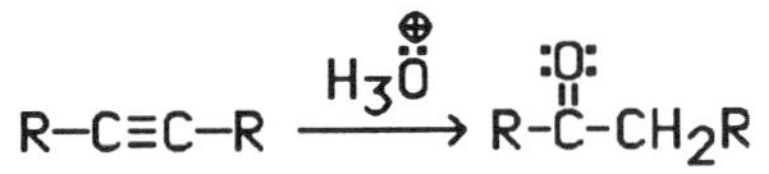

Understand the system. Source class: C≡C. Sink class: HA. Media: acidic. Acids: $H_3O^{\oplus}$ pK_a -1.7.

Find possible routes. The Ad_E2 and the Ad_E3 are reasonable for acidic media. The Ad_N2 is not a possibility since there is no electron-withdrawing group.

Ad_E2:

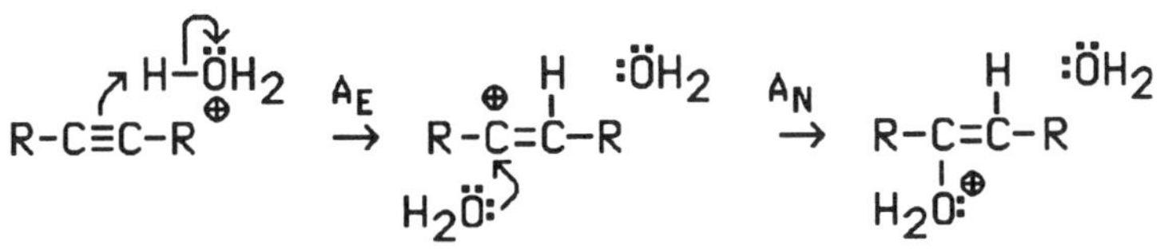

or Ad_E3:

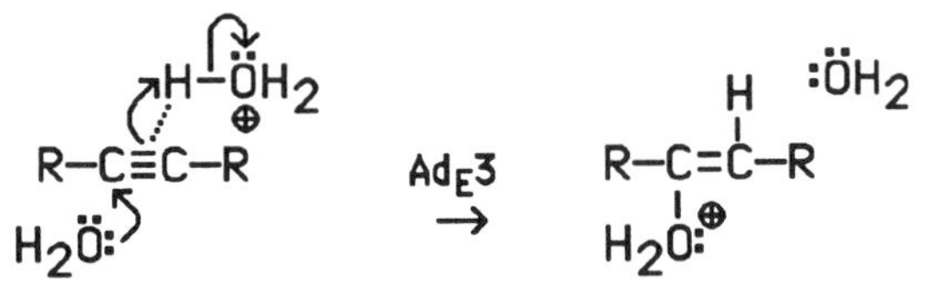

Evaluate and cross-check. The AdE3 is the most probable path since the AdE2 goes via the unstable vinyl carbocation. The cation produced has a pK_a of about -2 to -3 and is consistent with the acidity of the media.

Understand the system. The intermediate cation is a good acid.

Find possible routes. Proton transfer to water produces an enol (path p.t.).

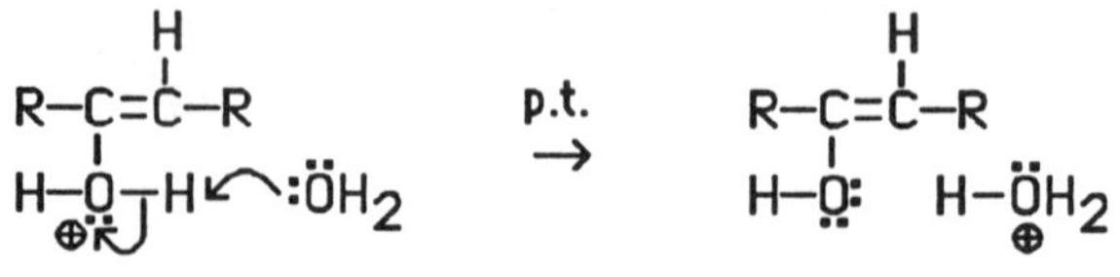

Evaluate and cross-check. The K_{eq} is about 10^1 and is favorable.

Understand the system. Enols are in equilibrium with the keto form via tautomerization. The product ketone is our desired end point.

Find possible routes. Tautomerization can be either base or acid catalyzed.

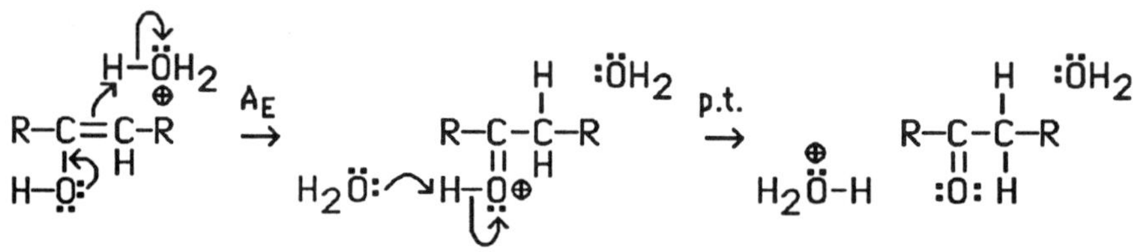

Evaluate and cross-check. The acid-catalyzed route is appropriate for the media.

Overview: An AdE3 occurs, then a proton transfer to water, giving the enol, which tautomerizes to the product ketone.

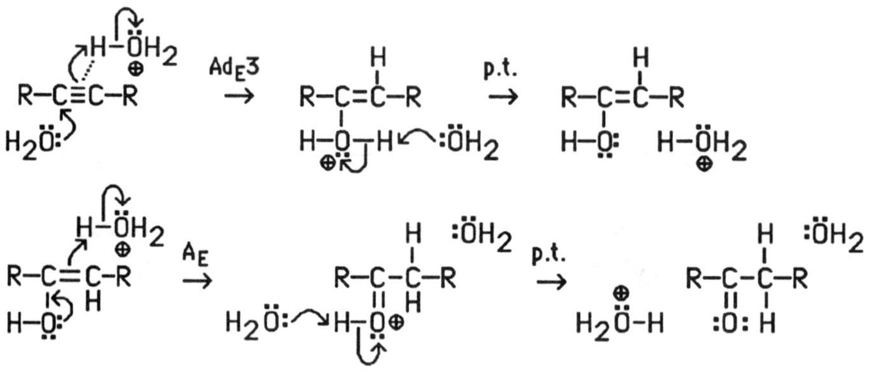

Problem 9.1d

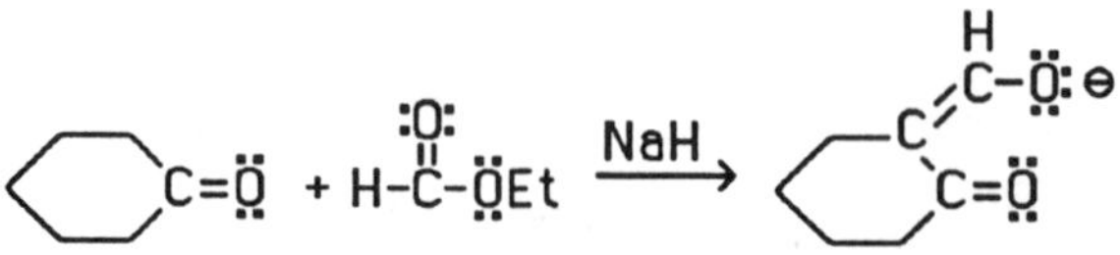

Understand the system. Source class: MH. Sink classes: C=Y and Y=C—L. Media: basic. Acids: the CH_2 next to the ketone, pK_a 19.2. Bases: NaH, pK_{abH} 35.

Find possible routes. Proton transfer should be rapid to generate the sodium enolate and hydrogen gas.

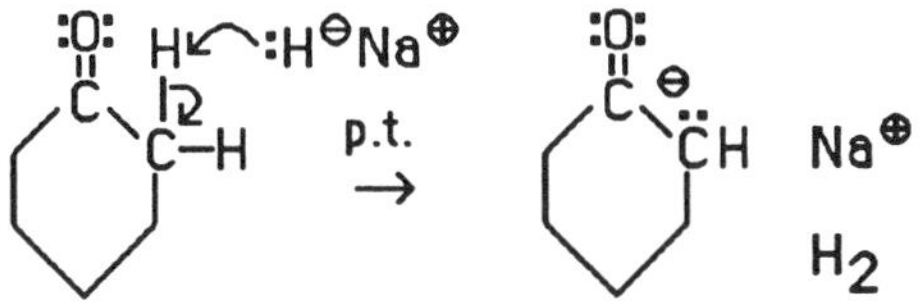

Evaluate and cross-check. The K_{eq} is $10^{15.8}$ and is therefore irreversible.
Understand the system. The enolate is a good allylic source. The ester is a Y=C—L and our only sink.
Find possible routes. The addition–elimination is the only route in basic media to substitute a leaving group on an sp^2 hybridized center.

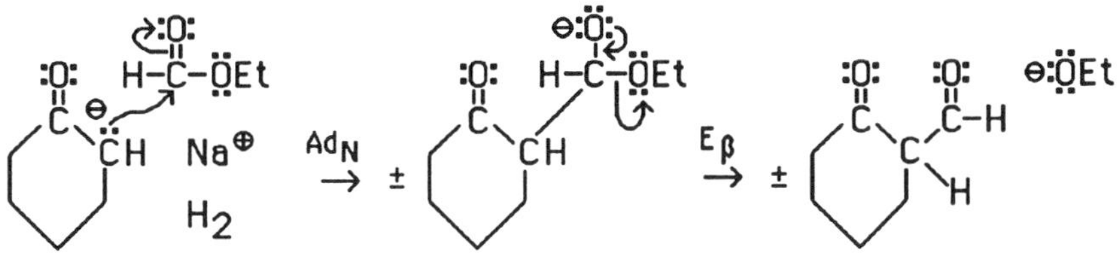

Evaluate and cross-check. The reaction has produced a weaker base going from pK_{abH} 19.2 to pK_{abH} 16.
Understand the system. The product is just a deprotonation away. The pK_a of the keto-aldehyde should be about 7, halfway between the value for the diketone ($pK_a = 9$) and that of the dialdehyde ($pK_a = 5$).
Find possible routes. Path p.t., deprotonation.

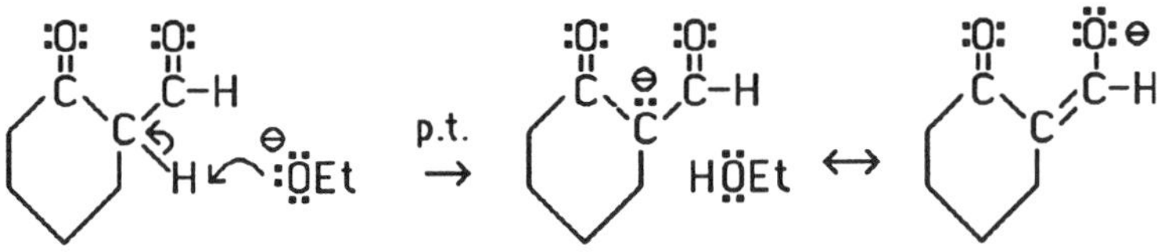

Evaluate and cross-check. The K_{eq} is 10^9, irreversible toward product.
Overview: Deprotonation, followed by addition–elimination, then another deprotonation gives us the product.

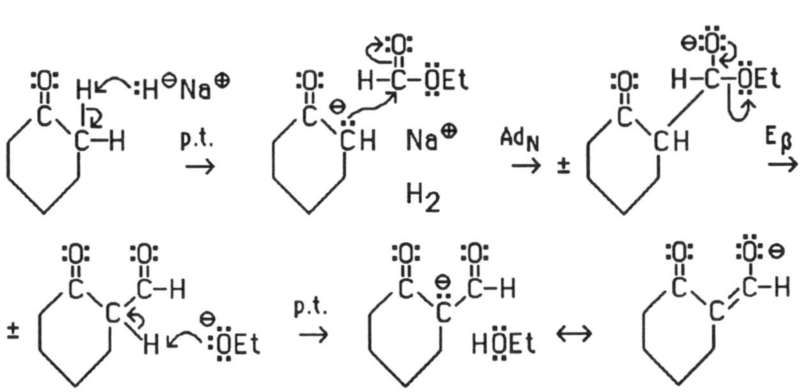

Problem 9.1e

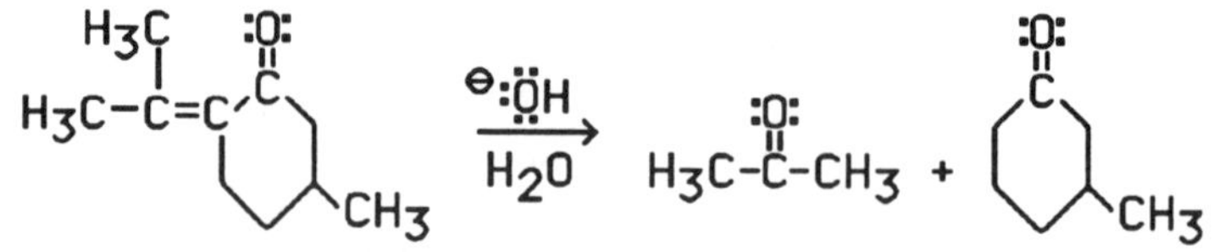

Understand the system. Source class: Z. Sink class: C=C−ewg. Media: basic. Acids: conjugated ketone, pK_a about 19. Bases: $OH^\ominus$, pK_{abH} 15.7. Objectives: Break the C=C and replace the left side with C=O and the right by two H's.

Find possible routes. The carbon bearing the two methyls is bonded to an oxygen in the product. This C−O bond is easily made with an Ad_N2 conjugate addition to the starting material (path Ad_N, then p.t.).

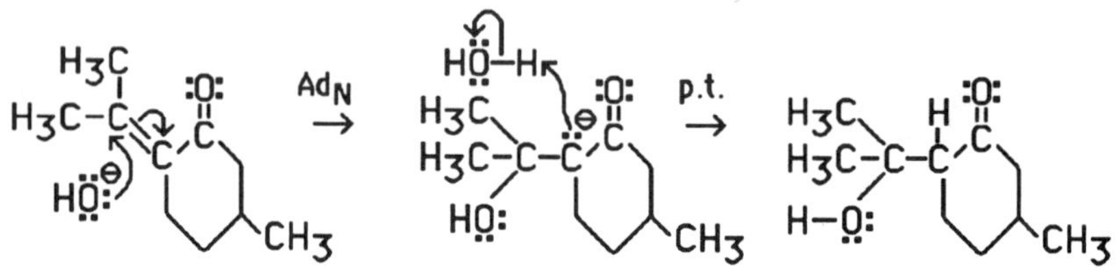

Evaluate and cross-check. All pK_a's are within reason.

Understand the system. We have achieved half of our objectives. The OH is acidic, pK_a 19, and we need to break off the three-carbon piece.

Find possible routes. The E1cB or the E2 are the possible elimination routes that are consistent with basic media.

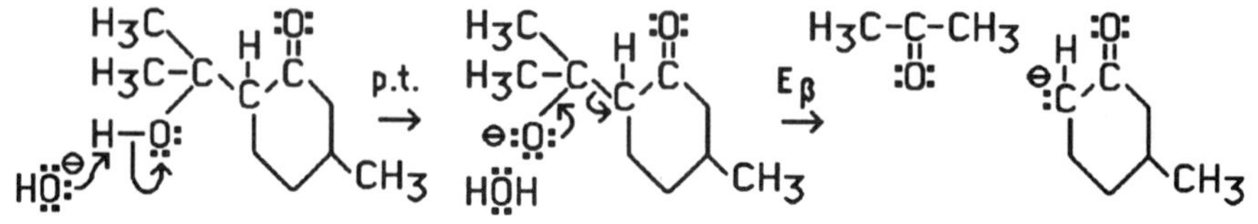

or

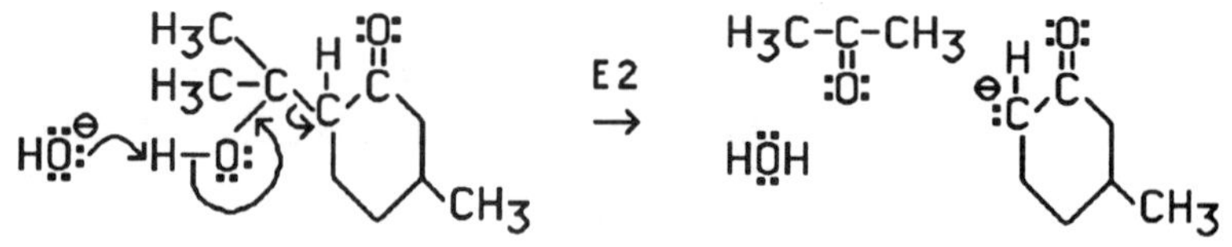

Evaluate and cross-check. Considering the leaving group is poor, pK_{aHL} 19.2, and the H is acidic, the E1cB is probably more likely.

Understand the system. Product is just a protonation away.

Find possible routes. Path p.t. protonation.

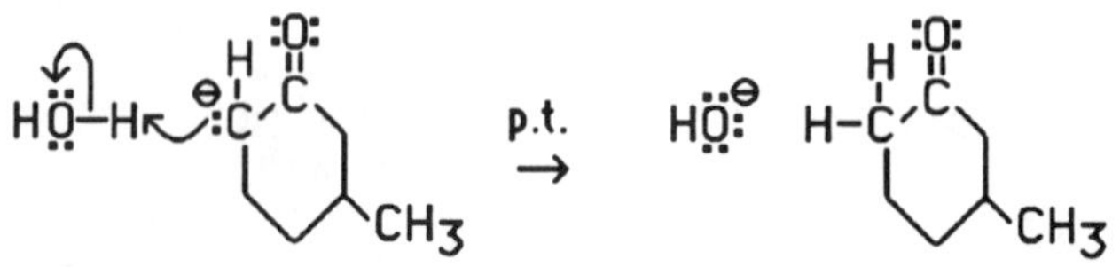

Evaluate and cross-check. Hydroxide is less basic than a ketone enolate, $K_{eq} = 10^{3.5}$.

Overview: We have done an Ad_N2, an E1cB, then a final proton transfer.

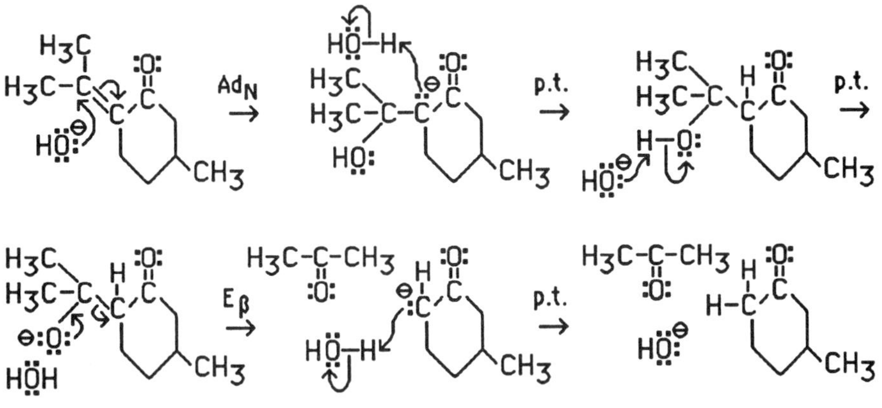

[Chemical reaction scheme with AdN, p.t., p.t. steps]

[Chemical reaction scheme with Eβ, p.t. steps]

Problem 9.1f

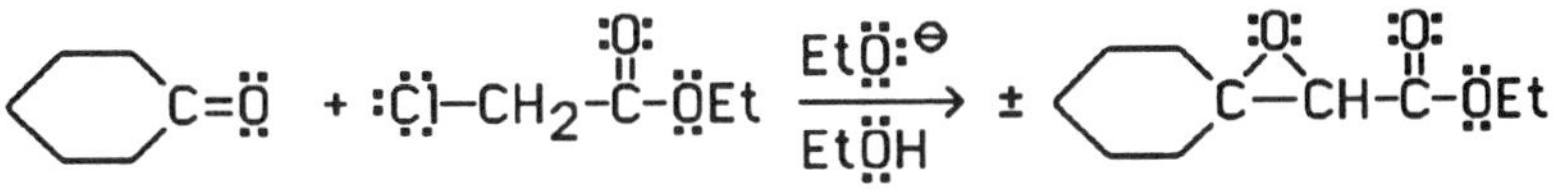

[Reaction: cyclohexanone $C=\ddot{O}$ + $:\overset{..}{C}l-CH_2-\overset{\overset{:O:}{\|}}{C}-\ddot{O}Et$ $\xrightarrow[Et\ddot{O}H]{EtÖ:^{\ominus}}$ ± product]

Understand the system. Source class: Z. Sink class: Y=C—L or C=Y or $\geqslant$C—L. Media: basic, ethoxide pK_{abH} = 16. Acids: CH_2 adjacent to the ketone, pK_a = 19.2, and the CH_2 between the chlorine and the ester, pK_a estimated at 21 (by analogy to a chloroketone, which is more acidic than a ketone by 3 units).

Find possible routes. With three possible sinks, we could use a good starting hint, and therefore let's look at the product. Since there are no additional EtO units on the product, ethoxide probably served as a base rather than a nucleophile. The CH_2's of the ring are undisturbed. The product has a three-membered ring, commonly formed in Nu—L reactions. Although we do not have an obvious Nu—L present, deprotonation of the chloroester would generate an Nu—L system.

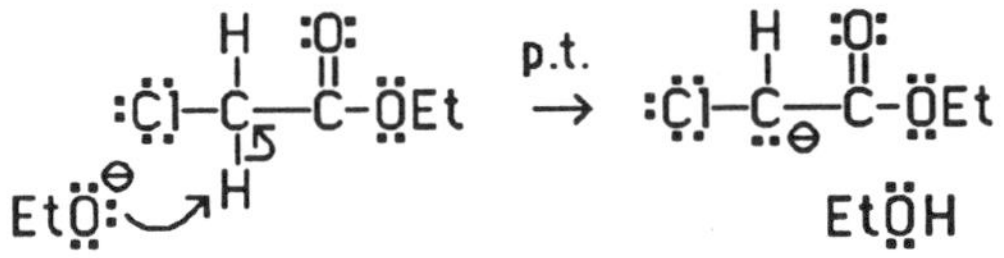

[Reaction scheme: deprotonation step with p.t., EtÖ: and EtÖH]

Evaluate and cross-check. The K_{eq} is about 10^{-5} but still within the useful range.

Understand the system. We now have a good Nu—L and a good C=Y sink in the ketone.

Find possible routes. Addition to the polarized multiple bond, path Ad_N.

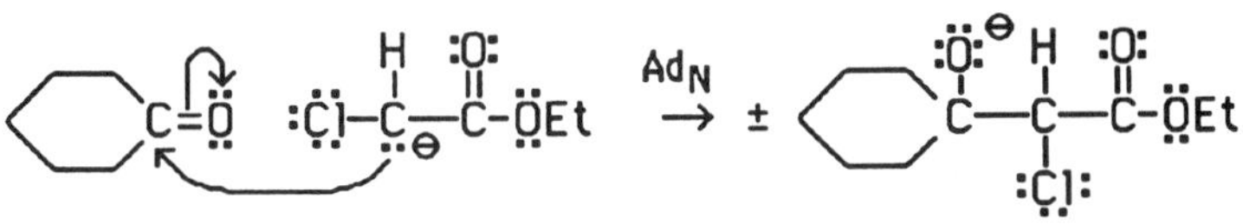

[Reaction scheme: AdN addition, cyclohexanone + enolate → product ±]

Evaluate and cross-check. We have made a slightly weaker base going from about pK_{abH} 21 to pK_{abH} 19.

Understand the system. The oxygen anion is an excellent nucleophile, and the chlorine is a good leaving group.

Find possible routes. Intramolecular S_N2 is expected to be rapid.

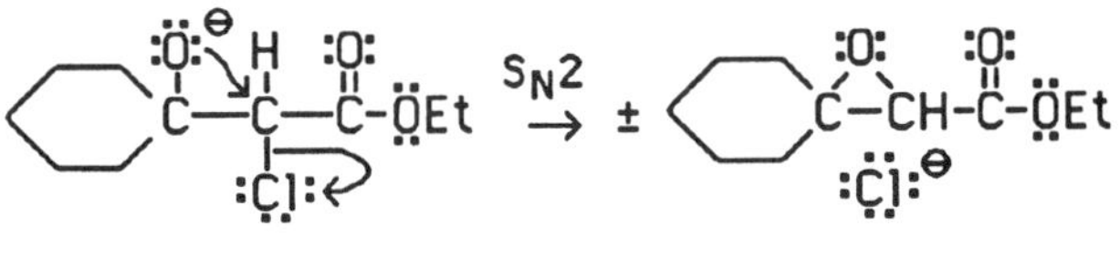

[Reaction scheme: S_N2 intramolecular → epoxide product ±]

Evaluate and cross-check. With the loss of chloride, pK_{abH} -7, the reaction has rationalized the formation of the product and definitely formed a weak base.

Overview: Proton transfer gave us a good nucleophile to attack the carbonyl, which was followed by an internal S_N2.

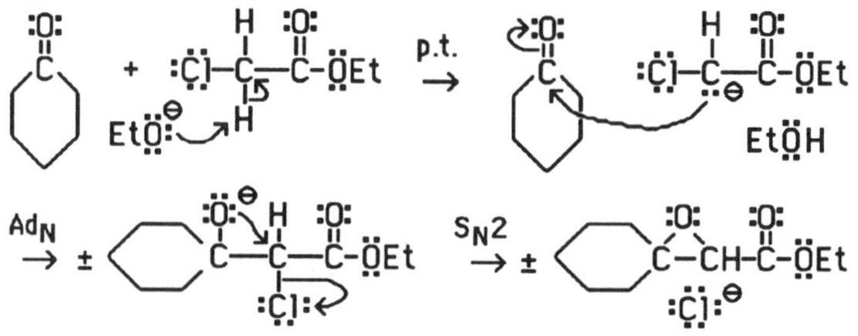

Problem 9.2a

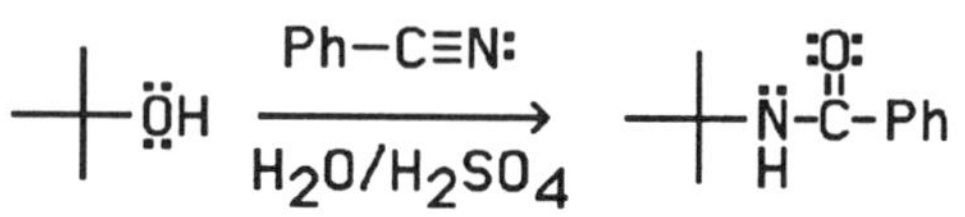

Understand the system. Source class: Z. Sink class: HA. Media: acidic.

Find possible routes. Proton transfer.

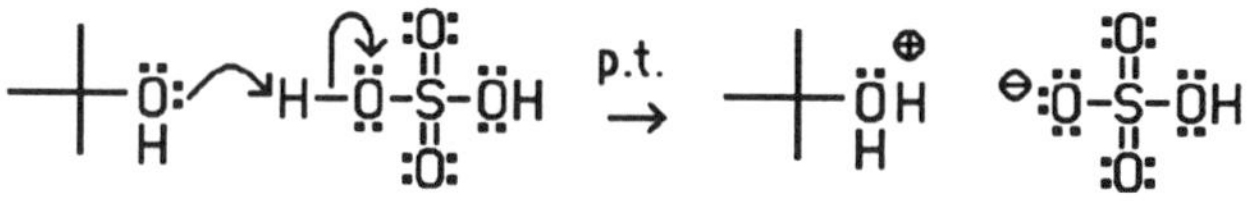

Evaluate and cross-check. We protonated the most basic lone pair in the reaction mixture (besides water). The K_{eq} is very favorable.

Understand the system. We have created a good leaving group on a tertiary carbon in a polar solvent. In the product this carbon is attached to the CH_3CN residue. A substitution is in order.

Find possible routes. S_N1, $D_N + A_N$, is expected to predominate under these conditions. Bisulfate can add reversibly by path A_N,

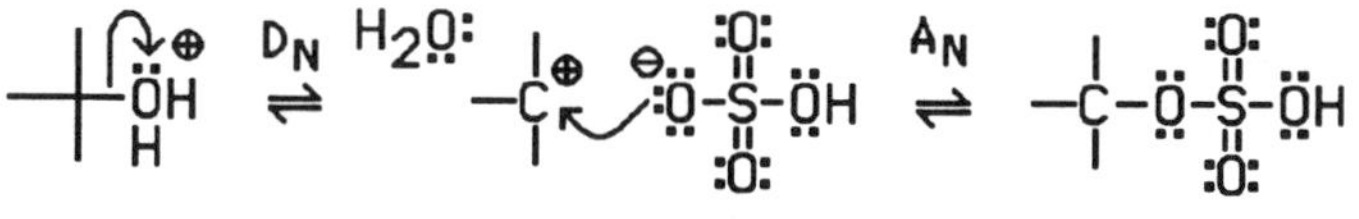

or the nitrile can add reversibly by path A_N.

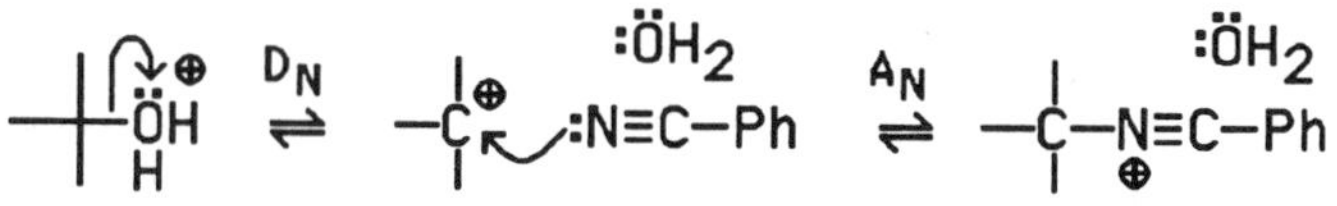

Evaluate and cross-check. A third possibility is to add water, but that just takes us back to reactants. The addition of bisulfate is highly reversible since bisulfate is a good leaving group, and the carbocation is reasonably stable. The addition of the nitrile creates a lone-pair-stabilized carbocation of similar stability to the tertiary carbocation.

Understand the system. The lone-pair-stabilized carbocation can be considered a highly polarized multiple bond, an excellent sink. The product has an oxygen on this carbon.

Find possible routes. Bisulfate can add reversibly by path AdN, or water can add by path Ad$_N$.

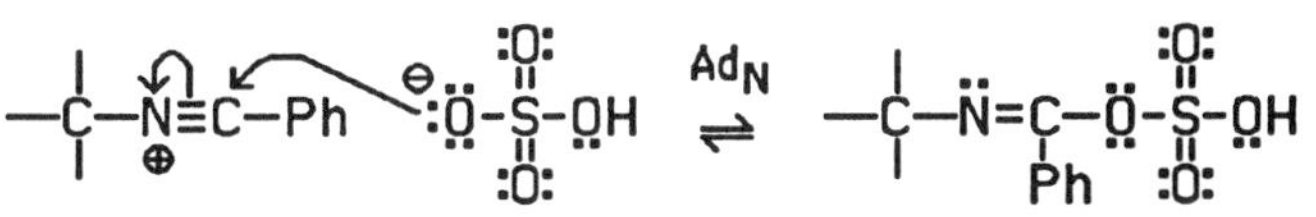

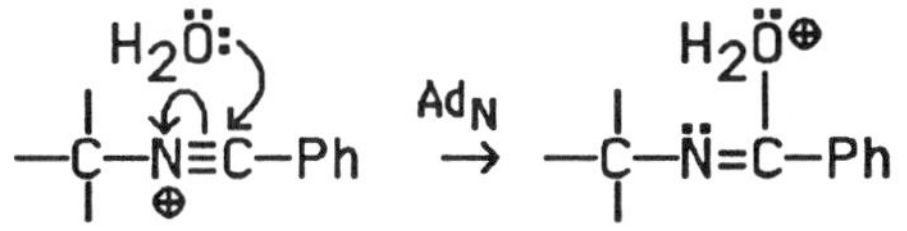

Evaluate and cross-check. Either could also be considered to be path A$_N$ trapping of a cation by a nucleophile. Again, the addition of bisulfate is highly reversible. With the addition of water, we are only a few proton transfers away from the final product.

Understand the system. The protonated oxygen should be a good acid. The most basic lone pair is the nitrogen lone pair.

Find possible routes. Proton transfer either internally or via solvent should be fast.

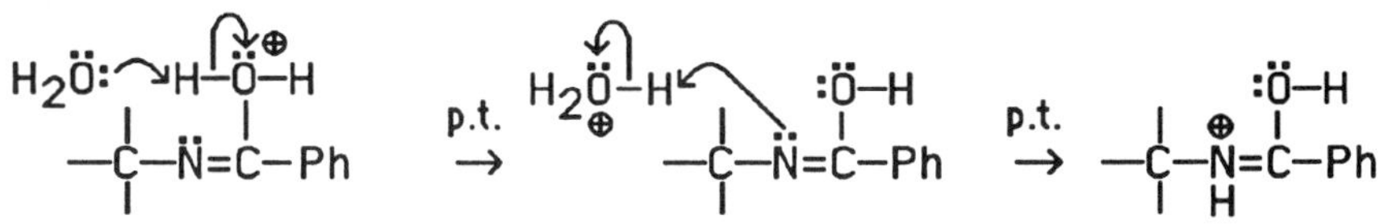

Evaluate and cross-check. This is just the protonated form of the final product.

Find possible routes. Deprotonation is all that is left.

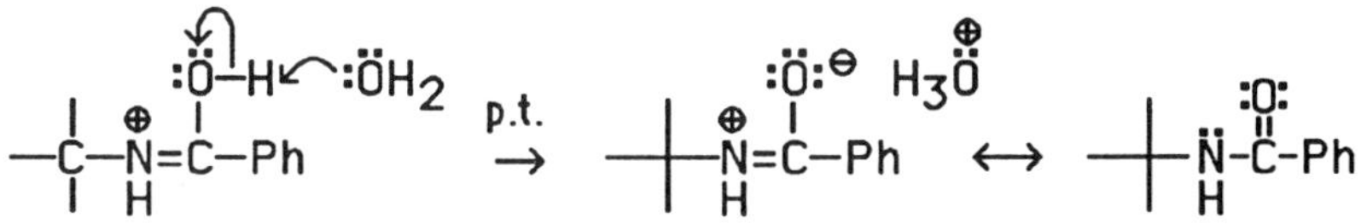

Evaluate and cross-check. The $K_{eq} = 10^{-1.2}$ is slightly uphill.

Overview: The overall process has been a proton transfer, an S$_N$1, an addition of water followed by a proton transfer, and an acid-catalyzed tautomerization.

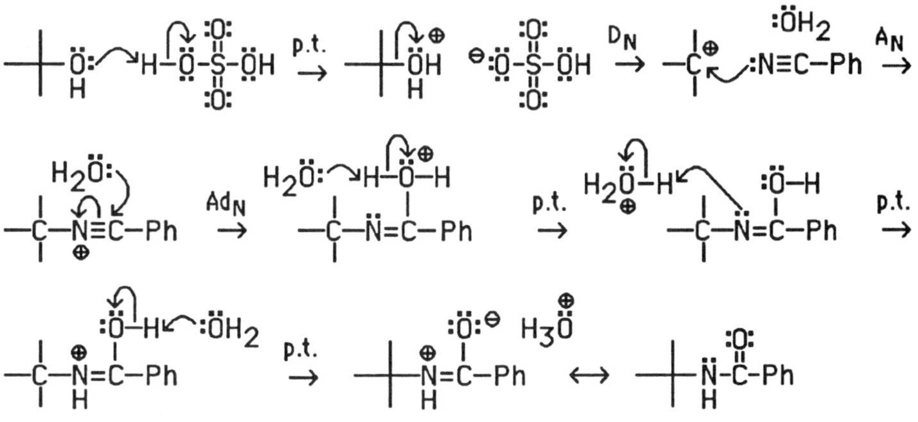

Problem 9.2b

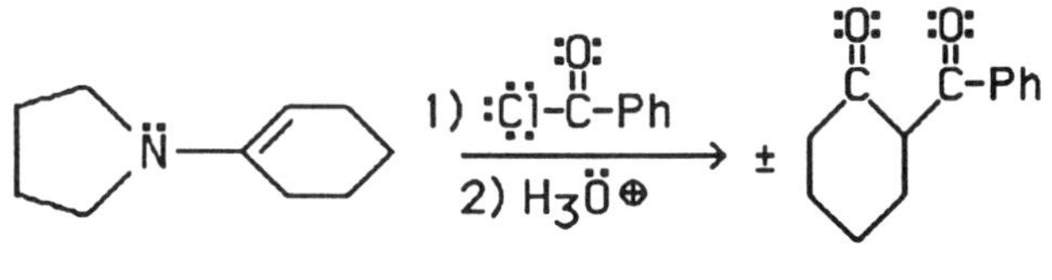

Understand the system. Source class: Z–C=C. Sink class: Y=C–L. Media: neutral, then acidic.

Find possible routes. Addition–elimination, path AdN, then Eβ.

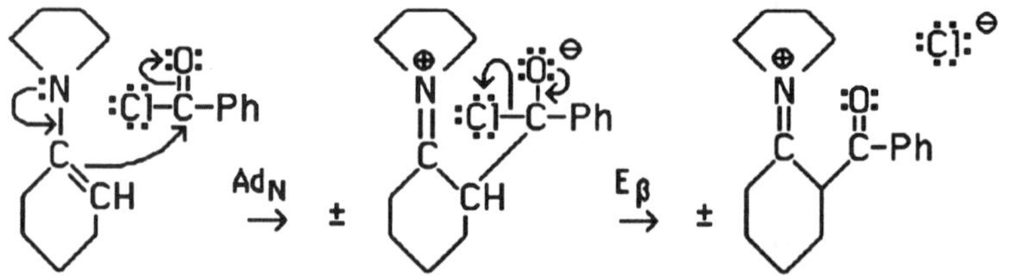

Evaluate and cross-check. The source is ambident, but the heteroatom is hindered and thus a poorer nucleophile. A glance at the product verifies that acylation went on carbon not nitrogen.

Understand the system. The C=N⊕ is an excellent sink, and we need to attack with an oxygen nucleophile. The water in step 2 can serve as the needed nucleophile.

Find possible routes. Addition to the polarized multiple bond, path AdN.

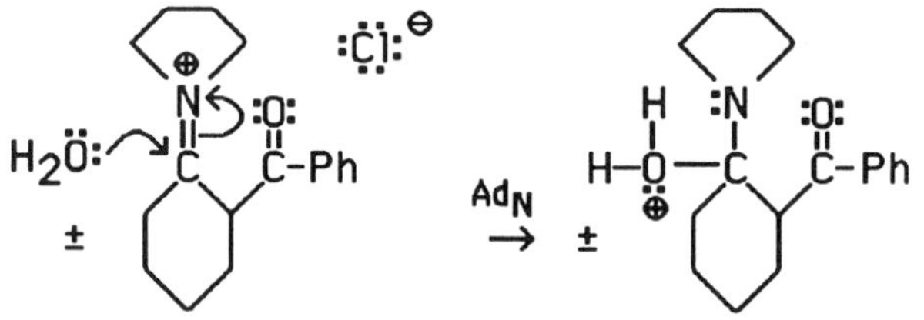

Evaluate and cross-check. This step should seem familiar. We covered the same ground from here to the ketone product in mechanism example 4 in Section 9.5.

Understand the system. The nitrogen lone pair is basic and the $OH_2^{\oplus}$ is acidic.

Find possible routes. Proton transfer can be direct but is most likely mediated by the water solvent.

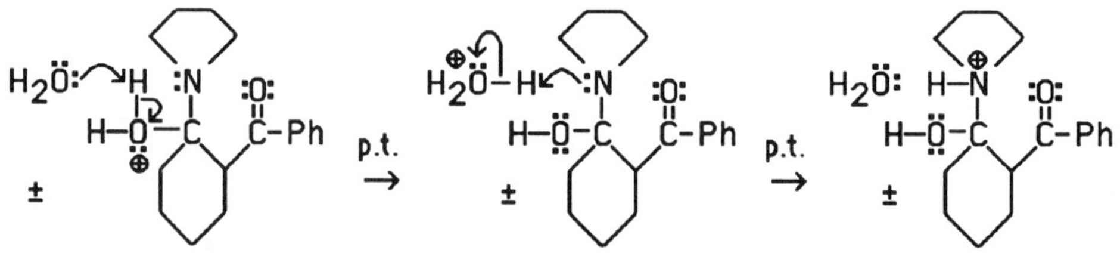

Evaluate and cross-check. The K_{eq} should be very good, greater than 10^{10}.

Understand the system. All that is needed to get to the product is an elimination.

Find possible routes. The E1, E2, or E1cB are all possible.

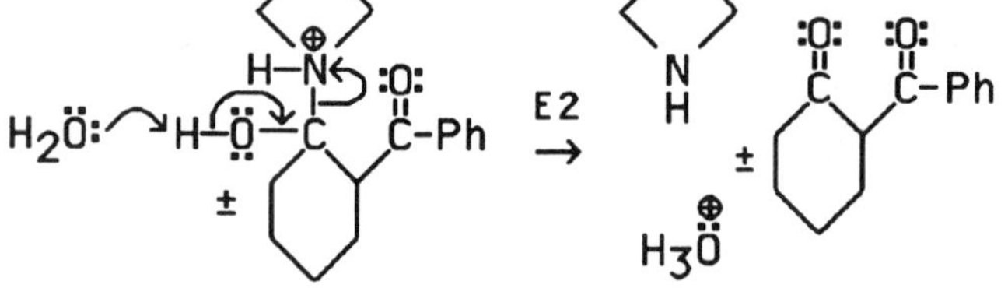

Evaluate and cross-check. The E2 is probably best, but see the discussion in mechanism example 4 in Section 9.5.

Overview: The overall process was an AdN + Eβ, an AdN, two p.t.s, and an E2.

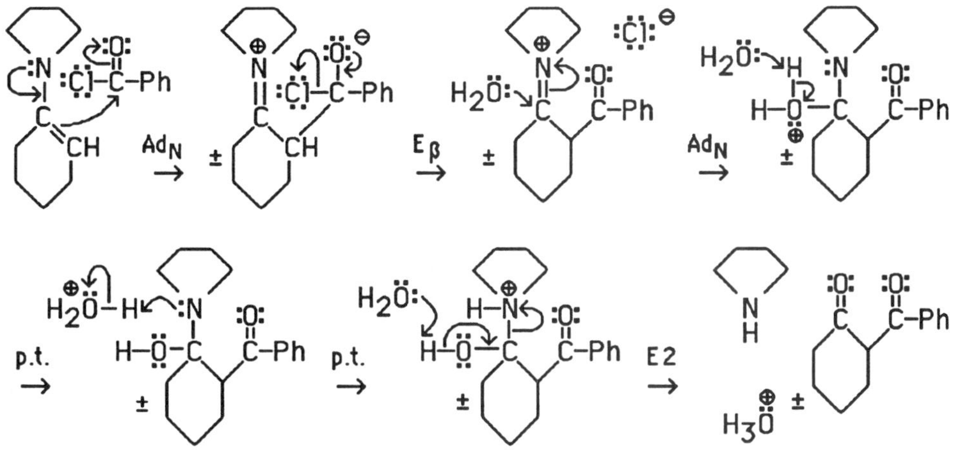

Problem 9.2c

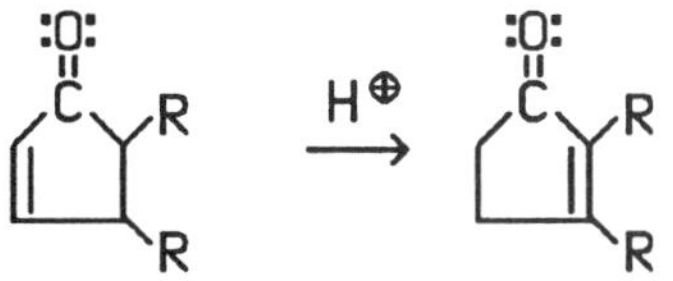

Understand the system. Source class: Z. Sink class: HA. Media: acidic. The product is an isomer of the starting material. Only the positions of the hydrogens and the double bonds have changed; suspect tautomerization since it does just that.

Find possible routes. Acid-catalyzed tautomerization to the enol, path p.t., then D_E.

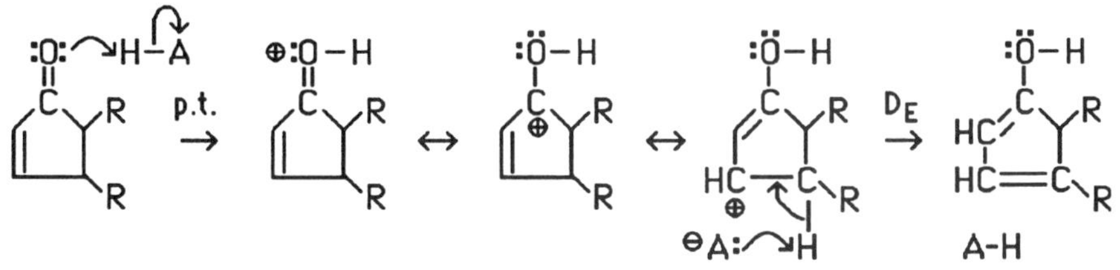

Evaluate and cross-check. The lone-pair-stabilized allylic carbocation intermediate is relatively stable.

Understand the system. The enol is energetically uphill by a ΔH calculation and can tautomerize back to the starting material or on to a new ketone.

Find possible routes. Acid-catalyzed tautomerization to a ketone, path A_E, then p.t.

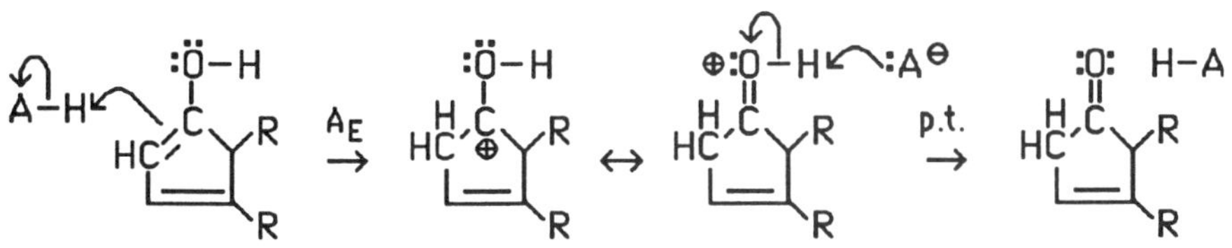

Evaluate and cross-check. This ketone, because it is not conjugated, should be a few kcal/mol (10 to 15 kJ/mol) higher in energy than the starting ketone.

Understand the system. Again we can enolize the ketone. One enol is our previous one, but a second enol is possible.

Find possible routes. Acid-catalyzed tautomerization to the enol, path p.t., then D_E.

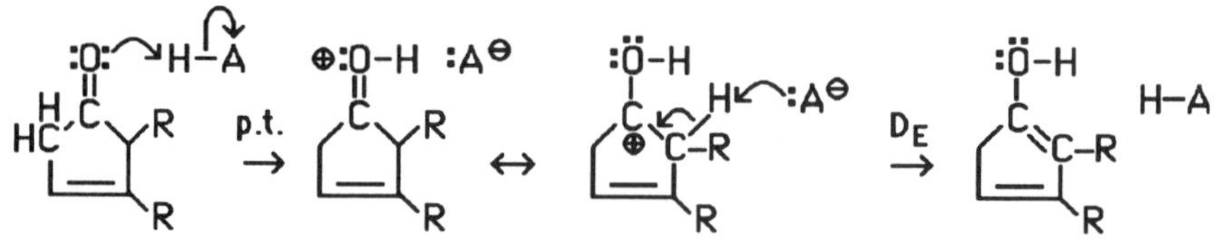

Evaluate and cross-check. The lone-pair-stabilized tertiary carbocation intermediate is relatively stable.

Understand the system. The enol is energetically uphill by a ΔH calculation and can tautomerize back to the unconjugated ketone or on to a new ketone.

Find possible routes. Acid-catalyzed tautomerization to a ketone, path A_E, then p.t..

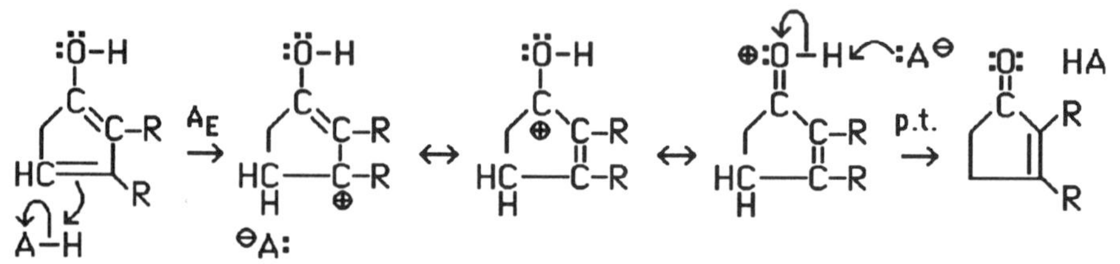

Evaluate and cross-check. This ketone is conjugated and the C=C is tetrasubstituted and therefore should be a few kcal/mol (10 to 15 kJ/mol) more stable than the starting ketone in which the C=C is only disubstituted.

Overview: The overall process has been four tautomerizations. This overview is slightly shorter because it skips the formation of the intermediate ketone (actually a side route, which was shown for completeness).

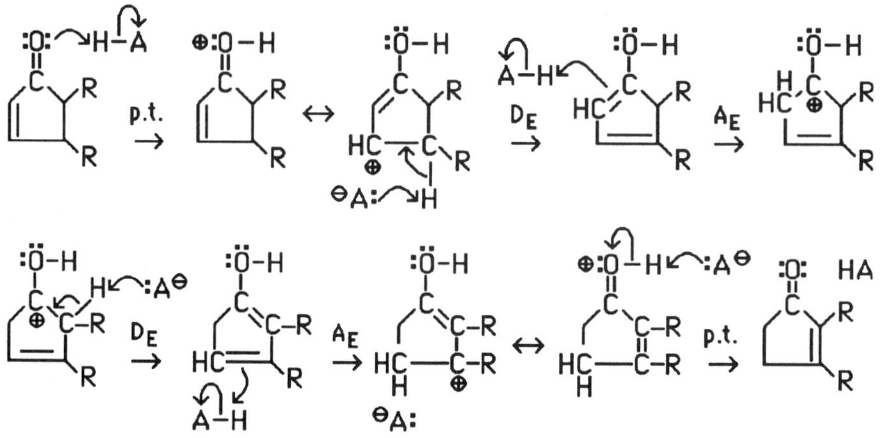

Problem 9.2d

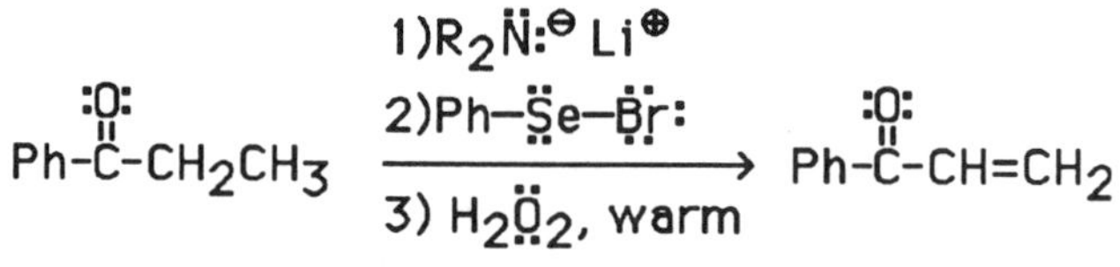

Understand the system. Source class: Z. $R_2N^\ominus$, pK_{abH} 36, is a strong base and a poor Nu. Sink class: C=Y. Media: basic. The ketone's CH_2 is the only acidic H, pK_a 19.2.

Find possible routes. Proton transfer.

$$Ph-\overset{\overset{\displaystyle :O:}{\|}}{C}-\overset{\underset{\displaystyle R_2\overset{\ominus}{\overset{..}{N}}:\curvearrowleft H}{|}}{\underset{\overset{\displaystyle Cl}{}}{C}}HCH_3 \quad \overset{p.t.}{\rightarrow} \quad Ph-\overset{\overset{\displaystyle :O:^{\ominus}}{\|}}{C}-\overset{..}{C}HCH_3 \qquad R_2\overset{..}{N}H \qquad Li^{\oplus}$$

Evaluate and cross-check. The $K_{eq} = 10^{+16.8}$, irreversible proton transfer.

Understand the system. The enolate is a good allylic source; PhSeBr is a Y—L sink.

Find possible routes. To displace the bromide on selenium we can use the S_N2.

$$Ph-\overset{\overset{\displaystyle :O:^{\ominus}}{\|}}{C}-\overset{\underset{\displaystyle Ph-\overset{..}{S}e\overset{}{\rightharpoonup}\overset{..}{B}r:}{}}{C}HCH_3 \qquad \overset{S_N2}{\rightarrow} \quad \pm \quad Ph-\overset{\overset{\displaystyle :O:}{\|}}{C}-\overset{\underset{\displaystyle Ph-\overset{..}{S}e:}{|}}{C}HCH_3 \quad :\overset{..}{B}r:^{\ominus}$$

Evaluate and cross-check. By the ΔpK_a rule this should be energetically downhill since we went from pK_{abH} of 19.2 to -7.

Understand the system. The selenium lone pairs are very nucleophilic. H_2O_2 is a Y—L sink.

Find possible routes. The S_N2 oxidizes the selenium.

$$Ph-\overset{\overset{\displaystyle :O:}{\|}}{C}-\overset{\underset{\displaystyle Ph-\overset{..}{S}e:\curvearrowright \overset{H\,\curvearrowleft}{:\overset{..}{O}-\overset{..}{O}H}}{|}}{C}HCH_3 \qquad \overset{S_N2}{\rightarrow} \quad Ph-\overset{\overset{\displaystyle :O:}{\|}}{C}-\overset{\underset{\displaystyle Ph-\overset{\oplus}{\underset{..}{S}e}-\overset{..}{O}-H}{|}}{C}HCH_3 \quad {}^{\ominus}:\overset{..}{O}H$$

Evaluate and cross-check. The reaction has broken a weak O—O bond.

Understand the system. The SeOH is acidic and the hydroxide is a good base.

Find possible routes. Proton transfer generates water and the selenoxide.

$$Ph-\overset{\overset{\displaystyle :O:}{\|}}{C}-\overset{\underset{\displaystyle Ph-\overset{\oplus}{\underset{..}{S}e}-\overset{..}{O}\underset{}{\rightharpoondown}H\curvearrowleft {}^{\ominus}:\overset{..}{O}H}{|}}{C}HCH_3 \qquad \overset{p.t.}{\rightarrow} \quad Ph-\overset{\overset{\displaystyle :O:}{\|}}{C}-\overset{\underset{\displaystyle Ph-\overset{\oplus}{\underset{..}{S}e}-\overset{..}{\underset{\ominus}{O}}:}{|}}{C}HCH_3 \quad H-\overset{..}{O}H$$

Evaluate and cross-check. The pK_a for the protonated selenoxide is not on the chart. We can guess that an oxygen anion bound to a positive atom is more stabilized than the oxygen anion of hydroxide (bound only to H). Recall that $R_3NOH^{\oplus}$ is about 11 pK_a units more acidic than water.

Understand the system. The oxygen anion is basic and the positive selenium can serve as a leaving group.

Find possible routes. This system is perfectly set up for the Ei elimination.

$$Ph-\overset{\overset{\displaystyle :O:}{\|}}{C}-\overset{\underset{\displaystyle Ph-\overset{\oplus}{\underset{\underset{\ominus}{\overset{..}{O}:}}{S}e}\curvearrowleft}{|}}{C}H\overset{\curvearrowleft H}{\underset{}{C}}H_2 \qquad \overset{Ei}{\rightarrow} \quad Ph-\overset{\overset{\displaystyle :O:}{\|}}{C}-CH=CH_2 \qquad \underset{\displaystyle \overset{..}{O}H}{Ph-\overset{..}{S}e}$$

Evaluate and cross-check. This intramolecular elimination reaction that produces our final product is expected to be fast.

Overview: The overall process has been a deprotonation, two S_N2 substitutions, a proton transfer, and an Ei elimination.

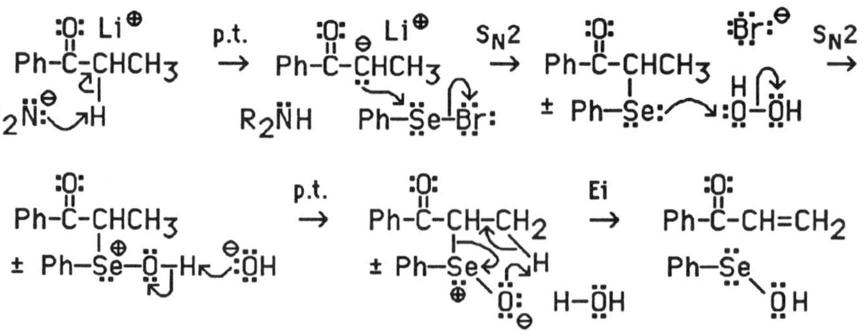

:O: Li⊕   p.t.   :O: Li⊕   S$_N$2   :O: :B̈r:⊖   S$_N$2
Ph-C-CHCH$_3$ → Ph-C-CHCH$_3$ → Ph-C-CHCH$_3$ →
R$_2$N̈:⤴H   R$_2$N̈H   Ph-S̈e-B̈r: ± Ph-S̈e: :O̤-ÖH

:O: p.t.   :O: Ei   :O:
Ph-C-CHCH$_3$ → Ph-C-CH-CH$_2$ → Ph-C-CH=CH$_2$
± Ph-S̈e-Ö-H⤴:ÖH   ± Ph-S̈e⊕   Ph-S̈e
           :Ö:⊖ H-ÖH   ÖH

Problem 9.2e

$$\text{R-C(=O)-CH}_2\text{-C(=O)-ÖEt} \xrightarrow[\text{heat}]{H_3O^⊕} \text{R-C(=O)-CH}_3 + \text{HÖEt} + CO_2$$

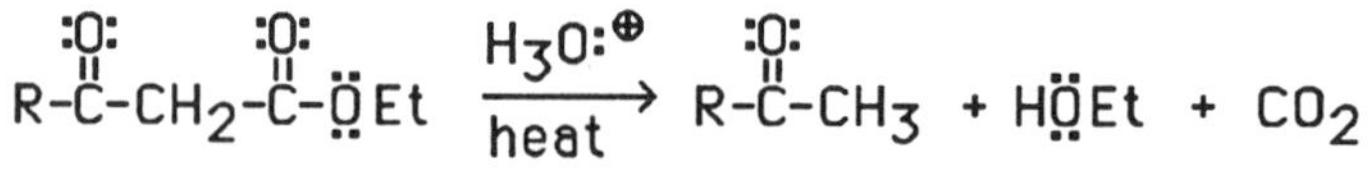

Understand the system. Source class: Z. Sink class: HA, C=Y, Y=C–L. Media: acidic, $H_3O^⊕$ pK_a is -1.7. Most basic site: the ester carbonyl lone pair (protonated ester pK_a is -6.5). Overall process: The alcohol portion of the ester has been dropped off, and the ester carbonyl has been lost as carbon dioxide. A common path to lose CO_2 that we know of is the decarboxylation of a carboxylic acid via path Ei. We can then set as an intermediate goal the conversion of the ester to the carboxylic acid. Therefore we will concentrate our first effort on replacing the OEt with OH by acidic media paths of adding H_2O then dropping off EtOH.

Find possible routes. Two possible routes are available to add water to the ester in acidic media: the hetero Ad$_E$2 (path p.t., then path Ad$_N$),

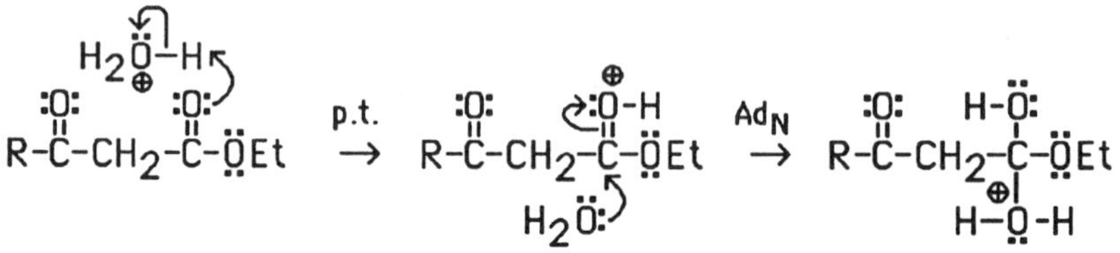

$$\text{R-C(=O)-CH}_2\text{-C(=O)-ÖEt} \xrightarrow{\text{p.t.}} \text{R-C(=O)-CH}_2\text{-C(=O-H)-ÖEt} \xrightarrow{\text{Ad}_N} \text{R-C(=O)-CH}_2\text{-C(-ÖEt)(-Ö H)-O-H}$$

(with $H_2\overset{⊕}{Ö}$-H and $H_2\ddot{O}:$ reagents)

or the hetero Ad$_E$3.

$$\text{R-C(=O)-CH}_2\text{-C(=O)-ÖEt} \xrightarrow{\text{Ad}_E3} \text{R-C(=O)-CH}_2\text{-C(-ÖEt)(-Ö H)-O-H}$$

(with $H_2\overset{⊕}{Ö}$-H and $H_2\ddot{O}:$ reagents)

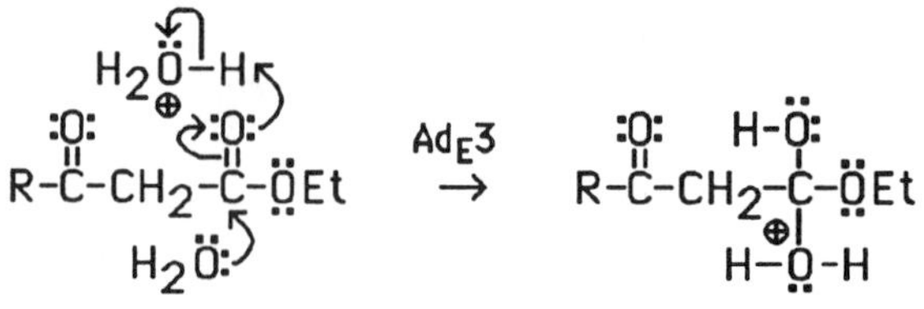

Evaluate and cross-check. Either route is reasonable; the Ad$_E$2 is commonly drawn in many texts.
Understand the system. The product should be a good acid; proton loss to water gives us the hydrate of the ester. Proton transfer between the various lone pairs of the hydrate should be fast.
Find possible routes. Proton transfer paths.

:O: H-Ö: p.t.   :O: H-Ö: ⟶H-ÖH₂ p.t.   :O: H-Ö: H :ÖH₂
R-C-CH₂-C-ÖEt   →   R-C-CH₂-C-ÖEt   ⊕   →   R-C-CH₂-C-ÖEt
H₂Ö:⟶H-O-H   :O-H   H-Ö: ⊕
        ⊕

Evaluate and cross-check. We have gone slightly uphill in energy since a protonated ether is slightly more acidic than a protonated alcohol.

Understand the system. Protonation of the ether oxygen is heading in the right direction since we want to drop off ethanol as a leaving group. Ethoxide (EtOH pK_{aHL} is 16) is a poor leaving group and protonation makes it much better (EtOH$_2^\oplus$ pK_{aHL} is -2.4).

Find possible routes. The lone-pair-assisted E1, path E$_\beta$, then path p.t.

:O: H-Ö: H   E$_\beta$:O: ⊕Ö-H⟶:ÖH₂   p.t.   :O: :O: H₃Ö⊕
R-C-CH₂-C-O: ⊕   →   R-C-CH₂-C   →   R-C-CH₂-C
        H-Ö: Et   :O-H HÖEt   :O-H

Evaluate and cross-check. The E2 is also a reasonable alternative (by the principle of microscopic reversibility) since it is the reverse of the Ad$_E$3 addition, which we found to be an acceptable alternative mode of addition in the first step.

Understand the system. We have achieved our intermediate goal, the carboxylic acid, and now can work on losing carbon dioxide.

Find possible routes. Path Ei, the cyclic decarboxylation,

:O: ⟶H-Ö: Ei   :Ö-H   :O:
R-C-CH₂-C   →   R-C=CH₂ +   C
         :O: :O:

or protonation of the carbonyl followed by the E2 with the enol as the leaving group.

H₂Ö-H⟶:O: :ÖH   p.t.   H-Ö⊕   :Ö⟶H⟶:ÖH₂ E2   H-Ö: :O: HÖH₂
⊕   R-C-CH₂-C   →   R-C-CH₂-C   →   R-C=CH₂   C
            :O: :O: :O:

Evaluate and cross-check. The intramolecular process is most likely faster. A ΔH calculation indicates that the reaction is endothermic by 20 kcal/mol (84 kK/mol), but that is why heat is necessary. The reaction is driven by the loss of carbon dioxide gas from the reaction mixture.

Understand the system. The product of the decarboxylation is the enol. The final product is just a tautomerization away.

Find possible routes. Acid-catalyzed tautomerization to a ketone, path A$_E$, then p.t.

:Ö-H   A$_E$ ⊕Ö-H⟶:ÖH₂   p.t.   :O: H₃Ö⊕
R-C=CH₂   →   R-C-CH₂   →   R-C-CH₃
    ⟶H-ÖH₂   H
        ⊕

Evaluate and cross-check. From a ΔH calculation, the ketone is downhill.

Overview: The overall process has been an Ad$_E$2, two proton transfers, a lone-pair-assisted E1, an Ei decarboxylation, then a tautomerization.

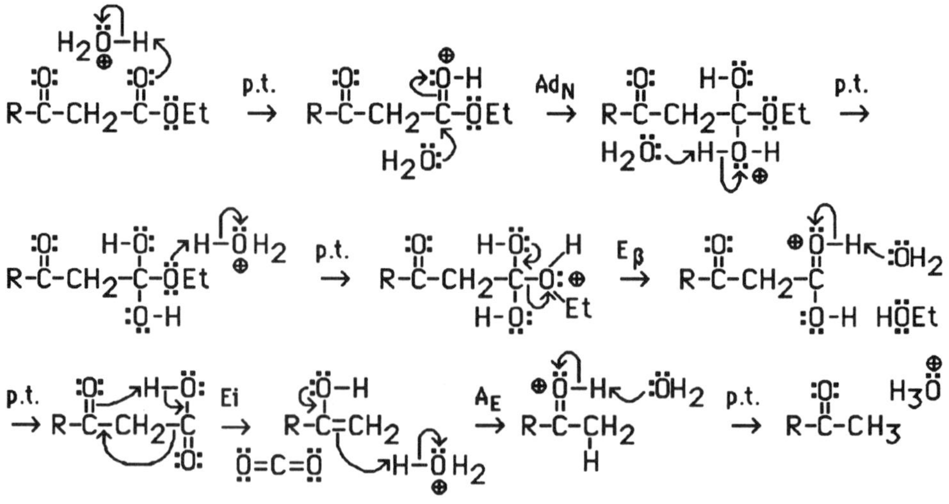

Problem 9.2f   (Robinson annulation)

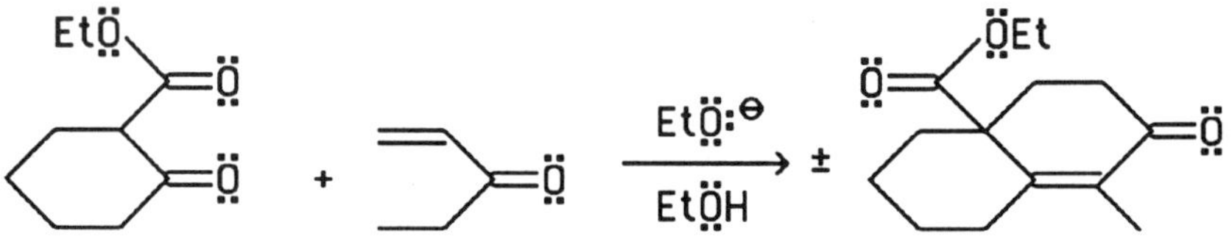

Because of the length of this example only the most probable paths off each point will be discussed.

Understand the system. Balanced? No, H_2O is missing from the right hand side. Medium: Basic, ethoxide conjugate acid pK_{abH} 16. Sources: Ethoxide. Sinks: The ketone is a polarized multiple bond, the ester is a polarized multiple bond with leaving group, and the enone is a conjugate acceptor. Acidic H's: The CH between the ester and the ketone is the most acidic H at pK_a 10.7. Leaving groups: No good ones, only the ethoxide on the ester. Resonance forms:

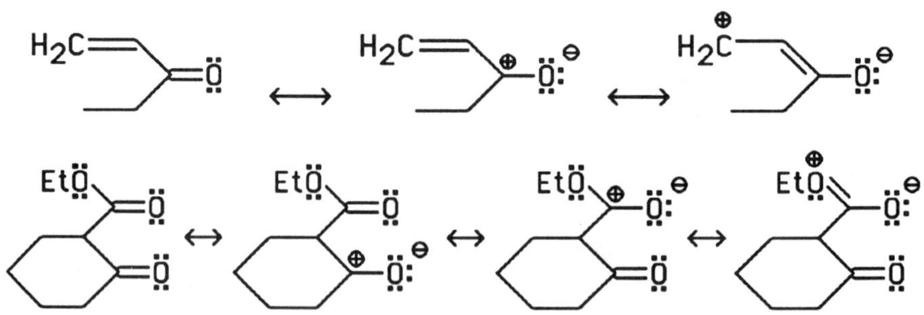

Find the pieces of the starting materials in the product and determine what bonds have been made. The new bonds are boldfaced in the following structure; they connect the two reactant pieces. Reasoning back from adjacent functionality, determine the possible nucleophile and electrophile pairs for each of the new bonds. Atoms next to electron-withdrawing groups can easily become anionic and therefore nucleophilic. The carbonyl and the conjugate acceptor carbons are already electrophilic. Generic process: An addition has occurred to produce the single bond and an addition and an elimination has occurred to create the double bond.

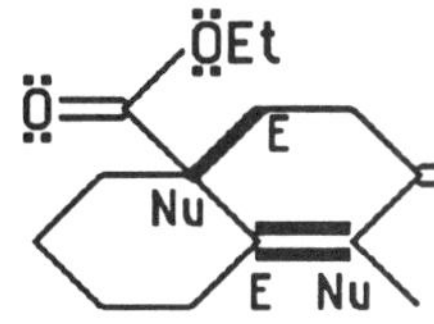

Find possible routes. The lone pair electrons of ethoxide are the best source; but if ethoxide had served as a nucleophile, we would expect an additional ethoxy group in the product. Since that is not the case, we should explore ethoxide as a base first, especially since there is an acidic H.

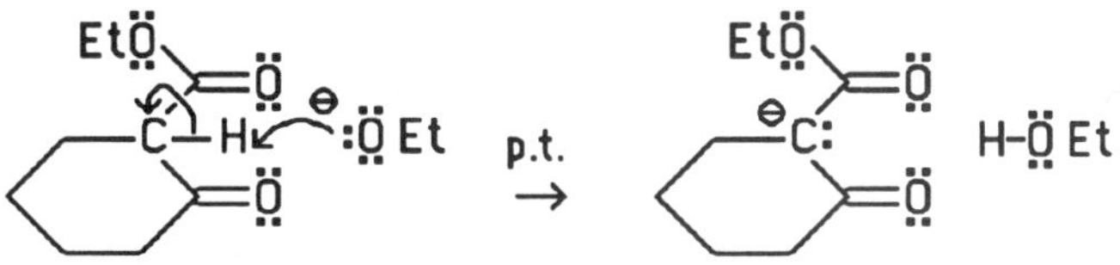

Evaluate and cross-check. The proton transfer has a very good K_{eq} of $10^{(16-10.7)} = 10^{+5.3}$. The reaction fits the medium and is downhill in energy.

Understand the system. Now there is a nucleophile where required for the single bond formation.

Find possible routes. The delocalized anion is a soft nucleophile and is expected to add conjugate to the enone (path Ad_N).

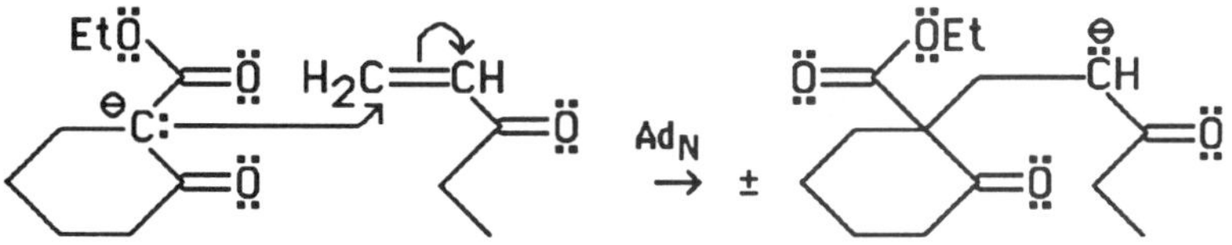

Evaluate and cross-check. If this stabilized nucleophile were to add to the ester or the ketone, the addition would simply reverse because the stabilized anion would also be the best leaving group. Although the pK_a of the product is about 9 pK_a units more basic than the nucleophile, a weak pi bond has been broken and a strong sigma bond has been formed, making this reaction's energetics hard to guess.

Understand the system. The anion from the conjugate addition can act as a nucleophile or as a base.

Find possible routes. The product has a CH_2 at the anionic carbon. Protonation of this anion gets us closer to product.

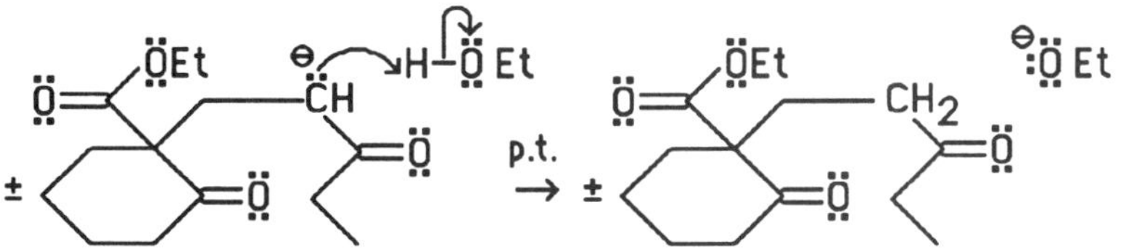

Evaluate and cross-check. Only one path was found that heads toward product. If the anion were to attack one of the carbonyls instead, it would form a strained four-membered ring, and therefore the addition would reverse. The K_{eq} for this proton transfer step is $10^{(19.2-16)} = 10^{+3.8}$ and will act to make the last two steps much more favorable, which overall compose an Ad_N2 process.

Understand the system. The only acidic H's within range of ethoxide's pK_{abH} are those next to the two ketones. The remaining bonds left to be made require a nucleophile at the CH_2 of the ethyl ketone.

Find possible routes. Proton transfer produces the anion shown. Even though the CH_2 next to the ring ketone has the same pK_a and is expected to be reversibly deprotonating under these reaction conditions, a glance at the product shows us that that CH_2 has not changed.

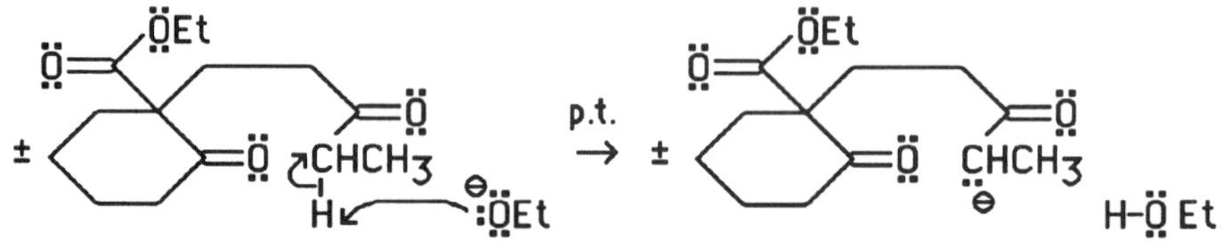

Evaluate and cross-check. Again, only one path headed towards product. The K_{eq} for this proton transfer is $10^{(19.2-16)} = 10^{+3.8}$ and is therefore somewhat uphill.

Understand the system. Now we have the nucleophile that we need to make the sigma portion of the double bond. The best electrophile is the ketone, not the ester.

Find possible routes. The ketone enolate can attack the ketone in an Ad_N2 process (path Ad_N, then by p.t.). We should feel we are on familiar ground; mechanism example 2 showed how to take an enolate and a carbonyl to an enone.

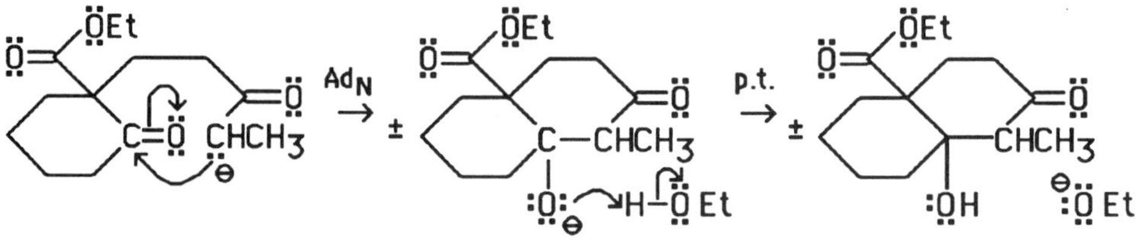

Evaluate and cross-check. Again, only one path headed towards product. The intermediate anion increases in basicity, but overall the Ad_N2 forms a weaker base.

Understand the system. Now all that is needed to get to product is to eliminate hydroxide to make the pi portion of the double bond.

Find possible routes. Either the E2 or the E1cB would be appropriate, but since the carbonyl makes the CH_2 acidic and the leaving group is poor, the most probable route would be an E1cB or E1cB-like process.

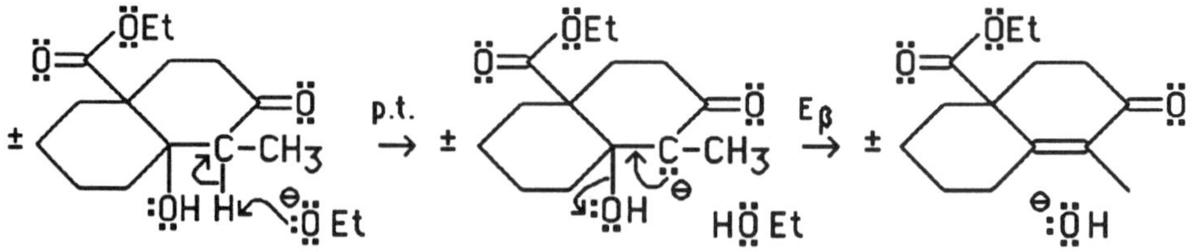

Evaluate and cross-check. The leaving group is no more basic than the incoming base and therefore checks by the ΔpK_a rule.

Overview: Let's look at how the mechanism for this transformation or "mechanistic sentence" can be made from our "mechanistic phrases" or path combinations: an Ad_N2 addition to a polarized multiple bond followed by a proton transfer, then another Ad_N2, and finally an E1cB elimination.

9.3a

$$CH_3CH_2CH_2\ddot{B}\ddot{r}: + CH_3\ddot{\underset{..}{S}}:^{\ominus} \rightarrow$$

Understand the system. Medium: Basic, $CH_3S^{\ominus}$ pK_{abH} is 10.6. Sources: $CH_3S^{\ominus}$, a lone pair source. Sinks: The primary sp^3 bound leaving group. Acidic H's: None. Leaving groups: Bromide is an excellent leaving group.

Find possible routes. Proton transfer is unreasonable because it would have a K_{eq} much greater than 10^{-20}. Substitution or elimination are the only routes left for this sink; the starting point is on the $S_N2/E2$ surface, not the $S_N1/E1$ surface, since the loss of the leaving group would produce an unstable primary carbocation.

S_N2:

$$CH_3\ddot{\underset{..}{S}}:^{\ominus} \qquad \xrightarrow{S_N2} \qquad CH_3\ddot{\underset{..}{S}}:$$
$$CH_3CH_2CH_2\!-\!\ddot{B}\ddot{r}: \qquad\qquad CH_3CH_2CH_2 \quad :\ddot{B}\ddot{r}:^{\ominus}$$

or the E2:

$$CH_3\ddot{\underset{..}{S}}:^{\ominus}\!\!\!-\!\!H \qquad \xrightarrow{E2} \quad CH_3\ddot{\underset{..}{S}}\!-\!H$$
$$CH_3\overset{|}{\underset{|}{C}}\!\!\!=\!\!CH_2\!-\!\ddot{B}\ddot{r}: \qquad\qquad CH_3\overset{|}{C}\!\!=\!\!CH_2 \quad :\ddot{B}\ddot{r}:^{\ominus}$$
$$H \qquad\qquad\qquad\qquad H$$

Evaluate and cross-check. We must now make a major decision, substitution versus elimination, as discussed in Section 8.5. Since the leaving group is on an unhindered primary site and the $CH_3S^{\ominus}$ is a good nucleophile and a weak base, substitution is preferred. The ΔpK_a rule checks. The nucleophile had a pK_{abH} of 10.6 and the leaving group -9; the reaction has definitely formed the weaker base. Since the product appears not to be able to react further, we are done.

9.3b

$$H_3C\!-\!\overset{:\ddot{O}:}{\underset{||}{C}}\!-\!\ddot{\underset{..}{O}}\!-\!\overset{:\ddot{O}:}{\underset{||}{C}}\!-\!CH_3 \quad + \quad \langle\!\!\!\bigcirc\!\!\!\rangle \quad \xrightarrow{AlCl_3}$$

Understand the system. Medium: Acidic; $AlCl_3$ is a strong Lewis acid and will generate HCl with any trace of moisture. Solvent: Benzene, nonpolar. Sources: The lone pairs of the anhydride and benzene, an aromatic source. Leaving groups: Acetate is a fair leaving group. Sinks: The anhydride is a polarized multiple bond with an attached leaving group, and the Lewis acid is excellent. The best sink is the electron-deficient Lewis acid. Resonance forms: The last two resonance forms are very minor, for the second carbonyl's partial plus destabilizes the positive oxygen.

$$\underset{H_3C}{\overset{:\ddot{O}:}{}}\overset{:\ddot{O}:}{\underset{\ddot{O}}{}}\underset{CH_3}{} \leftrightarrow \underset{H_3C}{\overset{\ominus:\ddot{O}:}{}}\overset{:\ddot{O}:}{\underset{\ddot{O}}{}}\underset{CH_3}{} \leftrightarrow \underset{H_3C}{\overset{:\ddot{O}:}{}}\overset{:\ddot{O}:\ominus}{\underset{\ddot{O}}{}}\underset{CH_3}{}$$

$$\leftrightarrow \underset{H_3C}{\overset{:\ddot{O}:\ominus\ :\ddot{O}:}{}}\underset{CH_3}{} \leftrightarrow \underset{H_3C}{\overset{:\ddot{O}:\ \ominus:\ddot{O}:}{}}\underset{CH_3}{}$$

Find possible routes. The Lewis acid is expected to complex with any lone pair source, path A_N. The anhydride is a lone pair source and can complex with the Lewis acid at two different sites.

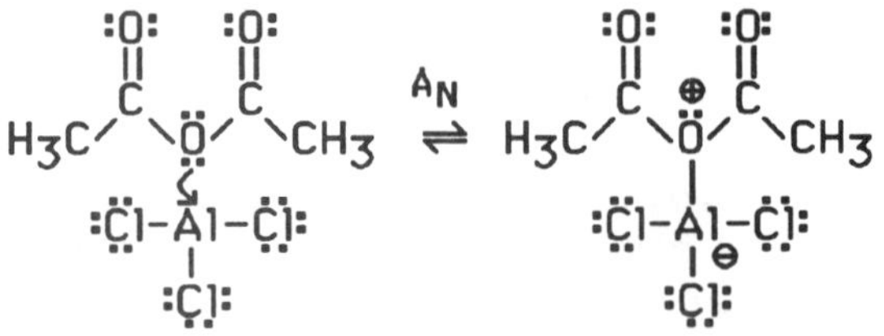

or

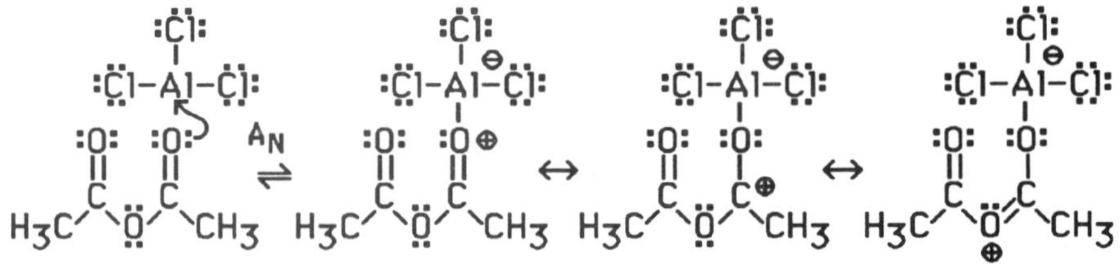

Evaluate and cross-check. The complexation with the ether oxygen is the less favored process since it creates an oxonium ion between two ewg's, the carbonyls. The best site for Lewis acid complexation is the carbonyl lone pair. The Lewis acid complex has a delocalized charge and is consistent with a strongly acidic medium.

Understand the system. The complex has several resonance forms. Complexation increases the leaving group's ability to ionize. The leaving group is now good to excellent, equivalent to a protonated acid, pK_{aHL} -6. The only source is the aromatic ring, which needs an excellent electrophile in order to react.

Find possible routes. By the criteria established in Section 8.3 we should expect ionization of the leaving group, path D_N.

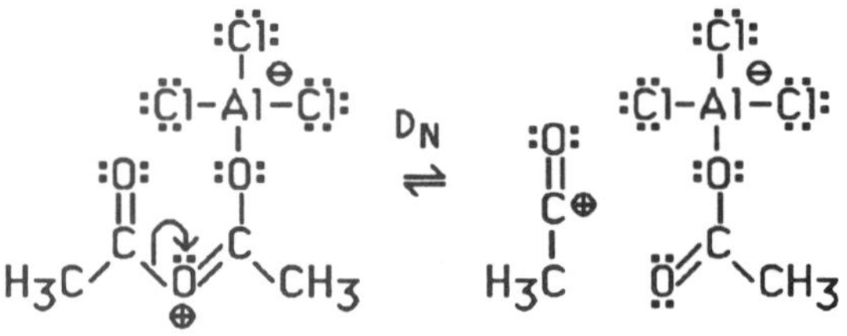

Evaluate and cross-check. The other path not discussed is the loss of the Lewis acid, but that returns to reactants. The ionization is probably uphill in energy because although the acylium ion is lone pair stabilized, it is also vinyl.

Understand the system. The acylium ion is an excellent electron sink.

$$CH_3-\overset{\oplus}{C}=\overset{..}{O}: \quad \leftrightarrow \quad CH_3-C\equiv\overset{\oplus}{O}:$$

Find possible routes. The aromatic ring can now undergo electrophilic aromatic substitution, path A_E followed by D_E.

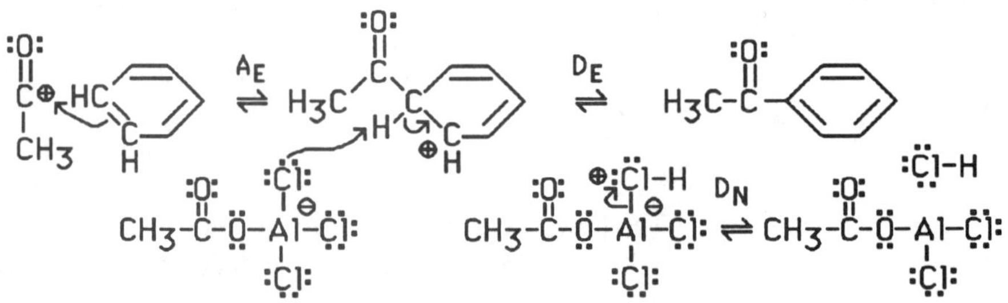

Evaluate and cross-check. The proton is most likely removed by the Lewis acid complex or by chloride to produce HCl, which in the nonpolar solvent does not ionize readily and will escape as a gas.

Overview: The carbonyl is an electron-withdrawing group and deactivates the aromatic ring to further attack. We have reached a logical end point.

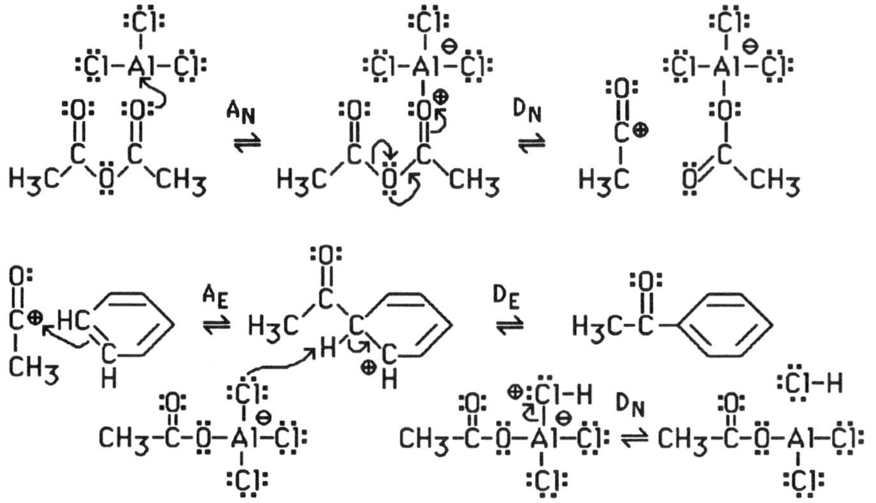

Problem 9.3c

$$CH_3CH_2CH_2CH_2CH_2-\ddot{B}r: \xrightarrow{Ph_3P:}$$

Understand the system. Source class: Z. Sink class: $\geqslant C-L$. Media: Neutral. Most basic site: Ph$_3$P lone pair (pK_{abH} is 2.7).

Find possible routes. Substitution or elimination. However, there is no group basic enough to do an elimination. The two possible substitution routes are S_N1 (path D_N, then A_N)

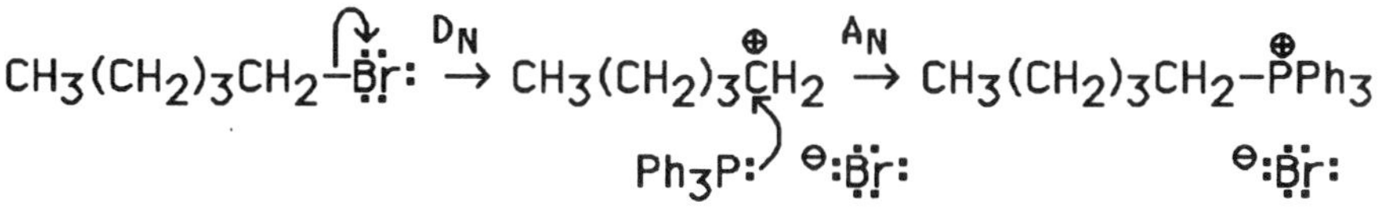

and S_N2.

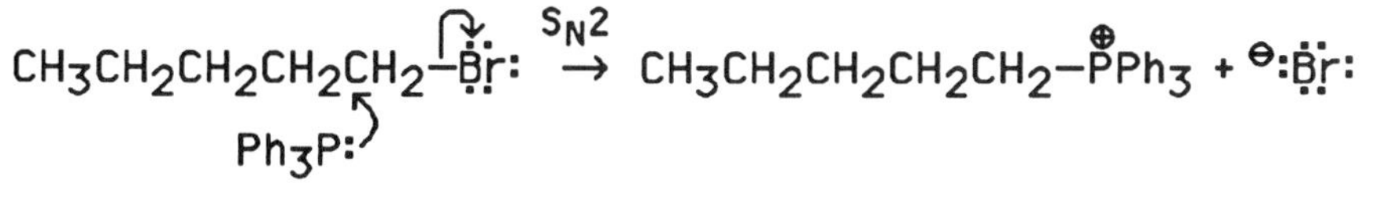

Evaluate and cross-check. The S_N1 goes via an unstable primary carbocation and thus is not reasonable. The S_N2 is the best route. By the ΔpK_a rule the reaction has gone downhill in energy (pK_{abH} dropping from 2.7 to -7).

Understand the system. The product is a salt, and it is not electron deficient. Although the ⊕PPh$_3$ is a fair leaving group, the only nucleophiles are bromide (which would take

us back to starting materials) and PPh₃, our original nucleophile. We are done.
Overview: The overall process has been an S_N2.

$$CH_3CH_2CH_2CH_2CH_2\text{--}\ddot{B}r\text{:} \xrightarrow{S_N2} CH_3CH_2CH_2CH_2CH_2\text{--}\overset{\oplus}{P}Ph_3 + {}^{\ominus}\text{:}\ddot{B}r\text{:}$$

$$Ph_3P\text{:}$$

Problem 9.3d

$$Ph\text{--}\overset{\overset{\displaystyle :O:}{\|}}{C}\text{--}\ddot{C}l\text{:} \xrightarrow[H_2\ddot{O}]{\text{:NH}_3}$$

Understand the system. Source class: Z. Sink class: Y=C—L. Media: Mildly basic, the
pK_a of NH₄⁺ is 9.2. Most basic site and best nucleophile: ammonia lone pair.
Find possible routes. Nucleophilic addition, path Ad_N.

$$Ph\text{--}\overset{\overset{\displaystyle :O:}{\|}}{C}\text{--}\ddot{C}l\text{:} \xrightarrow{Ad_N} \pm\ Ph\text{--}\overset{\overset{\displaystyle {}^{\ominus}:\ddot{O}:}{|}}{C}\text{--}\ddot{C}l\text{:}$$

$$H_3N\text{:} \qquad\qquad\qquad {}^{\oplus}NH_3$$

Evaluate and cross-check. This should be a relatively stable anion due to the electron
withdrawal of the positive nitrogen and the electronegative chlorine.
Understand the system. The chlorine is an excellent leaving group, and the protons on
the positive nitrogen are acidic.
Find possible routes. Path E_β, then deprotonation path p.t.,

$$Ph\text{--}\overset{\overset{\displaystyle {}^{\ominus}:\ddot{O}:}{|}}{C}\text{--}\ddot{C}l\text{:} \xrightarrow{E_\beta} Ph\text{--}\overset{\overset{\displaystyle :O:}{\|}}{C} \ \text{:}\ddot{C}l\text{:}^{\ominus} \xrightarrow{p.t.} Ph\text{--}\overset{\overset{\displaystyle :O:}{\|}}{C}\text{--}\ddot{N}H_2 + \text{:}\ddot{C}l\text{:}^{\ominus}$$

$$H_3\overset{\oplus}{N} \qquad\quad H_3N\text{:}\frown H\overset{\oplus}{-}NH_2 \qquad\qquad H_4N^{\oplus}$$

or deprotonation path p.t., then path E_β.

$$Ph\text{--}\overset{\overset{\displaystyle {}^{\ominus}:\ddot{O}:}{|}}{C}\text{--}\ddot{C}l\text{:} \xrightarrow{p.t.} Ph\text{--}\overset{\overset{\displaystyle {}^{\ominus}:\ddot{O}:}{|}}{C}\text{--}\ddot{C}l\text{:} \xrightarrow{E_\beta} Ph\text{--}\overset{\overset{\displaystyle :O:}{\|}}{C}\text{--}\ddot{N}H_2 + \text{:}\ddot{C}l\text{:}^{\ominus}$$

$$H_3N\text{:}\frown H\overset{\oplus}{-}NH_2 \qquad H_4N^{\oplus} \ H_2\ddot{N} \qquad\qquad H_4N^{\oplus}$$

Evaluate and cross-check. Loss of an excellent leaving group from an anionic
intermediate should be rapid, but this process produces the somewhat destabilized nitrogen
protonated amide. Proton transfer in water is often at the diffusion controlled rate. Both
routes are good and produce the same product, the amide.
Understand the system. Amides are one of the least reactive of the class Y=C—L; without
any powerful electrophiles or nucleophiles in the reaction mixture, this is a reasonable
place to stop.
Overview: The overall process has been an addition followed by (in either order)
elimination and proton transfer.

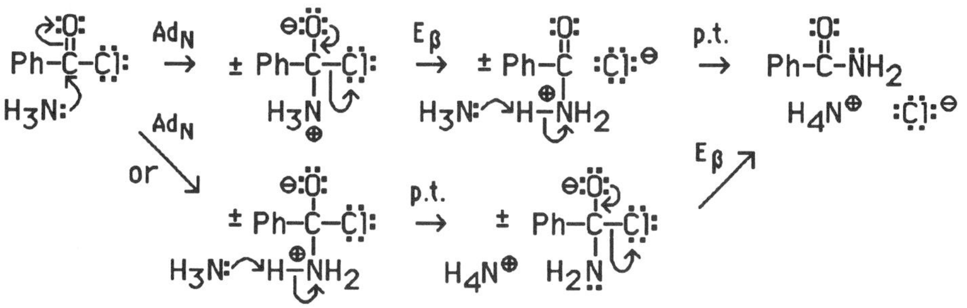

Problem 9.3e

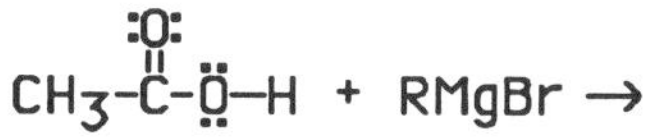

Understand the system. Source class: RM. Sink class: HA or Y=C–L. Media: Acidic, CH_3COOH pK_a is 4.8. Most basic site: the anionic carbon atom of the organometallic, pK_{abH} about 50.

Find possible routes. The two routes are nucleophilic addition (path Ad_N),

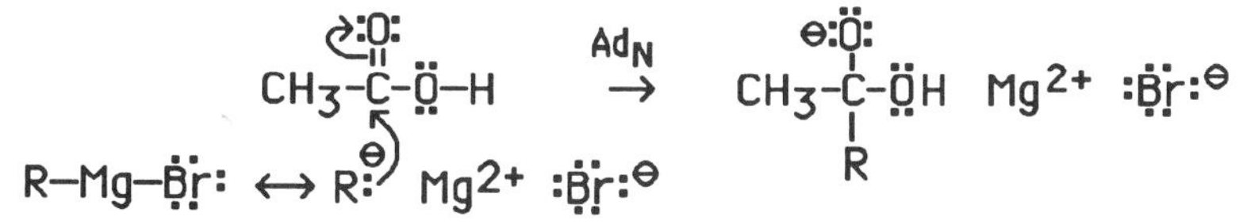

or protonation path p.t.

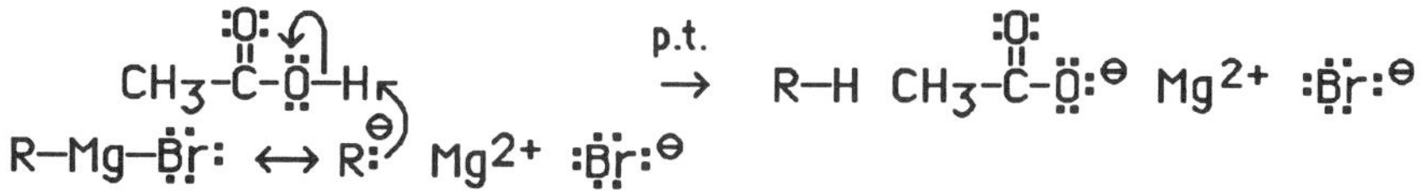

Evaluate and cross-check. From the large drop in pK_a's we can tell that both processes will be irreversible. Therefore the product will be determined by which reaction proceeds the fastest. Proton transfer from an acidic heteroatom H to the organometallic is almost always faster than nucleophilic attack.

Understand the system. Carboxylates are the least reactive of the class Y=C–L, and there are no nucleophiles present that will add to it.

Overview: The overall process has been the protonation of the organometallic.

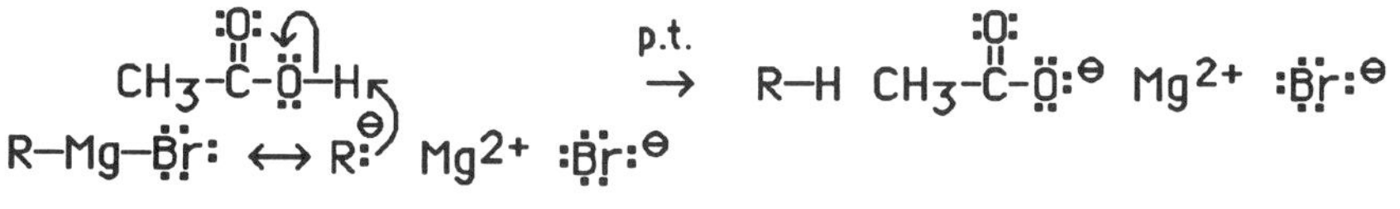

Problem 9.3f

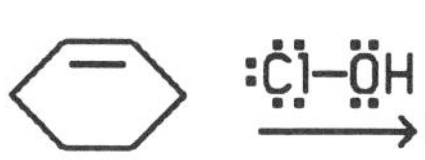

Understand the system. Source class: C=C. Sink class: Y–L or Nu–L (since oxygen is more electronegative than chlorine, chlorine is the ∂+ end and would be attacked by the electron source). Media: neutral.

Find possible routes. Two possible routes are electrophilic addition (path A$_E$),

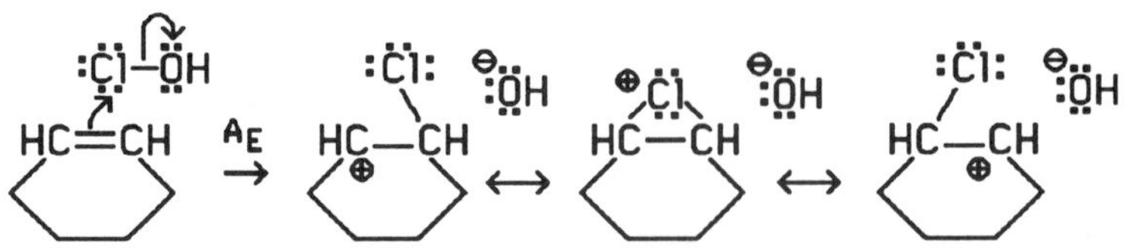

and Nu−L addition.

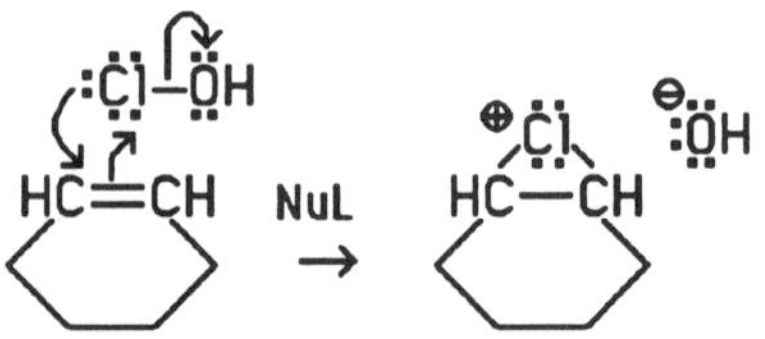

Evaluate and cross-check. Since both routes go to different resonance forms of the same bridged cation, they are essentially the same route. Since the secondary carbocation is not very stable, bridging is expected to predominate.

Understand the system. The bridged cation is a $\geqslant$C−L, and hydroxide is a good nucleophile.

Find possible routes. Substitution by S$_N$1 (path D$_N$, then A$_N$),

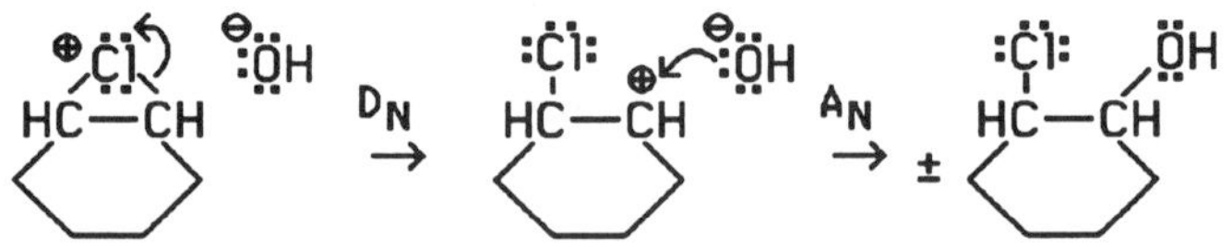

or S$_N$2.

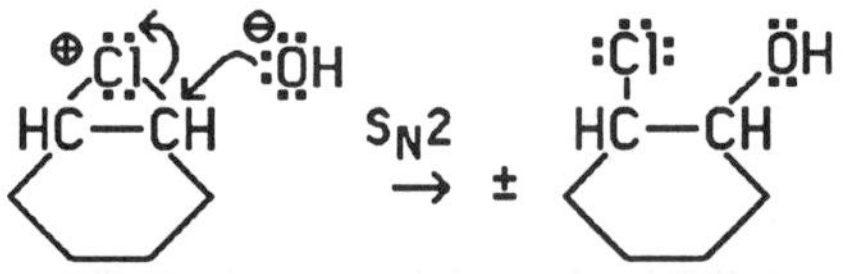

Evaluate and cross-check. The first step of the S$_N$1 route, ionization of a leaving group, gives us another resonance form of the bridged ion. If this open resonance form were predominant (which it is not because of the unstable secondary carbocation), we would expect the nucleophile to be able to attack from either face, giving a mixture of product stereoisomers. Since the bridged resonance form predominates, the S$_N$2 is most likely the best route.

Understand the system. The overall addition has been *anti* since attack would have to occur from the opposite face from the bridging chlorine for steric reasons. This is a good stopping point; a ΔH calculation indicates we have gone downhill in energy by 52 kcal/mol (218 kJ/mol) from our original reactants.

Overview: The overall process has been an Nu−L addition followed by an S$_N$2.

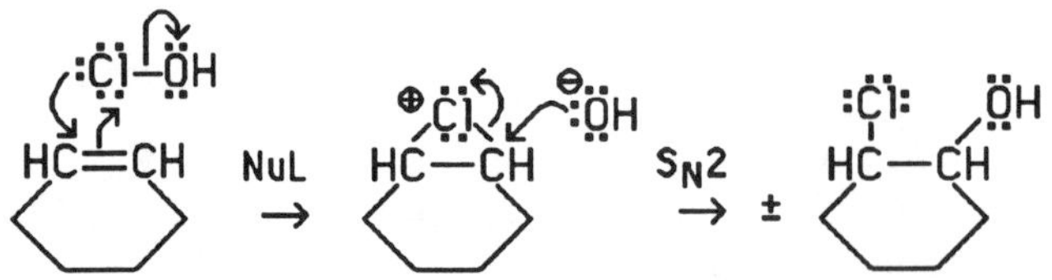

Problem   9.3g

$$\text{(cyclohexene oxide structure)} \quad \ddot{\text{O}} \quad \xrightarrow[H^{\oplus}]{R\ddot{\text{O}}H}$$

Understand the system. Source class: Z–C=C. Sink class: HA. Media: acidic. Basic sites: at either end of the Z–C=C.

Find possible routes. The two possible proton transfer routes are to oxygen,

$$\text{(structure) } HC{=}CH \quad \overset{H^{\oplus}}{\curvearrowright} \quad \xrightarrow[\to]{\text{p.t.}} \quad \text{(structure) } HC{=}CH \quad _{\oplus}\ddot{\text{O}}H$$

or to carbon.

$$\overset{H^{\oplus}}{\curvearrowright} \text{(structure) } HC{=}CH \quad \xrightarrow[\to]{\text{p.t.}} \quad \text{(structure) } \overset{H}{HC}{-}CH \quad \ddot{\text{O}}{:}_{\oplus} \quad \leftrightarrow \quad \text{(structure) } \overset{H}{HC}{-}\overset{\oplus}{CH} \quad {:}\ddot{\text{O}}{:}$$

Evaluate and cross-check. Both are strong acids; the protonated ether pK_a is -3.5, and the lone-pair-stabilized carbocation is even more acidic. Unless there is an excellent nucleophile present to attack the CH_2 next to the O, protonation on the heteroatom is a dead end. Since the reaction conditions indicate acid catalysis and a weak nucleophile, the protonation on carbon is the better route.

Understand the system. The lone-pair-stabilized carbocation is an excellent electron sink. It can deprotonate and return to reactants or add a nucleophile. Our only nucleophile is the lone pair of the alcohol.

Find possible routes. Nucleophilic addition to the highly polarized multiple bond (path Ad_N).

$$\text{(structure) } \overset{H}{HC}{-}\overset{R\ddot{\text{O}}H}{CH} \quad \ddot{\text{O}}{:}_{\oplus} \quad \xrightarrow[\pm]{Ad_N} \quad \text{(structure) } \overset{H}{HC}{-}\overset{R\overset{\oplus}{\ddot{\text{O}}}{-}H}{CH} \quad {:}\ddot{\text{O}}{:}$$

Evaluate and cross-check. If you drew the other resonance form of the carbocation, then this reaction would be the trapping of a carbocation by a nucleophile, path A_N. The protonated ether should be more stable than the lone-pair-stabilized carbocation.

Understand the system. The protonated ether is acidic, pK_a -3.5.

Find possible routes. Deprotonation (by solvent), path p.t.

$$\text{(structure) } \overset{H}{HC}{-}\overset{R\overset{\oplus}{\ddot{\text{O}}}{-}H\curvearrowright:A^{\ominus}}{CH} \quad {:}\ddot{\text{O}}{:} \quad \xrightarrow[\to]{\text{p.t.}} \quad \text{(structure) } \overset{H}{HC}{-}\overset{R\ddot{\text{O}}{:}\ \ H{-}A}{CH} \quad {:}\ddot{\text{O}}{:}$$

Evaluate and cross-check. A ΔH calculation indicates we have gone downhill in energy by 11 kcal/mol (46 kJ/mol) from our original reactants.

Understand the system. Nothing left to do but quit.

Overview: The overall process has been a lone-pair-stabilized version of the Ad_E2 followed by deprotonation of the nucleophile.

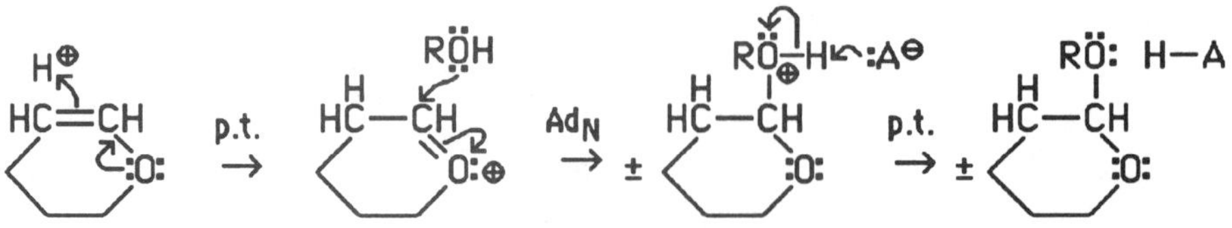

The organic reaction scheme at the top shows a series of steps (p.t., Ad_N, p.t.) involving an enol/carbonyl system with ROH and RÖ groups.

9.4a

$$Ph-\underset{\underset{CH_3}{|}}{\overset{\overset{H}{|}}{C}}-\ddot{N}(CH_3)_2 \xrightarrow[\text{2) Heat}]{\text{1) H}_2\text{O}_2}$$

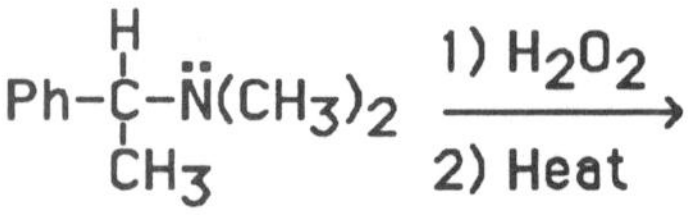

Understand the system. Medium: Mildly basic, because amines are weak bases. Sources: The amine lone pair. Leaving groups: Hydroxide. Sinks: Hydrogen peroxide has a very weak O–O bond and is a Y–L. Acidic H's: H_2O_2 has a pK_a of 11.6, close to the pK_{abH} of the amine, 10.7.

Find possible routes. Proton transfer and S_N2 substitution are possibilities.

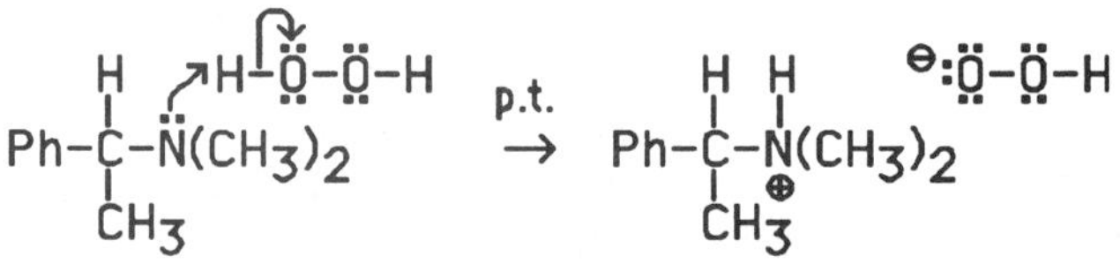

The reaction scheme shows the amine reacting with H_2O_2 via proton transfer (p.t.) to give the protonated amine with $^\ominus:\ddot{O}-\ddot{O}-H$.

or

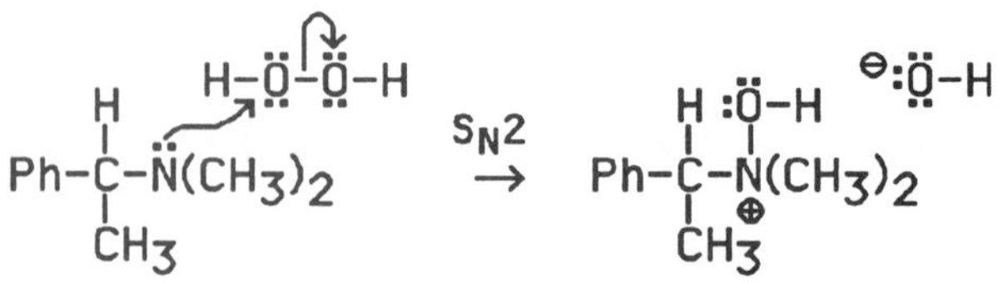

The reaction scheme shows the amine reacting with $H-\ddot{O}-\ddot{O}-H$ via S_N2 to give the amine oxide with $^\ominus:\ddot{O}-H$.

Evaluate and cross-check. The K_{eq} of the proton transfer step is calculated to be close to 0.1. A quick glimpse ahead down the proton transfer route shows that it is a dead end, although energetically quite reasonable. Amines generally are quite stable to acidic water. The S_N2 breaks a very weak bond and forms a stronger one. The ΔpK_a rule on the S_N2 shows a climb in pK_{abH} of five units (from 10.7 up to 15.7), but a weak bond has been broken and a stronger one formed so the reaction may be still energetically downhill.

Understand the system. Hydroxide is a relatively strong base, and the hydroxyl on the positive amine is definitely acidic.

Find possible routes. Proton transfer should be very fast, path p.t.

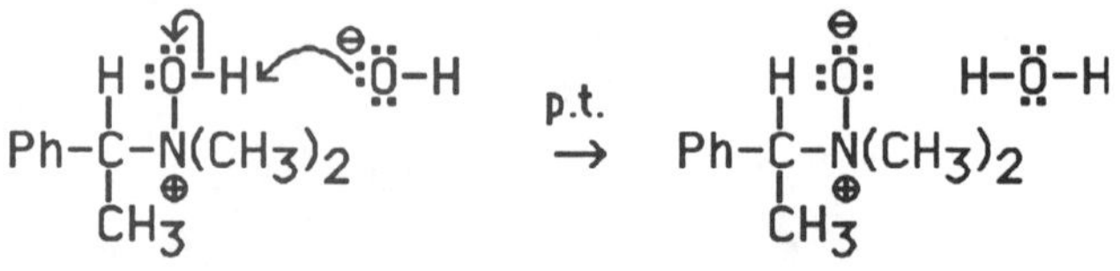

The reaction scheme shows the protonated amine oxide reacting with $^\ominus:\ddot{O}-H$ via proton transfer (p.t.) to give the amine oxide and $H-\ddot{O}-H$.

Evaluate and cross-check. The pK_a of the protonated amine oxide is 4.6, thus the K_{eq} is around $10^{+11.1}$, indicating irreversible reaction.

Understand the system. There are few paths left; let's consider step 2, heat.

Find possible routes. The amine oxide is set up for an internal *syn* elimination, path Ei. The oxygen anion serves as an internal base, and the positive nitrogen acts as a leaving group.

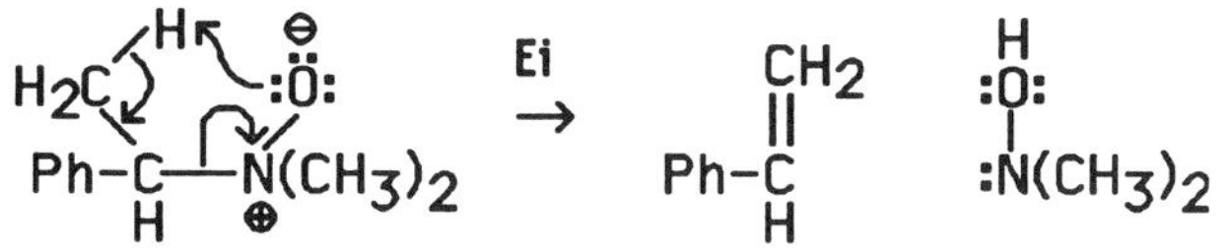

Evaluate and cross-check. A second path, the E2 elimination, would get to the same product but would be expected to be much slower than this intramolecular process. The reaction neutralizes charge and is energetically favorable. The products cannot react with each other, so this is a good stopping point.

Overview: The overall process is then an S_N2, a proton transfer, and a *syn* elimination.

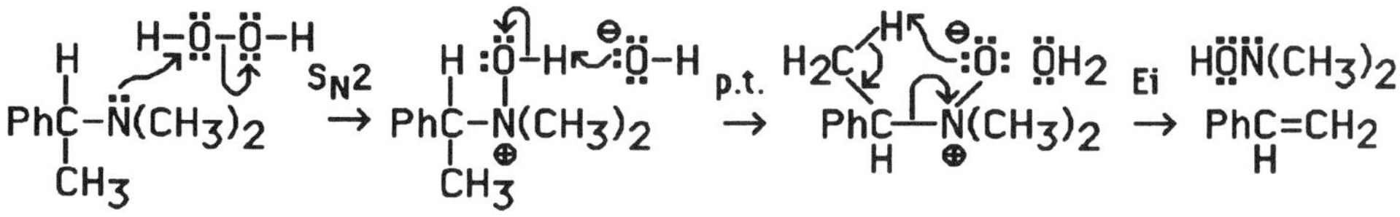

9.4b

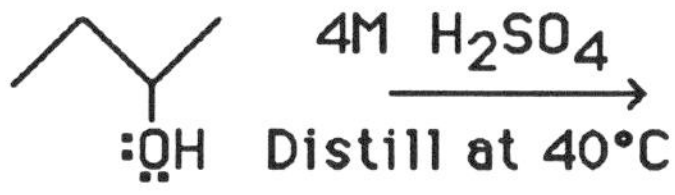

Understand the system. Medium: Strongly acidic. Sources: The alcohol lone pair. Leaving groups: The alcohol OH can get protonated and become a good leaving group. Sinks and acidic H's: Sulfuric acid.

Find possible routes. Since the leaving group is not good enough to ionize and sulfate is a poor nucleophile, the most probable path is proton transfer from the strong acid to the lone pair of the alcohol, path p.t.

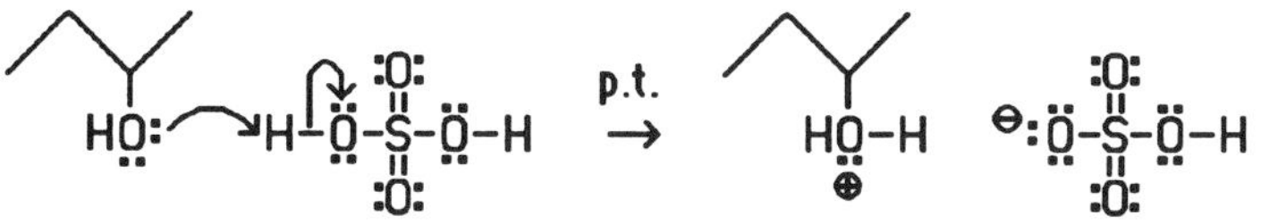

Evaluate and cross-check. Only one path found. The reaction has produced a weaker base.

Understand the system. Now we have generated a good leaving group bound to an sp^3 hybridized carbon. Since the reaction is in acidic medium and there is a good leaving group, we are at a well-known junction, the starting point of the $S_N1/E1$ energy surface.

Find possible routes. There are two routes available, the S_N1 (path D_N, then A_N),

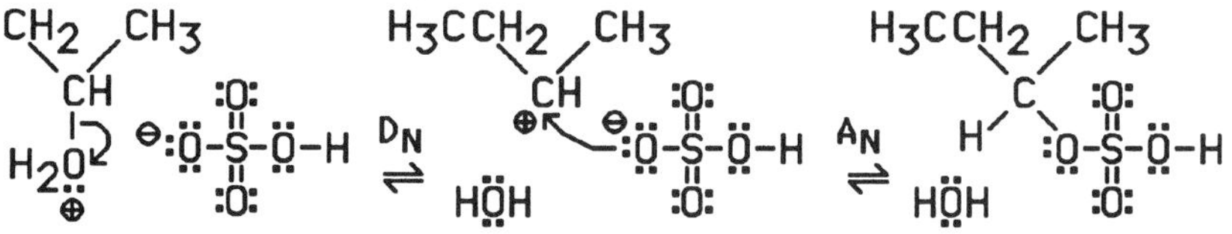

or the E1 (path D_N followed by D_E).

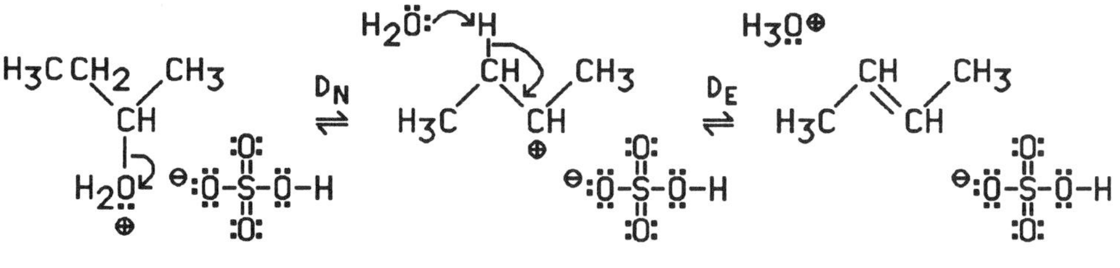

Evaluate and cross-check. As explained in Section 8.5, when the reaction is heated, the nucleophile is poor and also is a good leaving group, then the substitution easily reverses and elimination predominates. The alkene can reprotonate and return along these paths, but still the equilibrium will favor the most stable alkene, *trans*-2-butene. The reaction produces a weaker base. The overall process has been a protonation to give a better leaving group, followed by an E1 elimination. The low boiling alkene product is removed by distillation, driving the equilibria to produce more alkene.

Problem 9.4c

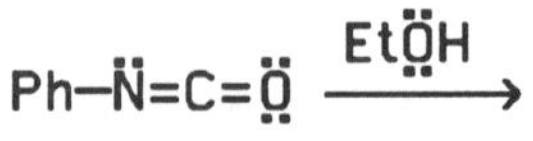

$$\text{Ph-}\ddot{\text{N}}\text{=C=}\ddot{\text{O}} \xrightarrow{\text{Et}\ddot{\text{O}}\text{H}}$$

Understand the system. Source class: Z. Sink class: Y=C=Y. Media: neutral. Overall process: addition.

Find possible routes. The addition surface tells us we have three possible routes: Ad_E2,

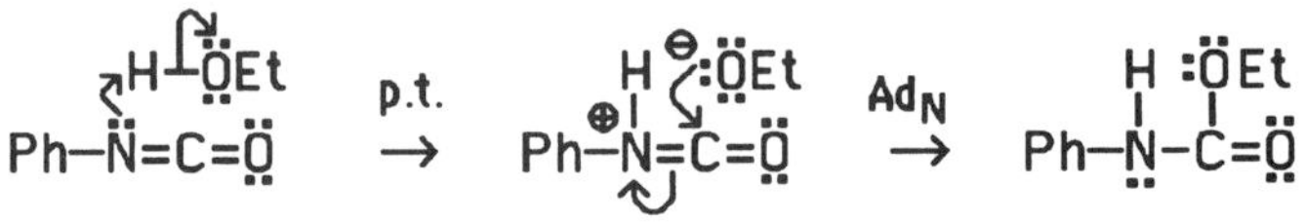

or Ad_E3,

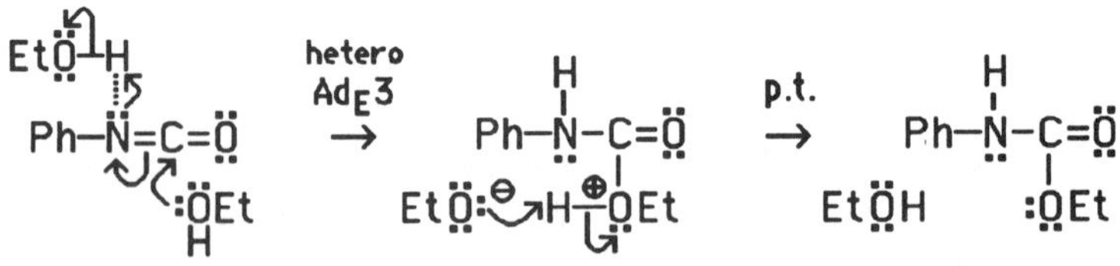

or Ad_N2, (Ad_N + p.t.).

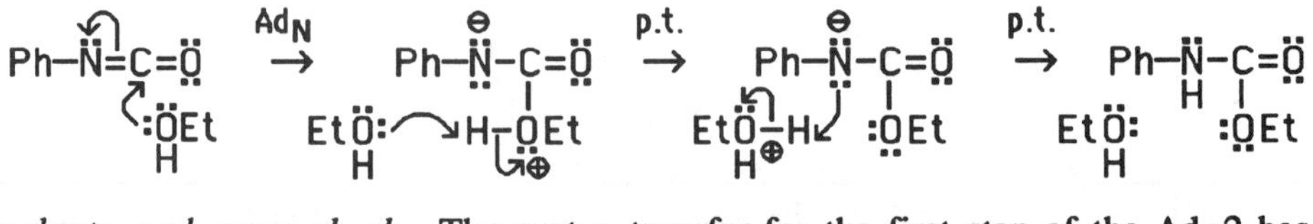

Evaluate and cross-check. The proton transfer for the first step of the Ad_E2 has a somewhat unfavorable K_{eq} crudely estimated at about 10^{-10}, but it produces a reactive ion pair that collapses directly to the carbamate. The Ad_E3 produces an ion pair that has an uncomfortably wide pK_a span of about 18 units, followed by a highly favorable proton transfer to yield the carbamate. The Ad_N2 produces a delocalized anion, whose pK_{abH} is difficult to estimate but may be above 8, and the pK_{abH} of the Nu was -2.4; we should suspect by the ΔpK_a rule that this step is uphill. We have generated two charges from neutral species, but the two proton transfers neutralize these charges giving the carbamate.

Understand the system. We have three almost equal routes to the same carbamate. We should note that the delocalized anion could have protonated on oxygen to give the tautomer of the carbamate. However, proton transfer would easily establish the tautomeric mixture regardless of which route formed the carbamate.

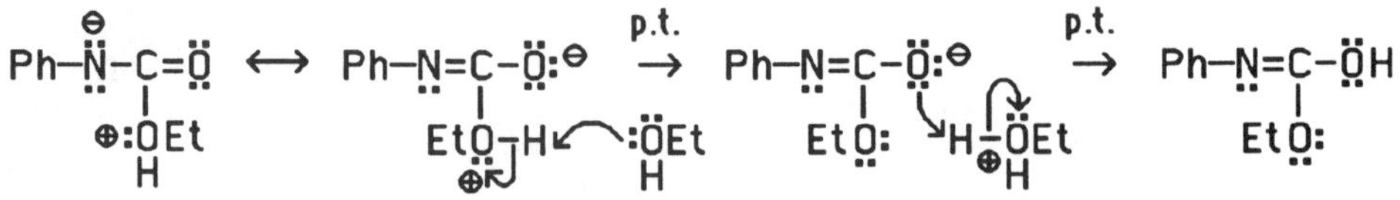

Evaluate and cross-check. A ΔH calculation tells us that the nitrogen protonated species has the strongest bonds but by only 1 kcal/mol (4 kJ/mol). Within the accuracy of our bond strength table, we should expect an equilibrium mixture of the tautomers.

Understand the system. This product carbamate is a highly deactivated Y=C−L and is stable to further attack under these mild conditions.

Overview: We seem to be at a good stopping point. Our addition surface in the first step did not appear to fold or tilt to prefer any route in particular, within the limitations of our crude tools to determine energetics. The mechanism that is commonly written is the $AdN2$. The overall process has been an addition followed by two proton transfers.

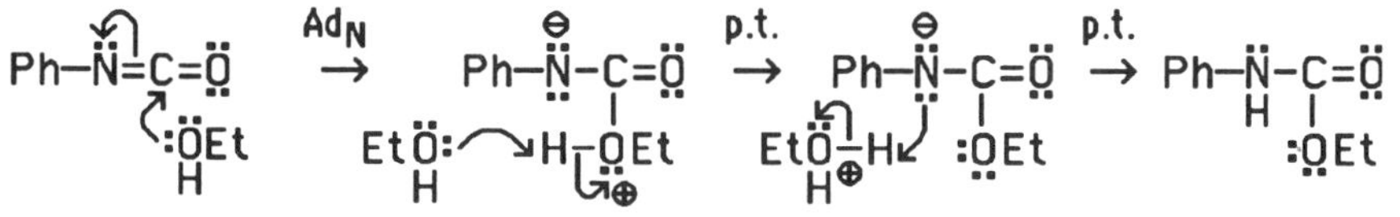

Problem 9.4d

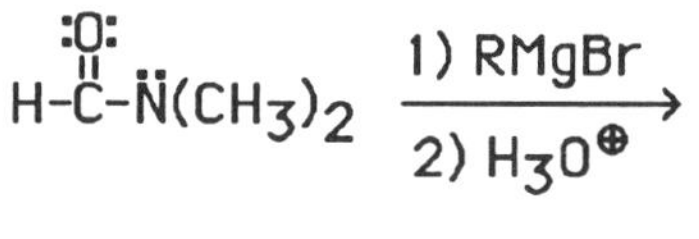

Understand the system. Source class: RM. Sink class: Y=C−L. Media: basic. Acidic H's: none.

Find possible routes. Nucleophilic addition to the polarized multiple bond, AdN.

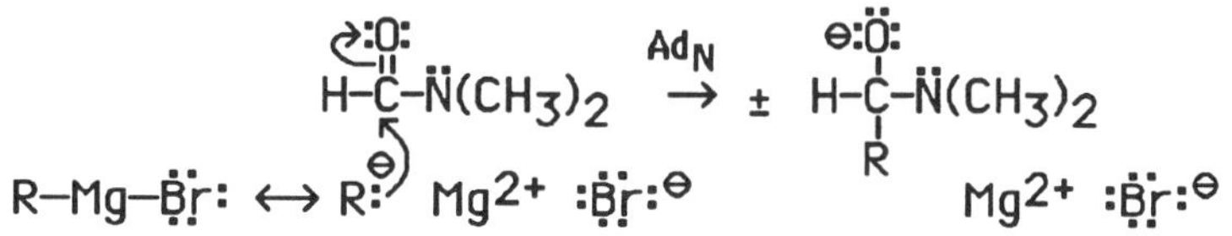

Evaluate and cross-check. This is certainly downhill in energy; the organometallic at approximately pK_{abH} 50 is forming an oxygen anion at about pK_{abH} 15.

Understand the system. One question that needs to be answered at this point is "does multiple addition occur?" Consulting Section 8.2, we find that the leaving group is so bad that it cannot be kicked out ($pK_{aHL} = 36$), and therefore only a single addition is expected. Since nothing obvious can happen, we proceed to step 2. The oxygen anion is a good base; the nitrogen lone pair is also basic.

Find possible routes. Proton transfer twice.

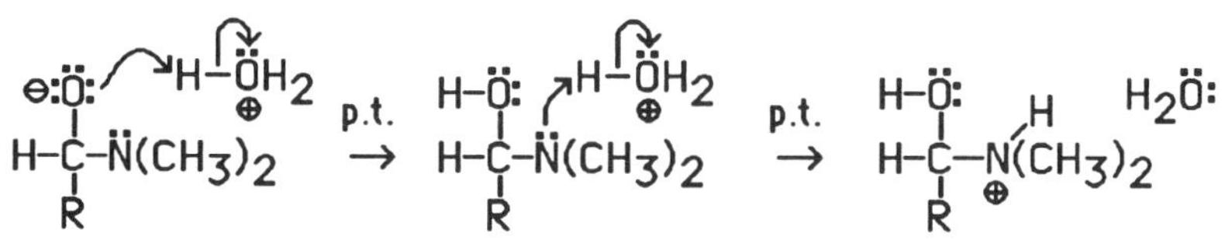

Evaluate and cross-check. The K_{eq} for the first proton transfer is approximately equal to $10^{[15 - (-1.7)]} = 10^{+16.7}$ and for the second K_{eq} is about $10^{[10.7 - (-1.7)]} = 10^{+12.4}$; therefore both are excellent processes. The product of the first proton transfer cannot do anything else since it bears no good leaving groups.

Understand the system. The protonated amine is a fair leaving group. We now are on familiar ground, for the last part of mechanism example 4 in Section 9.5 will take us to the ketone product.

Find possible routes. Three elimination routes, the E1cB (path p.t., then path E$_\beta$),

$$H_2\ddot{O}: \quad H-\ddot{O}: \quad \xrightarrow{\text{p.t.}} \quad H_3\overset{\oplus}{\ddot{O}} \quad :\overset{\ominus}{\ddot{O}}: \quad \xrightarrow{E_\beta} \quad :\overset{\cdot\cdot}{O}: \quad R-\overset{\|}{C}-H \quad H\ddot{N}(CH_3)_2$$

$$H-\overset{|}{\underset{R}{C}}-\overset{\oplus}{N}(CH_3)_2 \qquad H-\overset{|}{\underset{R}{C}}-\overset{\oplus}{N}(CH_3)_2$$

or the E2,

$$H_2\ddot{O}: \quad H-\ddot{O}: \quad \xrightarrow{E2} \quad H_3\overset{\oplus}{\ddot{O}} \quad R-\overset{\|}{C}-H \quad H\ddot{N}(CH_3)_2$$

$$H-\overset{|}{\underset{R}{C}}-\overset{\oplus}{N}(CH_3)_2$$

or the lone-pair-assisted E1 (path E$_\beta$, then p.t.).

$$H-\ddot{O}: \quad \xrightarrow{E_\beta} \quad H_2\ddot{O}: \quad H-\overset{\oplus}{\ddot{O}} \quad \xrightarrow{\text{p.t.}} \quad H_3\overset{\oplus}{\ddot{O}} \quad R-\overset{\|}{C}-H$$

$$H-\overset{|}{\underset{R}{C}}-\overset{\oplus}{N}(CH_3)_2 \qquad H-\overset{\|}{\underset{R}{C}} \quad H\ddot{N}(CH_3)_2$$

Evaluate and cross-check. The E1cB suffers from a poor proton transfer K_{eq} (estimated at about 10^{-12}), but the anion is a good enough source to kick out a fair to poor leaving group. The lone-pair-assisted E1 produces a very acidic cation, approximate pK_a -8, and usually requires a better leaving group. The E2 must have the O–H and C–N bonds coplanar for proper alignment. All routes have problems, but since the product of all three is the same, we can proceed.

Understand the system. The aldehyde will reversibly form the hydrate and the enol, as shown in Section 9.6, product prediction example 3. The aldehyde is the most stable, as demonstrated by a ΔH calculation.

Overview: The overall process has been an addition, two proton transfers, then an elimination. Finally, the amine will become protonated in this acidic medium.

$$:\overset{\cdot\cdot}{O}: \quad \xrightarrow{Ad_N} \quad \overset{\ominus}{\ddot{O}}: \quad H-\overset{\oplus}{O}H_2 \quad \xrightarrow{\text{p.t.}} \quad H-\ddot{O}: \quad H-\overset{\oplus}{O}H_2 \quad \xrightarrow{\text{p.t.}}$$

$$H-\overset{\|}{C}-\ddot{N}(CH_3)_2 \qquad \pm\ H-\overset{|}{\underset{R}{C}}-\ddot{N}(CH_3)_2 \qquad \pm\ H-\overset{|}{\underset{R}{C}}-\ddot{N}(CH_3)_2 \quad \rightarrow$$

$$R-Mg-\overset{\cdot\cdot}{B}r: \leftrightarrow R:^{\ominus} Mg^{2+} :\overset{\cdot\cdot}{B}r:^{\ominus} \qquad Mg^{2+} :\overset{\cdot\cdot}{B}r:^{\ominus}$$

$$H_2\ddot{O}: \quad H-\ddot{O}: \quad \xrightarrow{E2} \quad H_2\overset{\oplus}{\ddot{O}}-H \qquad :\overset{\cdot\cdot}{O}: \quad \xrightarrow{\text{p.t.}} \quad H_2\ddot{O}: \qquad :\overset{\cdot\cdot}{O}:$$

$$\pm\ H-\overset{|}{\underset{R}{\overset{\oplus}{C}}}-\ddot{N}(CH_3)_2 \qquad H\ddot{N}(CH_3)_2 \quad R-\overset{\|}{C}-H \qquad \overset{\oplus}{H_2}N(CH_3)_2 \quad R-\overset{\|}{C}-H$$

Problem 9.4e

$$\bigcirc C=\ddot{O}: \ + \ H\ddot{N}\bigcirc \quad \xrightarrow[-H_2O]{\overset{\oplus}{H}}$$

Understand the system. Source class: Z. Sink class: C=Y. Media: acidic. Basic sites: the ketone carbonyl lone pair (protonated ketone pK_a is -7) and the amine lone pair (protonated amine pK_a is 11). An important item to note is that this reaction is driven uphill in energy by the removal of water.

Find possible routes. Let's explore the best proton transfer first.

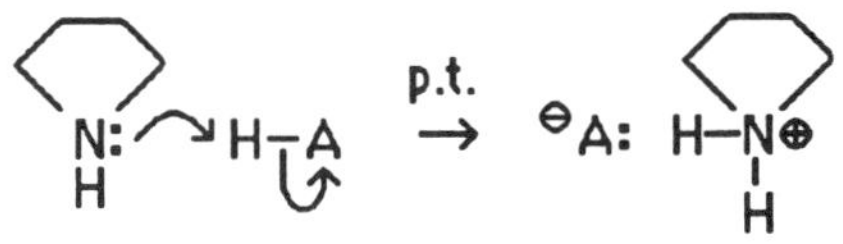

Evaluate and cross-check. This is a good process but the product is unable to react further since there are no good bases or nucleophiles present. In fact, we have just protonated our best nucleophile, so let's return to the start.

Understand the system. The ketone is a good site to add a nucleophile.

Find possible routes. Three ways to add a nucleophile to a carbonyl are the $AdE3$,

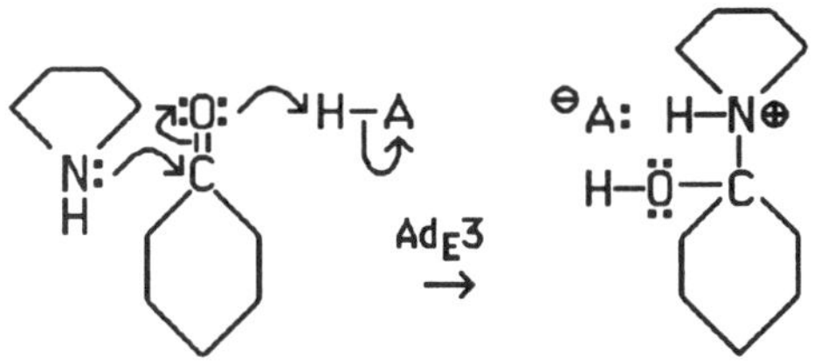

the $AdE2$ (path A_E, then A_N),

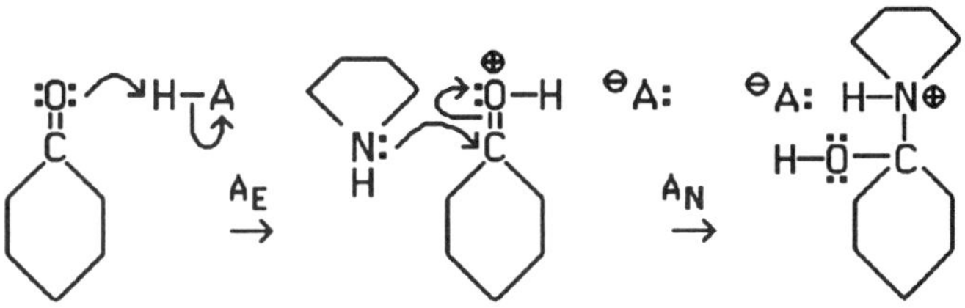

and the $AdN2$ (path AdN, then p.t.).

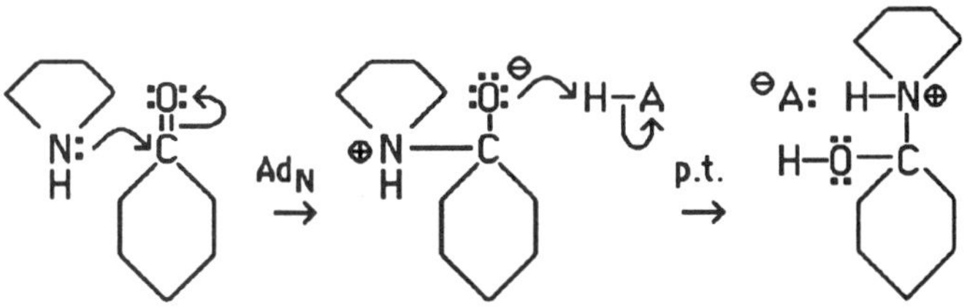

Evaluate and cross-check. The $AdE2$ is suspect because any media acidic enough to have a decent concentration of the protonated ketone will have very little unprotonated amine. The pK_a span is -7 (ketone) and 11 (amine) or 18 pK_a units. Both the $AdE3$ and the $AdN2$ are reasonable. The $AdN2$ would tend to predominate in media more basic than the $AdE3$.

Understand the system. Since the conditions indicate a loss of water, we need to convert the OH into a good leaving group.

Find possible routes. Proton transfer.

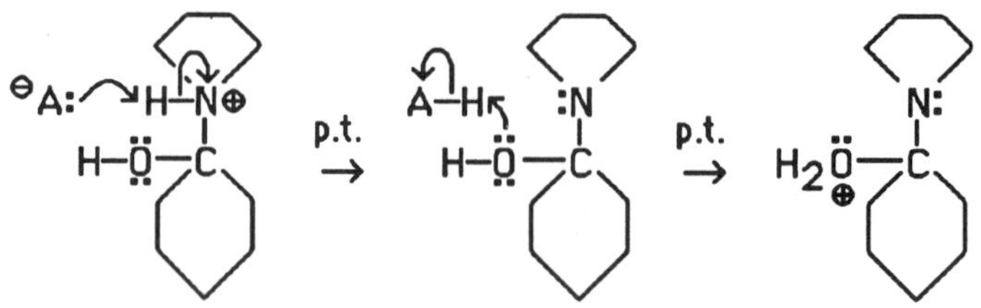

Evaluate and cross-check. These steps are uphill in energy since we are deprotonating the more basic site and protonating a less basic one.

Understand the system. We have a good leaving group, pK_a -1.7, and a lone pair to push it out with.

Find possible routes. Two elimination routes are possible, the E1,

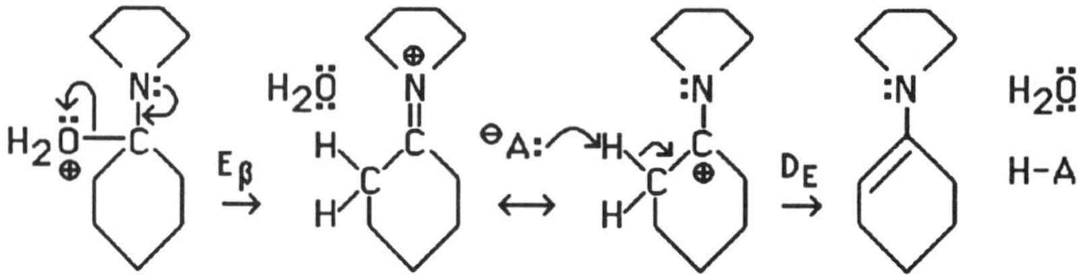

and the E2.

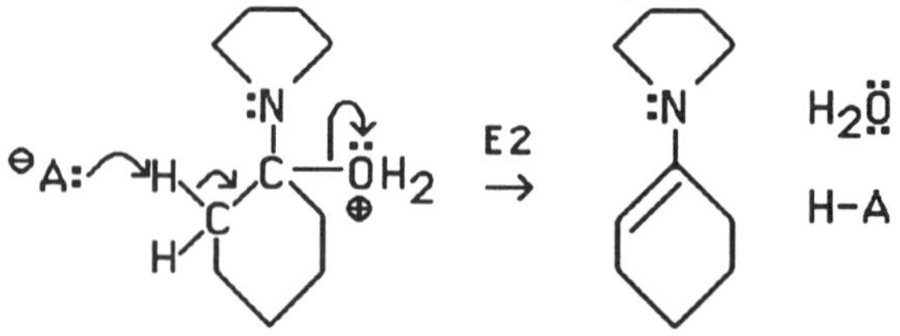

Evaluate and cross-check. Since the carbocation is stabilized and the leaving group is good, the E1 will probably be faster than the E2, but both are reasonable routes.

Understand the system. The enamine product is uphill from the reactants, but the equilibrium is driven by the removal of water.

Overview: The overall process has been an Ad_E3 addition, two proton transfers, and an elimination.

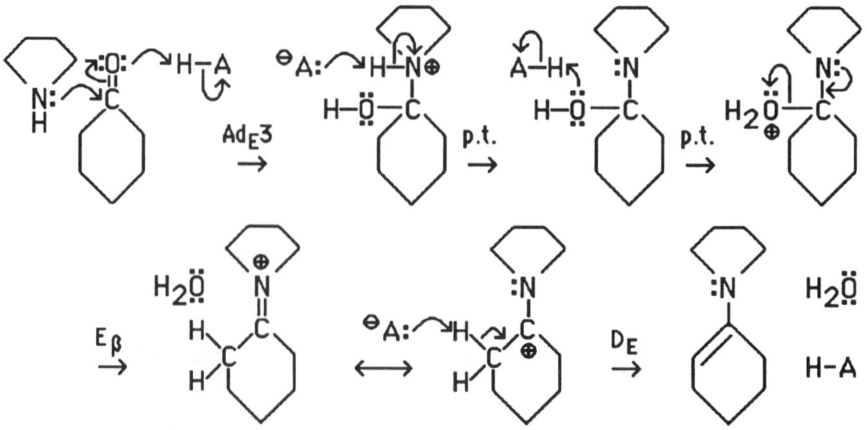

Problem 9.4f

$$H_3C-\!\!\langle \bigcirc \rangle \xrightarrow[H_2SO_4]{(CH_3)_3COH}$$

Understand the system. Source class: Z. Sink class: HA. Media: highly acidic. Most basic site: the alcohol lone pair (protonated alcohol pK_a is -2.4).

Find possible routes. The most obvious route is to protonate the alcohol and drop off water as a leaving group.

$$(CH_3)_3C\ddot{O}:\!\!\curvearrowright\!\!H\!-\!\ddot{O}\!-\!\overset{:\!O\!:}{\underset{:\!O\!:}{\overset{\|}{S}}}\!-\!\ddot{O}H \xrightarrow{\text{p.t.}} (CH_3)_3C\!-\!\overset{\curvearrowright}{\underset{H^{\oplus}}{\ddot{O}H}} \quad {}^{\ominus}:\!\ddot{O}\!-\!\overset{:\!O\!:}{\underset{:\!O\!:}{\overset{\|}{S}}}\!-\!\ddot{O}H \xrightarrow{D_N} (CH_3)_3\overset{\oplus}{C} \;\; \ddot{O}H_2$$

Evaluate and cross-check. The leaving group is good (pK_a -1.7) and the carbocation formed is tertiary and adequately stable.

Understand the system. The carbocation is an excellent electron sink. Possible sources are water, bisulfate, and the aromatic. Water is most likely protonated in media this acidic and would make a poor nucleophile (attack by water would return us to reactants anyway).

Find possible routes. Carbocation trap by bisulfate (path A_N),

$$(CH_3)_3\overset{\oplus}{C}\!\!\curvearrowright\!\!{}^{\ominus}:\!\ddot{O}\!-\!\overset{:\!O\!:}{\underset{:\!O\!:}{\overset{\|}{S}}}\!-\!\ddot{O}H \underset{}{\overset{A_N}{\rightleftharpoons}} (CH_3)_3C\!-\!\ddot{O}\!-\!\overset{:\!O\!:}{\underset{:\!O\!:}{\overset{\|}{S}}}\!-\!\ddot{O}H$$

or deprotonation of the carbocation (path D_E),

$$(CH_3)_2\overset{\oplus}{C}\!\!\overset{H}{\underset{CH_2}{\curvearrowleft}}\!\!{}^{\ominus}:\!\ddot{O}\!-\!\overset{:\!O\!:}{\underset{:\!O\!:}{\overset{\|}{S}}}\!-\!\ddot{O}H \overset{D_E}{\rightleftharpoons} (CH_3)_2C\!=\!CH_2 \quad H\ddot{O}\!-\!\overset{:\!O\!:}{\underset{:\!O\!:}{\overset{\|}{S}}}\!-\!\ddot{O}H$$

or electrophilic aromatic substitution (path A_E, then D_E).

$$H\ddot{O}\!-\!\overset{:\!O\!:}{\underset{:\!O\!:}{\overset{\|}{S}}}\!-\!\ddot{O}:^{\ominus} \quad \xrightarrow{A_E} \quad \xrightarrow{D_E}$$

Evaluate and cross-check. The carbocation trap by bisulfate completes an S_N1 process and should be quite reversible since bisulfate is also a good leaving group. The deprotonation of the carbocation completes an E1 process and should also be reversible since the strong acid can easily protonate the pi bond of the alkene. The electrophilic aromatic substitution is less reversible, for it makes a strong C—C bond and drops off an $H^{\oplus}$ to water, the alcohol, or sulfate.

Understand the system. The aromatic can add a second electrophile, but the addition would be appreciably more hindered; this is a good place to stop.

Overview: The overall process has been a proton transfer, an ionization of a leaving group, then an electrophilic aromatic substitution.

$$(CH_3)_3C\overset{..}{\underset{H}{O}}:\curvearrowright H-\overset{:\overset{..}{O}:}{\underset{:\overset{..}{O}:}{\overset{|}{\underset{|}{S}}}}-\overset{..}{O}H \xrightarrow{\text{p.t.}} (CH_3)_3C-\overset{\oplus}{\underset{H}{O}}H \quad :\overset{\ominus}{O}-\overset{:\overset{..}{O}:}{\underset{:\overset{..}{O}:}{\overset{|}{\underset{|}{S}}}}-\overset{..}{O}H \xrightarrow{D_N} (CH_3)_3\overset{\oplus}{C} \quad \overset{\oplus}{O}H_2$$

(reaction schemes with (CH₃)₃C⁺, sulfonate groups, cyclohexadiene ring — steps labeled A_E, D_E)

Problem 9.4g

$$Ph-\overset{:\overset{..}{O}:}{\underset{}{\overset{||}{C}}}-\overset{..}{O}CH_2CH_3 \xrightarrow{\text{LiAlH}_4}$$

Understand the system. Source class: MH₄. Sink class: Y=C—L. Media: basic. No acidic sites are present.

Find possible routes. The addition–elimination route is most likely (Ad$_N$, then E$_\beta$).

$$H_3\overset{\ominus}{Al}-H \; Li^{\oplus} \curvearrowright \overset{..}{O}: \quad Ph-\overset{}{\overset{||}{C}}-\overset{..}{O}CH_2CH_3 \xrightarrow{Ad_N} \pm \; Li^{\oplus} \; \overset{\ominus}{\overset{..}{O}}: \quad Ph-\overset{}{\overset{|}{C}}-\overset{..}{O}CH_2CH_3 \xrightarrow{E_\beta} Ph-\overset{:\overset{..}{O}:}{\overset{||}{C}}-H$$
$$H_3Al \; H:\overset{\ominus}{\frown} \qquad\qquad H_3Al \; H \qquad\qquad Li^{\oplus} \; \overset{\ominus}{\overset{..}{O}}:CH_2CH_3$$

Evaluate and cross-check. The reaction has certainly gone downhill in energy by the ΔpK_a rule (from a metal hydride at about pK_{abH} 35 to an alkoxide at pK_{abH} 16).

Understand the system. The aldehyde product is more reactive than the original ester so there will definitely be a second addition (Section 8.2).

Find possible routes. Nucleophilic addition to a polarized multiple bond, Ad$_N$.

$$H_3\overset{\ominus}{Al}-H \; Li^{\oplus} \curvearrowright \overset{..}{O}: \quad Ph-\overset{}{\overset{||}{C}}-H \xrightarrow{Ad_N} H_3Al \quad PhCH_2\overset{..}{O}:^{\ominus} \; Li^{\oplus}$$
$$H_3Al \; H:\overset{\ominus}{\frown}$$

Evaluate and cross-check. The reaction has gone downhill in energy by the ΔpK_a rule.

Understand the system. The H₃Al is a Lewis acid and will complex with the anionic product to form a Lewis salt, which is a good stopping point. The Lewis salt is hydrolyzed in the normal acidic aqueous workup of this reaction to give the alcohol and several water-soluble aluminum species.

Overview: The overall process is an addition–elimination followed by a second addition.

$$H_3\overset{\ominus}{Al}-H \; Li^{\oplus} \curvearrowright \overset{..}{O}: \quad Ph-\overset{}{\overset{||}{C}}-\overset{..}{O}CH_2CH_3 \xrightarrow{Ad_N} \pm \; Li^{\oplus} \; \overset{\ominus}{\overset{..}{O}}: \quad Ph-\overset{}{\overset{|}{C}}-\overset{..}{O}CH_2CH_3 \xrightarrow{E_\beta} Ph-\overset{:\overset{..}{O}:}{\overset{||}{C}}-H$$
$$H_3Al \; H:\overset{\ominus}{\frown} \qquad\qquad H_3Al \; H \qquad\qquad Li^{\oplus} \; \overset{\ominus}{\overset{..}{O}}:CH_2CH_3$$

$$H_3\overset{\ominus}{Al}-H \; Li^{\oplus} \curvearrowright \overset{..}{O}: \quad Ph-\overset{}{\overset{||}{C}}-H \xrightarrow{Ad_N} H_3Al \quad PhCH_2\overset{..}{O}:^{\ominus} \; Li^{\oplus}$$
$$H_3Al \; H:\overset{\ominus}{\frown}$$

9.4h

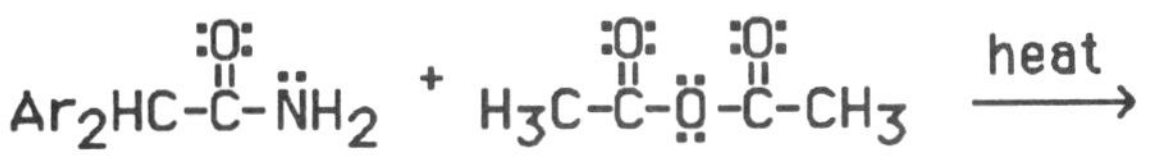

$$Ar_2HC\text{-}\overset{\overset{\displaystyle :O:}{\|}}{C}\text{-}\ddot{N}H_2 \;+\; H_3C\text{-}\overset{\overset{\displaystyle :O:}{\|}}{C}\text{-}\ddot{O}\text{-}\overset{\overset{\displaystyle :O:}{\|}}{C}\text{-}CH_3 \;\xrightarrow{\;heat\;}$$

Understand the system. Medium: Initially neutral; however, most reactions of carboxylic acid anhydrides form the carboxylic acid as a product. Sources: The oxygen lone pairs on the amide are the best source. Leaving groups: The carboxylate on the anhydride. Sinks: The anhydride is a Y=C–L. Acidic H's: The amide NH_2 has a pK_a of 17. Bases: None. Resonance forms:

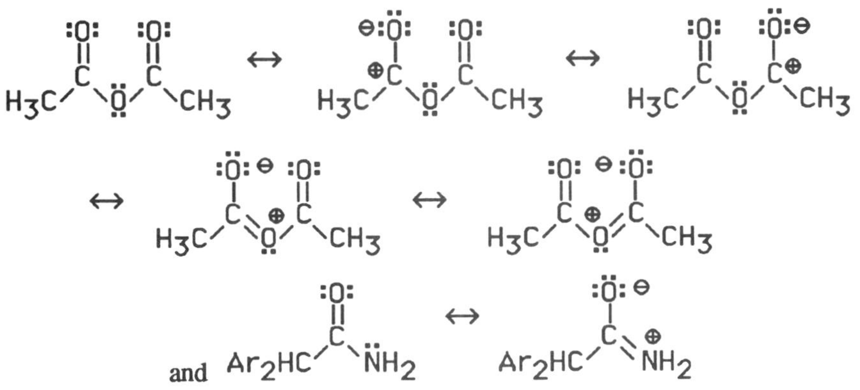

Find possible routes. There is only one route possible for this sink in this medium, addition–elimination (path Ad_N followed by path E_β).

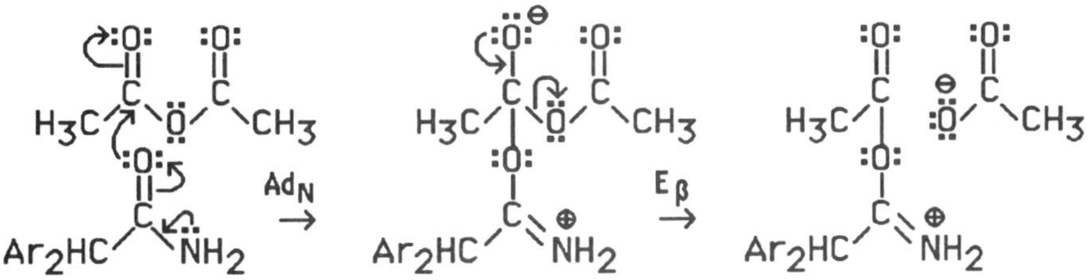

Evaluate and cross-check. The reaction is probably uphill in energy because it creates charge. The pK_{abH} of the nucleophile is -0.5 and the pK_{aHL} of the leaving group is 4.8; the reaction has made a slightly stronger base, but within the bounds of the ΔpK_a rule.

Understand the system. The O acylated amide is both a good site for nucleophilic attack and a source of acidic H's.

Find possible routes. The two possible routes are proton transfer,

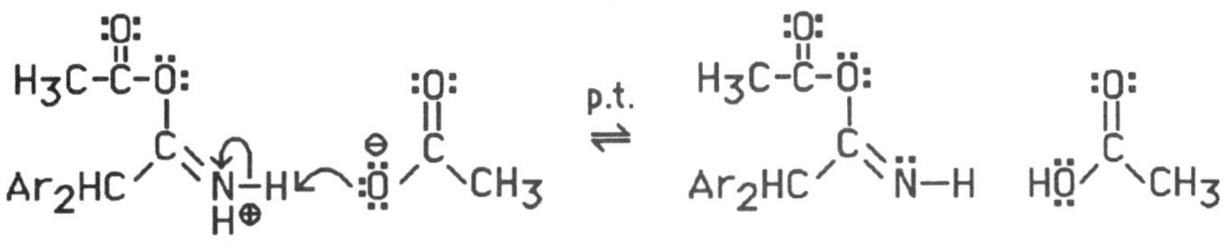

and addition to a polarized multiple bond (path Ad_N).

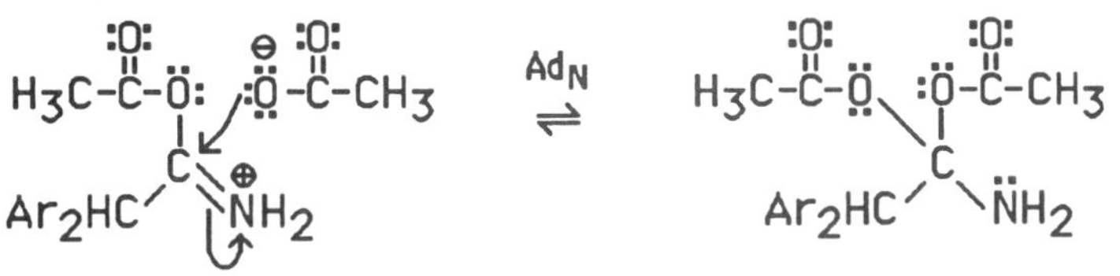

Evaluate and cross-check. The ΔH calculation gives that the nucleophilic attack is endothermic, slightly uphill, from initial reactants. The proton transfer step with an estimated $K_{eq} = 10^{+5.3}$ is favorable by a large margin. The K_{eq} of the proton transfer is good, and the product of nucleophilic attack has no new paths except to kick out the best leaving group, and that path returns to reactants.

Understand the system. The O acylated amide is downhill from the reactants, but we still need to explore to see whether there are any lower accessible points on the surface. The acetate can serve as a leaving group ($pK_{aHL} = 4.8$).

Find possible routes. We should explore two possible elimination routes that produce the same product, the E2,

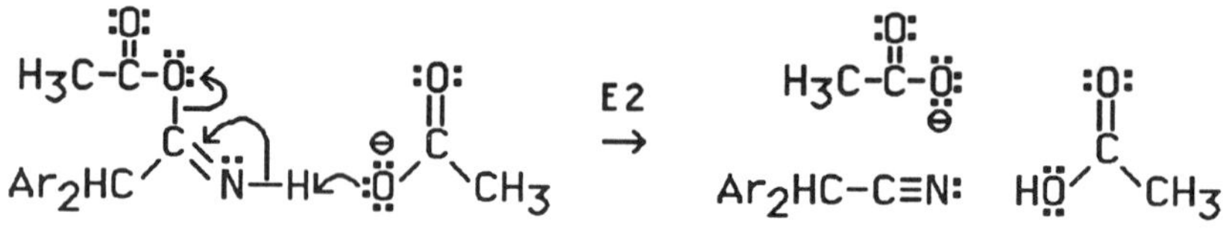

and the cyclic elimination, Ei.

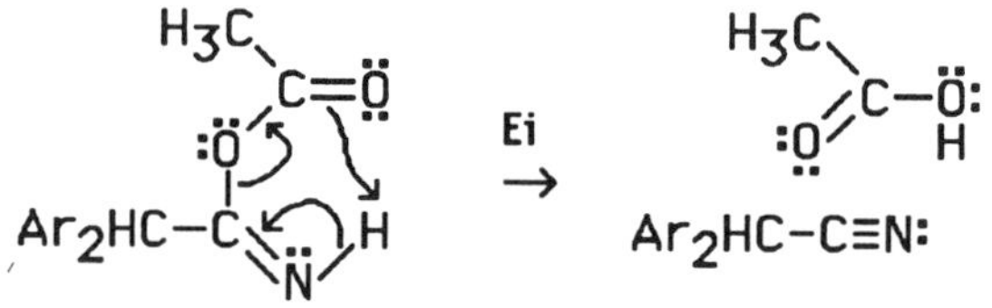

Evaluate and cross-check. Both routes are reasonable. Acetate is a better base than the carbonyl lone pair, and this favors the E2. However, the intramolecular process may be faster because of the greater number of effective collisions between the NH and the carbonyl oxygen. A ΔH calculation puts the nitrile only about 2 kcal/mol (8 kJ/mol) endothermic of the O acylated amide (within the error of a calculation based on values from an average bond strength table). Since we have three products from two reactants, entropy should favor the formation of the nitrile. Often a reaction like this is driven by distilling off the lower boiling acetic acid.

Overview: An addition–elimination was followed by a proton transfer, then an internal elimination (or E2).

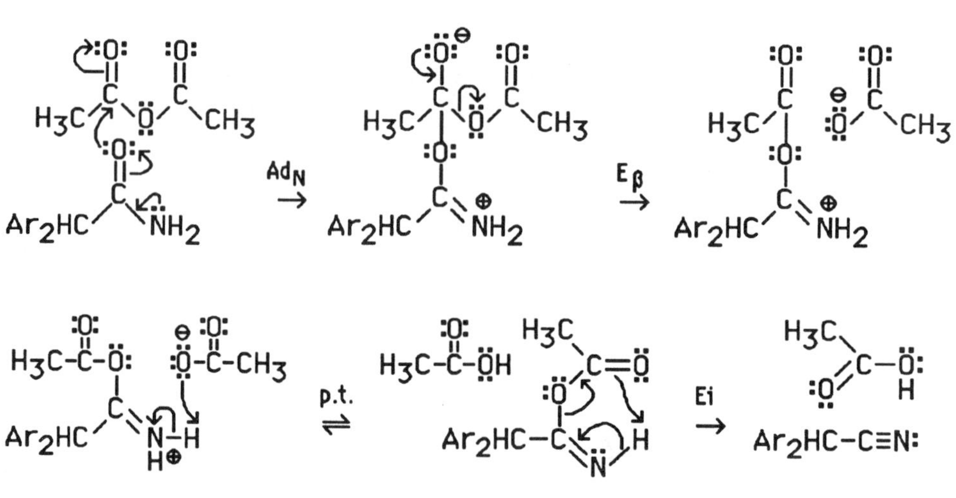

9.5a Similar systems run in ^{18}O water show no ^{18}O in the alcohol product.

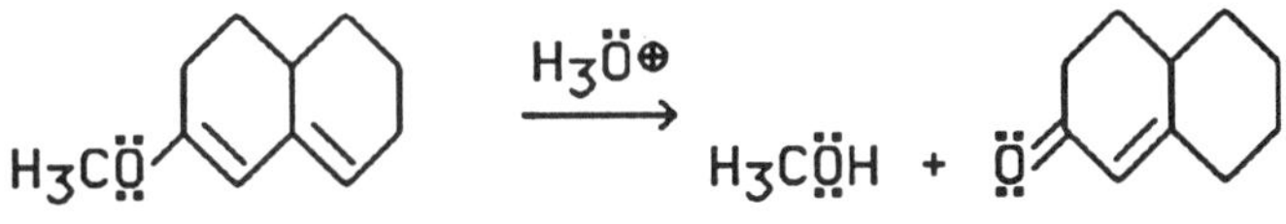

Understand the system. Source class: Z—C=C—C=C (vinylogous allylic source). Sink class: HA. Media: acidic. Best site for protonation: either end of the Z—C=C—C=C. *Find possible routes.* Protonation could occur on carbon by path A_E,

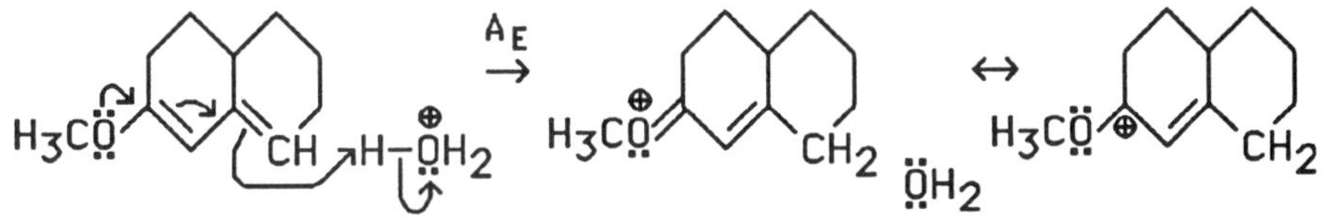

or on oxygen by path p.t. (which sets up the system for an S_N2).

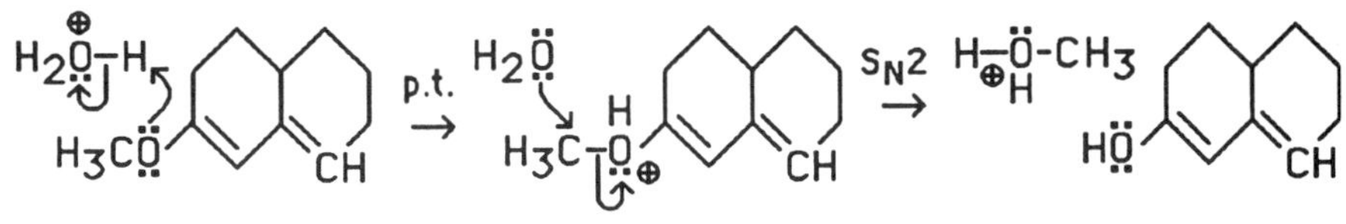

Evaluate and cross-check. Protonation of the CH puts an H in the right place on the right ring. The lone-pair-stabilized allylic carbocation is relatively stable and an excellent electron sink. Proton transfer to the oxygen makes a more stable oxonium ion, which has only one path, the S_N2, available to it (other than return to reactants). The S_N2 gets us closer to products but is ruled out by the $^{18}OH_2$ study, for the methanol formed would contain ^{18}O if the S_N2 occurred.
Understand the system. The sink is C=C—ewg, but from the product we can see that no conjugate addition has occurred. Water is our most abundant nucleophile.
Find possible routes. Addition of water to the polarized multiple bond,

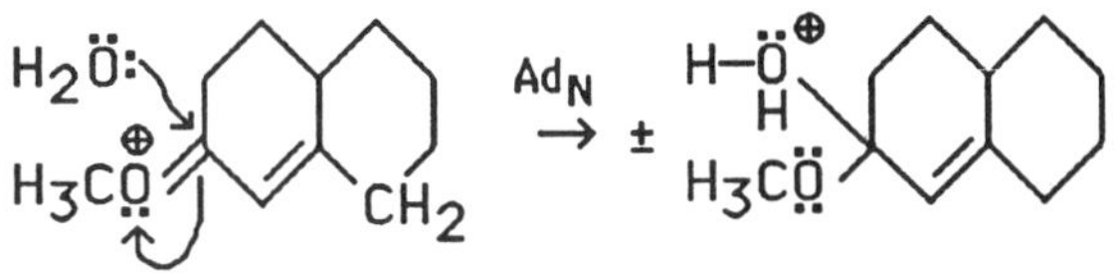

or S_N2 on the methyl.

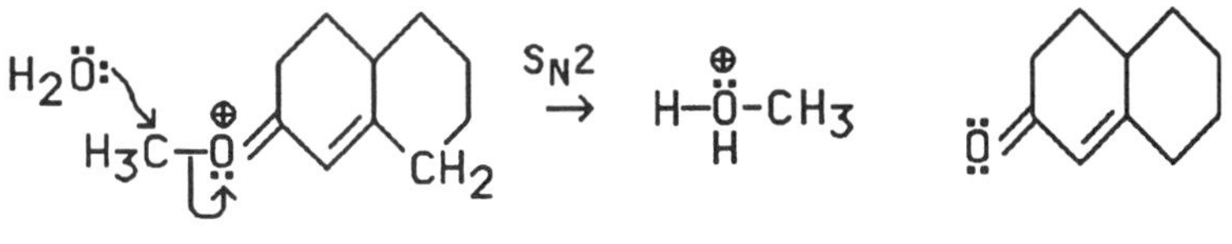

Evaluate and cross-check. This Ad_N could also be considered path A_N, trapping of a cation by a nucleophile. Again the S_N2 can be ruled out by the $^{18}OH_2$ study.
Understand the system. The product is a protonated hemiketal. Proton transfer between the lone pairs should be rapid. We need to make the methyl ether into a leaving group so that we can eliminate it.
Find possible routes. Proton transfer either internally or via solvent should be fast.

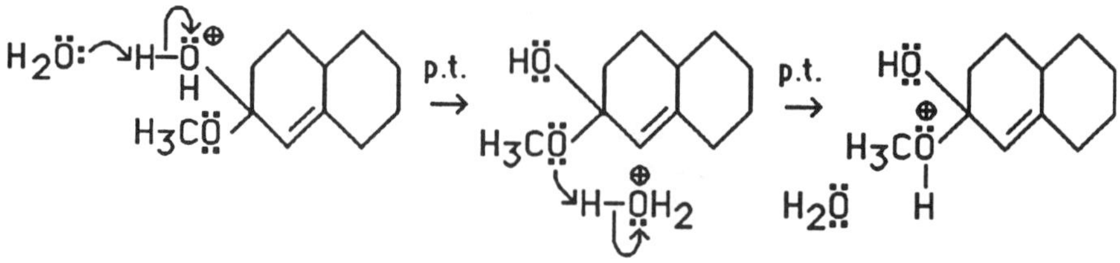

Evaluate and cross-check. The protonated ether is uphill; the K_{eq} is about 10^{-1}.
Understand the system. The protonated ether is a good leaving group, $pK_{aHL} = -2.4$.
Find possible routes. Considering the media, the good stability of the carbocation and the good leaving group, the lone-pair-assisted E1 is the best route.

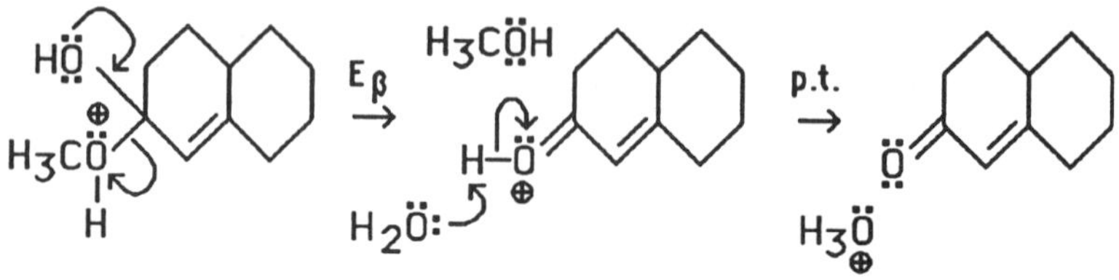

Evaluate and cross-check. The lone-pair-stabilized allylic carbocation intermediate is relatively stable. The methanol is not labeled, consistent with the $^{18}OH_2$ study.
Overview: Did this problem seem familiar? Look at the overview for product prediction example 3 in Section 9.6. We have gone from the enol ether via the hemiketal to the ketone. The grease has changed but the electron flow remains the same. The only difference was that this system had an extra double bond and started at the enol ether instead of the ketal.

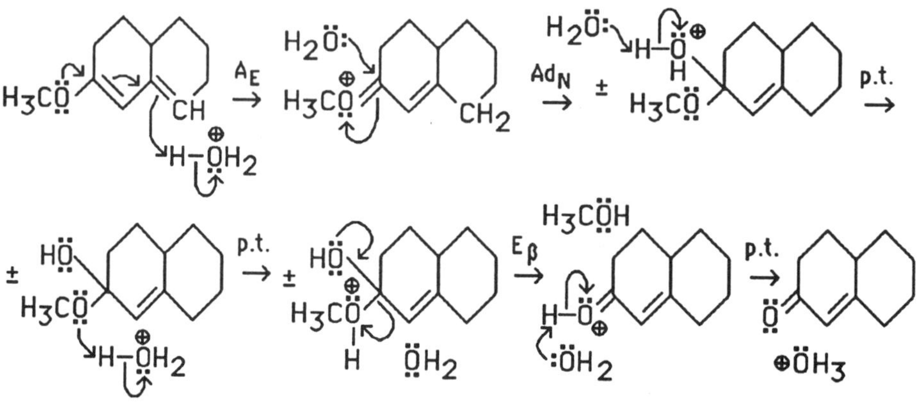

Problem 9.5b This reaction shows general acid and general base catalysis. The reaction is run in an acetate buffer, CH_3COOH and CH_3COONa. The rate decreases if the pH gets significantly more basic or more acidic.

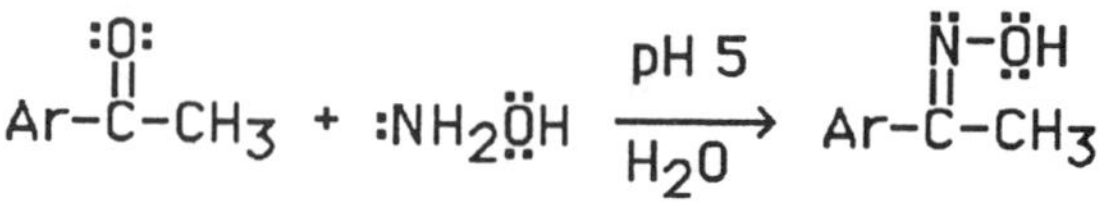

Understand the system. Balanced? No, H_2O is missing from the right-hand side. Generic process: An addition has occurred followed by the loss of water. Medium: Mildly acidic water, kept at pH 5 with a acetate buffer. Sources: Nitrogen lone pair and oxygen lone pair. Best source: The nitrogen lone pair. Sinks: The ketone is a polarized multiple bond. Resonance forms:

$$
\begin{array}{ccc}
:\ddot{O}: & & :\ddot{O}:\ ^{\ominus} \\
\| & & | \\
Ar\text{-}\overset{}{C}\text{-}CH_3 & \longleftrightarrow & Ar\text{-}\overset{\oplus}{C}\text{-}CH_3
\end{array}
$$

Find possible routes. Though the most obvious first step is to attack the carbonyl with the N of NH_2OH to make a C—N bond, proton transfer should be considered.

$$
A\text{-}H \curvearrowleft :NH_2\ddot{O}H \quad \overset{p.t.}{\rightleftharpoons} \quad A:^{\ominus} \quad H\text{-}\overset{\oplus}{N}H_2\ddot{O}H
$$

Evaluate and cross-check. The K_{eq} for acetic acid to protonate hydroxylamine is 10. We have just protonated our best nucleophile and removed it from reaction. If the medium becomes more acidic, the reaction to final product will slow down because the concentration of the needed nucleophile drops.

Understand the system. It is very important to recognize when you've been over similar territory before. This problem and the acidic medium part of example 4, Section 9.5, occur on similar mechanistic surfaces. In the worked example, the product of the organometallic addition was added to acidic water to produce upon the first proton transfer, the imine, $CH_3ArC=NH_2$. This imine is structurally similar to the product of the current example, $CH_3ArC=NHOH$. The product of the worked example, $CH_3ArC=O$, is the same as the current reactant.

Find possible routes. Instead of ending with the surface illustrated by Figure 6.13, this problem starts with that surface. We branch to three routes of addition that are the reverse of the elimination processes discussed in the example, the Ad_N2,

$$
\begin{array}{c}
\overset{:\ddot{O}:\ H\text{-}A}{\underset{\underset{H_3C}{|}}{Ar\text{-}\overset{\|}{C}}}\curvearrowleft :NH_2\ddot{O}H
\end{array}
\quad \overset{Ad_N}{\rightarrow} \pm \;
\begin{array}{c}
^{\ominus}:\ddot{O}:\curvearrowright H\text{-}A \\
| \\
Ar\text{-}\overset{\oplus}{C}\text{-}NH_2\ddot{O}H \\
| \\
H_3C
\end{array}
\quad \overset{p.t.}{\rightarrow} \pm \;
\begin{array}{c}
:\ddot{O}\text{-}H\ \ ^{\ominus}:A \\
| \\
Ar\text{-}\overset{}{C}\text{-}NH_2\ddot{O}H \\
\underset{H_3C}{|}\ \ ^{\oplus}
\end{array}
$$

the hetero Ad_E2,

$$
\begin{array}{c}
:\ddot{O}:\curvearrowright H\text{-}A \\
\| \\
Ar\text{-}\overset{}{C} \\
| \\
H_3C \quad :NH_2\ddot{O}H
\end{array}
\quad \overset{p.t.}{\rightarrow} \pm \;
\begin{array}{c}
^{\oplus}\overset{..}{O}\text{-}H\ ^{\ominus}:A \\
\| \\
Ar\text{-}\overset{}{C}\curvearrowleft :NH_2\ddot{O}H \\
| \\
H_3C
\end{array}
\quad \overset{Ad_N}{\rightarrow} \pm \;
\begin{array}{c}
:\ddot{O}\text{-}H \\
| \\
Ar\text{-}\overset{}{C}\text{-}NH_2\ddot{O}H \\
\underset{H_3C}{|}\ \ ^{\oplus}
\end{array}
$$

and the hetero Ad_E3.

$$
\begin{array}{c}
:\ddot{O}:\curvearrowright H\text{-}A \\
\| \\
Ar\text{-}\overset{}{C}\curvearrowleft \\
\underset{H_3C}{|}\ :NH_2\ddot{O}H
\end{array}
\quad \overset{Ad_E3}{\rightarrow} \pm \;
\begin{array}{c}
:\ddot{O}\text{-}H\ \ ^{\ominus}:A \\
| \\
Ar\text{-}\overset{}{C}\text{-}NH_2\ddot{O}H \\
\underset{H_3C}{|}\ \ ^{\oplus}
\end{array}
$$

Evaluate and cross-check. The Ad_N2 produces an anion of about pK_{abH} of 11, which is not that far away from the pH of the medium. The Ad_E2 forms a protonated ketone of about pK_a -7, which is much further away from the pH of the medium and therefore less likely. The Ad_E3 is the most probable route.

Understand the system. We need to lose the OH, and the H's on the N. Proton transfer not only can make the OH a better leaving group but can also give us a nitrogen lone pair to push it out with.

Find possible routes. Two proton transfer steps are possible.

$$\text{H-\ddot{O}:} \quad \underset{\text{Ar-C-N-\ddot{O}H}}{\overset{\oplus}{|}} \quad \overset{\text{p.t.}}{\rightleftharpoons} \quad \text{H-\ddot{O}:} \quad \text{Ar-C-N-\ddot{O}H} \quad \overset{\text{p.t.}}{\rightleftharpoons} \quad \overset{\oplus}{:}\text{OH}_2 \quad :A^{\ominus} \quad \text{Ar-C-N-\ddot{O}H}$$

Evaluate and cross-check. These steps are uphill in energy since we are deprotonating the more basic site and protonating a less basic one.

Understand the system. Either the O protonated or the neutral carbinolamine is a candidate for elimination.

Find possible routes. There are three possible elimination reactions consistent with a medium that contains a weak acid, the E2,

$$\underset{H_3C}{\overset{\oplus}{:}}\text{OH}_2 \quad \text{Ar-C-N-\ddot{O}H} \quad \overset{\text{E 2}}{\rightarrow} \quad \text{Ar-C=N-\ddot{O}H} \quad \text{H-A}$$

the E1,

$$\underset{H_3C}{\overset{\oplus}{:}}\text{OH} \quad \text{Ar-C-N-\ddot{O}H} \quad \overset{D_N}{\rightarrow} \quad \text{Ar-C=N-\ddot{O}H} \quad \overset{\text{p.t.}}{\rightarrow} \quad \text{Ar-C=N-\ddot{O}H} \quad \text{H-A}$$

or the general acid-catalyzed beta elimination, EgA, followed by proton transfer.

$$\text{A-H} \quad \ddot{O}\text{H} \quad \text{Ar-C-N-\ddot{O}H} \quad \overset{\text{EgA}}{\rightarrow} \quad :\text{OH}_2 \quad \text{Ar-C=N-\ddot{O}H} \quad \overset{\text{p.t.}}{\rightarrow} \quad :\text{OH}_2 \quad \text{Ar-C=N-\ddot{O}H} \quad \text{H-A}$$

Evaluate and cross-check. All are reasonable, but the general acid-catalyzed beta elimination, EgA, bypasses the energetically uphill O protonated carbinolamine and satisfies the kinetic observations. Since some form of OH protonation is required for any of these eliminations, the reaction will slow as the medium gets more basic.

Overview: A reasonable route is then the AdE3, a proton transfer, then the general acid-catalyzed beta elimination, EgA, followed by proton transfer.

$$:O: \quad \text{H-A} \quad \text{Ar-C} \quad :NH_2\ddot{O}H \quad \overset{\text{AdE3}}{\rightleftharpoons} \quad \pm \quad \text{H-\ddot{O}:} \quad \text{Ar-C-N-\ddot{O}H} \quad \overset{\text{p.t.}}{\rightleftharpoons} \quad \pm \quad \text{H-\ddot{O}:} \quad \text{Ar-C-N-\ddot{O}H}$$

$$\text{A-H} \quad \ddot{O}\text{H} \quad \pm \quad \text{Ar-C-N-\ddot{O}H} \quad \overset{\text{EgA}}{\rightleftharpoons} \quad :\text{OH}_2 \quad \text{Ar-C=N-\ddot{O}H} \quad \overset{\text{p.t.}}{\rightleftharpoons} \quad :\text{OH}_2 \quad \text{Ar-C=N-\ddot{O}H} \quad \text{H-A}$$

9.5c Only this stereoisomer of product is formed.

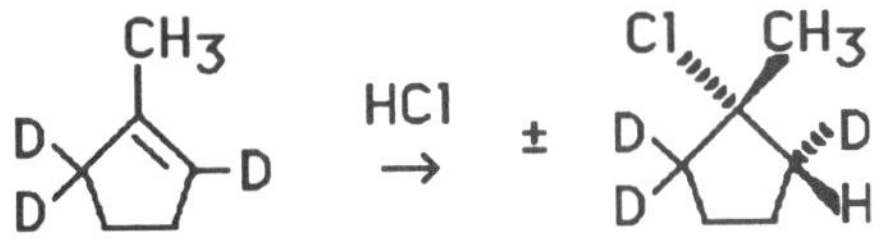

Understand the system. Source: C=C. Sink: HA. Media: acidic, HCl, pK_a -7.
Find possible routes. The Ad_E2 and the Ad_E3 are reasonable for acidic media. The Ad_N2
is not a possibility since there is no electron-withdrawing group.
Ad_E2:

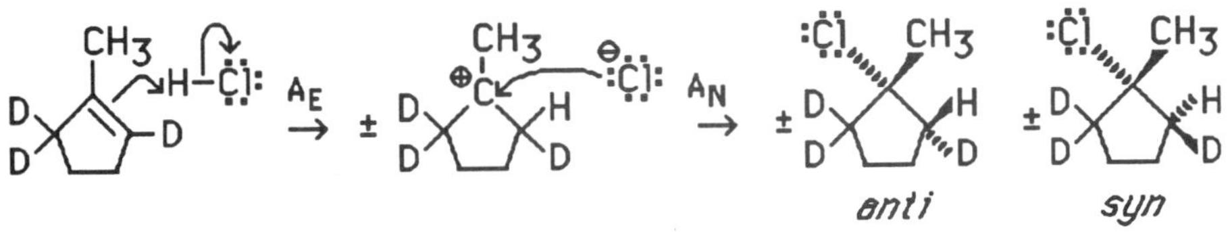

Ad_E3:

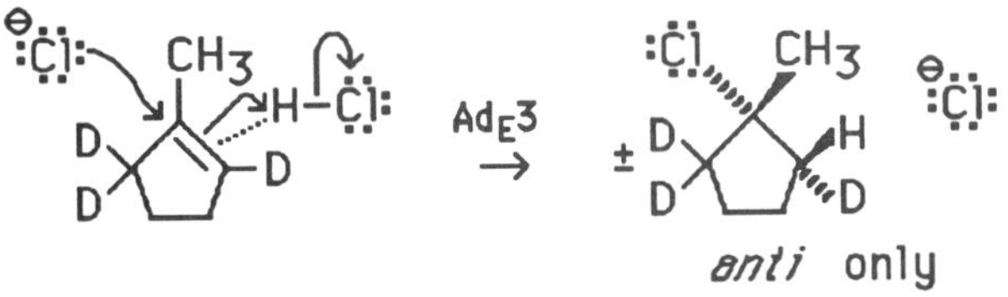

Evaluate and cross-check. The Ad_E2 goes via the tertiary carbocation, which if
symmetrically solvated should produce a 50:50 mixture of *anti* to *syn*. However, since
protonation forms a chloride ion on the same side of the molecule, *syn* addition can
predominate in Ad_E2 additions. The Ad_E3 is the most probable path since it is able to
explain the exclusive *anti* stereochemistry of the product.

9.5d Only this stereoisomer of product is formed at -72°C.

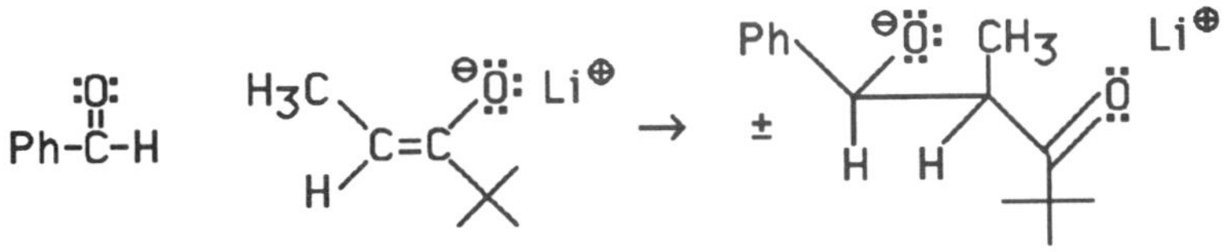

Understand the system. Source: C=C-Z. Sink: C=Y. Media: basic, enolate pK_{abH} is
about 19 to 20. No acidic H's. No leaving groups. Generic process: an addition.
Find possible routes. The Ad_N and the metal-chelated six-electron cyclic are good paths
for the basic medium. The Ad_N is expected to produce a product mixture.

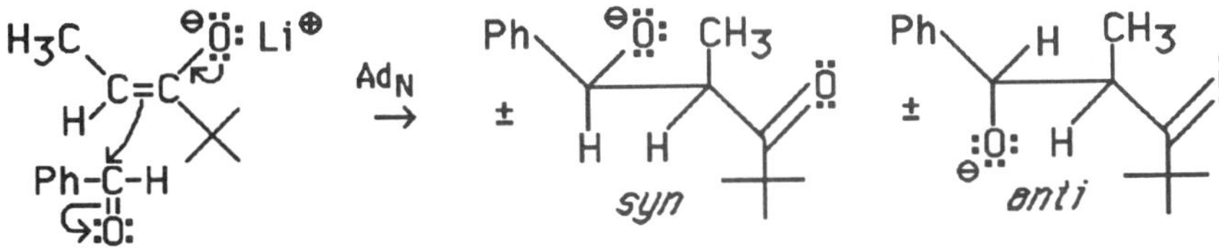

The metal-chelated 6e has two transition states that are of different energy.

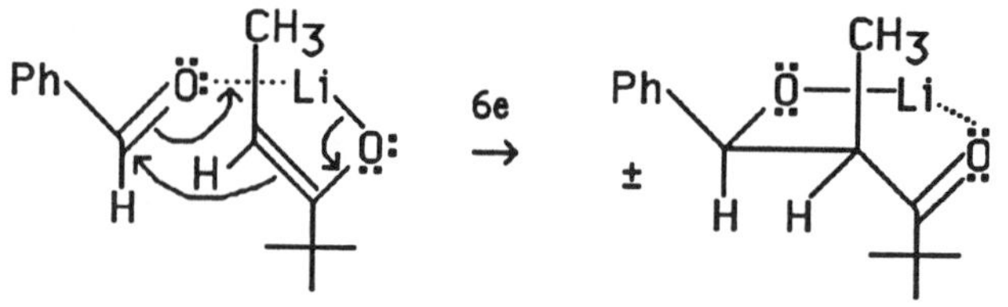

Owing to steric hindrance between the phenyl and *tert*-butyl groups, the following more hindered transition state is of much higher energy.

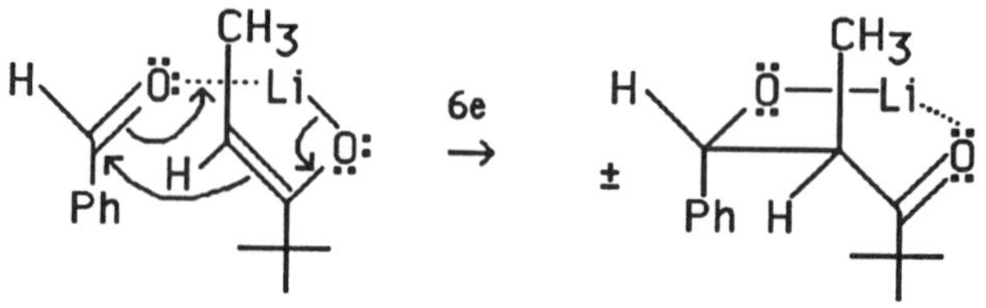

Evaluate and cross-check. The Ad$_N$ (no metal chelate) should produce a mixture of *anti* to *syn*. The less hindered metal-chelated six-electron cyclic path easily explains the exclusive *syn* stereochemistry of the product.

9.5e Chiral optically active reactant gives a racemic product with overall retention of stereochemistry.

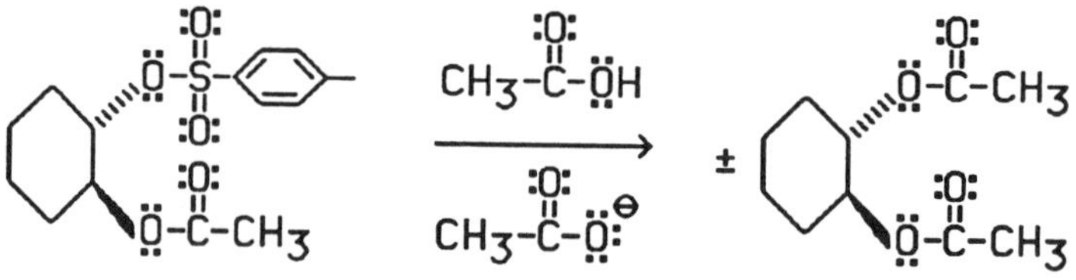

Understand the system. Medium: Acidic, CH_3COOH pK_a is 4.8. Best source: $CH_3COO^\ominus$, a lone pair source. Leaving groups: Tosylate is an excellent leaving group. Sinks: The tosylate is an sp^3 bound leaving group. Acidic H's: None.
Find possible routes. The expected alternatives are the S_N2,

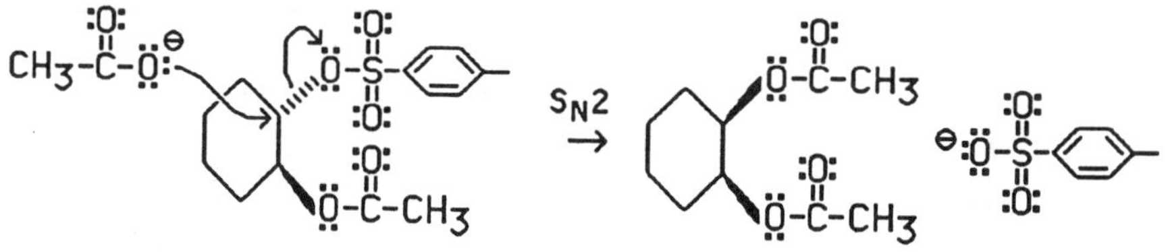

and the S_N1 (D_N, then A_N).

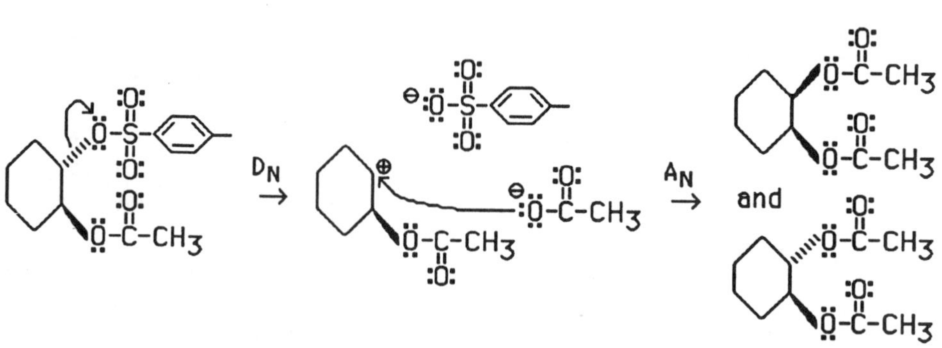

Evaluate and cross-check. The S_N2 would invert the stereochemistry and gives a product that is not chiral but is the wrong stereochemistry, *cis*. The S_N1 would give a mixture of *cis* and *trans* products. The *trans* product would still be optically active because the intermediate carbocation is chiral. Obviously neither route is acceptable; we must look for other routes.

Find possible routes. The carbonyl lone pair of the attached acetate can act as a nucleophile in an internal S_N2. A second S_N2 gives the product.

Evaluate and cross-check. The internal S_N2 creates a five-membered ring and should therefore be fast. The two sequential S_N2 reactions would invert the stereochemistry twice, giving the correct *trans* stereochemistry. The carbocation contains an internal mirror plane and is therefore not chiral; the product of acetate attack on this achiral intermediate is racemic. This route fits the experimental data.

CHAPTER 10: ADDITIONAL EXERCISES

10.1 The most stable radical is designated with the numeral 1.

$\cdot CH=CH_2$	$\cdot CH_2\text{-}CH=CH_2$	$\cdot CH_2\text{-}CH_3$	$\cdot CH(CH_3)_2$	$\cdot CH_3$
Rank: 5	1	3	2	4

10.2 Use radical stabilities to predict the regiochemistry of the radical addition chain reaction of HBr to propene (initiated by AIBN). Bromine radical adds to the alkene to give the more stable secondary radical, which then abstracts a hydrogen atom from HBr to give $BrCH_2CH_2CH_3$ as the product.

10.3 Calculate the ΔH for the abstraction of an H from ethane by a $\cdot CCl_3$ radical. ΔH = [bonds broken 98 kcal/mol (410 kJ/mol)] - [bonds made 96 kcal/mol (402 kJ/mol)] = +2 kcal/mol (8 kJ/mol).

10.4 Calculate the ΔH of an iodine radical adding to a carbon–carbon double bond. ΔH = [bonds broken 63 kcal/mol (263 kJ/mol)] - [bonds made 51 kcal/mol (213 kJ/mol)] = +12 kcal/mol (50 kJ/mol).

10.5 Check the ΔH of the propagation steps for the radical chain addition of HF to ethene to see whether it is an energetically reasonable process.

The two propagation steps are:

Step 1 $CH_2=CH_2 + \cdot F \rightarrow \cdot CH_2\text{-}CH_2\text{-}F$	$\Delta H = -53$	(-222)
Step 2 $\cdot CH_2\text{-}CH_2\text{-}F + H\text{-}F \rightarrow H\text{-}CH_2\text{-}CH_2\text{-}F + \cdot F$	$\Delta H = +36$	(+151)
Total: $CH_2=CH_2 + H\text{-}F \rightarrow H\text{-}CH_2\text{-}CH_2\text{-}F$	$\Delta H = -17$	(-71)

The radical chain addition of HF to an alkene is not an energetically reasonable process because the second propagation step is more than 20 kcal/mol (84 kJ/mol) endothermic, even though the entire process is exothermic.

10.6 The radical chain bromination of toluene yields $PhCH_2Br$. The abstraction of the methyl H by bromine radical gives the stabilized benzylic radical (abstraction of a ring H yields the less stable phenyl radical). The benzylic radical abstracts a bromine atom from Br_2 to continue the chain.

10.7 Radical chain bromination of an alkane would be more selective because the abstraction reaction is much more endothermic. Product radical stabilities would make much more of a difference in the abstraction transition state.

10.8 BHT is a radical chain terminator because H abstraction from oxygen yields a radical that is highly sterically hindered.

10.9 The product distribution of this radical reaction is dominated by the abstraction of H by bromine radical to form the most stable radical.

$$Br_2 \ + \ (CH_3)_2CHCH_2CH_2CH_3 \ \rightarrow \ (CH_3)_2CBrCH_2CH_2CH_3$$

10.10 A mechanism for the following radical chain reaction starts with initiation by light-induced homolytic cleavage. The radical chain has four propagation steps: decarboxylation, abstraction, addition, fragmentation.

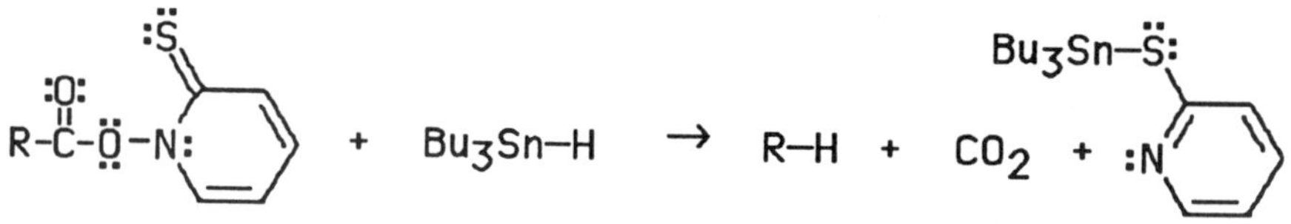

Initiation:

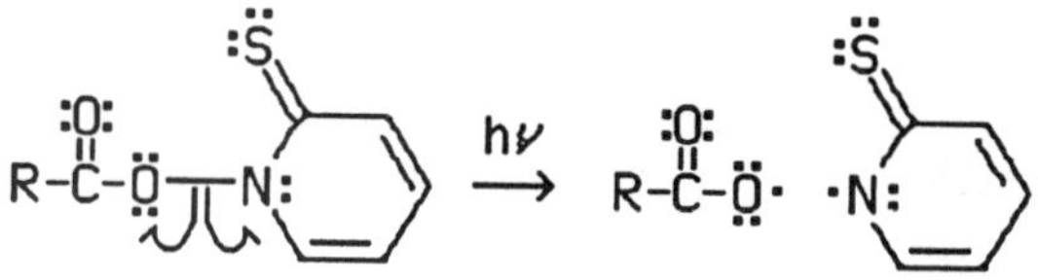

Propagation:

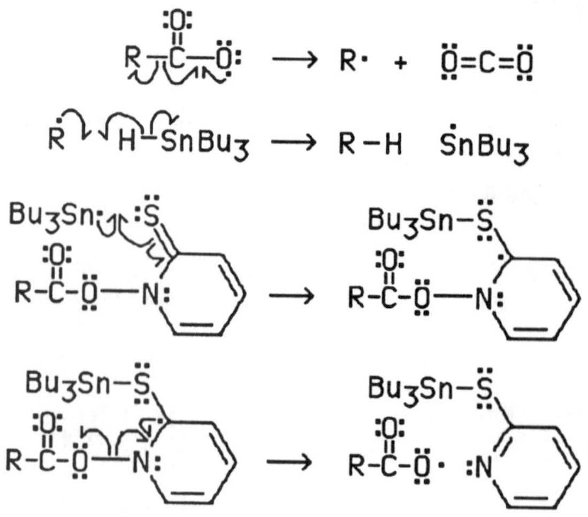

Index

A